T0263727

# Handbook of Glycomics

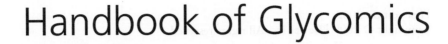

# Handbook of Glycomics

Edited by
Richard D. Cummings
*DEPARTMENT OF BIOCHEMISTRY,*
*EMORY UNIVERSITY SCHOOL OF*
*MEDICINE, ATLANTA, GEORGIA, USA*
and
J. Michael Pierce
*COMPLEX CARBOHYDRATE RESEARCH*
*CENTER AND UGA CANCER CENTER,*
*UNIVERSITY OF GEORGIA,*
*ATHENS, GEORGIA, USA*

AMSTERDAM • BOSTON • HEIDELBERG • LONDON • NEW YORK • OXFORD
PARIS • SAN DIEGO • SAN FRANCISCO • SINGAPORE • SYDNEY • TOKYO
Academic Press is an imprint of Elsevier

Academic Press is an imprint of Elsevier
32 Jamestown Road, London NW1 7BY, UK
30 Corporate Drive, Suite 400, Burlington, MA 01803, USA
525 B Street, Suite 1900, San Diego, CA 92101-4495, USA

First edition 2009
Copyright © 2009 Elsevier Inc. All rights reserved.

No part of this publication may be reproduced, stored in a retrieval system, or transmitted
in any form or by any means electronic, mechanical, photocopying, recording or otherwise
without the prior written permission of the publisher

Permissions may be sought directly from Elsevier's Science & Technology Rights Department
in Oxford, UK: phone: (+44) (0) 1865 843830, fax: (+44) (0) 1865 853333, e-mail:
permissions@elsevier.com. Alternatively visit the Science and Technology Books website at
www.elsevierdirect.com/rights for further information

Notice
No responsibility is assumed by the publisher for any injury and/or damage to persons or
property as a matter of products liability, negligence or otherwise, or from any use or operation
of any methods, products, instructions or ideas contained in the material herein. Because of
rapid advances in the medical sciences, in particular, independent verification of diagnoses
and drug dosages should be made

**British Library Cataloguing in Publication Data**
A catalogue record for this book is available from the British Library

**Library of Congress Cataloguing in Publication Data**
A catalogue record for this book is available from the Library of Congress

ISBN: 978-0-12-373600-0

For information on all Academic Press publications
visit our website at www.elsevierdirect.com

Working together to grow
libraries in developing countries

www.elsevier.com | www.bookaid.org | www.sabre.org

ELSEVIER      BOOK AID
              International      Sabre Foundation

# Contents

# List of Contributors

**Kazuhiro Aoki** (Chapter 13)
The Complex Carbohydrate Research Center, The University of Georgia, Athens, GA, USA

**Sarah Baas** (Chapter 13)
The Complex Carbohydrate Research Center, and Department of Biochemistry and Molecular Biology, The University of Georgia, Athens, GA, USA

**Sara Chalabi** (Chapter 12)
Division of Molecular Biosciences, Faculty of Natural Sciences, Imperial College London, SW7 2AZ, UK

**Paul R. Crocker** (Chapter 11)
College of Life Sciences, University of Dundee, Dundee, Scotland, UK

**Richard D. Cummings** (Chapters 4, 6, and 15)
Department of Biochemistry, Emory University School of Medicine, 1510 Clifton Road, Suite 4001, Atlanta, GA 30322, USA

**Eugene Davidson** (Chapter 14)
Department of Chemistry, Georgetown University, Washington, D.C., USA

**Anne Dell** (Chapter 12)
Division of Molecular Biosciences, Faculty of Natural Sciences, Imperial College London, SW7 2AZ, UK

**Erik Eklund** (Chapter 17)
Burnham Institute for Medical Research, Sanford Children's Health Research Center, La Jolla, California, USA

**Martin Frank** (Chapter 10)
Deutsches Krebsforschungszentrum (German Cancer Research Centre), Molecular Structure Analysis Core Facility–W160, Im Neuenheimer Feld 280, 69120 Heidelberg, Germany

**Hudson H. Freeze** (Chapter 17)
Burnham Institute for Medical Research, Sanford Children's Health Research Center, La Jolla, California, USA

**Kosuke Hashimoto** (Chapter 9)
Bioinformatics Center, Institute for Chemical Research, Kyoto University, Uji, Kyoto 611–0011, Japan

**Stuart M. Haslam** (Chapter 12)
Division of Molecular Biosciences, Faculty of Natural Sciences, Imperial College London, SW7 2AZ, UK

**Stephan Herget** (Chapter 10)
Deutsches Krebsforschungszentrum (German Cancer Research Centre), Molecular Structure Analysis Core Facility–W160, Im Neuenheimer Feld 280, 69120 Heidelberg, Germany

**Jun Hirabayashi** (Chapter 7)
Research Center for Medical Glycoscience, National Institute of Advanced Industrial Science and Technology, AIST Central 2, 1-1-1, Umezono, Tsukuba, Ibaraki 305-8568, Japan

**Minoru Kanehisa** (Chapter 9)
Bioinformatics Center, Institute for Chemical Research, Kyoto University, Uji, Kyoto 611–0011, Japan

**Krys J. Kochut** (Chapter 8)
Department of Computer Science, The University of Georgia, Athens, GA, USA

**R. J. Linhardt** (Chapter 3)
Department of Chemistry and Chemical Biology, RPI, 4005C BioTechnology Building, 110 8th Street, Troy, NY 12180-3590, USA

**Thomas Lütteke** (Chapter 10)
Deutsches Krebsforschungszentrum (German Cancer Research Centre), Molecular Structure Analysis Core Facility–W160, Im Neuenheimer Feld 280, 69120 Heidelberg, and Justus-Liebig-University Gießen, Institute of Biochemistry and Endocrinology, Frankfurter Str. 100, 35392 Gießen, Germany

**Yehia Mechref** (Chapter 1)
National Center for Glycomics and Glycoproteomics, Department of Chemistry, Indiana University, Bloomington, IN 47405, USA

**John A. Miller** (Chapter 8)
Department of Computer Science, The University of Georgia, Athens, GA, USA

**Kelley Moremen** (Chapter 5)
Complex Carbohydrate Research Center, The University of Georgia, 315 Riverbend Road, Athens, GA 30602, USA

**Alison Nairn** (Chapter 5)
Complex Carbohydrate Research Center, The University of Georgia, 315 Riverbend Road, Athens, GA 30602, USA

**Simon J. North** (Chapter 12)
Division of Molecular Biosciences, Faculty of Natural Sciences, Imperial College London, SW7 2AZ, UK

**Milos V. Novotny** (Chapter 1)
National Center for Glycomics and Glycoproteomics, Department of Chemistry, Indiana University, Bloomington, IN 47405, USA

**J. Michael Pierce** (Chapter 16)
The University of Georgia Cancer Center, Complex Carbohydrate Research Center, and Department of Biochemistry and Molecular Biology, The University of Georgia, Athens, GA 30602, USA

**Mindy Porterfield** (Chapter 13)
The Complex Carbohydrate Research Center, and the Department of Chemistry, The University of Georgia, Athens, GA, USA

**R. René Ranzinger** (Chapter 10)
Deutsches Krebsforschungszentrum (German Cancer Research Centre), Molecular Structure Analysis Core Facility–W160, Im Neuenheimer Feld 280, 69120 Heidelberg, Germany

**Pierre Redelinghuys** (Chapter 11)
College of Life Sciences, University of Dundee, Dundee, Scotland, UK

**Mary Sharrow** (Chapter 13)
The Complex Carbohydrate Research Center, The University of Georgia, Athens, GA, USA

**David F. Smith** (Chapters 4 and 6)
Department of Biochemistry, Emory University School of Medicine, 1510 Clifton Road, Suite 4001, Atlanta, GA 30322, USA

**Mark Sutton-Smith** (Chapter 12)
Division of Molecular Biosciences, Faculty of Natural Sciences, Imperial College London, SW7 2AZ, UK

**Michael Tiemeyer** (Chapter 13)
The Complex Carbohydrate Research Center and Department of Biochemistry and Molecular Biology, The University of Georgia, Athens, GA, USA

**Irma van Die** (Chapter 15)
Department of Molecular Cell Biology & Immunology, VU University Medical Center, van der Boechorststraat 7, 1081 BT Amsterdam, The Netherlands

**Lance Wells** (Chapter 2)
Complex Carbohydrate Research Center, and Department of Biochemistry and Molecular Biology, The University of Georgia, Athens, GA, USA

**William S. York** (Chapter 8)
Complex Carbohydrate Research Center, Department of Biochemistry and Molecular Biology, and Department of Computer Science, The University of Georgia, Athens, GA, USA

**Fuming Zhang** (Chapter 3)
Department of Chemistry and Chemical Biology, RPI, 4005C BioTechnology Building, 110 8th Street, Troy, NY 12180-3590, USA

**Zhenqing Zhang** (Chapter 3)
Department of Chemistry and Chemical Biology, RPI, 4005C BioTechnology Building, 110 8th Street, Troy, NY 12180-3590, USA

R. Räne Ravenger (Chapter 10)
Deutsches Krebsforschungszentrum (German Cancer Research Center), Molecular Structure Analysis Core Facility, W160, Im Neuenheimer Feld 280, 69120 Heidelberg, Germany.

Fiona Redelinghuys (Chapter 11)
College of Life Sciences, University of Dundee, Dundee, Scotland, UK.

Mary Sharrow (Chapter 15)
The Complex Carbohydrate Research Center, The University of Georgia, Athens, GA, USA.

David F. Smith (Chapters 4 and 6)
Department of Biochemistry, Emory University School of Medicine, 1510 Clifton Road, Suite 4001, Atlanta, GA 30322, USA.

Mari Sutton-Smith (Chapter 12)
Division of Molecular Biosciences, Faculty of Natural sciences, Imperial College London SW7 2AZ, UK.

Michael Tiemeyer (Chapter 13)
The Complex Carbohydrate Research Center and Department of Biochemistry and Molecular Biology, The University of Georgia, Athens, GA, USA.

Irma van Die (Chapter 15)
Department of Molecular Cell Biology or Immunology, VU University Medical Center van der Boechorststraat, 1081 BT Amsterdam, The Netherlands.

Lance Wells (Chapter 2)
Complex Carbohydrate Research Center, and Department of Biochemistry and Molecular Biology, The University of Georgia, Athens, GA, USA.

William S. York (Chapter 8)
Complex Carbohydrate Research Center, Department of Biochemistry and Molecular Biology, and Department of Computer Science, The University of Georgia, Athens, GA, USA.

Peiming Zhang (Chapter 3)
Department of Chemistry and Chemical Biology, RPI, 4025C BioTechnology Building, 110 8th Street, Troy, NY 12180-3590, USA.

Zhengjiang Zhang (Chapter 3)
Department of Chemistry and Chemical Biology, RPI, 4005C BioTechnology Building, 110 8th Street, Troy, NY 12180-3590, USA.

# Preface

*"L'esperienza de questa dolce vita."*
*The experience of this sweet life.*

Dante Alighieri, *La Divina Commedia*, *Paradiso*, Canto XIX

Many of us remember when complex carbohydrates were viewed by most scientists simply as nuisances whose structures were hopelessly complicated and whose functions at best were obscure. Arguments were made that cells synthesized their glycans more for sport or even "decoration" than for necessity, and that differences in glycan structures between cells might be epigenetic irrelevancies, rather than genetically controlled essentials. If Dante had been a scientist in the latter part of the twentieth century, he surely would have assigned a level in his *Inferno* to those who found themselves mired in understanding glycan structure and function, and who understood well the warning at *Inferno's* entrance, "Abandon hope all ye who enter." With the shrinking number of open reading frames (ORFs) identifiable in genomes from humans to yeast, however, no one would now hypothesize that a cell makes a protein or any product, especially those made by the action of several proteins within a pathway, which serves no particular function. Thus, with new times comes new knowledge; surely now Dante would have the glycoscientists ascend to *Paradiso* to join the anointed ones.

The list of recent discoveries showing critical functions of glycans is impressive:

- Regulation of lysosomal enzyme biosynthesis and the roles of glycosylation in molecular targeting
- Regulation of the initial steps of the inflammatory response and leukocyte adhesion and turnover, as well as lymphocyte homing
- Mediation of the recognition and infection by many pathogens, including HIV, via dendritic cell lectins
- Identification of the family of congenital disorders of glycosylation that cause birth defects
- Demonstration that mutations in glycosyltransferases cause a form of muscular dystrophy known as the muscle–eye–brain disease
- A Ser/Thr modification by N-acetylglucosamine, O-linked GlcNAc, serves to regulate diverse signaling pathways, from transcription factor binding to insulin resistance
- Understanding that protein N-glycosylation fundamentally controls glycoprotein quality control and biosynthesis in the endoplasmic reticulum/Golgi apparatus
- Multiple activities of glycosaminoglycans, including heparan sulfate, regulate cell signaling and angiogenesis, as well as blood clotting

- Glycans are recognized by antigen-presenting cells, including macrophages and dendritic cells, via numerous glycan-binding proteins that regulate immune responses

- Sialic acid-binding proteins in leukocytes control cellular differentiation.

Moreover, throughout the field of glycoscience there is excitement and anticipation because new glycan functions are constantly being discovered.

All of these biological processes are controlled by specific glycan structures that are biosynthetically regulated and that are often recognized by specific glycan-binding proteins or lectins. A scientific hallmark of the first part of the twenty-first century, therefore, has been the development of technologies and methodologies to identify the details of post-translational modifications, in particular glycan structures, thus accelerating the pace of discovery of glycan function. These accomplishments have led to the establishment of glycomics as one of the fundamental areas of biological and biochemical study, and the ambitious goal of defining human and animal glycomes as one of the twenty-first century's most important quests. The definition of glycomics used in this book is very broad, encompassing all those molecules that contain glycans and which regulate or whose structure and function are regulated by glycans. Thus, this quest to understand the glycome is extraordinarily challenging, since it requires an understanding of both glycan structures and the molecules to which glycans are linked, and requires defining their altered expression in different cells during development and disease.

The *Handbook of Glycomics* contains descriptions of state-of-the-art technologies that are in current use to describe and quantify changes in glycomes, as well as the application of these technologies to the study of specific organisms, cells, and diseases. For a comprehensive background on glycobiology and glycoscience, readers are referred to the *Essentials of Glycobiology,* 2nd edition, 2009, Cold Spring Harbor Press.

The chapters in the *Handbook of Glycomics* were written by experts in the field with experience in studying glycan structures using newly developed methods for the analysis of many different types of glycoconjugates, including glycoproteins, glycolipids, and glycosaminoglycans/proteoglycans. Moreover, these authors are involved in integrating this knowledge into a systems approach to the field, incorporating glycan structure, gene expression, and glycobioinformatics to propel our understanding of the functions of glycans in physiology, development, and disease. The book is organized into six sections. In Section I, Glycoconjugate Structural Analysis, Meschref and Novotny describe *N*-glycan analysis and Wells describes *O*-glycan analysis using a glycoproteomics approach. The complex nature of glycosaminoglycan structural analysis is addressed by Zhang, Zhang, and Linhardt. In Section II on Glycotranscriptomics, the use of isotope labeling approaches to aid in mass spectrometry of glycans and the rapid identification of structural differences of glycans compared between different sources are described by Cummings and Smith. Then Nairn and Moremen describe the approaches used to analyze transcription of genes that regulate and recognize the glycomes. In Section III, Protein–Glycan Interactions, Smith and Cummings describe the use of glycan microarrays to identify glycan recognition by glycan-binding proteins and lectins, and Hirabayashi describes how chromatographic and mass spectrometric techniques are used to define glycan–protein interactions. Section IV, Glycobioinformatics, includes chapters by York, Kochut, and Miller on integrating glycomic information and databases, and by Hashimoto and Kanehisa on the development and use of the KEGG GLYCAN for integrating biosynthetic pathways, genetics, and glycan structures. Ranzinger, Herget, Lütteke, and Frank describe the European Glycomics Portal, which promotes integration of glycan-related data from many sources. Section V gives overviews of different glycomes and their relationships to functional glycomics and immunity, as described by Redelinghuys and Crocker in relation to the mouse and human immune systems; North, Chalabi, Sutton-Smith, Dell, and Haslam in regard to humans and mice; Sharrow, Aoki, Baas, Porterfield, and Tiemeyer in *Drosophila*; Davidson in malaria;

and van Die and Cummings in parasitic worms. The final section of the book, Section VI, is on Disease Glycomics, where cancer glycomics is discussed by Pierce, and glycans in human disorders of glycosylation are discussed by Freeze and Eklund.

The *Handbook of Glycomics* serves as a portal to these broad subjects on glycomics, since the field is expanding at a stunning pace, with information pouring in daily on glycomes, glycan-binding proteins, enzymes (glycosyltransferases and glycosidases), and transporters in microbes, parasites, plants, and animals. There are clearly many areas of research on glycomics that will need to be included in updated editions of the *Handbook*. The techniques and conceptual approaches described in this edition of the *Handbook of Glycomics* will no doubt be a valuable reference in this field for years to come, and will provide the interested reader both depth and diversity in this rapidly expanding and exciting component of biology.

We are indebted to those who have helped make this book possible, including the many authors who graciously responded to our requests and cajoling, along with the expert staff at Elsevier, including acquisitions editor Christine Masahina and developmental editor Rogue Schindler.

*The most beautiful experience we can have is the mysterious.*

—Albert Einstein

The Editors:
Richard D. Cummings
Atlanta, GA
J. Michael Pierce
Athens, GA

# Acknowledgments

The glycan analyses were performed by the Analytical Glycotechnology Core of the Consortium for Functional Glycomics (GM62116). Anne Dell was a Biotechnology and Biological Sciences Research Council (BBSRC) Professorial Fellow. We also acknowledge the efforts of various personnel from the CFG Mouse Transgenic Core F and M-Scan INC, 606, Brandywine Parkway, West Chester, PA 19380 USA. We also wish to acknowledge the efforts of Lisa Stocks, Executive Director of Lifesharing, the organ procurement organization of San Diego and her professional staff that undertook the collection and initial processing of the human tissue samples for this research.

# Glycoconjugate Structural Analysis

# 1

# High-Sensitivity Analytical Approaches to the Analysis of *N*-Glycans

Yehia Mechref and Milos V. Novotny
NATIONAL CENTER FOR GLYCOMICS AND GLYCOPROTEOMICS,
DEPARTMENT OF CHEMISTRY, INDIANA UNIVERSITY,
BLOOMINGTON, USA

## Introduction

The importance of knowing the structural details of biologically important oligosaccharides has been increasingly emphasized throughout the current glycobiology literature, as it seems that most processes in living cells are in one way or another associated with some form of carbohydrate interaction. During the last 10–15 years, there have been significant strides in the development of powerful structural tools to probe even the most intricate structural attributes of signaling and regulations in and between various cells. Various investigations on glycosylation of proteins and the interactions of glycans and other biomolecular entities are not stimulated by the scientific curiosity alone. The far-reaching consequences of these biomolecular interactions touch on many areas of modern biology and medicine. Any alterations in a glycan composition or structure could potentially be either a cause or consequential attribute of the biochemical imbalance that we recognize as a "disease." Potentially, new clinical diagnostic procedures, as well as the development of potent pharmaceuticals, can be aided significantly through precise, repeatable, and sensitive glycomic measurements.

The difficulties of structural analysis in glycobiology mainly originate from (a) structural complexity of glycoconjugates featuring various forms of branching and other types of isomerism; (b) microheterogeneities at the sites of glycosylation; (c) extensive occurrence of glycosylation in eukaryotic proteins; (d) distribution of glycosylated proteins in different parts of living cells; and (e) considerably wide dynamic ranges in which different glycoproteins are encountered in biological mixtures (e.g., human blood). To address these issues at different degrees of complexity, analytical strategies have gradually evolved.

**Handbook of Glycomics**
Copyright © 2009 by Elsevier Inc. All rights of reproduction in any form reserved.

This chapter addresses primarily the contemporary analytical methodologies used in molecular profiling and mass spectroscopy (MS)-based structural elucidation of N-glycans cleaved from glycoproteins. The tasks and analytical attributes of these determinations in different aspects of glycobiology may vary in terms of difficulty, complexity of materials, and the amounts of available glycoproteins. It is fair to state that under the current situation in this field, the glycomic analysis of asparagine-linked sugars is viewed as a more facile target of investigation than the analysis of O-glycans. This situation is largely due to: (a) the availability of endoglycosidase enzymes which relatively easily remove N-glycans from their protein backbone; and (b) existence of the uniform core structure with its extensively researched biosynthetic attributes.

This chapter emphasizes primarily the approaches based on chemical/analytical measurements such as MS and laser-induced fluorescence (LIF) detection following the use of a capillary separation technique. Identification and quantification of glycans from the physico-chemical viewpoint is thus the primary focus of this chapter, while the other approaches that may otherwise be valuable in glycobiology research (e.g., lectin or glycan arrays) will not be discussed here and are described in chapters 6 and 7.

Traditionally, either N- or O-linked oligosaccharides have been determined by various analytical techniques after their cleavage from the previously isolated, individual glycoproteins. It was still customary during the early 1990s that at least milligram quantities of isolated glycoproteins were required for the glycan analysis. Since the first demonstrations of capillary electrophoresis (CE)-LIF [1–7] detecting fluorescently labeled sugars at subattomole levels, there has been growing incentive for development of structurally informative tools and methodologies to match such sensitive, albeit structurally uninformative, measurements. With the gradual development of biomolecular MS, this gap has been substantially decreased. It is now becoming customary to work with microgram or submicrogram quantities of glycoproteins. With the recent emphasis on the direct glycomic measurements from complex biological materials at very high sensitivity, analytical glycobiology has become an important scientific field, as will be demonstrated here through various applications to N-glycans.

We will first emphasize that the glycan release methods and small-scale purification protocols are a very important adjunct to the following chemical measurements. The role of capillary separations in dealing with complex oligosaccharide mixtures will be dealt with as the next subject. The huge importance of MS in this field has become evident through its ability to provide unique structural information at very high sensitivity. Recent advances in the use of MS tandem techniques will be discussed in some detail here, reviewing important research avenues and potential directions.

# Release of N-Glycans from Glycoproteins

## Chemical Release

Using a classical method of carbohydrate chemistry [8] or its more recently improved versions [9], N- and O-glycans can be chemically cleaved from glycoproteins with hydrazine. Using this approach, O-glycans are claimed to be specifically released at

60°C, while 95°C is needed to release *N*-linked glycans [9]. However, this chemical release approach generally suffers from some major disadvantages. Since the reagent cleaves amidic bonds, including the linkage between the *N*-glycans and asparagine, the analyzed protein samples are significantly altered. Consequently, any information pertaining to the site of glycosylation and its microheterogeneity is lost. Secondly, under the reaction conditions used, the acyl groups of *N*-acetylamino sugars and sialic acids are hydrolyzed, calling for a reacetylation step. As a result, any information related to the acylation or glycolyation of the sialic acids is not available. Thirdly, the residual hydrazide or amino groups are often incorporated in the reducing terminus of some glycans, thus altering their natural state. In addition, conducting the reaction at high temperatures may result in a loss of the reducing terminal GlcNAc. Finally, this chemical release requires strictly anhydrous conditions, which may not be easily controlled or achieved. As a result of these disadvantages, hydrazinolysis has now become secondary to other chemical release approaches or the even more favored enzymatic release.

The release of glycans from glycoproteins in alkaline aqueous media provides yet another commonly used approach, the so-called β-elimination [10]. More recently, Novotny and co-workers introduced a modified β-elimination procedure which cleaves the *N*- and *O*-glycans with the reducing ends intact. This reaction is attained under very mild conditions, thus making the glycans amenable to MS analysis [11]. This so-called ammonia-based β-elimination procedure provides a viable alternative to the previously used hydrazinolysis and other chemical release methods. Its simplicity, the lack of peeling reactions and deacetylation byproducts supplement its effectiveness at microscale. A minimum sample handling and an easy removal of reactants make it possible to cleave effectively the glycan chains from low-microgram quantities of glycoproteins. This procedure is particularly effective in conjunction with the highly sensitive techniques of MS and capillary separation methodologies. Moreover, this chemical method results in the formation of both *O*- and *N*-linked glycans with their intact reducing ends, thus rendering them amenable to subsequent labeling with a fluorophore or any other structural entity that may facilitate their better separation, detection or ionization capabilities.

## Enzymatic Release

*N*-Glycans can be released enzymatically from glycoproteins by several commercially available enzymes. Among those, peptide *N*-glycosidase F (PNGase F or *N*-glycanase) has become the most widely used. The enzyme cleaves off the intact glycans as glycosylamines, which are readily converted to regular glycans. This enzyme initially releases *N*-linked glycans as their 1-amino form, which are easily hydrolyzed to form ammonia and reducing oligosaccharides [12]. This type of hydrolysis is fast in neutral and acidic media, but slow in alkaline media. Therefore, it is possible to release the glycans as having either a reducing terminus, or a 1-amino terminus by choosing the appropriate buffer. We have demonstrated that using 10mM ammonium bicarbonate buffer at pH 7.8 (with 1% mercaptoethanol) results in the formation of 1-amino glycans [13]. The enzymatic release of glycans also results in the conversion of asparagine to aspartic acid at the N-glycosylation site of a protein. PNGase F has a very broad specificity, cleaving all N-glycans except those having fucose monosaccharide moiety attached to the reducing terminal GlcNAc through $\alpha$(1–3) linkage. Glycans possessing such moiety are commonly found on plant and insect proteins, and can be cleaved using a different enzyme, PNGase A [14].

There are other endoglycosidases that are more specific, such as endoglycosidase H (Endo H) known to cleave high-mannose and most hybrid structures [14]. Moreover, Endo H hydrolyzes the bond between the two GlcNAc residues of the chitobiose core, leaving a GlcNAc attached to the protein. This is somewhat disadvantageous, since the information related to the presence of fucose residues at the reducing end of the glycans is lost following such a cleavage. Other endoglycosidases with variable degrees of specificity are also available (for detailed description of endoglycosidases see [14]).

## Purification of N-Glycans

Mass spectrometric analysis of glycans is often more effective upon their purification from crude hydrolyzed samples. This is due to the fact that carbohydrates are generally less tolerant than proteins to salts and other contaminants which may adversely affect ionization. This is despite the fact that small amounts of sodium or other alkali metals are actually required for an efficient MS ionization of carbohydrate structures. It has been observed that metals cause clustering between the matrix and sample, thus adversely affecting resolution [15]. Therefore, many methods have been developed for the removal of salts and buffers to enhance the MS analyses of glycans.

### Dialysis

Although salts and buffers can be removed from glycoproteins simply by washing the dried sample spots with cold water [16] or a dilute trifluoroacetic acid aqueous solution [17], this procedure cannot be generally recommended for glycans because of their high water solubility. However, glycans can be desalted by drop dialysis on a 500 Da cut-off dialysis membrane [18,19]. This step simply involves spotting the glycan sample on a small piece of the dialysis membrane floating on the surface of water. The evaporation of sample is prevented by covering the water container, while the sample is removed after 10–60 min. Since the ion yield and crystal formation in matrix-assisted laser desorption/ionization-MS (MALDI-MS) analysis of glycans are adversely influenced by the presence of salts and buffers, this time is sufficient for a partial desalting of the sample, thus ultimately ensuring effective glycan crystallization for MALDI-MS analyses.

### Solid-Phase Extraction

Another approach to the purification of glycans has been the use of short columns of ion-exchange or hydrophobic resins packed into disposable pipette tips. Kussmann et al. [20] constructed disposable micro-liquid chromatography (LC) columns with

1–2 μL volume using gel loader tips packed with porous reversed-phase materials. Such disposable tips were found very effective in removal of sample contaminants. Currently, these disposable tips with various packing materials (such as cation- and anion-exchange, reversed-phase, and normal-phase resins) are commercially available from Millipore (Bedford, MA, USA) and Harvard Apparatus (Holliston, MA, USA).

Packer and co-workers [21] investigated extensively the use of graphitized carbon as a solid extraction medium. It was demonstrated that this process allows purification of glycans from the solutions containing one or more of the following contaminants: salts, monosaccharides which do not interact with the graphitized carbon, detergents (sodium dodecylsulfate and Triton X-100), proteins (including enzymes), and the reagents used for glycan release (such as hydrazine and sodium borohydride). These authors demonstrated the utilization of graphitized carbon in cartridges for (a) purification of *N*-linked glycans after their enzymatic or chemical removal from glycoproteins; (b) desalting of *O*-linked glycans after alkaline removal from glycoproteins; (c) online desalting of chromatographically separated oligosaccharides; and (d) purification of oligosaccharides from urine. Such disposable tips with various amounts of packing materials are commercially available from Millipore and Harvard Apparatus.

Currently, many solid-phase extractions are easily performed using SpinColumn™, a trade mark of Harvard Apparatus. SpinColumns can be pre-filled with a wide selection of chromatographic materials, including gel-filtration, ion-exchange, silica-based reversed- and normal-phase materials, as well as specific materials, such as charcoal or cellulose confined by frits and/or caps. The SpinColumn is simply placed in a centrifuge tube and a sample is centrifuged briefly to separate sample components. The column material binds and purifies the sample according to size and shape, chemical composition, charge or other physico-chemical properties according to the protocol outlined in Figure 1.1.

PrepTip (Harvard Apparatus) or ZipTip (Millipore) are alternatives to SpinColumn. These alternatives are now routinely used to purify small sample amounts. Chromatographic media, in this case, are either packed in small fritted pipetting tips or the interior walls of the tip are coated with the sample-binding material. In the case of the latter, the sample can flow freely through the opening without back pressure, since the tip opening is not plugged by the interior coating process. The same protocol used with SpinColumns (outlined in Figure 1.1) is employed with PrepTips or Ziptips.

## Other Purification Protocols

Perfluorosulfonated ionomer (Nafion-117) membranes have also been used for purification of glycans [22]. This membrane was first pretreated by heating at 80°C in nitric acid or hydrochloric acid for 2 h to extensively protonate the sulfate groups, after which the membrane was washed with water. This membrane not only allows desalting of glycan samples, but it also adsorbs proteins and peptides.

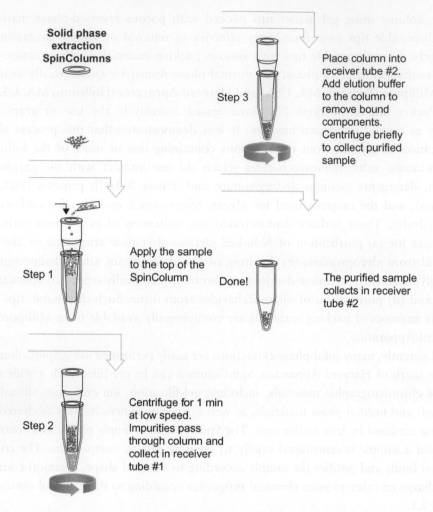

Solid phase
extraction
SpinColumns

Step 3 — Place column into
receiver tube #2.
Add elution buffer
to the column to
remove bound
components.
Centrifuge briefly
to collect purified
sample

Step 1 — Apply the sample
to the top of the
SpinColumn

Done! — The purified sample
collects in receiver
tube #2

Step 2 — Centrifuge for 1 min
at low speed.
Impurities pass
through column and
collect in receiver
tube #1

**FIGURE 1.1**    SpinColumn™ solid-phase extraction protocol for free glycans. (See color plate 1.)

The membrane combined with ion-exchange chromatography was also used for cleaning the glycans released by hydrazinolysis [23].

Worrall et al. [24] replaced the conventional stainless-steel probes with activated synthetic membranes. Contaminated samples were directly spotted onto the activated synthetic membranes (such as polyethylene and polypropylene membranes), while various impurities, including salts, glycerol, and detergents were washed from the sample.

Jacobs and Dahlman [25] have demonstrated a substantial enhancement in the quality of MALDI mass spectra for highly acidic oligosaccharides by coating the MALDI probe surface with a film consisting of a perfluorosulfonated ionomer (Nafion). The effectiveness of this coating in improving the quality of MALDI spectra of highly acidic oligosaccharides was compared with that for the uncoated and nitro-cellulose-coated target plates. While the MALDI spectra of oligouronates (oligomers containing mannuronic/glucuronic and galacturonic acid residues) obtained through the use of uncoated or nitrocellulose-coated probes consisted of a series of broad and multiple peaks, the corresponding spectra obtained with Nafion-coated probes contained only a single series of sharp peaks of nondissociated oligomers exhibiting chain lengths of up to 15 uronic acid residues. This improvement was attributed to the ability of Nafion coating to remove the sodium counterions remaining in the deposit of the sample/matrix mixture on the probe.

Rouse and Vath [26] developed an on-the-probe sample clean-up technique in which the small amounts of resins were co-deposited with a matrix on the sample spot on the probe. The technique involved first drying the sample on the sample plate. Next, the matrix was added, followed by the addition of appropriate adsorbing media, depending on which types of contaminants were present. The sample was dried again and the adsorbent was loosened with a microspatula before being removed by an air stream. Detergents were removed with Extracti-gel-D resin, cations with AG-50W-X8, and anions with Mono-Q resin. However, the authors recommended removal of acetate ions by drop dialysis prior to their on-probe clean-up technique.

Novotny and co-workers [27] reported a simple microscale procedure that effectively removes detergents from the small quantities of oligosaccharides recovered after the common endoglycosidase or exoglycosidase enzymatic digestion of glycoproteins. This involved the addition of a small quantity of a hydrophobic resin (Supelco SP20SS) directly into the digest solutions. The resin adsorbs detergents, remaining deglycosylated proteins, and possible hydrophobic contaminants while leaving the glycans intact in the solution. This contaminant removal enhances substantially the sensitivity of oligosaccharide detection and sequencing by MALDI-MS. The effectiveness of this technique was demonstrated with submicrogram quantities of several model glycoproteins as well as the sperm-binding protein isolated from an extremely small biological sample originating from the egg vitelline envelope of a frog species. The effectiveness of this resin in removing the contaminants was determined to be superior to that of $C_8$- and $C_{18}$-bonded silica materials.

# Analysis of Glycan Mixtures

## Capillary Electrophoresis-Based Approaches

While glycan pools released from different glycoproteins (e.g., recombinant glycoprotein products and affinity-isolated individual proteins), or even unfractionated glycoprotein biological mixtures can be profiled through MALDI-MS, fluorescently tagged mixtures separated by high-performance capillary electrophoresis (HPCE) and detected by laser-induced fluorescence detection (LIF) are often viewed as somewhat complementary to MALDI-MS. Different forms of HPCE are often capable of resolving isomers, which are otherwise indistinguishable by MS. Several types of capillary electromigration techniques have been occasionally utilized for glycomic analysis, in either capillary zone electrophoresis (CZE) employing either free-buffer media or gels, or through micellar electrokinetic chromatography (MEKC). However, high-resolution and fast analysis inherently offered by these techniques can only be appreciated when used in conjunction with sufficiently sensitive detectors. While still employed in certain applications [28], the sensitivity of UV absorbance is marginal for carbohydrates, and fluorescence-tagging procedures for LIF detection are generally recommended in the absence of MS capabilities. Moreover, the tagging procedure is also expected to introduce ionic groups needed for electromigration, since not all N-glycans are inherently charged. However, the MEKC or CZE modes using a borate buffer can also allow the electromigration of glycans tagged with neutral tagging reagents.

In comparison to the miniaturized LC-based techniques discussed below, CZE suffers from a limited capability to inject large aliquots of biological samples. While sample stacking and the use of other means of preconcentration can somewhat reduce this problem, CZE performs best at small column diameters and, correspondingly, very small solute concentrations and peak volumes [29–31]. The detection challenges of CZE, and to a various degree also the other capillary electromigration techniques, can increasingly be met through the inherently sensitive LIF and high-sensitivity MS techniques.

LIF detection necessitates the use of fluorogenic or fluorescence-tagging reagents, whose structures are best utilized in conjunction with reliable laser technologies. Typically, argon-ion lasers are preferred over helium/cadmium (He-Cd) lasers. In 1991, the first carbohydrate analysis by CE-LIF combination was demonstrated, using 3-(4-carboxybenzoyl)-2-quinolinecarboxaldehyde (CBQCA) as a derivatizing agent, for amino sugars [32], and glycoprotein-derived N-glycans [1,2,32]. With the detection limits reaching subattomole levels, the oligosaccharide constituents of bovine fetuin (CBQCA-labeled N-glycans) were resolved into four major (expected) peaks and some minor components. Since then, additional efforts have been made to exploit the potential of various derivatizing reagents for high-sensitivity glycomic analysis.

The effect of several inherently charged fluorophores, including 8-amino-naphthalene-1,3,6-trisulfonic acid (ANTS), 7-aminonaphthalene-1,3-disulfonic acid (ANDS), and 2-aminonaphthalene-1-sulfonic acid (ANS) on the electrophoretic separation of glycans has been investigated [33–35]. As expected, a greater charge caused faster analyses and higher resolution, making ANTS one of the most effective derivatizing agents among the aminonaphthalene derivatives for the CE analysis of N-glycans derived from glycoproteins. This reagent was successfully employed for characterization of glycans derived from various glycoproteins, including human immunoglobulin G [36], ovalbumin [37,38], fetuin [37,39], recombinant HIV gp120 [39], and $\alpha_1$-acid glycoprotein [39].

Although it was demonstrated that ANTS is effective in glycomic analysis, the instability and high cost of the required He-Cd laser prompted the need to exploit the potential of alternative fluorophores compatible with a more convenient light source such as the argon-ion laser. As a result, the most commonly used fluorescence-tagging reagent for CE-LIF today is 1-aminopyrene-3,6,8-trisulfonic acid (APTS) [40]. Using this reagent, Chen and Evangelista [41] devised a nearly complete method for the analysis of N-glycans derived from glycoproteins. It is based on a combination of specific chemical and enzymatic conversions coupled with CE-LIF. First, *N*-glycans were released enzymatically from their glycoproteins and derivatized with APTS under mild reductive amination conditions to preserve sialic acid and fucose residues. The method was successful in profiling heavily sialylated *N*-glycans. Using similar methodologies, additional authors reported the analyses of *N*-glycans derived from various glycoproteins, including ribonuclease B [4,5,40–44], fetuin [41,42], recombinant human erythropoietin [41], kallikrein [41], and a chimeric recombinant monoclonal antibody [6].

Guttman devised a method for multistructure sequencing with *N*-glycans by CE and exoglycosidase digestions [5]. It involves a carefully designed exoglycosidase matrix with a subsequent comparison of the positions of the separated exoglycosidase digest fragments to maltooligosaccharides of a known size, in terms of relative migration shifts. Accordingly, the appropriate linkage information could be deduced from the positions of separated peaks and combined with the shifts resulting from treatment with a specific exoglycosidase.

Ma and Nashabeh [6] extended the use of CZE of APTS-labeled glycans to monitor certain variations in the glycosylation of rituximab, a chimeric recombinant monoclonal antibody, during production. The *N*-glycans derived from rituximab are neutral, complex biantennary oligosaccharides with zero, one and two terminal galactose residues (G0, G1, and G2, respectively). The method was based on releasing *N*-glycans from the glycoprotein via PNGase F, then derivatizing them with APTS prior to a CE mapping. All observed glycans were fully resolved, including the positional isomers of G1. The two G1 positional isomers were identified by comparing CE profiles obtained from sequential enzymatic digestions. *N*-glycans were first released enzymatically by treating the glycoprotein with PNGase F; as an example, a CE profile of APTS-derivatized *N*-glycans is shown in Figure 1.2a. The *N*-glycans were further incubated with β-*N*-acetylglucosaminidase to remove the terminal GlcNAc residues on the G0 and G1 isomers, as seen in Figure 1.2b. The migration time of G2 remained unchanged, since it does not possess a terminal GlcNAc. On the other hand, G0 exhibited a larger change in migration time than G1 isomers because of the removal of two GlcNAc residues from G0, relative to one from G1 (see Figure 1.2b). Finally, the use of α(1–2, 3)-mannosidase, which specifically cleaves terminal α(1–3) linkage, in conjunction with a change in the CE profile, indicated the predominance of G1 with a galactose residue on the mannose α(1–6) arm. The method has several advantages over other schemes, including simplicity, accuracy, precision, high throughput, and robustness.

The use of APTS in conjunction with CZE-LIF permitted a baseline separation of closely related structural isomers [7]. The CE-LIF trace (Figure 1.3a) of the fluorescently labeled standards (core-fucosylated biantennary, asialylated, and mono- and disialylated glycans) clearly illustrates the ability of this technique to separate structural isomers. Monosialylated structures, which differ only in the attachment of a terminal sialic acid residue to 1–6 or 1–3 antenna, were partially resolved in this CE-LIF analysis scheme. The CE-LIF analysis of the *N*-glycans derived from a monoclonal antibody (mAb)

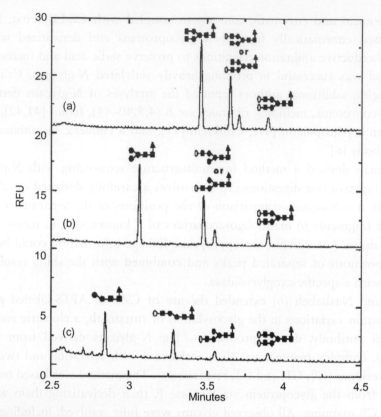

**FIGURE 1.2** Electropherograms of glycans obtained from a chimeric recombinant antibody (rituximab) after sequential enzymatic digestion steps: (a) PNGase F; (b) β-N-acetylhexosaminidase; (c) α1–2,3 mannosidase. Samples were derivatized before capillary electrophoresis analysis. Reproduced from [6], with permission.

revealed the presence of seven major components in addition to many minor components (Figure 1.3b) [7]. In this figure, the resolving power of CZE is also evident from the ability to baseline-resolve monofucosylated biantennary glycan structures lacking one terminal galactose residue.

The application of CE-MS and tandem MS to glycoscreening in the biomedical field has been highlighted in several reviews [28,45–49]. Off-line coupling of CE to MALDI-MS for the structural characterization of APTS-labeled oligosaccharides and N-glycans derived from a glycoprotein was initially described by Suzuki et al. [44]. The CE-resolved components were first isolated using an automated high-resolution fraction collector. The negative ion mode detection of the APTS-labeled oligosaccharides and N-glycans derived from glycoproteins was attained using an on-probe sample clean-up with a cation-exchange resin and a matrix mixture of 6-hydroxypicolinic and 3-hydroxypicolinic acids (1:1 ratio). The authors reported the detection of singly deprotonated ions with a detection limit of about 30 fmol for APTS-labeled maltoheptaose. They also noted that for highly sialylated structures (tri-, tetra-, pentasialylated, and higher) the sensitivity was not as high as for asialylated or mono- and disialylated structures. This might be due to the in-source cleavage of the highly labile sialic acid, especially in the case of heavily sialylated glycans.

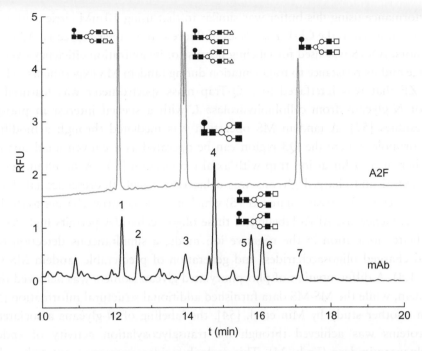

**FIGURE 1.3** Capillary electrophoresis profile of APTS-labeled glycans derived from a monoclonal antibody. The upper trace represents standard core-fucosylated biantennary/disialylated, monosialylated, and asialylated glycans. Conditions: column, polyacrylamide-coated 50/365 μm ID/OD; length, 50.5 cm total, 40.5 cm effective length; temperature, 25°C; injection pressure, 0.5 psi for 5.0 s; voltage, 15 kV anodic electroosmotic flow; λex 488 nm, λem 520 nm. Reproduced from [7], with permission.

Recently, a sheathless electrospray ionization (ESI) coupling of CE to a quadrupole time-of-flight (QTOF)-MS was utilized for glycoscreening of a complex mixture of O-glycosylated peptides purified from the urine of patients suffering from N-acetylhexosaminidase deficiency [50]. The arrangement utilized in this study was based on coating the outlet of the CE separation capillary with a conductive metal using a homemade metallic clenching device. The data-dependent acquisition capability of the QTOF instrument allowed on-the-fly switching between the negative-ion mode MS scans and low-energy collision-induced dissociation tandem MS on selected precursor ions, thus permitting the sensitive identification of the different glycopeptides present in urine samples. The sensitivity offered by this approach reached the values corresponding to the low detection levels of glycans typically derived from biological matrices. Sialylated O-glycosylated peptides, undetectable in complex mixtures by direct ESI MS, were identified using CE-ESI-MS.

In another study, Harvey and co-workers utilized an ESI sheath-flow interface to couple CE to an ion-trap mass spectrometer for the analysis of negatively charged labeled and native glycans [51]. The interface was similar to the ones described above, using 60–100% methanol solution as a coaxial sheath liquid and nitrogen sheath gas to sustain a stable spray. This arrangement was employed to analyze ANTS-labeled N-glycans derived from model glycoproteins such as ribonuclease B and fetuin. A highly efficient CE separation was attained in this study using 20 mM 6-aminocaproic acid.

CE performance using this buffer was similar to that using 10 mM citric acid, which is commonly utilized in the CE-LIF analyses of fluorescently labeled glycans. According to the authors, ANTS was used for labeling because of its ionization efficiency over a wide pH range and its resistance to fragmentation during tandem MS experiments [51].

CZE that was interfaced to a Q-Trap mass spectrometer was featured in the study of N-glycans from cellobiohydrolase I, with a special interest in phosphorylated residues [52]. A tandem MS operation was mediated through a modified triple quadrupole, where the Q3 region can be operated as a conventional quadrupole mass filter or as a linear ion trap with axial ion ejection [53]. A simultaneous analysis of neutral and charged glycans was achieved through fluorescence labeling with 8-aminopyrene-1,3,6-trisulfonate (APTS) which imparts negative charges to the labeled structures. Generally, APTS labeling of these oligosaccharides permits high-resolution CE, a better ionization in the negative ion mode, a simultaneous detection of neutral and charged oligosaccharides, and generation of predictable tandem MS spectra (Figure 1.4). A differentiation of phosphorylated glycan isomers was achieved through this system, while the MS-MS data furnished additional structural information [52].

In another study by Min et al. [54], the labeling of N-glycans associated with glycoproteins was achieved through the transglycosylation activity of endo-β-N-acetylglucosaminidase (Endo M). This endoglycosidase possesses not only a hydrolytic activity toward the glycosidic bond in the N,N9-diacetylchitobiose moiety of the N-glycans of glycopeptides, but also a transglycosylation activity to transfer both the complex type and high-mannose type oligosaccharides of the N-linked sugar chains from glycopeptides to suitable acceptors having an N-acetylglucosamine residue. A fluorescent acceptor such as NDA-Asn-GlcNAc was used in conjunction with Endo M to the fluorescently labeled N-glycans derived from glycoproteins prior to their CE-TOF-MS analysis. All N-glycans derived from ovalbumin were utilized to evaluate and validate the method. Although the derivatization was deemed to reach completion in 30 min, the overall approach is lengthy, since pronase digestion of glycoproteins is required prior to treatment with Endo M [54].

Gennaro and Salas-Solano described an online CE-LIF-MS technology for the direct characterization of N-linked glycans from therapeutic antibodies [55]. A schematic of the online CE-LIF-MS is shown in Figure 1.5. Their system consisted of an Agilent CE instrument equipped with a CE-MS capillary cartridge, allowing for external detection. The online LIF detection was made possible by fitting a PVA-coated capillary with an ellipsoid from Picometrics (Toulouse, France) and using a separate Picometrics Zeta-LIF Discovery detection system consisting of a 488 nm laser unit and an LIF detection cell connected to the detector via a fiber-optic cable, as shown in Figure 1.5. The detection cell is supported by an adjustable arm, approximately 20 cm from the ESI tip. MS was achieved using a mass spectrometer (MicrOTOF, Bruker Daltonics, Bremen, Germany) equipped with an orthogonal, grounded ESI source. A 50:50 isopropanol:water (0.2% ammonia) sheath liquid was added coaxially to the separation capillary at 4 μL/min flow rate, using a CE-MS sprayer (Agilent Technologies). This configuration allowed the identification of major and minor

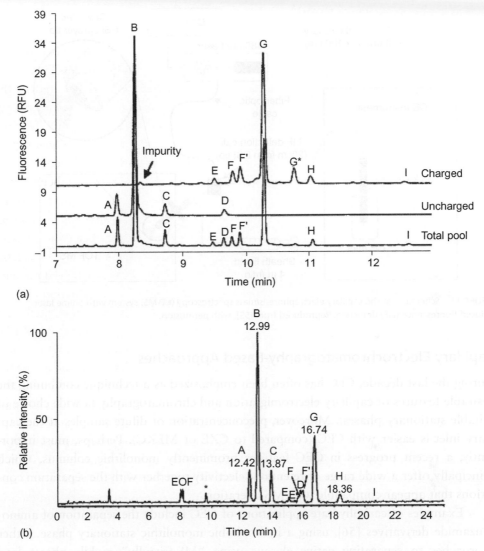

**FIGURE 1.4**  Capillary electrophoresis/laser-induced fluorescence (CE-LIF) electropherograms of the APTS-derivatized total cellobiohydrolase I *N*-glycan pool and of the uncharged and charged glycans (a). CE-MS base peak electropherogram of the total CBH I *N*-glycan pool (b). The neutral glycans are labeled A–D, the charged ones E–I. Unlabeled neutral glycans are detected in the electroosmotic flow (EOF); unlabeled phosphorylated glycans migrate more slowly and are indicated by an asterisk. Reproduced from [52], with permission.

glycan species by providing their accurate masses. This configuration also allowed for the first time the ability to attain CE-MS separation with a sensitivity and component resolution comparable to that of CE-LIF [55]. Figure 1.6 displays expanded-scale views of (a) a conventional CE-LIF electropherogram using a 60 cm capillary, (b) a CE-LIF trace obtained online with MS detection, and (c) a CE-MS base-peak electropherogram. The four early migrating species can be seen clearly in both the online LIF and MS ion electropherograms. However, the authors noticed that the MS-based assays (b and c) suffer from a shift in migration time and a minor loss in resolution. This was believed to be due to the addition of their MS detector.

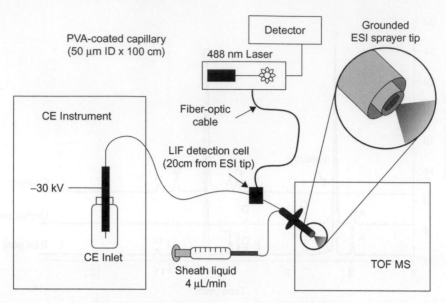

**FIGURE 1.5**   Schematic of the capillary electrophoresis/mass spectroscopy (CE-MS) system with online laser-induced fluorescence (LIF) detection. Reproduced from [55], with permission.

## Capillary Electrochromatography-Based Approaches

During the last decade, CEC has often been emphasized as a technique combining the desirable features of capillary electromigration and chromatography (a wide choice of suitable stationary phases). Moreover, preconcentration of dilute samples at the capillary inlet is easier with CEC (compared to CZE or MEKC). Perhaps, most importantly, a recent progress in CEC features prominently monolithic columns, which principally offer a wide range of retention selectivity together with the separation conditions that appear compatible with MS operation.

Examples of carbohydrate applications of CEC include the separation of aminobenzamide derivatives [56], using a hydrophobic monolithic stationary phase. Other approaches to separating native glycans using "MS-friendly" mobile phases lead to the use of columns featuring hydrophilic interactions. Among these efforts, CEC monolithic columns were shown to be capable of dealing with very complex pools of glycans because of the high resolving power associated with this approach [57–59]. This is clearly illustrated for the case of bile salt-stimulated lipase (BSSL) from human breast milk, which is a relatively large glycoprotein consisting of 722 amino acid residues [60], with numerous O-glycosylation sites near the C-terminus. The CEC/ion-trap MS profile of chemically cleaved N- and O-glycans using a hydrophilic monolithic column is depicted in Figure 1.7. The high selectivity of these columns assisted a partial or complete resolution of several structural isomers, as suggested by the detection of several m/z values at different retention times. These CEC columns were shown to be very effective in conjunction with ESI-MS, using both the ion trap [57,59], and Fourier transform MS (FT-MS) [58] mass analyzers, as well as in MALDI-MS using a micro-deposition device prior to the MS analysis [61].

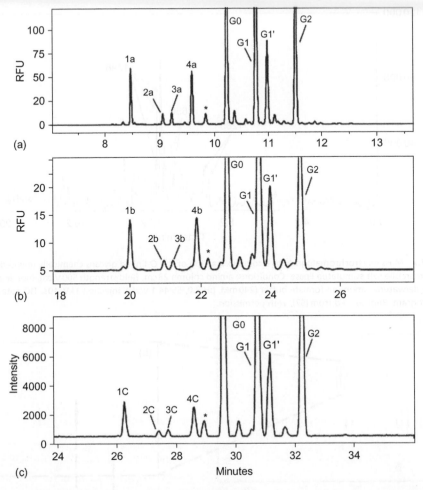

**FIGURE 1.6**   Expanded-scale electropherograms of APTS-labeled *N*-glycans derived from rMAb 1: (a) standard capillary electrophoresis/laser-induced fluorescence (CE-LIF) electropherogram using a 60 cm capillary, (b) CE-LIF trace obtained online with MS detection and (c) CE-MS base-peak electropherogram. Reproduced from [55], with permission.

Novotny and co-workers demonstrated the advantage of interfacing CEC to FT-MS in the analysis of closely related glycan structures [58]. The mixture of *O*-glycans chemically released from mucin was resolved by the CEC/FT-MS system, as shown in Figure 1.8a. The average mass measurement accuracy for FT-MS with external calibration was 3.9 ppm. The mass resolution and accuracy of FT-MS for saccharides was demonstrated in this study by providing characterization of two glycan structures that differ from each other by 1 *m/z* unit. These were an acidic glycan with *m/z* 756 (dot at 10.5 min in Figure 1.8b) and a neutral glycan with *m/z* 757 (dot at 12 min in Figure 1.8b). The average resolution of the spectrum illustrated in Figure 1.8a is 30 000. Previously, these structures could not be distinguished by MALDI-TOF-MS, because of its inability to isolate either ions for the following post-source decay studies. In contrast, these two oligosaccharides could be electrochromatographically separated here from each other, while the high mass accuracy of FT-MS allowed their accurate

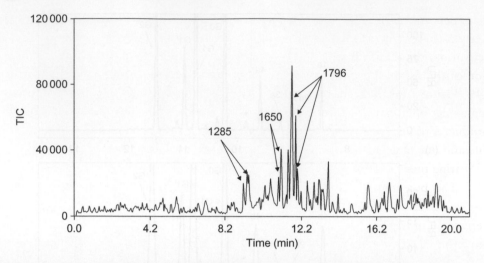

**FIGURE 1.7**    Mass electrochromatogram of a complex fraction of the *O*-linked glycans chemically released from human bile salt-stimulated lipase. Conditions: amino column 28 cm, field strength 500 V/cm, mobile phase acetonitrile/water/ammonium formate buffer (240 mM, pH 3.0, 55:44:1 v/v/v), injection 1 kV, 10 s. TIC, total ion chromatogram. Reproduced from [57], with permission.

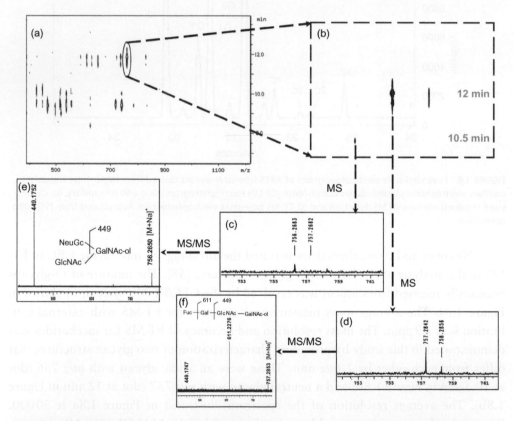

**FIGURE 1.8**    (a) Two-dimensional (2D) contour plot of a mixture of *O*-glycans cleaved from mucin; (b) zoomed 2D contour plot; (c and d) spectra of glycans with *m/z* values of 756 and 757 and their corresponding tandem mass spectra (e and f). An average resolution of ~30000 is demonstrated. Experimental conditions: cyano column, 26 cm; mobile phase, 2.4 mM ammonium formate buffer (240 mM, pH 3) and 0.2 mM sodium acetate in the mixture of acetonitrile/water (71:29, v/v); field strength, 600 V/cm; injection, 12 kV, 30 s. Reproduced from [58], with permission. (See color plate 2.)

mass determination (756.2683 *m/z* and 757.2851 *m/z*, Figure 1.8c and 1.8d, respectively). These structures were further characterized by tandem MS. Upon collisional activation, the acidic glycan (*m/z*) 756 with the GlcNAc(NeuGc)GalNAc-ol composition easily lost its acidic residue at the non-reducing end to form the fragment at *m/z* 449 (Figure 1.8e). The glycosidic bonds of sialic acids were readily cleaved from the remainder of glycan structures. Conversely, the neutral glycan (*m/z* 757) produced two product ions with *m/z* values of 449 and 611. These fragments result from the loss of Fuc and Fuc plus Hex from the non-reducing end of the structures (Figure 1.8f), thus indicating that the composition of this glycan is FucHexHexNAcGalNAc-ol.

The potential of CEC-MALDI-MS was exploited in the analysis of *N*- and *O*-glycans derived from BSSL. The high microheterogeneity of the glycan structures derived from this glycoprotein is evident in Figure 1.9, which depicts three- and two-dimensional CEC-MALDI-TOF-MS recordings. Approximately 50 distinct peaks were

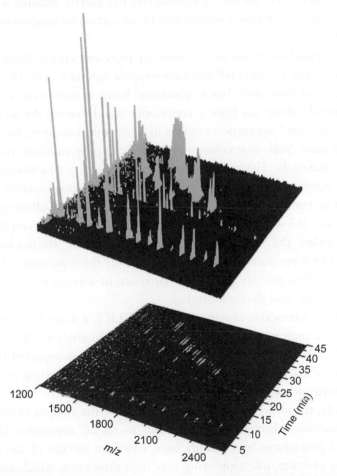

**FIGURE 1.9**  Three-dimensional electrochromatogram of the mixture of *N*-linked and *O*-linked glycans derived from human bile salt-stimulated lipase. Experimental conditions: cyano capillary column, 28 cm; mobile phase, 2.4 mM ammonium formate buffer in a 60/40 mixture of acetonitrile/water; field strength, 535 V/cm; injection, 10 kV, 10 s; matrix, 20 mg/mL DHB in 80/20 methanol/water. Reproduced from [61], with permission.

displayed in this figure. Many of the depicted peaks differ by a few mass units, while some other peaks appear to represent structural isomers [61].

## Liquid Chromatography-Based Approaches

While the MS-based technologies combined with bioinformatic tools have become most crucial to the final structural identifications of particular glycoconjugates, analytical separations are extremely important in securing optimum amounts of materials to be analyzed through MS. These important approaches range from the currently common uses of lectin chromatography isolation, which allows the enrichment of trace amount of glycoproteins from biological samples [62]; or two-dimensional gel electrophoresis in conjunction with glycoprotein-specific stains, which permits the so-called global monitoring of glycosylation changes [63]; to the sophisticated uses of multidimensional microcolumn chromatography in the online systems coupled to different mass analyzers. Modern glycoanalysis has clearly become a multimethodological task in which various combinations of separation techniques and MS become applicable [28].

Most analytical situations in glycobiology represent sample-limited applications which necessitate the use of small chromatographic columns, with the corresponding drastic reduction of flow rates, but a substantial boost in sensitivity of most detectors [64]. Consequently, there has been a continuous trend toward the so-called "micro-flow" and "nano-flow" separation systems in biochemical analysis, including the separations of glycans. Such new column types are typically capillary tubes filled with small-particle materials. The overall separation efficiencies of chromatographic columns can be significantly enhanced further through the combination of very small particles with increased column length, leading to the use of ultrahigh pressures in microcolumn LC [65–68]. Finally, the properties of monolithic capillary columns, operating in either the electromigration mode (capillary electrochromatography) [56–59,69–71] or a pressure-driven mode [72,73] for the separation of both peptides and oligosaccharides, provide an additional route to solving the problems of peak overlap in proteomic and glycomic studies.

Hydrophilic-interaction chromatography (HILIC), a name first coined by Alpert [74], has become increasingly popular in carbohydrate analysis. Although some LC separations of carbohydrates on unmodified silica have been reported [75], it is more effective to use silica packings modified through chemical bonding of suitable polar functional groups such as amine [76–79] or amide [80–83].

Commonly, the utility of the HILIC approach for the analysis of oligosaccharides involves fluorescence labeling of glycans prior to an LC separation [28]. However, most described procedures have mainly been based on the use of the common analytical columns (4.6 mm, i.d.) employing very high flow rates, which adversely affects detection sensitivity. This problem can be overcome through the use of capillary amide columns operating at a flow rate of 300 nL/min, which are also more compatible with MS [84]. The use of this nano-LC format permitted the detection of oligosaccharide mixtures at low-femtomole sensitivity. The approach was validated with the *N*-glycans derived from keyhole limpet hemocyanin and horseradish peroxidase [84].

Although hydrophilic-phase LC columns using amine- and amide-bonded stationary phases have widely been considered most effective for the oligosaccharide separations [28], graphitized carbon columns (GCC) are recently gaining popularity because of their ease of use, high sample capacity, adequate efficiency, and more recently, their commercial availability as microcolumns (from SGE, Ringwood, Australia). Carbohydrate retention in GCC is mainly based on adsorption, while some hydrophobic interactions are also expected. GCC is a unique adsorbent with some capabilities to resolve isomeric and other closely related carbohydrates.

The retentive properties of GCC are greater than what is typically observed with the reversed-phase materials. Thus, a greater percentage of organic solvent in the mobile phase is required to assist the separation of compounds as hydrophilic as carbohydrates. Another major advantage of GCC is their ruggedness, which apparently allows a repeated use without a loss of performance or retention reproducibility. Moreover, these columns are unaffected by strongly acidic or alkaline conditions and can be used throughout the entire pH range with a wide range of solvents. Thus far, these columns have been utilized in conjunction with MS for the analysis of glycans derived from different samples, including egg jelly surrounding *Xenopus laevis* eggs [85–87], human bronchial epithelial cell cultures [88], membrane proteins from premature aging Huchinson-Gilford progeria syndrome fibroblasts [89], human tear-fluid [90], and human blood plasma [91]. Moreover, microliter flow rates were employed in all these applications, allowing only the detection of structures present in abundance. This situation has substantially improved recently as a result of the introduction of nanocolumns, which are utilized at much lower flow rates.

The use of graphitized carbon nano-LC-MS has allowed *N*- and *O*-glycans analysis at a substantially higher sensitivity than with other column types, thus allowing the analysis to be performed on low-abundance glycoproteins. Three orders of magnitude increase in sensitivity was observed as a result of using nanoflow LC-MS instead of microflow LC-MS (Figure 1.10) [92]. As a result, negative-ion MS in conjunction with graphitized carbon nano-LC-MS allowed the simultaneous detection of femtomole amounts of both acidic and neutral oligosaccharides. Although baseline resolution of several structural isomers is common with GCC, many of these structures were not detected using the larger column format at microflow rates. Moreover, *N*- and *O*-glycans derived from glycoprotein mixtures could be isolated from mucosal surfaces and ovarian cancer cells [92].

In the picomole concentration range, anion-exchange chromatography of glycans with pulsed-amperometric detection has become one of the standard methodologies in glycobiology because of its simplicity and quantitative reliability. While mass spectrometers are capable of detecting glycans at the levels that is many orders of magnitude lower than those measured by amperometric detectors, interfacing ion-exchange LC to MS is not convenient due to a limited compatibility of the mobile phases with the MS instrumentation. Several technical modifications [93–97] were made to address this problem, but neither approach was particularly successful.

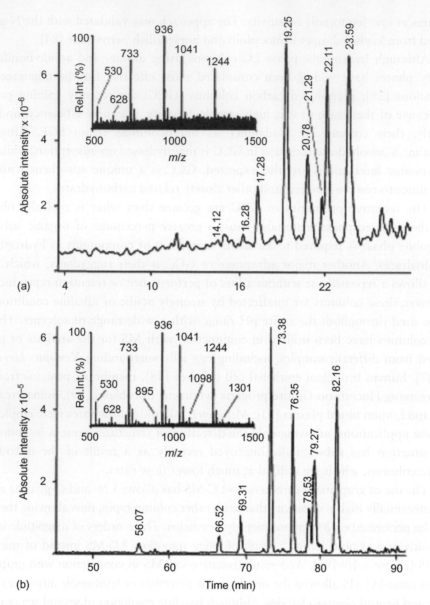

**FIGURE 1.10** Comparison of negative ion (a) micro liquid chromatography/mass spectroscopy (LC-MS) versus (b) nano-LC-MS analysis of neutral O-linked oligosaccharides (5.5 ng) using graphitized carbon chromatography (base-peak chromatograms). Combined MS1 mass spectra of the region where oligosaccharides were eluted are shown as insets. Reproduced from [92], with permission.

## Mass Spectrometry

### Mass Spectrometry of Native N-Glycans

MALDI-MS of native N-glycans is most widely achieved using 2,5-dihydroxybenzoic acid (DHB) as a matrix [98]. Since the intensities of MALDI-MS signals are very susceptible to the matrix effects and the method of sample preparation, crystallization of the DHB matrix and its influence on the signal of analyzed glycans were thoroughly investigated [99,100]. DHB, prepared in a mixture of acetonitrile or methanol and water, typically crystallizes as long needle-shaped crystals that are formed at the periphery of the spot and project toward the center. The central region of the spot is commonly formed of an amorphous mixture of the analyte, contaminants, and salts. Stahl et al. [100] reported that in a mixture of glycans and glycoproteins, glycans were fractioned in the central region of the spot while the glycoproteins were in the periphery, as suggested by the spectral intensities. Recrystallization of the spot using dry ethanol often results in the formation of a more even film of crystals, which is expected to generate stronger signals [99]. This increase in sensitivity by an order of magnitude is attributed to a more efficient mixing of a matrix and analyte. The formation of an even film of crystals is also attained through vacuum drying.

An alternative to the DHB matrix is the one that is commonly referred to as "super-DHB" [101] This matrix consists of DHB and other substituted benzoic acids or related compounds, of which 2-hydroxy-5-methoxybenzoic acid seems to be the most effective. The inclusion of this additive causes a disorder of the crystal lattice, thus allowing a "softer" desorption. The mixture was shown to achieve a two- to three-fold increase in sensitivity for the dextrin standard as well as a resolution increase attributed to the reduction in metastable ion formation [101].

Although other matrices have been tested for the analysis of carbohydrates, none have enjoyed the popularity of the aforementioned ones. Different matrices that have been successfully used for the analysis of carbohydrates include 3-aminoquinoline [102], potassium hexacyanoferrate and glycerol [103], hygroscopic tetrabutylammonium bromide, chloride, or acetate [104], 5-chloro-2-mercaptobenzothiazole (CBMT) [105], β-carbolines [106], ferulic acid [107], and 2,5-dihydroxyacetophenone [108]. Generally, DHB is currently considered as the matrix of choice for this analytical task.

Negatively charged *N*-linked glycans, such as the sialic acid-containing structures, generally give poor MALDI spectra when ionized with matrices such as DHB. Sialylated and some sulfated glycans are very labile and often endure fragmentation, resulting in the loss of either sialic acid or carbon dioxide from the sialic acid, or a sulfate, respectively. This problem can be somewhat remedied by alternative substances. Accordingly, it has been demonstrated that the use of 6-aza-2-thiothymine (ATT) as a matrix gives a significant increase in sensitivity for acidic glycans over what is produced by the more common matrices [109]. This matrix improved detection of *N*-linked glycans by approximately tenfold over DHB. However, it did not prevent fragmentation and the loss of sialic acid in the linear mode of detection. The loss of sialic acid in the reflectron mode was determined to be even more substantial.

A mixture of 2,4,6-trihydroxyacetophenone (THAP) and ammonium citrate has been shown to be effective in the analysis of sialylated glycans [109]. This mixture gives a single negative ion peak from sialylated *N*-linked glycans and allows their detection at 10 fmole level in the linear mode with no evidence of fragmentation (Figure 1.11). However, like the other matrices, a major loss of sialic acid in the reflectron mode was observed. The signal intensity observed with the THAP matrix was dependent on sample

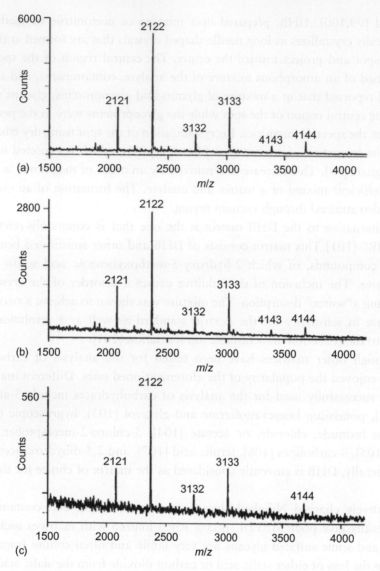

**FIGURE 1.11** Negative-ion matrix-assisted laser desorption/ionization (MALDI) mass spectrum of 2.5% of the *N*-linked oligosaccharides released from (a) 1 mg, (b) 0.5 mg, and (c) 0.1 mg of recombinant tissue-type plasminogen activator. The spectra was acquired using 2,4,6-trihydroxyacetophenone (THAP) as the matrix and smoothed with a 19-point Savitzky–Golay function. Reproduced from [109], with permission.

preparation conditions. The best results are attained through vacuum-drying of the sample spot, preventing formation of large crystals. Moreover, the formation of small crystals using this matrix is achieved by allowing the sample to absorb water from the atmosphere, which is critical for sample desorption and ionization.

The use of the base spermine as a DHB co-matrix has been demonstrated as an effective means for the determination of sialylated glycans [110]. Since the carboxylic acid groups of sialic acids have a tendency to form salts with alkali metals and distributing the signal intensity among multiple peaks in the spectra, the addition of spermine was found to reduce (if not eliminate) sodium salt formation. Moreover, spermine also provides a good crystalline surface. Therefore, ca. 50 fmol detection limits in the negative-ion mode become attainable using spermine/DHB mixture. The matrix was effective in analyzing standard sialylated *N*-linked glycans as well as the glycans derived from fetuin and $\alpha_1$-acid glycoprotein [110].

Due to the sensitivity problems associated with large glycans, excessive sample consumption and mass discrimination, ESI-MS of the native glycans did not originally match the popularity of MALDI-MS. However, this is now changing because of the advent of nanospray. In 1992, the potential of ESI-MS in the characterization of native *N*-linked glycans derived from glycoproteins was exploited for the first time. Sialylated oligosaccharides were detected best in the negative-ion mode, while neutral glycans were detected best in the positive-ion mode [111]. The primary structural features of the studied oligosaccharides were also determined by tandem MS. The MS-MS characterization of the native complex oligosaccharides yielded fragment ions that resulted from the cleavages of glycosidic bonds, while no linkage determination was feasible. The composition and relative abundances of the ovalbumin-derived oligosaccharides, as analyzed by ESI-MS, matched the results obtained by other analytical techniques such as HPAEC-PAD and gel permeation chromatography [111]. Since 1992, substantial advances in ESI-MS have occurred.

Harvey [112] investigated the influence of alkali metals on both the ionization and fragmentation of *N*-linked glycans examined with an ESI-QTOF mass spectrometer. The glycans were ionized most effectively as the adducts of alkaline metals, with lithium providing the most abundant signal and cesium the least. Although significant effort was initially aimed at direct analysis of *N*-glycans through ESI-MS, most successful work focused on using this mode of ionization in conjunction with chromatography or electrically driven separations, as we have described already. The effectiveness of a particular ionization technique (i.e., the use of ESI vs. MALDI) must be considered in view of the relatively inefficient ionization of oligosaccharides when compared to peptides. While the MALDI process forms exclusively the sodium adduct ions of oligosaccharides in the positive-ion mode, multiply charged states are formed in ESI, causing a loss in sensitivity as a result of dividing the signal of a single analyte over several charge states.

## Mass Spectrometry of Derivatized N-Glycans

Derivatization of glycan structures has been demonstrated to improve substantially the sensitivity of MS measurements. Among different approaches, permethylation of glycans prior to an MS measurement has significant analytical merits such as better ionization and fragmentation as well as the stabilization of sialylated (acidic) oligosaccharides.

Linsley et al. [113] demonstrated early on the enhanced sensitivity attained when analyzing permethylated *N*- and *O*-linked glycans derived from recombinant erythropoietin. The glycans were permethylated and profiled through ESI-MS without desialylation. On the other hand, Okamoto et al. [114] achieved an improvement in sensitivity of ESI-MS measurements through derivatization of glycans with trimethyl(*p*-aminophenyl)-ammonium chloride (TMAPA). The TMAPA derivatives exhibited extremely high sensitivity in the positive-ion ESI-MS and provided as Y- and Z-fragment ions by ESI-MS-MS. Since then, the utility in ESI-MS for the analysis of glycoprotein glycans has been growing continuously.

ESI-MS was used in conjunction with reversed-phase LC to characterize permethylated oligosaccharides [115]. Weiskopf et al. [116,117], Viseux et al. [118], and Sheeley and Reinhold [119] described the use of ESI/quadrupole ion trap (QIT)-MS, with its capacity to perform the multiple stages of fragmentation ($MS^n$), for the characterization of permethylated N-linked glycans derived from glycoproteins. Collisional activation of the permethylated oligosaccharide molecular ions ($MS^2$) produced abundant fragments from the glycosidic bond cleavages which depicted composition and sequence. However, these fragment ions do not furnish any information related to linkages. Consequently, selected fragments were trapped and further dissociated (during $MS^n$), now permitting interpretation and confirmation of cross-ring cleavage products. The mixtures of isobaric oligosaccharides, which were ionized and introduced into the trap simultaneously, were resolved and further examined in isolation by selection of $MS^2$ or $MS^n$ specific to only one glycomer. This potential of ion trap was demonstrated for the $GlcNac_2Man_5$ oligosaccharide from ribonuclease B and two isobaric HexNAc5 oligosaccharides from ovalbumin [116]. The higher order MS experiments further illustrated the potential of ESI-QIT-MS for oligosaccharide analysis, as $MS^8$ was used to illustrate a significant branching information for an oligosaccharide from ovalbumin (Figure 1.12) [117]. ESI-QIT-MS was also effective in characterization of monosialylated N-linked glycans [119]. The applicability of this technique was further extended to characterize a series of subunits generated from fucosylated and sialylated oligosaccharides [118].

Although permethylation of oligosaccharides has been effectively utilized for the characterization of glycans as discussed above, until recently [120,121], this time-honored derivatization technique of carbohydrate chemistry appeared unreliable for quantitative measurements due to incomplete methylation and degradation of some structures in the derivatization process. Consequently, a solid-phase permethylation of oligosaccharides in mixtures was developed at microscale [121], overcoming the quantitation difficulties with previous procedures. In using MALDI-MS as the glycomic profiling approach, it has become feasible to include both the neutral and acidic glycans in one profile, as shown in Figure 1.13. The figure compares representative glycomic maps of a cancer patient to an individual free of the disease [122]. Importantly, these samples were derived from only 10-μL aliquots of human blood serum, demonstrating very high sensitivity for today's analytical glycobiology.

Naven and Harvey [123] devised a cationic derivatization of oligosaccharides with Girard's T reagent for improved performance in ESI-MS. The oligosaccharides were derivatized to form hydrazones in order to introduce a cationic site needed for detection by ESI-MS. The derivative exhibited a high yield and did not require extensive clean-up prior to MS examination, which was an advantage over reductive amination. Such derivatives offered a tenfold increase in detection sensitivity over the underivatized oligosaccharides and provided intense spectra in the positive-ion ESI without the need to add cations to the solvent.

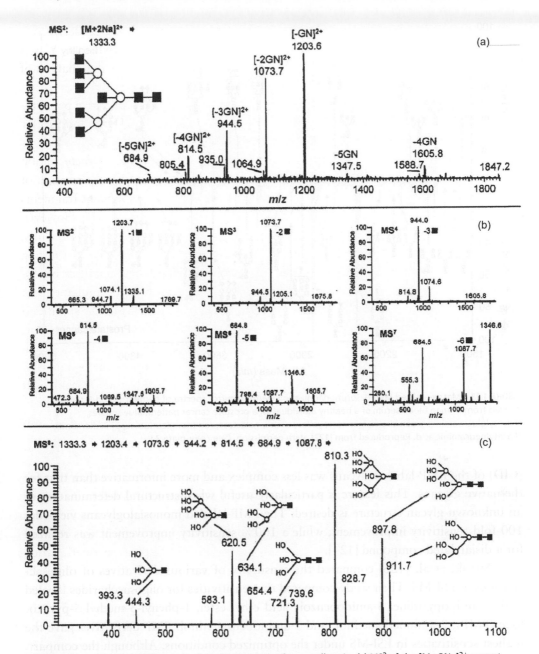

**FIGURE 1.12** High-order MS$^n$ analysis of the GlcNAc$_8$Man$_3$ from ovalbumin: (a) MS$^2$ of the [M+2Na]$^{2+}$ parent ion, *m/z* 1333.3; MS$^2$–MS$^7$ stepwise removal of *N*-acetylglucosamine residues; (c) MS$^8$ analysis of the exposed saccharide core for determination of branching patterns. Filled squares, GlcNAc; open circles, Man. Reproduced from [117], with permission.

Perreault and co-workers [124,125] discussed the advantages of labeling free *N*-linked glycans with 1-phenyl-3-methyl-5-pyrazolone (PMP), for HPLC and ESI-MS. The studies focused on some asialo and sialylated glycans, comparing HPLC and ESI-MS behaviors of the PMP-labeled glycans against the native counterparts. PMP-asialo glycans did not yield a significant increase in sensitivity relative to the native structures; however, fragmentation produced by in-source collision-induced dissociation

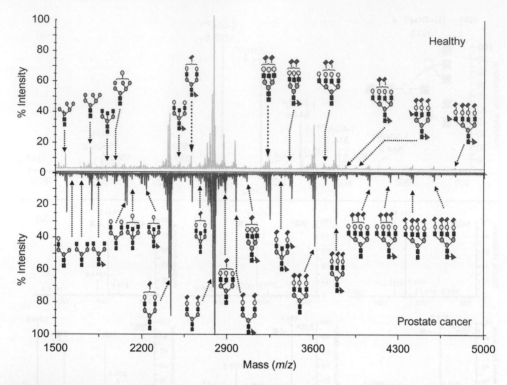

**FIGURE 1.13**   Matrix-assisted laser desorption/ionization (MALDI) mirror spectra of permethylated *N*-glycans derived from human blood serum of a healthy individual vs. a prostate cancer patient. Blue squares, *N*-acetylglucosamine; green circles, mannose; yellow circles, galactose; red triangles, fucose; pink diamonds, *N*-acetylneuraminic acid. Reproduced from [122], with permission. (See color plate 3.)

(CID) of the PMP-labeled glycans was less complex and more informative than that for the native glycans. This feature is particularly useful when structural determination of an unknown glycan structure is desired. The PMP-labeled monosialoglycans yielded a 100-fold sensitivity improvement, while a 100% sensitivity improvement was realized for a disialylated compound [124].

Suzuki et al. [126] compared the sensitivity of various derivatives of oligosaccharides in ESI-MS. Their study compared the sensitivities for oligosaccharides labeled with α-aminopyridine, 4-aminobenzoic acid ethyl ester, 1-phenyl-3-methyl-5-pyrazolone (PMP), 2-aminoethanethiol, and 2-aminobenzenethiol. PMP derivatives gave the highest sensitivities in ESI-MS under the optimized conditions. Although the comparison study was based on linear oligosaccharides, PMP labeling allowed the microanalysis of *N*-linked glycans derived from ovalbumin and porcine thyroglobulin.

## Mass Spectrometry of N-Glycans Derivatized with Stable Isotopes

An inherent advantage of MS measurements is the capability of including stable isotope labels for the sake of quantification. The use of isotopic labeling for glycan identification is further discussed in Chapters 2 and 4. Although stable isotopic labeling has found widespread use in the proteomics field, its application to carbohydrate quantification has been limited and is only beginning to emerge. Yuan et al. [127] and Hitchcock et al. [128] were the first to demonstrate the feasibility of stable isotope glycan tags in which isotopically varied reductive amination tag was used to quantify the relative amounts of carbohydrate present in their samples through LC-MS. In addition, Hsu et al. [129] used a synthesized affinity tag that incorporates a biotin moiety and allows for the incorporation of a single deuterium atom via reduction with sodium borodeuteride; however, the mass shift of 1 Da may not suffice for quantitation due to the overlapping isotopic envelopes.

Bowman and Zaia reported the design, synthesis, and application of a novel series of compounds that allow for the incorporation of isotopic variation within the glycan structures [130]. The novel feature of the compounds is the ability to incorporate the isotopes in a controlled manner, allowing for the generation of four tags that vary only in their isotopic content. This allows for the direct comparisons of three samples or triplicate measurements with an internal standard during one mass spectral analysis. Deuterated tags were designed for use with the widely accepted reductive amination chemistry. The modules are connected with amide bonds that are less labile than the glycosidic bonds and are unlikely to fragment under the conditions used for glycan tandem MS. They incorporate a fluorophore useful for UV or fluorescence-based detection and absolute quantification. Three distinct modules create four different isotopic variations of the tags (Figure 1.14). A reference glycan mixture is labeled with the light form of the tag, and three unknown glycan mixtures are labeled with the heavy forms. The MS dimension of tagged tetraplex glycan mixtures determines the monosaccharide compositions of the ions and their quantities relative to the standard mixture. Each

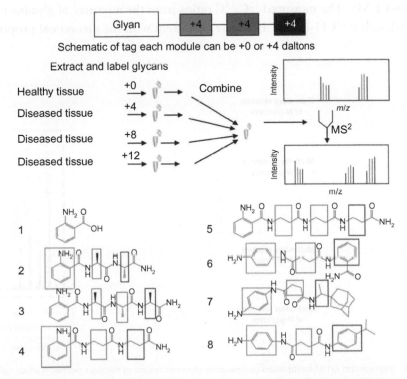

**FIGURE 1.14**   Representation of tetraplex tags and experimental flowchart and synthesized compounds indicating potential sites for isotopic modification. Reproduced from [130], with permission. (See color plate 4.)

ion in the MS mode reflects a mixture of isobaric glycoforms. The abundances of tandem MS product ions reflect the glycoforms that compose each precursor ion. Thus, the tandem mass spectra provide a means of following the expression of glycoform fine structure as a function of biological change. For the glycan classes where well-defined standards exist, the product ion abundances determine the glycoform mixture percentages. Fragmentation characteristics of glycoconjugate glycans vary significantly, as based on a compound class. In particular, the presence of acidic functional groups strongly influences the product ion pattern. Quantitation of partially depolymerized glycosaminoglycan mixtures, as well as $N$-linked glycans released from fetuin, was used to demonstrate the utility of the tetraplex tagging strategy.

An alternative approach is to permethylate glycans using different isotopically labeled methyl iodide reagents. Recently, we have expanded our solid-phase permethylation procedure, which commonly uses methyl iodide as the permethylation agent, to use deuteromethyl iodide, thus allowing us to simultaneously compare normal and pathological samples in a single MALDI-MS acquisition [131]. This approach was referred to as comparative glycomic mapping and abbreviated as C-GlycoMAP (Figure 1.15). This approach was recently expanded using singly, doubly, and triply deuteriated methyl iodide agent, thus allowing tetraplexing. The use of other isotopically labeled methyl iodide reagents, such as $^{13}CH_3I$, was also investigated [132]. This procedure was used to analyze $N$-linked glycans released from various mixtures of glycoproteins, such as $\alpha$-1 acid glycoprotein, human transferrin, and bovine fetuin. The MS techniques that were used included MALDI-MS and ESI with ion cyclotron resonance-FT-MS. The measured $^{12}C:^{13}C$ ratios from the mixtures of glycans permethylated with either $^{12}CH_3I$ or $^{13}CH_3I$ were consistent with the theoretical proportions.

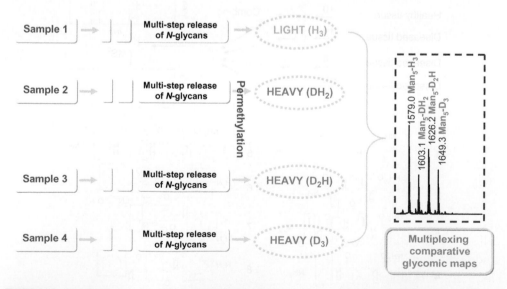

**FIGURE 1.15** Representation of multiplexed comparative glycomic mapping through permethylation using stable isotopic methyl iodide reagents. This approach is an extension of C-GlycoMAP [131]. (See color plate 5.)

## Tandem Mass Spectrometry

While mass determination through MALDI-MS can often lead to compositional data (in terms of isobaric monosaccharides) additional information must be secured through other methodologies. Monosaccharide sequences, branching and, in some cases, linkages can be determined through fragmentation that a glycan may experience in either a post-source decay or a collision-induced dissociation. However, the amounts of material, which are available, may not be sufficient for a full characterization and conclusions can often be vague.

Fragmentation of glycans observed in MALDI-MS is similar to that observed in FAB-MS and ESI-MS and is dependent on many factors such as ion formation, its charge state, the energy deposited into an ion, and the time available for fragmentation. In general, glycans undergo two types of cleavages: (a) glycosidic cleavages that result from breaking the bond linking two sugar residues, and (b) cross-ring cleavages that involve rupturing two bonds on the same sugar residue. The former provide information pertaining mainly to the sequence and branching, while the latter may reveal some details on a linkage.

Domon and Costello [133] introduced the nomenclature pertaining to fragmentation of carbohydrates. According to this nomenclature, the ions retaining the charge at the reducing terminus are designated as X for cross-ring cleavages, and Y and Z for glycosidic cleavages. Those retaining a charge at the non-reducing terminus are designated A for cross-ring cleavages, and B and C for glycosidic cleavages. Ions are designated by a subscript number that follows the letter showing the fragment type. Sugar rings are numbered from the nonreducing end for A, B, and C ions and from the reducing end for the others. Greek letters are used to distinguish fragments from branched-chain glycans, with the letter $\alpha$ representing the largest branch. In the case of ring cleavages, superscript numbers are given to show the ruptured bonds.

Ngoka et al. [134] and Cancilla et al. [135] determined that protonated species decompose much more readily than metal-cationized species, and that the order in which cationized species decompose is $Li^+>Na^+>K^+>Cs^+$. This is prompted by the exothermicity of cation binding, which follows the same order, so that protonated and lithiated species are formed with the excess energy that can be utilized to initiate fragmentation. Moreover, it was also observed that linear carbohydrates produce more fragmentation than the branched ones [135]. This has also been attributed to coordination of the metal ions.

Fragmentations in MALDI-MS can result from (a) the post-source decay (PSD) which designates the fragments formed after ion extraction from the ion source, (b) in-source decay (ISD) which designates the fragments formed within the ion source, and (c) collision-induced dissociation (CID) which designates the fragments formed in a collision cell filled with a gas. PSD spectra of sodiated ions from neutral carbohydrates tend to be dominated by the glycosidic and internal cleavages with very weak cross-ring ions. This abundance of internal and glycosidic cleavages makes spectral interpretation somewhat difficult. Major ions are usually the result of B- and Y-cleavages and the information related to sequence and branching. A lack of abundant cross-ring cleavages limits the linkage information to be deduced. The lack of cross-ring fragmentation in PSD is attributed to their high-energy requirements.

The most useful cross-ring cleavage ions in the spectra of N-linked glycans are the $^{3,5}A$ and $^{0,4}A$ ions produced by cleavages of the core-branching mannose residues. These ions contain only the antenna attached to the 6-position, providing a wealth of information on the composition of each antenna. Spengler et al. [136] reported $^{0,4}A$ ions in the PSD spectra of bi- and triantennary N-linked glycans. These two cross-ring cleavage ions were used to determine the antenna configuration in high-mannose N-linked glycans derived from hemoglobin extracted from tubeworms [137].

The incorporation of delayed ion extraction (DE) technique into MALDI-MS caused a dramatic improvement in sensitivity, mass resolution, and mass accuracy of the precursor ions up to ~10 kDa. However, under DE the loss of total PSD fragment ion yield can be as large as one order of magnitude [138]. A part of this loss was balanced by a better signal-to-noise ratio, which resulted from a significantly improved mass resolution of the PSD fragment ions. While this compensating effect was true for the middle to high mass range of the PSD fragment ions, it gradually diminished toward the low mass scale. As shown in the case of linear peptides, some important information can be lost [138]. Although DE adversely affects PSD fragmentation, it improves the overall quality of PSD spectra, while any loss of analytical information can be compensated for by CID.

Lemoine et al. [139] demonstrated that derivatization can significantly enhance the quality of PSD mass spectra and thus the information they facilitate. Peracetylated and, to some extent, permethylated oligosaccharides yielded useful preliminary information. However, deuteroacetylation allowed a clear distinction between the hexose and N-acetylhexosamine residues. The best results were obtained with benzylamino derivatives, which offered a basic site allowing the formation of protonated ions (which, as discussed above, easily fragment), thus providing sequence and branching information [139].

Rouse et al. [140] devised a strategy to characterize unknown isomeric N-linked glycan structures that was referred to as "knowledge-based." The strategy was based on comparing specific fragment ion types and their distributions in the unknown PSD spectrum to those in the PSD spectra of standards possessing similar structural features. A precursor ion selection device was employed to isolate the component of interest from the mass profile without any chromatographic isolation steps. This PSD knowledge-based isomeric differentiation strategy permits to distinguish the respective isomers for certain high-mannose and asialo complex N-glycan standards that were characterized previously by NMR spectroscopy. This strategy also provides the means to identify oligosaccharide isomer types, to pinpoint the locations of structural variations, to recognize isomerically pure oligosaccharides from isomeric mixtures, and to estimate the major and minor oligosaccharide isomer species in a mixture.

More recently, Yamagaki and Nakanishi exploited the potential of a MALDI mass spectrometer with the curved-field reflectron for PSD fragmentation analyses of the linkage isomers of oligosaccharides [141,142]. A reflectron of the curved voltage gradient type, which was first introduced by Cornish and Cotter [143,144], allows to acquire a total fragment spectrum in a single experiment. This type of reflectron employs a modified single-stage reflector; the gradient differential increases of its axial voltage produce an alignment of energy focal points for product ions. Therefore, the parent ion and all the fragment ions are detected under the same measurement conditions. A major advantage of this reflectron is an accurate comparison of the intensity of the differentially observed fragments, since they are all detected under the same conditions. As in the case of a conventional reflectron, glycosidic cleavage ions are predominant, while cross-ring cleavage ions are weak or absent.

Generally, CID fragmentation spectra contain more abundant cross-ring fragments than are normally produced under ISD and PSD conditions [145]. The very abundant $^{1,5}X$-ions are of particular significance. Their masses could be used to determine the branching patterns of glycans such as the high-mannose sugars.

The effect of the reducing-terminal substituents on the high-energy CID MALDI spectra of oligosaccharides has been investigated [146]. The study compared the CID spectra of oligosaccharides bearing different, commonly encountered, reducing-terminal modifications including hydroxyl, 2-aminobenzamide-labeled, and asparagine- and tetrapeptide-linked substances. All compounds formed abundant sodiated molecular and fragment ions, the latter corresponding to glycosidic and cross-ring cleavages as well as to internal fragment ions. However, the nature of the modification at the reducing end considerably influenced CID behavior. The strongest and most complete series of glycosidic cleavage ions were observed for the underivatized oligosaccharides, while most cross-ring fragment ions (diagnostic of a linkage) were observed for the glycopeptides. A-type cross-ring cleavage ions were abundant in the spectrum of asparagine derivatives. Reductive amination of the reducing end with 2-aminobenzamide resulted in suppression of the cross-ring fragment ions, prompted by the presence of an opened reducing terminal sugar as a result of labeling.

Penn et al. [147] reported MALDI-CID spectra of milk sugars and *N*-linked glycans using a Fourier transform mass spectrometer fitted with an external ion source. A fucose loss was observed as the favorable glycosidic cleavage with the lowest relative dissociation threshold. This threshold increased when cesium was used as the cationization species, being attributed to complexation between cesium and fucose. Dissociation thresholds for the appearance of cross-ring cleavage ions were determined to be higher than those due to glycosidic cleavages, correlating well with the appearance of these ions in the CID spectra. The ability to measure dissociation threshold values with the FT MS instrument can be employed to distinguish between different monosaccharide residues, thus allowing the determination of glycan composition. This type of instrument also permits a storage of ions for considerable periods of time, making it possible to carry $MS^n$.

In a different communication, the same group proposed a method for obtaining the relative dissociation thresholds using a multiple-collision CID [148]. The multiple-collision dissociation thresholds (MCDT) values were used to examine the relative affinities of the alkali metal ions for oligosaccharides and the effect of alkali metal ions on oligosaccharide fragmentation (including glycosidic bond and cross-ring cleavages). According to these MCDT values, the activation barriers for glycosidic bond cleavages were found to be dependent on the size of alkali metal ions, as was determined previously for MALDI-TOF-MS (see above). Moreover, cross-ring cleavages were found to be independent of the alkali metal ion, but dependent on a linkage type. The results also suggested that glycosidic bond cleavages were charge-induced, while cross-ring cleavages were a charge-remote process.

The application of MALDI-FT-MS was extended to characterization of permethylated oligosaccharides [149]. The study evaluated several aspects of MALDI-FT-MS for carbohydrate structural analyses including sustained off-resonance irradiation (SORI), CID, quadrupolar axialization, multiple stages of isolation and dissociation ($MS^n$), and ion remeasurement. It was determined that SORI/CID internal energies were adequate for the linkage analysis of permethylated glucose oligomers. In FT-MS, ions continually drift out of the trap as a result of the exponential increase in the off-axis displacement of the center of an ion cyclotron orbit (i.e., the magnetron radius) with time. This magnetron radial expansion is also observed during sequential dissociation. Nevertheless, magnetron radial expansion is rectified by ion axialization through cyclotron-resonant quadrupolar excitation in the presence of collision damping. Therefore, ion axialization greatly improves measurement efficiency, being a crucial component of the ion control required for $MS^n$. Accordingly, ion remeasurement and axialization techniques enhance the sensitivity of ion fragmentation analysis.

Harvey et al. [150] reported for the first time the utility of QTOF-MS fitted with the MALDI ionization source for ionization and fragmentation of complex glycans, particularly the N-linked glycans derived from glycoproteins. Positive-ion spectra of sialylated glycans obtained on this instrument were simpler than those obtained with a conventional MALDI instrument because of the absence of ions resulting from metastable fragmentations occurring in the flight tube. MS-MS spectra of the sodiated ions from all compounds were far superior to MALDI-PSD spectra recorded with a conventional MALDI instrument. Fragmentation of the parent ion was dominated by B- and Y-type glycosidic cleavages, leading to both primary and internal fragment ions. Less abundant cross-ring cleavage ions providing linkage information were also detected. The utility of this instrument was demonstrated for the native and derivatized N-linked glycans cleaved from ribonuclease B and human secretory IgA and ApoB100 from human low-density lipoproteins.

A suitable general platform for functional glycomics must, first of all, consider the best possible structural technique that is capable of distinguishing different forms of isomerism at high sensitivity. During the recent past, distinctly different approaches and modern MS technologies were applied to satisfy this requirement. The use of sequential fragmentation in tandem MS (for example, $MS^n$ in an ion trap or quadrupole ion trap) was found to be effective in distinguishing isobaric structures among methylated branched oligosaccharides [151–157], albeit at the expense of a need for considerable sample quantity and limited compatibility with a prior LC separation. The effectiveness of a particular ionization technique (i.e., the use of ESI vs. MALDI) must also be considered in view of the relatively inefficient ionization of oligosaccharides when compared to peptides. While MALDI forms exclusively the sodium adduct ions of oligosaccharides in the positive-ion mode, multiply charged states are formed in ESI that cause a loss in sensitivity as a result of splitting the signal of a single analyte.

For the branched structures and differently linked residues in the oligosaccharides from typical glycoproteins, it is not sufficient to generate secondary fragments due to the cleavages at the sites of glycosidic bonds; sufficient energies must be available to the analyte sugar molecules to form the more informative cross-ring fragments. As the low-energy CID in the ion trap analyzers [57,59,158] or the post-source decay in MALDI-TOF-MS instruments [159] is generally insufficient to accomplish this fragmentation task, MALDI-TOF-TOF-MS with high-energy CID capabilities is now routinely employed to furnish extensive fragmentation of model neutral [160–164] and acidic [160,165] glycans. An alternative approach pursued for extensive fragmentation of glycans involves Fourier transform ion cyclotron resonance MS used in conjunction with SORI [166], CID [86,166], IR multiphoton dissociation (IRMPD) [167–169], or electron capture dissociation (ECD) [167,168]. It has been demonstrated that in MALDI-TOF-TOF experiments, the distinguishing between a subtle form of isomerism in the mannose-rich glycans [161] (Figure 1.16) is attainable from the ratios of diagnostic cross-ring fragments as well as different linkages of sialic acid residues and/or their attachment to a specific antenna within a branched structure [165]. Much of this structurally/functionally important information had previously

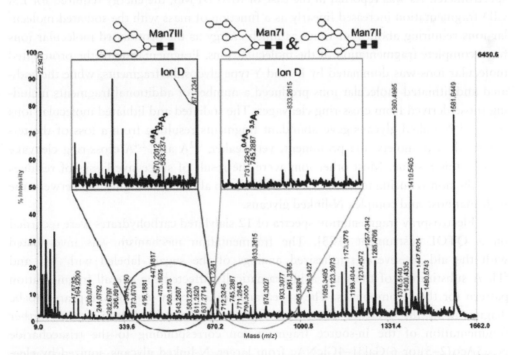

**FIGURE 1.16**   Tandem mass spectroscopy recording of GlcNAc$_2$Man$_7$. Inset represents a zoomed-in section of the spectrum. Reproduced from [161], with permission.

been available only through NMR spectrometry which generally requires highly concentrated samples.

A hybrid quadrupole orthogonal acceleration time-of-flight mass spectrometer (QTOF) in the precursor ion scanning and the MS-MS modes was utilized for the analysis of 2-aminoacridone-labeled N-linked glycan mixtures. The use of a precursor ion scanning strategy on this instrument provides a rapid and sensitive method for screening glycan mixtures without a prior separation by chromatographic methods. Although this will not furnish information related to linkages between the various sugar residues, it allows a facile and preliminary characterization of glycans into different classes such as high-mannose or complex types. Using the MS-MS mode of this instrumentation permits, with selected glycans, to determine sequences and provide branching information. This methodology allowed characterization of the N-linked glycans derived from ovomucoid, ovalbumin, and porcine thyroglobulin. The same instrument was also utilized for the analysis of 2-aminobenzamide-labeled N-linked glycans from ovalbumin and related glycoproteins from egg white [113], N-(2-diethylamino)ethyl-4-aminobenzamide-labeled standard complex, hybrid, and high-mannose N-linked glycans [170], and 2-aminoacridone-labeled standard high-mannose, hybrid and complex N-linked glycans [171], and 4-aminobenzoic acid 2-(diethylamino)ethyl ester-labeled N-linked glycans from IgY and thyroglobulin [172].

Similar to what Ngoka et al. [134] and Cancilla et al. [135] reported using MALDI mass spectrometers, Harvey [112] investigated the influence of alkali metals on the fragmentation of N-linked glycans examined with an ESI-QTOF mass spectrometer. As was reported in the case of MALDI-MS, the energy required for ESI CID fragmentation increased linearly as a function of mass with the sodiated molecular ions requiring about four times as much energy as the protonated molecular ions for a complete fragmentation of the molecular ions. Fragmentation of the protonated molecular ions was dominated by B- and Y-type glycosidic fragments, while the sodiated and lithiated molecular ions produced a number of additional fragments including those derived from cross-ring cleavages. The sodiated and lithiated molecular ions from all N-linked glycans gave abundant fragments resulting from a loss of the terminal GlcNAc moiety and prominent, yet weaker, $^{0,2}$A and $^{2,4}$A cross-ring cleavage ions of this residue. Most other ions were the result of successive losses of residues from the non-reducing terminus. MS-MS data also allowed distinguishing between the high-mannose and complex N-linked glycans.

Electrospray fragmentation spectra of 12 sialylated carbohydrates were recorded on a QTOF instrument [173]. The fragmentation mechanism was investigated with the aid of several synthesized analogs of the sugars labeled with $^{13}$C and $^2$H. A substitution of the sialic acid dramatically affected the overall fragmentation pattern for these compounds. The appearance of an ion at m/z 306 appeared to be diagnostic of the presence of an α(2–6)-linked sialic acid. Selection and further fragmentation of the in-source fragment ion corresponding to the trisaccharide Neu5Acα(2–3)(or 6)Galβ1–4GlcNAc from larger, N-linked glycans, ionized by electrospray, gave fragmentation patterns identical to those of the reference trisaccharides, thus furnishing a method for confirming a sialic acid linkage.

## Conclusions

During the last 10–15 years, there has been remarkable proliferation of sophisticated mass spectrometric techniques and instrumentation in the biomedical field. While these developments have largely been driven by the needs and expectations in the area of protein science and proteomic measurements, contemporary glycobiology benefits substantially from these developments as well. As described in this chapter, there is a dazzling array of new instrumental approaches and MS tools that can address the key structural and functional aspects of glycoconjugates in ways probably unforeseen by a previous generation of scientists. The new types of mass analyzers and tandem MS techniques now work at unprecedented sensitivity and mass resolution to determine sugar sequences, chain branching, linkages and additional aspects of the oligosaccharide structure. Mass spectrometers are relatively expensive instruments that generate vast amounts of useful analytical data. Their justification for future glycobiology research, and perhaps even routine measurements in biology and medicine, will increasingly rely on the needs for high-quality structural work and high analytical throughput.

Mass spectrometers alone are unlikely to provide all needed structural information in future glycomic and glycoproteomic investigations. As emphasized previously, a complete structural analysis of glycoproteins often becomes a complex multimethodological task. It is thus interesting to observe the current fusion of compatible approaches in the field for the sake of sensitive analysis: miniaturization of the well-established chemical and biomedical principles and the modified techniques for sample treatment, preconcentration and fractionation; the use of modern separation tools such as capillary liquid chromatography or capillary electrophoresis; and the most definitive measurements of the glycan masses and their structural attributes through mass spectrometry. Whether working on the structures of isolated glycoproteins or on complex mixtures of biomedically relevant glycoproteins, these approaches must be utilized in the integrated fashion with each other. For example, imperfect sample treatment and preconcentration cannot be bottlenecks for the following MS measurements.

While there seem currently multiple options in approaching different analytical tasks in glycobiology, as shown in this chapter, the current procedural optimization efforts may lead to more stabilized and unified analytical platforms in the next few years. The instrumental design and improvements in MS are likely to continue in future years with respect to reaching high sensitivity and further gains in mass spectral resolution. The unique advantage of MS-based quantification is the use of stable isotopic labels for glycans, as exemplified here with the sugar permethylation techniques.

The propensity of glycan structures to exist and function as different isomeric forms is clearly one of the most analytically challenging themes of our field and among the overwhelming considerations for the use of separation methodologies that combine well with mass spectrometry. There has been progress in separating some sugar isomers through the techniques of both capillary LC and CE, albeit not without difficulties. The current intensive research aiming at the optimization of CE-MS coupling is a good example of these efforts. Far from routine, this is one area where significant gains can be achieved in the near future.

While our chapter has been primarily centered around new analytical methods/ instrumentation and applications to *N*-glycans, high-sensitivity analysis of *O*-linked oligosaccharides has become the next important target of investigations where new chemical treatment approaches and powerful analytical technologies must meet for further advances.

## Acknowledgments

This work was supported by grant No. CA128535–01 from the National Cancer Institute, U.S. Department of Health and Human Services. Further support was provided by the Indiana Metabolomics and Cytomics Initiative (METACyt), grant No. GM24349 from the National Institute of General Medical Sciences, U.S. Department of Health and Human Services, and by NIH/NCRR–National Center for Glycomics and Glycoproteomics (NCGG), grant No. RR018942.

# References

1. Liu J, Shirota O, Novotny M. *J Chromatogr A* 1991;**559**:223.

2. Liu J, Shirota O, Novotny MV. *Anal Chem* 1992;**64**:973.

3. Guttman A. *Nature* 1996;**380**:461.

4. Guttman A, Chen F-T, Evangelista RA. *Electrophoresis* 1996;**17**:412.

5. Guttman A. Multistructure sequencing of N-linked fetuin glycans by capillary gel electrophoresis and enzyme matrix digestion. *Electrophoresis* 1997;**18**:1136.

6. Ma S, Nashabeh W. *Anal Chem* 1999;**71**:5185.

7. Mechref Y, Muzikar J, Novotny MV. *Electrophoresis* 2005;**26**:2034.

8. Takasa S, Misuochi T, Kobata A. *Methods Enzymol* 1982;**83**:263.

9. Patel T, Bruce J, Merry A, Bigge C, Wormald M, Jaques A, Parekh R. *Biochemistry* 1993;**32**:679.

10. Carlson DM. *J Biol Chem* 1968;**243**:616.

11. Huang Y, Mechref Y, Novotny MV. *Anal Chem* 2001;**73**:6063.

12. Tarentino AL, Plummer TH. *Method Enzymol* 1994;**230**:44.

13. Muzikar J, Mechref Y, Huang Y, Novotny MV. *Rapid Commun Mass Spectrom* 2004;**18**:1513.

14. O'Neill RA. *J Chromatogr A* 1996;**720**:201.

15. Chen P, Baker AG, Novotny MV. *Anal Biochem* 1997;**244**:144.

16. Beavis RC, Chait BT. *Anal Chem* 1990;**62**:1836.

17. Strub J-M, Goumon Y, Lugardon K, Capon C, Lopez M, Moniatte M, van Dorsselaer A, Aunis D, Metz-Boutigue M-H. *J Biol Chem* 1996;**271**:28533.

18. Gorisch H. *Anal Biochem* 1988;**173**:393.

19. Marusyk R, Sergeant A. *Anal Biochem* 1980;**105**:403.

20. Kussmann M, Nordhoff E, Rahbek-Nielsen H, Haebel S, Rossel-Larsen M, Jakobsen L, Gobom J, Mirgorodskaya E, Kroll-Kristensen A, Palm L, Roepstroff P. *J Mass Spectrom* 1997;**32**:593.

21. Packer NH, Lawson AM, Jardine DR, Redmond JW. *Glycoconjugate J* 1998;**15**:737.

22. Bornsen KO, Mohr MD, Widmer HM. *Rapid Commun Mass Spectrom* 1995;**9**:1031.

23. Harvey DJ, Kuster B, Wheeler SF, Hunter AP, Bateman RH, Dwek RA. *Mass Spectrometry in Biology and Medicine*. Totowa, NJ: Humana Press; 1999.

24. Worrall TA, Cotter RJ, Woods AS. *Anal Chem* 1998;**70**:750.

25. Jacobs A, Dahlman O. *Anal Chem* 2001;**73**:405.

26. Rouse JC, Vath JE. *Anal Biochem* 1996;**238**:82.

27. Huang Y, Mechref Y, Tian J, Gong H, Lennarz WJ, Novotny MV. *Rapid Commun Mass Spectrom* 2000;**14**:1233.

28. Mechref Y, Novotny MV. *Chem Reviews* 2002;**102**:321.

29. Delinger SL, Davis JM. *Anal Chem* 1992;**64**:1947.

30. Li SFY. *Capillary Electrophoresis. Principles, Practice and Applications*. Amsterdam: Elsevier; 1992.

31. Liu J, Dolnik V, Hsieh Y-Z, Novotny MV. *Anal Chem* 1992;**64**:1328.

32. Liu J, Shirota O, Wiesler D, Novotny MV. *Proc Natl Acad Sci USA* 1991;**88**:2302.

33. Chiesa C, O'Neill RA. *Electrophoresis* 1994;**15**:1132.

34. Stefansson M, Novotny M. *Carbohydr Res* 1994;**258**.

35. Stefansson M, Novotny MV. *Anal Chem* 1994;**66**:1134.

36. Klockow A, Widmer HM, Amado R, Paulus A. *Fresenius J Anal Chem* 1994;**350**:415.

37. Klockow A, Amado R, Widmer HM, Paulus A. *J Chromatogr A* 1995;**716**:241.

38. Klockow A, Amado R, Widmer HM, Paulus A. *Electrophoresis* 1996;**17**:110.

39. Guttman A, Starr CM. *Electrophoresis* 1995;**16**:993.

40. Guttman A, Pritchett T. *Electrophoresis* 1995;**16**:1906.

41. Chen F-T, Evangelista RA. *Electrophoresis* 1998;**19**:2639.

42. Evangelista RA, Chen F-T, Guttman A. *J Chromatogr A* 1996;**745**:273.

43. Guttman A, Herrick S. *Anal Biochem* 1996;**235**:236.

44. Suzuki H, Mueller O, Guttman A, Karger BL. *Anal Chem* 1997;**69**:4554.

45. Campa C, Coslovi A, Flamigni A, Rossi M. *Electrophoresis* 2006;**27**:2027.

46. Mechref Y, Novotny MV. *J Chromatogr B* 2006;**841**:65.

47. Novotny MV, Mechref Y. *J Sep Sci* 2005;**28**:1956.

48. Novotny MV, Soini HA, Mechref Y. *J Chromatogr B* 2008;**866**:26.

49. Zamfir A, Peter-Katalinic J. *Electrophoresis* 2004;**25**:1949.

50. Zamfir A, Peter-Katalinic J. *Electrophoresis* 2001;**22**:2448.

51. Gennaro LA, Delaney J, Vouros P, Harvey DJ, Domon B. *Rapid Commun Mass Spectrom* 2002;**16**:192.

52. Sandra K, van Beeumen J, Stals I, Sandra P, Claeyssens M, Devreese B. *Anal Chem* 2004;**76**:5878.

53. Sandra K, Devreese B, Stals I, Claeyssens M, van Beeumen J. *J Am Soc Mass Spectrom* 2004;**15**:413.

54. Min JZ, Toyo'oka T, Kato M, Fukushima T. *Chem Commun* 2005:3484.

55. Gennaro LA, Salas-Solano O. *Anal Chem* 2008;**80**:3838.

56. Palm A, Novotny MV. *Anal Chem* 1997;**69**:4499.

57. Que A, Novotny MV. *Anal Bioanal Chem* 2003;**375**:599.

58. Que AH, Mechref Y, Huang Y, Taraszka JA, Clemmer DE, Novotny MV. *Anal Chem* 2003;**75**:1684.

59. Que AH, Novotny MV. *Anal Chem* 2002;**74**:5184.

60. Mechref Y, Chen P, Novotny MV. *Glycobiology* 1999;**9**:227.

61. Tegeler TJ, Mechref Y, Boraas K, Reilly JP, Novotny MV. *Anal Chem* 2004;**76**:6698.

62. Mechref Y, Zidek L, Ma W, Novotny MV. *Glycobiology* 2000;**10**:1.

63. Kanninen K, Goldsteins G, Auriola S, Koistinaho AIJ. *Neurosci Lett* 2004;**367**:235.

64. Novotny M. Microcolumn liquid chromatography in biochemical analysis. *Methods Enzymol* 1996;**270**:101–33.

65. MacNair JE, Lewis KC, Jorgenson JW. *Anal Chem* 1997;**69**:983.

66. MacNair JE, Patel KD, Jorgenson JW. *Anal Chem* 1999;**71**:700.

67. Wu N, Collins DC, Lippert JA, Xiang Y, Lee ML. *J Microcol Sep* 2000;**12**:462.

68. Xiang Y, Yan B, McNeff CV, Carr PW, Lee ML. *J Chromatogr A* 2003;**1002**:71.

69. Allen D, El Rassi Z. *J Chromatogr A* 2004;**1029**:239.

70. Huber CG, Choudhary G, Horvath C. *Anal Chem* 1997;**69**:4429.

71. Peters EC, Petro M, Svec F, Frechet JMS. *Anal Chem* 1997;**69**:3646.

72. Ishizuka N, Minakuchi H, Nakanishi K, Soga N, Nagayama H, Hosoya K, Tanaka N. *Anal Chem* 2000;**72**:1275.

73. Motokawa M, Kobayashi H, Ishizuka N, Minakuchi H, Nakanishi K, Jinnai H, Hosoya K, Ikegami T, Tanaka N. *J Chromatogr A* 2002;**961**:53.

74. Alpert AJ. *J Chromatogr* 1990;**499**:177.

75. Churms SC. *J Chromatogr A* 1996;**720**:75.

76. Birrell H, Charlwood J, Lynch I, North S, Camilleri P. *Anal Chem* 1999;**71**:102.

77. Charlwood J, Birrell H, Organ A, Camilleri P. *Rapid Commun Mass Spectrom* 1999;**13**:716.

78. Morelle W, Strecker G. *J Chromatogr B* 1998;**706**:101.

79. Thomsson KA, Karlsson H, Hansson GC. *Anal Chem* 2000;**72**:4543.

80. Guile GR, Rudd PM, Wing DR, Prime SB, Dwek RA. *Anal Biochem* 1996;**240**:210.

81. Rudd PM, Mattu TS, Masure S, Bratt T, van den Steen PE, Wormald MR, Kuster B, Harvey DJ, Borregaard N, van Damme J, Dwek RA, Opdenakker G. *Biochemistry* 1999;**38**:13937.

82. Rudd PM, Wormald MR, Harvey DJD, McAlister MMSB, Brown MH, Davis SJ, Barclay AN, Dwek RA. *Glycobiology* 1999;**9**:443.

83. Taverna M, Tran NT, Valentin C, Level O, Merry T, Kolbe HVJ, Ferrier D. *J Biotech* 1999;**68**:37.

84.  Wuhrer M, Koeleman CAM, Deelder AM, Hokke CH. *Anal Chem* 2004;76:833.

85. Xie Y, Liu J, Zhang J, Hedrick JL, Lebrilla CB. *Anal Chem* 2004;**76**:5186.

86. Zhang J, Lindsay LL, Hedrick JL, Lebrilla CB. *Anal Chem* 2004;**76**:5990.

87. Zhang J, Xie Y, Hedrick JL, Lebrilla CB. *Anal Biochem* 2004;**334**:20.

88. Holmen JM, Karlsson NG, Abdullah LH, Randell SH, Sheehan JK, Hansson GC, Davis CW. *Am J Physiol* 2004;**287**:L824.

89. Robinson LJ, Karlsson GN, Weiss AS, Packer NH. *J Proteome Res* 2003;**2**:556.

90. Schulz BL, Oxley D, Packer NH, Karlsson GN. *Biochem J* 2002;**366**:511.

91. Wilson NL, Schulz BL, Karlsson GN, Packer NH. *J Proteome Res* 2002;**1**:521.

92. Karlsson GN, Wilson NL, Wirth H-J, Dawes P, Joshi H, Packer NH. *Rapid Commun Mass Spectrom* 2004;**18**:2282.

93. Conboy JJ, Henion JD. *Biol Mass Spectrom* 1992;**21**:397.

94. Conboy JJ, Henion JD, Martin MW, Zweigenbaum JA. *Anal Chem* 1990;**62**:800.

95. Simpson RC, Fenselau CC, Hardy MR, Townsend RR, Lee YC, Cotter RJ. *Anal Chem* 1990;**62**:248.

96. Torto N, Hofte AJP, van der Hoeven RAM, Tjaden UR, Gorton L, Marko-Varga G, Bruggink C, van der Greef J. *J Mass Spectrom* 1998;**33**:334.

97. van der Hoeven RAM, Hofte AJP, Tjaden UR, van der Greef J, Torto N, Gorton L, Marko-Varga G, Bruggink C. *Rapid Commun Mass Spectrom* 1998;**12**:69.

98. Stahl B, Steup M, Karas M, Hillenkamp F. *Anal Chem* 1991;**63**:1463.

99. Harvey DJ. *Rapid Commun Mass Spectrom* 1993;**7**:614.

100. Stahl B, Thurl S, Zeng J, Karas M, Hillenkamp F, Steup M, Sawatzki G. *Anal Biochem* 1994;**223**:218.

101. Strupat K, Karas M, Hillenkamp F. *Int J Mass Spectrom, Ion Processes* 1991;**111**:89.

102. Metzger JO, Woisch R, Tuszynski W, Angermann R. *Fresenius J Anal Chem* 1994;**349**:473.

103. Zollner P, Schmid ER, Allmaier G. *Rapid Commun Mass Spectrom* 1996;**10**:1278.

104. Breuker K, Knochenmuss R, Zenobi R. *Int J Mass Spectrom, Ion Processes* 1998;**176**:149.

105. Xu N, Huang Z-H, Watson JT, Gage DA. *J Amer Soc Mass Spectrom* 1997;**8**:116.

106. Nonami H, Tanaka K, Fukuyama Y, Erra-Balsells R. *Rapid Commun Mass Spectrom* 1998;**12**:285.

107. Kim SH, Shin CM, Yoo JS. *Rapid Commun Mass Spectrom* 1998;**12**:701.

108. Kraus J, Stoeckli M, Schlunegger UP. *Rapid Commun Mass Spectrom* 1996;**10**:1927.

109. Papac DI, Wong A, Jones AJS. *Anal Chem* 1996;**68**:3215.

110. Mechref Y, Novotny MV. *J Amer Soc Mass Spectrom* 1998;**9**:1292.

111. Duffin KL, Welply JK, Huang E, Henion JD. *Anal Chem* 1992;**64**:1440.

112. Harvey DJ. *J Mass Spectrom* 2000;**35**:1178.

113. Linsley KB, Chan S-Y, Chan S, Reinhold BB, Lisi PJ, Reinhold VN. *Anal Biochem* 1994;**219**:207.

114. Okamoto M, Takahashi K-I, Doi T. *Rapid Commun Mass Spectrom* 1995;**9**:641.

115. Viseux N, de Hoffmann E, Domon B. *Anal Chem* 1997;**69**:3193.

116. Weiskopf AS, Vouros P, Harvey DJ. *Rapid Commun Mass Spectrom* 1997;**11**:1493.

117. Weiskopf AS, Vouros P, Harvey DJ. *Anal Chem* 1998;**70**:4441.

118. Viseux N, De Hoffmann E, Domon B. *Anal Chem* 1998;**70**:4951.

119. Sheeley DM, Reinhold VN. *Anal Chem* 1998;**70**:3053.

120. Ciucanu I, Costello CE. *J Am Chem Soc* 2003;**125**:16213.

121. Kang P, Mechref Y, Klouckova I, Novotny MV. *Rapid Commun Mass Spectrom* 2005;**19**:3421.

122. Kyselova Z, Mechref Y, Al Bataineh MM, Dobrolecki LE, Hickey RJ, Vinson J, Sweeney CJ, Novotny MV. *J Proteome Res* 2007;**6**:1822.

123. Naven TJP, Harvey DJ. *Rapid Commun Mass Spectrom* 1996;**10**:829.

124. Saba JA, Shen X, Jamieson JC, Perreault H. *Rapid Commun Mass Spectrom* 1999;**13**:704.

125. Shen X, Perreault H. *J Mass Spectrom* 1999;**34**:502.

126. Suzuki S, Kakehi K, Honda S. *Anal Chem* 1996;**68**:2073.

127. Yuan J, Hashii N, Kawasaki N, Itoh S, Kawanishi T, Hayakawa T. *J Chromatogr A* 2005;**1067**:145.

128. Hitchcock AM, Costello CE, Zaia J. *Biochemistry* 2006;**45**:2350.

129. Hsu J, Chang SJ, Franz AH. *J Am Soc Mass Spectrom* 2006;**17**:194.

130. Bowman MJ, Zaia J. *Anal Chem* 2007;**79**:5777.

131. Kang P, Mechref Y, Kyselova Z, Goetz JA, Novotny MV. *Anal Chem* submitted 2007.

132. Alvarez-Manilla III G, Warren NL, Abney T, Atwood J, Azadi P, York WS, Pierce M, Orlando R. *Glycobiology* 2007;**17**:677.

133. Domon B, Costello CE. *Glycoconjugate J* 1988;**5**:397.

134. Ngoka LC, Gal J-F, Lebrilla CB. *Anal Chem* 1994;**66**:692.

135. Cancilla MT, Penn SG, Carroll JA, Lebrilla CB. *J Amer Chem Soc* 1996;**118**:6736.

136. Spengler B, Kirsch D, Kaufmann R, Lemoine J. *J Mass Spectrom* 1995;**30**:782.

137. Zal F, Kuster B, Green BN, Harvey DJ. *Glycobiology* 1998;**8**:663.

138. Kufmann R, Chaurand P, Kirsch D, Spengler B. *Rapid Commun Mass Spectrom* 1996;**10**:1199.

139. Lemoine J, Chirat F, Domon B. *J Mass Spectrom* 1996;**31**:908.

140. Rouse JC, Strang A-M, Yu W, Vath JE. *Anal Biochem* 1998;**256**:33.

141. Yamagaki T, Nakanishi H. *J Mass Spectrom* 2000;**35**:1300.

142. Yamagaki T, Nakanishi H. *Proteomics* 2001;**1**:329.

143. Cornish TJ, Cotter RJ. *Rapid Commun Mass Spectrom* 1993;**7**:1037.

144. Cornish TJ, Cotter RJ. *Rapid Commun Mass Spectrom* 1994;**8**:781.

145. Harvey DJ, Bateman RH, Green MR. *J Mass Spectrom* 1997;**32**:167.

146. Kuster B, Naven TJP, Harvey DJ. *Rapid Commun Mass Spectrom* 1996;**10**:1645.

147. Penn SG, Cancilla MT, Lebrilla CB. *Anal Chem* 1996;**68**:2331.

148. Cancilla MT, Wong AW, Voss LR, Lebrilla CB. *Anal Chem* 1999;**71**:3206.

149. Solouki T, Reinhold BB, Costello CE, O'Malley M, Guan S, Marshall AG. *Anal Chem* 1998;**70**:857.

150. Harvey DJ, Bateman RH, Bordoli RS, Tyldesley R. *Rapid Commun Mass Spectrom* 2000;**14**:2135.

151. Muhlecker W, Gulati S, McQuillen DP, Ram S, Rice PA, Reinhold VN. *Glycobiology* 1999;**9**:157.

152. Reinhold VN, Sheeley DM. *Anal Biochem* 1998;**259**:28.

153. Sheeley DM, Reinhold VN. *Anal Chem* 1998;**70**:3053.

154. Viseux N, de Hoffmann E, Domon B. *Anal Chem* 1997;**69**:3193.

155. Viseux N, de Hoffmann E, Domon B. *Anal Chem* 1998;**70**:4951.

156. Weiskopf AS, Vouros P, Harvey DJ. *Rapid Commun Mass Spectrom* 1997;**11**:1493.

157. Weiskopf AS, Vouros P, Harvey DJ. *Anal Chem* 1998;**70**:4441.

158. Baker AG, Alexander A, Novotny MV. *Anal Chem* 1999;**71**:2945.

159. Mechref Y, Baker AG, Novotny MV. *Carbohydr Res* 1998;**313**:145.

160. Lewandrowski U, Resemann A, Sickmann A. *Anal Chem* 2005;**77**:3274.

161. Mechref Y, Novotny MV, Krishnan C. *Anal Chem* 2003;**75**:4895.

162. Morelle W, Slomianny M-C, Diemer H, Schaeffer C, van Dorsselaer A, Michalski J-C. *Rapid Commun Mass Spectrom* 2004;**18**:2637.

163. Spina E, Sturiale L, Romeo D, Impallomeni G, Garozzo D, Waidelich D, Glueckmann M. *Rapid Commun Mass Spectrom* 2004;**18**:392.

164. Stephens E, Maslen SL, Gree LG, Williams DH. *Anal Chem* 2004;**76**:2343.

165. Mechref Y, Kang P, Novotny MV. *Rapid Commun. Mass Spectrom* 2006;**20**:1381–9.

166. Solouki T, Reinhold BB, Costello CE, O'Malley M, Guan S, Marshall AG. *Anal Chem* 1998;**70**:857.

167. Hakansson K, Cooper HJ, Emmett MR, Costello CE, Marshall AG, Nilsson CL. *Anal Chem* 2001;**73**:4530.

168. Hakansson K, Chalmers MJ, Quinn JP, McFarland MA, Hendrickson CL, Marshall AG. *Anal Chem* 2003;**75**:3256.

169. Zhang J, Schubothe K, Li B, Russell S, Lebrilla CB. *Anal Chem* 2005;**77**:208.

170. Harvey DJ. *Rapid Commun Mass Spectrom* 2000;**14**:862.

171. Charlwood J, Langridge J, Tolson D, Birrell H, Camilleri P. *Rapid Commun Mass Spectrom* 1999;**13**:107.

172. Mo W, Sakamoto H, Nishikawa A, Kagi N, Langridge JI, Shimonishi Y, Takao T. *Anal Chem* 1999;**71**:4100.

173. Wheeler SF, Harvey DJ. *Anal Chem* 2000;**72**:5027.

# 2

# O-Glycan Complexity and Analysis

Lance Wells

*COMPLEX CARBOHYDRATE RESEARCH CENTER, AND
DEPARTMENT OF BIOCHEMISTRY AND MOLECULAR BIOLOGY,
THE UNIVERSITY OF GEORGIA, ATHENS, GA, USA*

## Introduction

O-Glycans are usually attached to serine and threonine residues of the polypeptide, though modification of hydroxyproline and hydroxylysine is observed in limited cases. This chapter will focus on the non-proteoglycan (non-xylose initiated) O-glycans, the first of which were described nearly 150 years ago. The O-glycans are structurally diverse (i.e., they do not contain a uniform core structure like that of the N-linked glycans) and have been functionally implicated in a plethora of biological processes. Here we will first delineate the various types of O-glycans in terms of their initiating reducing end glycan (Figure 2.1), detail the various structural features of each class, and briefly review their biological function. As there is no equivalent of N-glycanase for O-glycans, we will then focus on the chemical release of O-linked glycans. Characterization of the released glycans with a focus on tandem mass spectrometry (MS) approaches including identification and relative quantification will follow. The recent development of technological platforms for the direct analysis of O-glycans from complex mixtures at high sensitivity will be illustrated. Finally, we will highlight efforts to analyze O-glycopeptides directly and illustrate some of the future endeavors for the glycobiology field as it relates to O-linked glycans.

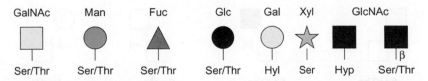

**FIGURE 2.1** Mammalian O-glycan initiating structures. Hydroxylysine (Hyl). CFG cartoon representations for glycans are used for this and all subsequent figures.

**Handbook of Glycomics**
Copyright © 2009 by Elsevier Inc. All rights of reproduction in any form reserved.

# O-Glycan Complexity

## O-GalNAc-Initiated Mucin Type

O-Linked N-acetylgalactosamine modification in an alpha linkage to the hydroxyl of serine and threonine residues is often referred to as mucin-type O-glycosylation, as the mucins are heavily O-GalNAc modified [1]. Interestingly, mammals have genes encoding for approximately 20 different polypeptide-N-acetylgalactosaminyltransferases (ppGalNAcTs), all of which transfer GalNAc from UDP-GalNAc to a hydroxyl-containing amino acid [2]. O-GalNAc modification of the polypeptide begins in the cis-Golgi. There are four common core structures and four less common core structures (Figure 2.2). Each of these core structures can be extended and branched to generate a diversity of structures. In the mucins, literally hundreds of sites on the polypeptide can be decorated with a variety of O-GalNAc-initiated extended core structures. This provides for a significant analytical challenge that will be discussed further in the chapter.

O-GalNAc-initiated glycoproteins appear to play a variety of essential roles. Among these is the ability of the mucins to hydrate and protect tissues by trapping bacteria [3]. These O-glycans can also significantly alter the conformation of the protein and on the heavily modified proteins may protect the polypeptide from proteolytic digestion [1]. O-GalNAc structures also appear to play an essential role in sperm–egg interactions [4]. From a pathophysiological perspective, O-GalNAc modification appears to play a critical role in the immune system, cell–cell interactions, and cancer [5].

## O-Mannose Initiated

The addition of O-mannose in an alpha linkage to the hydroxyl of serine and threonine residues was originally considered a modification that occurred only in fungi [6].

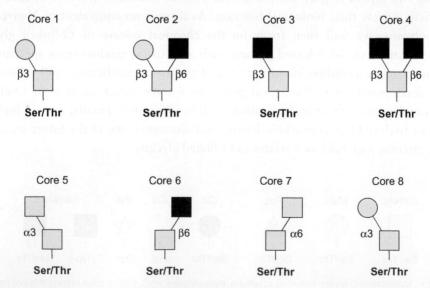

**FIGURE 2.2**  The eight core structures for O-GalNAc-initiated glycans.

However, Margolis and co-workers established O-Man-modified proteins in mammals more than 20 years ago now [7]. To date, only an extremely limited number of modified proteins, most notably alpha-dystroglycan, have been demonstrated to carry O-Man structures, though studies have demonstrated that approximately one-third of all released O-glycans from rodent brain are O-Man initiated [8]. The initial attachment of O-Man to the polypeptide occurs in the endoplasmic reticulum and is catalyzed by POMT1/POMT2 that uses dolichol-Man as the donor [9]. The O-Man can be extended to generate a Galβ4GlcNAcβ2-Manα-O-Ser/Thr structure that can be further modified by fucose, GalNAc, and sialic acid (Figure 2.3, [10]). Furthermore, a branched structure can be created by the addition of a GlcNAc in a β6 linkage to the core mannose that can also be extended [10].

While α-dystroglycan is the only well-studied O-Man modified protein, the importance of this modification has come to light due to its role in human disease. A number of congenital muscular dystrophies arise from defects in the enzymes that build the O-Man structure and result in hypoglycosylation of α-dystroglycan [10,11]. For instance, some patients with Walker–Warburg syndrome have been shown to have defects in POMT1 while patients with muscle–eye–brain disease have been shown to have defects in POMGnT1 (the enzyme responsible for extending the mannose with a β2-GlcNAc). In all, there have been six glycosyltransferases or putative glycosyltransferases implicated in hypoglycosylation of α-dystroglycan and congenital muscular dystrophy [10,11]. Elucidating the functional structures involved in the disease phenotype, the function of the putative glycosyltransferases implicated in disease and O-mannosylation, and determining other protein targets for O-mannosylation are current areas of focus in the field.

## O-Fucose Initiated

O-Fucose modification in an α linkage to the hydroxyl of serine and threonine residues can occur on proteins with an epidermal growth factor-like repeat and/or a thrombospondin type I repeat [12]. Initial modification occurs in the endoplasmic reticulum where GDP fucose is the donor sugar nucleotide [12]. Unlike other types

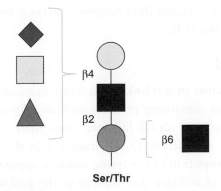

**FIGURE 2.3** O-Mannose-initiated glycan complexity.

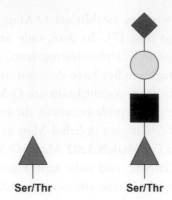

**FIGURE 2.4**  The two observed *O*-Fuc-initiated *O*-glycan structures.

of *O*-glycosylation discussed so far, the structural diversity is highly limited with only *O*-Fuc or an extended linear tetrasaccharide so far observed (Figure 2.4). Furthermore, there appears to be a strict primary sequence constraint upon *O*-fucosylation [12].

Strong experimental evidence implicates a functionally important role for *O*-fucosylation in development. For example, extension of the *O*-fucose by Fringe (a β3 GlcNAc transferase) and subsequent extension to the mature tetrasaccharide modulates Notch's affinity for different ligands [13]. Understanding the precise regulation of *O*-fucosylation and the role for this modification in modulating protein:protein interactions remains a central area of study in this field.

## *O*-Glucose Initiated

The addition of *O*-glucose to the hydroxyl of serine and threonine residues has recently been described and appears to be restricted to proteins with epidermal growth factor-like repeats [14]. The enzyme responsible for this modification, Rumi, was cloned in 2008 and appears to modify proteins in the endoplasmic reticulum at precise primary amino acid sequence restricted sites [15]. The *O*-Glc structure is then extended to Xylα3Xylα3Glc*O*-Ser/Thr. The trisaccharide structure is the only observed form of *O*-glucosylation to date. The exact functional role of this modification is still under study but recent data suggest it plays a role, similar to *O*-Fuc, in modulating Notch signaling [15].

## *O*-Galactose Initiated

The addition of *O*-galactose in a β linkage to hydroxylysine is observed in collagen and collagen-like domain containing proteins such as adiponectin [16]. This *O*-Gal can be extended by glucose in an α2 linkage. Thus, there are two observed *O*-Gal initiated structures that are initially *O*-Gal modified in the endoplasmic reticulum. Hydroxylation and glycosylation of the lysine residues appears to key for triple helix formation and stability of collagen and defects in the pathway are associated with connective tissue disorders [17].

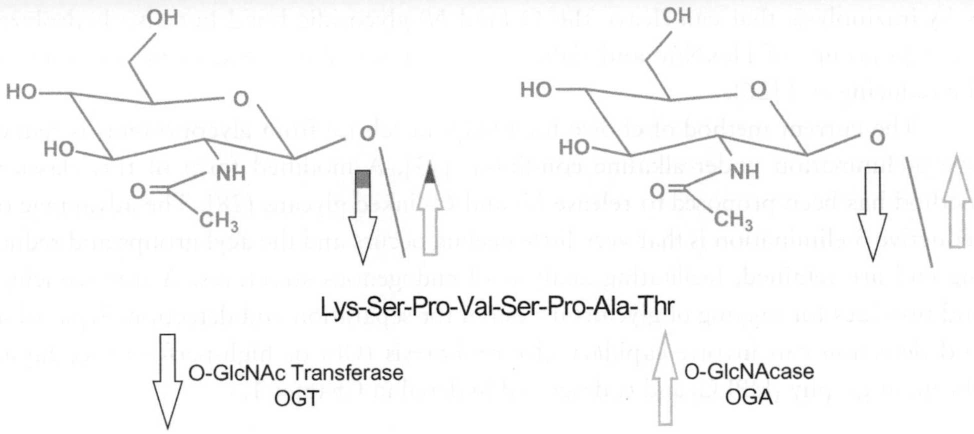

**FIGURE 2.5** The regulatory O-GlcNAc modification of nuclear and cytosolic proteins.

## O-β-N-Acetylglucosamine Initiated

The addition of O-β-N-acetylglucosamine to the hydroxyl group of serine and threonine is observed on nuclear and cytoplasmic proteins in multicellular organisms [18]. There also is an O-GlcNAc extended structure attached to hydroxyproline that appears to be evolutionary restricted to *Dictoyostelium* and *Toxoplasma* [19]. The rest of this section will focus only on the mammalian O-GlcNAc modification. Unlike other forms of glycosylation, the O-GlcNAc transferase and OGA, the enzyme that removes the modification, are found in the nucleus and cytosol [20]. To date, more than 400 proteins have been identified that carry this modification though the number of mapped sites remains under 100 [21]. No extensions of O-GlcNAc have been observed, making it exclusively a monosaccharide modification on proteins (Figure 2.5). A regulatory role for this modification has been suggested, given that the modification appears to be inducible and dynamic [22]. A clear role has been established for this modification in the regulation of insulin resistance, a hallmark of type 2 diabetes, and it appears the modification may serve as a nutrient sensor in the cell [23,24]. Of particular interest, this modification appears to have extensive interplay with serine/threonine phosphorylation and may be regulating a variety of signal transduction cascades [25]. Current efforts in the field are to elucidate the functional impact of O-GlcNAc modification at specific sites on particular proteins.

# Analysis of Released O-Glycans

## Release

Unlike N-linked glycans, no universal enzyme can remove the majority of O-linked glycans. The so-called O-glycanase enzyme only removes unsubstituted disaccharide core 1 (and possibly core 8) O-GalNAc initiated glycans. Therefore chemical release is the method of choice for O-glycan release unless specific subsets of O-glycans are to be focused upon in analysis. A classical approach that has lost favor in recent years is hydrazinolysis that can cleave the *O* (and *N*)-glycosidic bond but also hydrolyzes the acyl groups of HexNAc and sialic acid residues and can lead to modification of the reducing end [26].

The current method of choice for *O*-glycan release from glycoproteins is reductive β-elimination under alkaline conditions [27]. A modified form of this classical method has been proposed to release *N*- and *O*-linked glycans [28]. The advantage of reductive β-elimination is that very little peeling occurs and the acyl groups and reducing end are retained, facilitating analysis of endogenous structures. A free reducing-end provides for tagging of glycans if desired for separation and detection. Separation and detection can involve capillary electrophoresis (CE) or high-performance liquid chromatography (HPLC) and is described in detail in Chapter 1.

## Mass Spectrometry of Derivatized *O*-Glycans

Following reductive β-elimination released *O*-glycans can be permethylated with iodomethane to facilitate analysis by MS [27]. Permethylation aids in analyzing acidic glycans and enhances ionization, sensitivity, and informative fragmentation in the mass spectrometer. While a considerable amount of *O*-glycan characterization has been performed by matrix-assisted laser desorption/ionization-time-of-flight (MALDI-TOF) analysis, tandem MS is now the method of choice as it allows for more robust and reproducible fragmentation and when done on an ion trap instrument allows for MS$^n$ analysis. Unlike *N*-glycan analysis, determination of not only the non-reducing end structures but also the reducing end structure is informative.

A variety of approaches have been used for *O*-glycan profiling, such as direct infusion and liquid chromatography (LC)-based (often using graphitized carbon columns) approaches. One such direct infusion approach is total ion monitoring in combination with full MS profiling (Figure 2.6, [29]). In this approach, small *m/z* windows over the entire *m/z* range of the instrument are sequentially subjected to collision-activated dissociation (CAD) fragmentation and recorded facilitating a non-biased accounting of the *O*-glycans observed in a sample. A disadvantage of this approach is the large data set acquired (often as many 1000 individual MS-MS scans from a single sample) and the lack of automated software for the interpretation of spectra. LC-based approaches have been primarily performed on the often larger *N*-linked glycans and are still underutilized in *O*-glycan analysis. Regardless of approach, the initial fragmentation of *O*-glycans generates primarily glycosidic bond fragments that provide almost exclusively compositional and positional information without providing linkage, anomeric configuration, or monosaccharide identification. The Reinhold group has pioneered methods using MS$^n$ approaches to generate cross-ring cleavages in order to assign the complete structure [30,31]. Also, Harvey and co-workers as well as others have used various alkali metals on samples to facilitate cross-ring cleavages [32]. While elegant, these approaches require significant quantities of material as well as instrument and analysis time. Also the use of selective exoglycosidases facilitate assignments [27]. Many workers have compromised and use a limited MS$^n$ approach along with limited enzymatic digestion of the glycans and

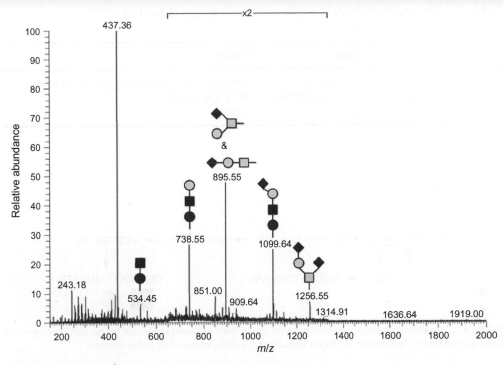

**FIGURE 2.6**  Full mass spectometry analysis of permethylated *O*-glycans released from a purified protein analyzed on a linear ion trap tandem mass spectrometer.

biological rules about the diversity and rules of linkage for *O*-glycans to facilitate a more high-throughput analysis that does not fully address isobaric structures but is much quicker and more sensitive than the full MS$^n$ analysis of each individual glycan in limited-material biological samples. Regardless of the degree of data acquisition, data analysis remains a significant bottleneck in the high-throughput analysis of both *O*- and *N*-linked glycans. Current efforts are underway to fully automate the analysis and will benefit greatly from existing efforts such as GlycoWorkBench [33,34].

## Quantitative Mass Spectrometry Approaches

Similar to the comparative proteomics field, glycomic efforts often seek to achieve relative quantification between a set of samples. For MS-based quantification in proteomics both non-isotope and stable isotope approaches have been developed, validated, and applied in an array of biological settings [35]. Recent work in glycomics has sought to develop similar quantification techniques that can be used for both *N*- and *O*-linked released glycans (Table 2.1). The use of isotopic labeling for glycan quantification is also discussed in Chapters 1 and 4.

Full mass spectrum upon direct infusion of glycan samples allows for calculation of area under or height of the peak that can be normalized and compared between various samples similar to the non-isotopic approaches used in proteomics [36]. This method is limited in that it cannot quantify minor glycans, requires the analysis of the samples separately adding to technical variability, and cannot differentiate between isobaric structures.

## Table 2.1 Quantification techniques that can be used for both N- and O-linked glycans

|  | Techniques | Examples |
| --- | --- | --- |
| Non-isotopic | Spectral counts | Total ion monitoring; normalized peak height/area |
| In vitro isotopic labeling | $^{18}O$-$H_2O$, ICAT, BEMAD | $^2H/^{13}C$-$CH_3I$; QUIBL |
| In vivo isotopic labeling | SILAC | IDAWG |

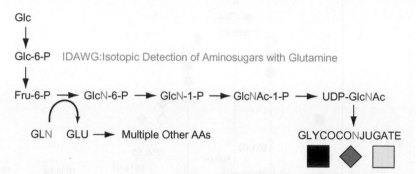

**FIGURE 2.7**   The IDAWG (isotopic detection of amino sugars with glutamine) approach for isotopic labeling in cell culture allowing for the addition of 1 Da for each GlcNAc, GalNAc, or sialic acid in the structure.

Several stable isotope-based approaches have been developed for quantitative glycomics. Several groups have sought to take advantage of in vitro chemical modification of the glycan. For example, reductive amination can be used to tag the reducing end with a light and heavy tags [37]. Another approach that has recently been applied is the use of $^{13}C$ or deuterated iodomethane for permethylation [36,38,39]. Orlando and colleagues have developed a method, QUIBL, based on combining the two different permethylation approaches to quantify isobaric glycan structures in mixtures [40].

Finally, the IDAWG (isotopic detection of amino sugars with glutamine) method has recently been developed. This takes advantage of the fact that the amide side chain of Gln is the sole source of nitrogen in GlcNAc, GalNAc, and sialic acid (Figure 2.7, [41]). Thus, cells can be efficiently labeled in culture by replacing normal Gln supplement with amide-$^{15}N$-Gln resulting in a +1 Da increase in mass for each GlcNAc, GalNAc, and sialic acid in the structure. This approach has the advantage, similar to SILAC (stable isotope labeling with amino acids in cell culture) for proteomics [42], that the cells can be added together before further processing to reduce technical variability. However, with regard to O-glycans this method suffers in that it will fail to label O-Gal and O-Glc initiated structures as well as the non-extended O-Fuc glycan.

While several approaches have recently been developed for quantitative glycomics, more testing of these methods with biological specimens will further validate this method as well as reveal shortcomings that need to be solved.

## Analysis of *O*-Glycopeptides

While glycomics provides a considerable wealth of biologically relevant information, ultimately glycoproteomics is where the field needs to move. Over the last several years methods have been developed for site mapping of both *N*-linked and *O*-linked glycosylation sites on proteins [43–45]. However, actually studying glycopeptides directly has been a great challenge. Enrichment strategies using lectins and/or glyco-specific antibodies can facilitate detection. Several studies on specific *N*-linked glycoproteins, such as IgGs, have been successful using tandem MS approaches [46]. For *O*-linked glycosylation there are a variety of attractive approaches. Unfortunately in the tandem mass spectrometer, CAD fragmentation of *O*-linked glycopeptides almost always results in the loss of part or all of the glycan from the peptide and results in poor fragmentation of the peptide backbone for identification. However, using an ion trap instrument one can take advantage of pseudo-neutral losses of the glycan structure to isolate and successfully fragment the peptide backbone [47]. Such neutral loss triggered $MS^n$ approaches provide a method in which to identify specific glycan structures on assigned peptides but often fail to map the exact site of modification if there are multiple hydroxyl-containing amino acids in the peptide. It should be noted that unlike phosphorylation that usually takes the hydroxyl oxygen with it when it undergoes neutral loss resulting in a dehydro amino acid at the site of modification, *O*-glycans preferentially cleave so as to leave the serine or threonine residue unmodified. One promising approach to overcoming this shortcoming is the use of electron transfer dissociation (ETD) or electron capture dissociation (ECD) for fragmentation as opposed to CAD. Most posttranslational modifications are stable upon ETD/ECD fragmentation facilitating assignment of the site of modification and the structure (Figure 2.8, [48]).

Of course with both techniques there is very little structural information for the glycan (i.e. no linkage, anomericity, or isobaric monosaccharide can be elucidated). Thus, the standard approach is to release and characterize the *O*-glycans present in the sample as well as perform standard shotgun proteomics to determine the protein(s) present in the sample first so as to define the possible modifications to be observed on hydroxyl amino acid-containing peptides. Given the abundance of serine and threonines and the lack of a primary sequence motif for many types of *O*-glycosylation this still provides a daunting task for even simple mixtures of proteins.

## Conclusions

In this chapter, we have illustrated the great diversity of *O*-linked glycans observed in mammals and highlighted just a few of their many biological functions. Unlike *N*-linked glycans, these structures have no common core structure from which they are elaborated nor do many of them use a defined consensus sequence on the polypeptide that they modify. Further, there is no single enzyme that can remove all, or even most, of the *O*-linked glycans from a complex mixture. For all of these reasons, *O*-linked glycan analysis has lagged behind *N*-linked glycan analysis. However, robust qualitative

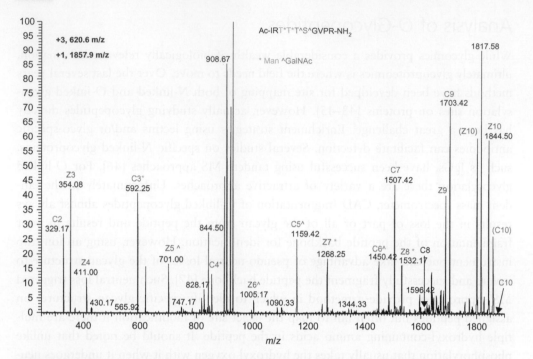

**FIGURE 2.8** Electron transfer dissociation fragmentation of a synthetic peptide containing two sites of *O*-Man modification and two sites of *O*-GalNAc modification facilitating assignment of all structures and sites.

and quantitative methods have been developed for their analysis. Furthermore, given their relatively small size compared with *N*-linked glycans for many *O*-glycans, promising approaches have and are being developed for the direct analysis of glycopeptides. Currently these methods require purified glycoproteins or extremely simple mixtures that are defined first by glycomics and proteomics. Future developments in sample handling, tagging, enrichment, and data analysis are needed if glycoproteomics is going to become a shotgun, high-throughput field. Progress in the field of proteomics and glycomics has been staggering in the last ten years, the next decade will hopefully see the merging of these two fields so that glycoproteomics can become routine so as to facilitate the elucidation of the biological functions of glycoproteins.

## Acknowledgments

This work was supported in part by the following grants: 1R01DK075069 from NIDDK, 1U01CA128454 from NCI, and 5P41RR018502 from NCRR. The author would like to thank Dr. David Live for providing the synthetic *O*-Man/*O*-GalNAc modified peptide used in ETD studies in this chapter.

# References

1. Carraway KL, Hull SR. *O*-glycosylation pathway for mucin-type glycoproteins. *Bioessays* 1989;**10**:117–21.

2. Ten Hagen KG, Fritz TA, Tabak LA. All in the family: the UDP-GalNAc:polypeptide N-acetylgalactosaminyl transferases. *Glycobiology* 2003;**13**:1R–16R.

3. Linden SK, Sutton P, Karlsson NG, Korolik V, McGuckin MA. Mucins in the mucosal barrier to infection. *Mucosal Immunol* 2008;**1**:183–97.

4. Van den Steen P, Rudd PM, Dwek RA, Opdenakker G. Concepts and principles of *O*-linked glycosylation. *Crit Rev Biochem Mol Biol* 1998;**33**:151–208.

5. Hanisch FG. *O*-glycosylation of the mucin type. *Biol Chem* 2001;**382**:143–9.

6. Strahl-Bolsinger S, Immervoll T, Deutzmann R, Tanner W. PMT1, the gene for a key enzyme of protein *O*-glycosylation in *Saccharomyces cerevisiae*. *Proc Natl Acad Sci USA* 1993;**90**:8164–8.

7. Krusius T, Finne J, Margolis RK, Margolis RU. Identification of an *O*-glycosidic mannose-linked sialylated tetrasaccharide and keratan sulfate oligosaccharides in the chondroitin sulfate proteoglycan of brain. *J Biol Chem* 1986;**261**:8237–42.

8. Chai W, Yuen CT, Kogelberg H, Carruthers RA, Margolis RU, Feizi T, Lawson AM. High prevalence of 2-mono- and 2,6-di-substituted manol-terminating sequences among *O*-glycans released from brain glycopeptides by reductive alkaline hydrolysis. *Eur J Biochem* 1999;**263**:879–88.

9. Manya H, Chiba A, Yoshida A, Wang X, Chiba Y, Jigami Y, Margolis RU, Endo T. Demonstration of mammalian protein *O*-mannosyltransferase activity: coexpression of POMT1 and POMT2 required for enzymatic activity. *Proc Natl Acad Sci USA* 2004;**101**:500–5.

10. Martin PT. Congenital muscular dystrophies involving the *O*-mannose pathway. *Curr Mol Med* 2007;**7**:417–25.

11. Muntoni F, Brockington M, Blake DJ, Torelli S, Brown SC. Defective glycosylation in muscular dystrophy. *Lancet* 2002;**360**:1419–21.

12. Haltiwanger RS, Lowe JB. Role of glycosylation in development. *Annu Rev Biochem* 2004;**73**:491–537.

13. Moloney DJ, Panin VM, Johnston SH, Chen J, Shao L, Wilson R, Wang Y, Stanley P, Irvine KD, Haltiwanger RS, Vogt TF. Fringe is a glycosyltransferase that modifies Notch. *Nature* 2000;**406**:369–75.

14. Kozma K, Keusch JJ, Hegemann B, Luther KB, Klein D, Hess D, Haltiwanger RS, Hofsteenge J. Identification and characterization of a beta1,3-glucosyltransferase that synthesizes the Glc-beta1,3-Fuc disaccharide on thrombospondin type 1 repeats. *J Biol Chem* 2006;**281**:36742–51.

15. Acar M, Jafar-Nejad H, Takeuchi H, Rajan A, Ibrani D, Rana NA, Pan H, Haltiwanger RS, Bellen HJ. Rumi is a CAP10 domain glycosyltransferase that modifies Notch and is required for Notch signaling. *Cell* 2008;**132**:247–58.

16. Schegg B, Hulsmeier AJ, Rutschmann C, Maag C, Hennet T. Core glycosylation of collagen is initiated by two beta(1-O)galactosyltransferases. *Mol Cell Biol* 2009;**29**:943–52.

17. Bann JG, Peyton DH, Bachinger HP. Sweet is stable: glycosylation stabilizes collagen. *FEBS Lett* 2000;**473**:237–40.

18. Snow DM, Hart GW. Nuclear and cytoplasmic glycosylation. *Int Rev Cytol* 1998;**181**:43–74.

19. West CM, van der Wel H, Blader IJ. Detection of cytoplasmic glycosylation associated with hydroxyproline. *Methods Enzymol* 2006;**417**:389–404.

20. Iyer SP, Hart GW. Dynamic nuclear and cytoplasmic glycosylation: enzymes of *O*-GlcNAc cycling. *Biochemistry* 2003;**42**:2493–9.

21. Copeland RJ, Bullen JW, Hart GW. Cross-talk between GlcNAcylation and phosphorylation: roles in insulin resistance and glucose toxicity. *Am J Physiol Endocrinol Metab* 2008;**295**:E17–28.

22. Wells L, Vosseller K, Hart GW. Glycosylation of nucleocytoplasmic proteins: signal transduction and *O*-GlcNAc. *Science* 2001;**291**:2376–8.

23. Dentin 3rd R, Hedrick S, Xie J, Yates J, Montminy M. Hepatic glucose sensing via the CREB coactivator CRTC2. *Science* 2008;**319**:1402–5.

24. Wells L, Vosseller K, Hart GW. A role for Nacetylglucosamine as a nutrient sensor and mediator of insulin resistance. *Cell Mol Life Sci* 2003;**60**:222–8.

25. Wang Z, Gucek M, Hart GW. Cross-talk between GlcNAcylation and phosphorylation: site-specific phosphorylation dynamics in response to globally elevated OGlcNAc. *Proc Natl Acad Sci USA* 2008;**105**:13793–8.

26. Merry T, Astrautsova S. Chemical and enzymatic release of glycans from glycoproteins. *Methods Mol Biol* 2003;**213**:27–40.

27. Morelle W, Michalski JC. Analysis of protein glycosylation by mass spectrometry. *Nat Protoc* 2007;**2**:1585–602.

28. Huang Y, Mechref Y, Novotny MV. Microscale nonreductive release of *O*-linked glycans for subsequent analysis through MALDI mass spectrometry and capillary electrophoresis. *Anal Chem* 2001;**73**:6063–9.

29. Aoki K, Porterfield M, Lee SS, Dong B, Nguyen K, McGlamry KH, Tiemeyer M. The diversity of O-linked glycans expressed during *Drosophila melanogaster* development reflects stage- and tissue-specific requirements for cell signaling. *J Biol Chem* 2008;**283**:30385–400.

30. Ashline D, Singh S, Hanneman A, Reinhold V. Congruent strategies for carbohydrate sequencing. 1. Mining structural details by MSn. *Anal Chem* 2005;**77**:6250–62.

31. Ashline DJ, Lapadula AJ, Liu YH, Lin M, Grace M, Pramanik B, Reinhold VN. Carbohydrate structural isomers analyzed by sequential mass spectrometry. *Anal Chem* 2007;**79**:3830–42.

32. Harvey DJ. Collision-induced fragmentation of underivatized N-linked carbohydrates ionized by electrospray. *J Mass Spectrom* 2000;**35**:1178–90.

33. Ceroni A, Maass K, Geyer H, Geyer R, Dell A, Haslam SM. GlycoWorkbench: a tool for the computer-assisted annotation of mass spectra of glycans. *J Proteome Res* 2008;**7**:1650–9.

34. Royle L, Mattu TS, Hart E, Langridge JI, Merry AH, Murphy N, Harvey DJ, Dwek RA, Rudd PM. An analytical and structural database provides a strategy for sequencing *O*-glycans from microgram quantities of glycoproteins. *Anal Biochem* 2002;**304**:70–90.

35. Aebersold R, Mann M. Mass spectrometry-based proteomics. *Nature* 2003;**422**:198–207.

36. Aoki K, Perlman M, Lim JM, Cantu R, Wells L, Tiemeyer M. Dynamic developmental elaboration of N-linked glycan complexity in the *Drosophila melanogaster* embryo. *J Biol Chem* 2007;**282**:9127–42.

37. Bowman MJ, Zaia J. Tags for the stable isotopic labeling of carbohydrates and quantitative analysis by mass spectrometry. *Anal Chem* 2007;**79**:5777–84.

38. Alvarez-Manilla 3rd G, Warren NL, Abney T, Atwood J, Azadi P, York WS, Pierce M, Orlando R. Tools for glycomics: relative quantitation of glycans by isotopic permethylation using 13CH3I. *Glycobiology* 2007;**17**:677–87.

39. Kang P, Mechref Y, Kyselova Z, Goetz JA, Novotny MV. Comparative glycomic mapping through quantitative permethylation and stable-isotope labeling. *Anal Chem* 2007;**79**:6064–73.

40. Atwood 3rd JA, Cheng L, Alvarez-Manilla G, Warren NL, York WS, Orlando R. Quantitation by isobaric labeling: applications to glycomics. *J Proteome Res* 2008;**7**:367–74.

41. Orlando R, Lim JM, Atwood JA, 3rd, Angel PM, Fang M, Aoki K, Alvarez-Manilla G, Moremen KW, York WS, Tiemeyer M, Pierce M, Dalton S, Wells L. IDAWG: Metabolic incorporation of stable isotope labels for quantitative glycomics of cultured cells. *J Proteome Res* 2009.

42. Ong SE, Blagoev B, Kratchmarova I, Kristensen DB, Steen H, Pandey A, Mann M. Stable isotope labeling by amino acids in cell culture, SILAC, as a simple and accurate approach to expression proteomics. *Mol Cell Proteomics* 2002;**1**:376–86.

43. Koles K, Lim JM, Aoki K, Porterfield M, Tiemeyer M, Wells L, Panin V. Identification of N-glycosylated proteins from the central nervous system of Drosophila melanogaster. *Glycobiology* 2007;**17**:1388–403.

44. Kristiansen TZ, Bunkenborg J, Gronborg M, Molina H, Thuluvath PJ, Argani P, Goggins MG, Maitra A, Pandey A. A proteomic analysis of human bile. *Mol Cell Proteomics* 2004;**3**:715–28.

45. Wells L, Vosseller K, Cole RN, Cronshaw JM, Matunis MJ, Hart GW. Mapping sites of *O*-GlcNAc modification using affinity tags for serine and threonine posttranslational modifications. *Mol Cell Proteomics* 2002;**1**:791–804.

46. Huhn C, Selman MH, Ruhaak LR, Deelder AM, Wuhrer M. IgG glycosylation analysis. *Proteomics* 2009;**9**:882–913.

47. Greis KD, Hayes BK, Comer FI, Kirk M, Barnes S, Lowary TL, Hart GW. Selective detection and site analysis of *O*-GlcNAc-modified glycopeptides by beta-elimination and tandem electrospray mass spectrometry. *Anal Biochem* 1996;**234**:38–49.

48. Mikesh LM, Ueberheide B, Chi A, Coon JJ, Syka JE, Shabanowitz J, Hunt DF. The utility of ETD mass spectrometry in proteomic analysis. *Biochim Biophys Acta* 2006;**1764**:1811–22.

# Glycosaminoglycans

## Fuming Zhang[a], Zhenqing Zhang[b], and Robert J. Linhardt[a,b,c]

*DEPARTMENT OF [a]CHEMICAL AND BIOLOGICAL ENGINEERING,
[b]CHEMISTRY AND CHEMICAL BIOLOGY,
[c]BIOLOGY, CENTER FOR BIOTECHNOLOGY AND
INTERDISCIPLINARY STUDIES, RENSSELAER POLYTECHNIC
INSTITUTE, TROY, NEW YORK, USA*

## Introduction

Glycosaminoglycans (GAGs) are a family of highly sulfated, complex, polydisperse linear polysaccharides that display a variety of important biological roles. Based on the difference of repeating disaccharide units comprising GAGs, they can be categorized into four main groups: heparin (HP)/heparan sulfate (HS), chondroitin sulfate (CS)/dermatan sulfate (DS), keratan sulfate (KS), and hyaluronan (HA) (Figure 3.1). GAGs are found in all animals from *Caenorhabditis elegans* to human, in the extracellular matrix and basement membranes as a structural scaffold [1].

HP and HS are mixtures of sulfated linear polysaccharides having a molecular weight (MW) range from 5000 to 40000 with an MW average of 10000 to 25000. The major repeating disaccharide sequence of heparin (75–90%) is [→4)-2-$O$-sulfo-α-L-idopyranosyluronic acid (IdoA$_p$) (1→4)-2-deoxy-2-$N$-,6-$O$-sulfo-α-D-glucopyranosylsamine (GlcN$_p$) (1→], while minor sequences contain β-D-glucopyranosyluronic acid (GlcA$_p$) residues, a reduced content of sulfo groups as well as $N$-acetylation. HP proteoglycan (PG) is $O$-linked to serine residues of the core protein serglycin and is found intracellularly in the granules of mast cells [2]. HS is $O$-linked to serine residues of a number of core proteins resulting in a number of PGs, including glypican and syndecan. HS has the similar structure with heparin, primarily containing nonsulfated disaccharide [→4) β-D-GlcA$_p$ (1→4)-α-D-GlcN$_p$Ac (1→] and monosulfated disaccharides, such as [→4) β-D-GlcA$_p$ (1→4)-6-sulfo-α-D-$N$-GlcN$_p$Ac].

HP, commonly used as a clinical anticoagulant, has other biological activities with potential clinical applications, such as effects on lipoprotein lipase, effects on smooth muscle proliferation, inhibition of complement activation, anti-inflammatory activity, angiogenic and antiangiogenic activities, anticancer activity, and antiviral activity [2–8]. HS has a number of important biological activities. It has polysaccharide components that bind antithrombin, supporting blood flow within blood vessels [5]. It is also

**Handbook of Glycomics**
Copyright © 2009 by Elsevier Inc. All rights of reproduction in any form reserved.

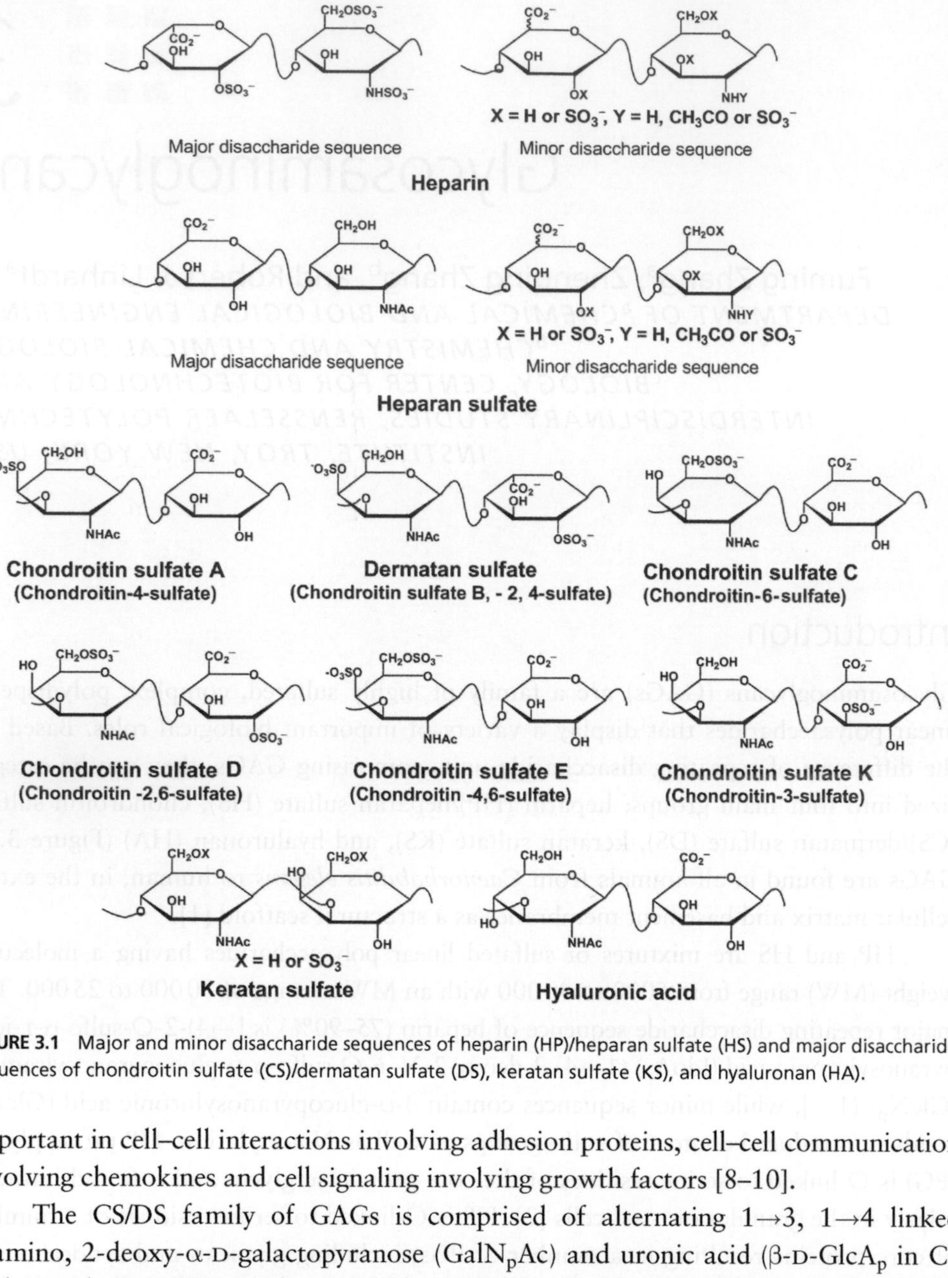

**FIGURE 3.1** Major and minor disaccharide sequences of heparin (HP)/heparan sulfate (HS) and major disaccharide sequences of chondroitin sulfate (CS)/dermatan sulfate (DS), keratan sulfate (KS), and hyaluronan (HA).

important in cell–cell interactions involving adhesion proteins, cell–cell communication involving chemokines and cell signaling involving growth factors [8–10].

The CS/DS family of GAGs is comprised of alternating 1→3, 1→4 linked 2-amino, 2-deoxy-α-D-galactopyranose (GalN$_p$Ac) and uronic acid (β-D-GlcA$_p$ in CS and α-L-IdoA$_p$ in DS) residue [11]. The molecular weights of CS/DS range from 2 to 50 kDa. There are multiple forms of CS named A, B (also known as DS), C, D, E, and K, differing based on sulfo group substitution and the type of uronic acid that each contains (Figure 3.1) [11]. This family is the most common type of GAG found in extracellular matrix PGs. CS is important in cell–cell interaction and communication, and DS exhibits important venous antithrombotic activity [11–13]. CS and DS also have intriguing functions in infection, inflammation, neurite outgrowth, growth factor signaling, morphogenesis, and cell division [13–15].

KS has two different forms, KS-I and KS-II, originally designated based on differences between KS from cornea and cartilage. Currently, the term KS-I includes all Asn-linked KS molecules, and KS-II is used to refer to all KS linked to protein through GalNAc-O-Ser/Thr [16,17]. A third type of KS linkage (mannose-O-Ser) has been identified that has been called KS-III [18]. KS is comprised primarily of 6-O-sulfo-GlcN$_p$Ac and galactose (Gal) (which may contain 6-O-sulfo groups) [17]. KS-PGs are the major class of PG in the corneal stroma and are thought to play an important role in corneal structure and physiology, particularly in the maintenance of corneal transparency [19]. KS-containing molecules have been identified in numerous epithelial and neural tissues in which KS expression responds to embryonic development, physiological variations, and to wound healing [17]. Evidence has also been presented supporting functional roles of KS in cellular recognition of protein ligands, axonal guidance, cell motility, and in embryo implantation [17].

HA is a homocopolymer made of repeating disaccharide units of [→3)-D-GlcN$_p$Ac(1→4)-β-D-GlcA$_p$(1→] [20]. It is an unsulfated GAG with very high MW of up to 2000 kDa. HA is widely distributed in cartilage, skin, eye, and most body liquids. It is not only an important structural component of extracellular matrices but also interacts instructively with cells during embryonic development, healing processes, inflammation, and cancer [21].

GAGs perform a variety of biological functions and play an important role in a number of different diseases. Their activities are mainly triggered by interactions with a wide range of proteins [4]. Experiments have shown that specific sequences within heparin and HS act as protein binding sites [22]. This suggests that polysaccharides may work as informational molecules and suggests the importance of developing analytical techniques to sequence GAG molecules [23,24]. The primary structure determination of biopolymers, such as proteins and nucleic acids, are commonly solved by automated sequencing of amino acid residues and nucleotide residues, respectively. In contrast with current state-of-the-art methods, the sequencing of polysaccharides with a high level of structural complexity is still extremely challenging.

## Isolation of GAGs

The long linear chain and high density of negative charges in GAGs distinguish this class of glycans from the short-branched oligosaccharides of glycoproteins and glycolipids. The high density of negatively charged structure of GAGs is the basis for their physical separation. This property facilitates GAG isolation/purification by strong anion-exchange (SAX) chromatography and precipitation with cetylpyridinium chloride and methanol/ethanol [25]. GAGs are generally attached to core proteins but free chains can accumulate because of proteolysis and endoglycosidic cleavage of the PG. GAGs can be integrated in the extracellular matrix by non-covalent interactions or be associated with the cell surface by hydrophobic interaction or by ionic binding with other cell surface molecules. GAGs can also be sequestered in intracellular compartments, such as in storage or secretory granules, or in pre-lysosomal or endosomal compartments [26]. Fragmentation of matrix PGs often occurs and gives rise to GAG chains attached to short peptides [25]. The principle of GAG isolation is similar to the methods for PG extraction except that proteolysis is used for GAG isolation. Chaotropic reagents, such as 4 M guanidine hydrochloride or 8 M

## Table 3.1  Glycosaminoglycans (GAGs) isolated from different animal tissues, cell cultures, and biological fluids

| | Sources | GAGs | Ref. |
|---|---|---|---|
| Organ tissues | Porcine intestine, bovine lung | HP, HS | 27 |
| | Bovine/porcine kidney, pancreas, lung livers | HS | 28 |
| | Dromedary intestine | HP/HS | 29 |
| | Human liver | HS | 30 |
| | Human amyloid A and fibrils | HS, DS | 31 |
| | Brains from rat, monkey, chicken, sheep, rabbit | HA, CS, HS | 32 |
| | Human non-epithelial tumors | CS, HA | 33 |
| | Chick embryo | CS | 34 |
| | Human placenta | DS | 35 |
| | Bovine, human cartilage | CS, KS | 36 |
| | Mouse organs | HS, CS | 37 |
| Cell culture | Stem cells | HS | 38 |
| | Chinese hamster ovary (CHO) cells | HS | 39 |
| Biological fluids | Plasma, urine | HP, CS, HA | 40,41 |
| | Human follicular fluid | HS | 42 |
| | Bovine follicular fluid | HS, CS | 43 |
| Insects | Drosophila | | 44 |
| | Mosquitoes | | 45 |
| Fish/marine/mulluscs | Zebra fish | HS, CS | 46 |
| | Clams | HS, CS | 47 |
| | Eel skin | DS, KS | 48 |
| | Africa giant snail | Acharan sulfate | 49 |

HP, heparin; HS, heparan sulfate; CS, chondroitin sulfate; DS, dermatan sulfate; KS, keratan sulfate; HA, hyaluronan.

urea, are effective in recovering of GAGs (or PGs) from cell matrix compartments. These reagents denature proteins and dissociate most non-covalent interactions. Detergents, such as Triton X-100, 2% of CHAPS, are often used to dissociate hydrophobic interactions [26]. The general procedures for the isolation/purification GAGs include: (i) sample pre-treatment, such as homogenization, freeze-drying and de-fatting, (ii) proteolysis, (iii) high-salt or chaotropic reagent/detergent extraction; (iv) anion-exchange chromatography; and (v) methanol or ethanol precipitation. GAGs have been recovered from a variety of different tissues, cell culture samples, and biological fluids (Table 3.1).

The following detailed procedure provides an example for isolation of GAGs from large organ or tissue samples (i.e., pig liver, bovine brain, human liver):

1. Wash the tissue in cold phosphate buffered saline (PBS) at 4°C.
2. Cut tissue into small pieces (2 × 2 cm).
3. Freeze-dry, and weigh the dried tissue.
4. Grind dry tissue into powder.

5. Remove fat by extracting the tissues with three solvent mixtures: chloroform: methanol (2:1, 1:1, 1:2 (v:v)) each left overnight at room temperature, dry defatted tissue in fume hood. Weigh the dry, defatted tissue.

6. Suspend defatted sample (5–10% (w/v)) in water and proteolyze at 55°C with 10% (w/w) of a non-specific protease such as actinase E (20 mg/mL) for 18 h.

7. Add 8 M urea containing 2% CHAPS to supernatant.

8. Remove insoluble residue by centrifugation (5000 rpm, for 30 min).

9. Equilibrate a SAX column with 8 M urea, 2% CHAPS at pH 8.3, load sample and wash the column with 3 column volumes of 200 mM aqueous NaCl. GAGs are then released by washing the column with 1 column volume of 16% NaCl.

10. Methanol precipitate (80%vol methanol) the released GAGs dissolved in 16% NaCl at 4°C overnight. Recover the precipitated GAGs by centrifugation (5000×$g$ for 30 min).

## Purification of Individual GAGs

Individual GAGs can be purified by: (i) selective enzymatic digestion with polysaccharide lyases; (ii) selective degradation with nitrous acid; (iii) SAX chromatography purification; and (iv) selective precipitation.

Polysaccharide lyases cleave specific glycosidic linkages present in acidic polysaccharides and result in depolymerization [50]. These enzymes act through an eliminase mechanism resulting in unsaturated oligosaccharide products that have UV absorbance at 232 nm. This class of enzymes includes heparin lyases (heparinases), heparan sulfate lyases (heparinases or heparitinases), chondroitin lyases (chondroitinases), and hyaluronate lyases (hyaluronidases) (Table 3.2). These enzymes can be used, alone or in combinations, to degrade the undesired GAGs, or to confirm the presence of GAGs in a sample as well as to distinguish between different GAGs. Chondroitin lyase ABC can depolymerize all forms of CS as well as HA, chondroitin lyase AC can depolymerize CS-A and CS-C and HA but not DS, and chondroitin lyase B can only degrade DS (CS-B). Heparin lyase III degrades HS but does not act on HP or other GAGs while heparin lyase I degrades HP but only acts in a very limited extent on HS. Each lyase enzyme has different optimal buffer and reaction conditions and can be inhibited by the presence of other GAGs [50]. Enzyme activity should be assayed prior to using it in an experiment to ensure it is active and has been stored properly.

A unique structural feature of HP and HS is that a large proportion of their GlcNp residues contain N-sulfo groups (85–90% in heparin; 30–60% in a typical heparan sulfate) and the remainder is modified with N-acetyl groups [50]. The glycosidic bonds of these N-sulfo GlcN$_p$ residues can be cleaved rapidly at room temperature with nitrous acid at pH 1.5 to afford a mixture of oligosaccharides. Nitrous acid susceptibility distinguishes HP and HS from all other GAGs (i.e., CS, DS, HA, and KS) as these contain only N-acetyl substituted amino sugars, which are not cleaved by nitrous acid [51].

GAGs also differ based on their average charge densities and these differences can be exploited to identify and separate different GAG classes by using anion-exchange

**Table 3.2 Polysaccharidases used in glycosaminoglycan analysis**

| Enzyme | Source | Action pattern | Specificity |
|---|---|---|---|
| Chondroitin lyase ACI | *Flavobacterium heparinum* | CS (endo) | →3)α-D-GalNpAc(4S/OH,6S/OH)(1→4)-β-D-GlcAp (1→<br>Primary to disaccharide products |
| Chondroitin lyase ACII | *Arthrobacter aurescens* | CS (exo) | →3)α-D-GalNpAc(4S/OH,6S/OH)(1→4)-β-D-GlcAp (1→<br>Primary to disaccharide products |
| Chondroitin lyase ABC | *Proteus vulgaris* | CS/DS/HA (endo and exo) | →3)α-D-GalNpAc(4S/OH,6S/OH)(1→4)HexAp (1→<br>Primary to disaccharide products |
| Chondroitin lyase ABC | *Bacteroides thetaiotaomicron* | CS/DS/HA (exo) | →3)α-D-GalNpAc(4S/OH,6S/OH)(1→4)HexAp (1→<br>Primary to disaccharide products |
| Chondroitin lyase B | *Flavobacterium heparinum* | DS (endo) | →3)α-D-GalNpAc(4S/OH)(1→4)-α-L-IdoAp (1→<br>Primary to disaccharide products |
| Heparin lyase I | *Flavobacterium heparinum* | Hp (endo) | →4)α-D-GlcNpS(6S/OH)(1→4)-α-L-IdoAp2S(1→<br>Primary to disaccharide products |
| Heparin lyase II | *Flavobacterium heparinum* | Hp/HS (endo) | →4)α-D-GlcNp(S, Ac)(6S/OH)(1→4)-α-L-IdoAp/-β-D-GlcAp 2S/OH)(1→<br>Primary to disaccharide products |
| Heparin lyase III | *Flavobacterium heparinum* | HS (endo) | →4)α-D-GlcNp(Ac/S)( OH/6S)(1→4)-β-D-GlcAp-α-L-IdoAp(1→<br>Primary to disaccharide products |
| Hyaluronate lyase | *Streptomyces hyalurolyricus* | HA (endo) | →3)α-D-GlcNpAc(1→4)-β-D-GlcAp(1→<br>Primary to disaccharide products |
| Hyaluronidase | Bovine testicular | HA/CS (endo) | →3)α-D-GlcNpAc(1→4)-β-D-GlcAp(1→<br>Primary to saturated tetrasaccharide products |

chromatography or selective precipitation. GAGs with different charge densities can resolve, at least partially, into separate peaks with a continuous salt gradient release. Many different types of ion exchangers are now available in high-performance liquid chromatography (HPLC) columns and membrane cartridges. For example, KS and DS were purified from a GAG-containing mixture (extracted from eel skin) by anion-exchange chromatography on a DEAE Sephacel column eluted with 2 column volumes of 0.5 M NaCl, 1 M, 1.2 M, 1.4 M, and 1.6 M NaCl [48].

# Structure Analysis

## Molecular Weight Analysis of GAGs

Polyacrylamide gel electrophoresis (PAGE) analysis can be conveniently applied to analyze the molecular weight of sulfated GAGs. Gels on which GAGs have been

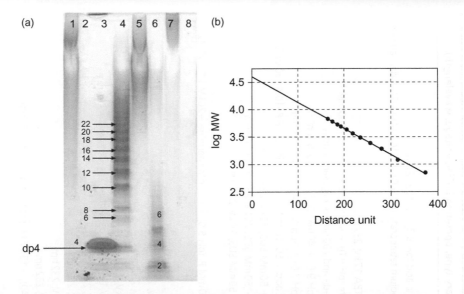

**FIGURE 3.2** (a) Gradient polyacrylamide gel electrophoresis (PAGE) analysis with Alcian Blue staining of glycosaminoglycan samples before and after treatment with heparin lyases. Lane 1 is intact porcine intestinal heparan sulfate (HS); lane 2 is porcine intestinal HS after treatment with heparin lyase 1, 2, and 3; lane 3 is a hexasulfated tetrasaccharide standard derived from heparin (HP) indicated by the arrow; lane 4 is a mixture of HP-derived oligosaccharide standards enzymatically prepared from bovine lung HP—the numbers indicate their degree of polymerization (i.e., 4 is a tetrasaccharide); lane 5 is intact porcine HP; lane 6 is porcine intestinal HP after heparin lyase 1 treatment; lane 7 is human liver HS; lane 8 is human liver HS after heparin lyase 3 treatment. (b) A plot of log molecular weight of bovine lung HP-derived oligosaccharide standards as a function of migration distance of each oligosaccharide from which the average molecular weight of HP and HS can be calculated. (See color plate 6.)

fractionated can be visualized with Alcian Blue with or without silver staining and the bands can be scanned and digitized. The average MW of a GAG is then calculated based on a mixture of HP-derived oligosaccharide standards prepared through the partial enzymatic depolymerization of HP. PAGE analysis of HS purified from human placenta is shown in Figure 3.2 [30]. The polydispersity of GAGs is observed as a broad smear in PAGE and a numerical value for the dispersity can be calculated.

Gel permeation chromatography (GPC), which separates molecule solely on the basis of differences in molecular size has been used for the MW analysis of GAGs. Dextrans, dextran sulfates, or GAGs of different MWs can be used as standards in a GPC column to calibrate the MW of GAGs. Refractive index detection is typically used in this method [52].

## Disaccharide Analysis

Disaccharide analysis is one of the most important ways to characterize a GAG, which consists of repeating disaccharide units composed of different monosaccharide residues, linkages, and sulfation patterns. Disaccharide analysis typically follows the complete enzymatic digestion of GAGs by corresponding lyases. There are several techniques used to measure the resulting disaccharide composition.

Capillary electrophoresis (CE) is often used in disaccharide analysis because of its high resolving power and sensitivity. The methods for CE disaccharide analysis are presented in Table 3.3 [53–76]. CE can be used with UV, fluorescence, or MS for detection of analyte.

Table 3.3 Capillary electrophoresis methods and operating condition used for analysis of glycosaminoglycan disaccharides

| Analytes | Operating conditions (capillary type, operating buffer, voltage, wavelength) |
| --- | --- |
| **High pH, normal polarity** | |
| HA- and CS-derived-Δ-disaccharides [53] | UFS, 10 mM borate-50 mM boric acid (pH 8.8), 10 kV, 232 nm |
| HA- and CS-derived-Δ-disaccharides [54] | UFS, 100 mM borate-25 mM sodium tetraborate (pH 9.0), 15 kV, 214 nm (PMP derivatives) |
| **High pH + additives, normal polarity** | |
| HA-derived-Δ-di-, tetra- and hexasaccharides; CS-derived-Δ-disaccharides [55] | UFS, alkaline borate (pH 9.0)+TBA 17 kV, 232 nm |
| HA-derived-Δ-disaccharides, saturated oligosaccharides and CS-derived-Δ-disaccharides [56] | UFS, (1) 40 mM Pi-10 mM borate (pH 9.0) + 40 mM SDS, 15 kV, (2) 200 mM Pi (pH 3.0) − 15 kV, 232 nm for Δ-di- and 200 nm for oligosaccharides |
| CS-derived-Δ-disaccharides [57] | UFS, 40 mM Pi-10 mM borate (pH 9.0) + 40 mM SDS, 15 kV, 232 nm |
| CS-derived-Δ-disaccharides [58] | UFS, 18 mM borate-30 mM Pi (pH 7.0) + 50 mM CTAB, −20 kV, 232 nm |
| DS- and HS-derived-Δ-disaccharides [59] | UFS, 50 mM borate 10 mM boric acid + 50–100 mM triethylamine (pH 8.8–10.4), 30 kV, 214 nm |
| HS-derived-Δ-disaccharides and tetra-, hexa- or higher oligosaccharides [60] | UFS, 10 mM borate (pH 8.81) + 50 mM SDS, 20 kV, 232 nm |
| Acetylated heparin-derived trisulfated disaccharides [61] | UFS, 10 mM borate (pH 8.81) + 50 mM SDS, 20 kV, 232 nm or 206 nm |
| HS-derived-Δ-disaccharides [62] | UFS, 10 mM borate (pH 8.5) + 50 mM SDS, 20 kV, 231 nm |
| Coated capillary | |
| CS-derived-Δ-disaccharides [63] | Polyether-coated fused silica, 50 mM Pi (pH 5.0) + 0.1 mM spermine − 20 kV, LIF (325/425) (ANDSA derivatives) |
| **Low pH, reversed polarity** | |
| CS- and HS-derived-Δ-disaccharides, saturated and -Δ-oligosaccharides [64] | UFS, 25 mM Pi (pH 3.0), −15 kV, 254 nm (AMAC derivatives) |
| CS-, HS-, and heparin-derived-Δ-disaccharides [65] | UFS, 20 mM Pi (pH 3.48), −8 kV, 232 nm |
| HA- and CS-derived-Δ-disaccharides [66] | UFS, 15 mM Pi (pH 3.00), −20 kV, 232 nm |
| CS-derived-Δ-di- and oligosaccharides [67,68] | UFS, 15 mM Pi (pH 3.0), −20 kV, LIF (AMAC derivatives) |
| HP-derived-Δ-di- and oligosaccharides [69] | UFS, 20 mM Pi (pH 3.48) −8 kV, 232 nm |
| HS-/heparin-derived-Δ-disaccharides [70] | UFS, 15 mM Pi (pH 3.5) −20 kV, 232 nm |
| HP-derived-Δ-di-and oligosaccharides [71] | UFS, 20 mM Pi (pH 3.48) −18 kV, 232 nm |
| HS-derived-Δ-disaccharides [72] | UFS, 20 mM Pi (pH 3.5) −15 kV, 232 nm |
| HP-derived-Δ-disaccharides [73] | UFS, 60 mM formic acid (pH 3.40), −15 kV (+ pressure gradient), 231 nm |
| Microemulsion electrokinetic capillary chromatography | |
| CS-derived-Δ-disaccharides [74] | UFS, microemulsion buffer, 30 kV, 260 nm |

[a] Δ is a disaccharide or oligosaccharide having a Δ4,5-deoxy-α-ʟ-*threo*-hex-4-enopyranosyluronic acid (ΔUAp) residue at its non-reducing end as a result of the action of a polysaccharide lyase.

Strong ion-exchange (SAX) high-performance liquid chromatography (HPLC) has also been used for disaccharide analysis [77,78]. This method is also widely used in oligosaccharide mapping. SAX-HPLC relies on UV detection at 232 nm and thus has rather low sensitivity, limiting its utility for microanalysis of GAGs prepared from small tissue of cell culture samples.

Reversed phase (RP)-HPLC is another method to analyze disaccharide composition. This method often utilizes fluorescence derivatization of the disaccharides with 2-aminoacridone or some other fluorescent tag [79]. RP-HPLC has been used for the disaccharide analysis of both HS and CS/DS obtained from biological samples. The pre-analysis derivatization of disaccharides, without sample clean-up, is followed by RP-HPLC separation and can detect as little as approximately 100 pg (approximately $10^{-13}$ mol) of each disaccharide present in the mixture, thereby requiring >10 ng of total GAG for analysis.

Reversed-phase ion-pairing (RPIP)-HPLC is also widely used recently to analyze disaccharide composition. Different conditions are listed in Table 3.4 [80–95]. This analysis is generally performed with post-column fluorescent derivatization of disaccharides. The detection sensitivity of this method is picomolar per each disaccharide. Unfortunately, post-column derivatization requires a custom built, temperature-controlled, post-column reactor with two additional HPLC pumps. Recently, in-line electrospray ionization (ESI) mass spectrometry (MS) has been used in place of fluorescence detection to obtain comparable analytical sensitivity while simplifying this method. RPIP-HPLC with ESI-MS detection has also been used for oligosaccharide mapping.

ESI-MS and ESI-MS-MS and multidimensional mass spectrometry ($MS^n$) can be relied on for the quantification of the isomeric disaccharides of GAGs without the use of chromatography or prior separation. The compositional analysis of disaccharide constituents of HP/HS can be achieved from a full-scan $MS^1$ spectra using an internal standard and a calculated response factor for each disaccharide [96]. Diagnostic product ions from $MS^n$ spectra of isomeric disaccharides can provide quantitative analysis of the relative amounts of each of the isomers in mixtures. This protocol was validated using several quality control samples and showed satisfactory accuracy and precision. Using this quantitative analysis procedure, percentages of disaccharide compositions for heparins from porcine and bovine intestinal mucosa and heparan sulfate from bovine kidney were determined.

CS/DS and KS disaccharides were also identified using a combination of electrospray ionization MS and tandem MS [97,98].

Reverse phase ion-pairing (RPIP) HPLC-MS is the most recent method to analyze disaccharide qualitatively and quantitatively. RPIP is the primary method to separate disaccharides and MS is used to detect and confirm the structure of each disaccharide. Selected methods are listed in Table 3.4. The extracted ion chromatogram (EIC) of the HPLC-MS analysis of 8 HP/HS disaccharides and the MS of each disaccharide is shown in Figure 3.3 [90]. The composition is obtained by integration of the peak area of the EIC and the structure of the disaccharides present in this mixture is unambiguously established from the MS. Based on the improvement in

## Table 3.4 Ion-pairing reverse-phase HPLC methods and conditions for disaccharide analysis and oligosaccharide mapping

| Column | Ion-pairing reagent | Sample | Detector |
|---|---|---|---|
| Regular C18 column [80] | Tetrabutylamium hydroxide | HA derived oligosaccharides | Post-column derivatization by 2-cyanoacetamide monitored at 270 nm |
| Supelcosil LC18 column [81] | Tetrabutylamium | Twelve HP/HS disaccharides | 232 nm |
| Cosmosil packed ODS column [82] | Tetrabutylammonium phosphate | PNP labeled HP/HS disaccharides | 390 nm |
| Regular C18 column [83] | Triethylamium, dibutylamium, tributylamium, tripentylamium, tetrapropylamium, and tetrabutylamium acetate | HP/HS disaccharides, oligosaccharides, heparosan oligosaccharides | 232 nm and MS |
| Regular C18 column [84,85] | Tributylammonium acetate | HP-derived oligosaccharides | 232 nm and MS |
| TSKgel Super-Octyl column [86–88] | Tetrabutylammonium hydrogensulfate | HP/HS disaccharides CS/DS disaccharides | Post-column derivatization by 2-cyanoacetamide monitored at 346 nm of excitation and 410 nm of emission with a fluorescence detection |
| XTerra MS C18 column [89] | Tripropyl ammonium acetate and butyl ammonium acetate | HP oligosaccharides | 232 nm and MS |
| Agilent capillary C18 column [90,91] | Tributylammonium acetate | HP/HS disaccharides CS/DS disaccharides | MS |
| Agilent capillary C18 column [92,93] | Tributylammonium acetate | HA, heparosan and CS oligosaccharides | MS |
| Regular C18 column [94] | Tributylammonium acetate | Non-sulfated CS oligosaccharides | MS |
| Acquity UPLC BEH C18 column [95] | Tributylammonium acetate | HP/HS disaccharides | 232 nm and MS |

separation method and application of an internal standard, the HP/HS disaccharide composition was quantified [95]. The method offers the advantage of rapid analysis with minimum sample consumption and without the need for sample preparation or further purification. This fast and reliable method is suitable for structural characterization and quantification of pharmaceutical HP preparations and samples of HP/HS isolated from a variety of biological sources.

## Oligosaccharide Mapping and Analysis

Oligosaccharide mapping techniques have been applied to GAGs to understand the structural features responsible for their activity differences. Oligosaccharide mapping

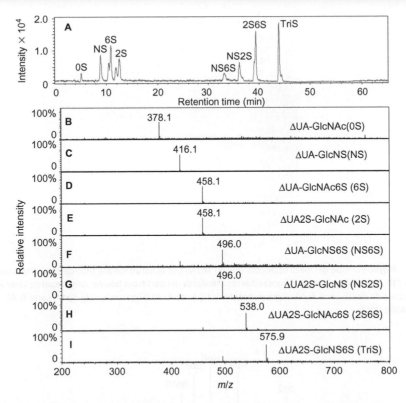

**FIGURE 3.3** Heparin (HP)/heparan sulfate (HS) disaccharide analysis by liquid chromatography/mass spectroscopy (LC-MS). (A) Extracted ion chromatography; (B–I) mass spectra of each disaccharide in negative mode. (See color plate 7.)

of GAGs represents an approach comparable to the peptide mapping of proteins. Oligosaccharide mapping can provide an estimate of molecular weight by detection of oligosaccharides arising from chain ends and gives information on the presence and distribution of discrete oligosaccharide sequences within a GAG [99]. Oligosaccharide mapping is a valuable method for the analysis of fine structure and to better understand the sequence of complex GAGs. Oligosaccharide mapping is also useful for making rapid assessments of the molecular distinctions between GAGs from different tissues and species [100–102]. The method involves specific enzymatic or chemical (i.e., through the use of nitrous acid) scission of polysaccharide chains followed by high-resolution separation of the degradation products by gradient PAGE (Figure 3.4) [103], GPC, CE (Figure 3.5) [103], SAX-HPLC (Figure 3.6a) [104], and other chromatographic methods. The molecular structural information of purified oligosaccharides (such as mass, degree of sulfation, and sequence) can be elucidated using ESI-MS (Figure 3.6b) [104]. Preparative separations can provide sufficient pure oligosaccharide for detailed structural analysis using nuclear magnetic resonance (NMR) spectroscopy (Figure 3.6c) [104].

## Sequence Analysis

Tandem MS has been recently applied to GAG [105]. Successful ionization and detection of GAG-derived oligosaccharides by tandem MS, including HA [106–109], KS

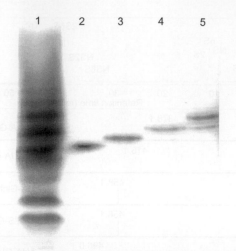

**FIGURE 3.4**  Polyacrylamide gel electrophoresis (PAGE) analysis of acharan sulfate (AS)-derived oligosaccharides visualized with Alcian Blue. Lane 1: Oligosaccharide standards derived from bovine lung heparin; Lane 2: AS oligosaccharide dp10; Lane 3: AS oligosaccharide dp12; Lane 4: AS oligosaccharide dp14; Lane 5: AS oligosaccharide dp16.

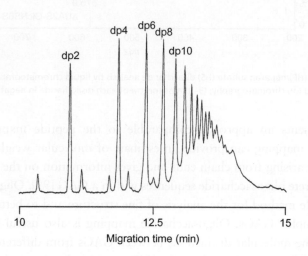

**FIGURE 3.5**  Capillary electrophoresis (CE) analysis of acharan sulfate after 40% digestion with heparin lyase II. Absorbance at 230 nm is plotted as a function of migration time. The peaks corresponding to the disaccharide (2mer) through octasaccharide were confirmed using authentic standards that had been previously prepared and characterized in our laboratory. The peaks of decasaccharide (10 mer) and higher oligosaccharides (not labeled) were also separated with high resolution.

[110–112], CS [113–116], DS [104], and HP/HS [96,117–118] oligosaccharides have been reported.

Since there is no sequence heterogeneity in HA, the structure is confirmed by chain length. Fragment ions can be useful for determining the terminal sugars in the HA oligosaccharides. Fragment ions show difference of $m/z = 174.8$ ($GlcA_p$-$H_2O$) or $m/z = 201.8$ ($GlcN_pAc$-$H_2O$). Glycosidic cleavage fragment ions were also observed in MS/MS and tandem MS of KS, CS/DS and HS/HP oligosaccharides, which confirmed

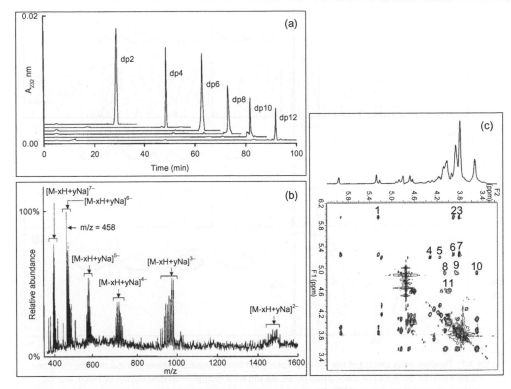

**FIGURE 3.6**  (a) Analytical strong ion-exchange (SAX) high-performance liquid chromatography (HPLC) analysis of individual dermatan sulfate-derived oligosaccharides. (b) Electrospray ionization mass spectrometry (MS-ESI) analysis of dermatan sulfate dodecasaccharide. Six ion clusters are marked with charges ranging from −7 to −2. The number of protons lost (x) ranged from 1 to 12 and the number of sodium atoms added (y) ranged from 0 to 9. (c) Two-dimensional $^1$H-NMR TOCSY spectrum of dermatan sulfate dodecasaccharide is presented. The cross-peaks are assigned as: 1, $\Delta$UAp H1/H4; 2, $\Delta$UAp H3/H4; 3, $\Delta$UAp H2/H4; 4, $\alpha$-GalpNAc H1/H2; 5, $\alpha$-GalpNAc H1/H3,H6; 6, $\Delta$UAp H1/H3; 7, $\Delta$UAp H1/H3; 8, IdoAp H1/H4; 9, IdoAp H1/H3; 10, IdoAp H1/H2; and 11, $\beta$-GalpNAc H1/H2,H3.

the residues and composition. The cross-ring cleavage fragment ions in tandem MS can distinguish GAG oligosaccharides with different linkage and sulfo patterns.

Most recently, electron detachment dissociation (EDD) has been applied to analyze the sequence of HS oligosaccharides, including substituent position and epimerized residues of HS oligosaccharides [119,120]. Compared to collisionally activated dissociation (CAD) and infrared multiphoton dissociation (IRMPD) MS, EDD provides improved cross-ring fragmentation important for determining the pattern of sulfation, acetylation, and hexuronic acid stereochemistry (GlcA$_p$ and IdoA$_p$) on a GAG oligosaccharide. The MS/MS spectra on [M-2H]$^-$ precursor ion of $\Delta$UA$_p$-GlcN$_p$S-IdoA$_p$-GlcN$_p$Ac6S in different dissociation are shown in Figure 3.7. Also this technology was applied to sequence DS oligosaccharides. The exact sequence of DS tetrasaccharides was assigned by the detail fragmentation in EDD-MS/MS [121,122]. Based on the extensive fragmentation observed in EDD, these complicated sulfated oligosaccharides can be sequenced unambiguously.

Electrospray ionization Fourier transform-ion cyclotron resonance mass spectrometry (ESI-FTICR-MS) was first applied to identify several major components

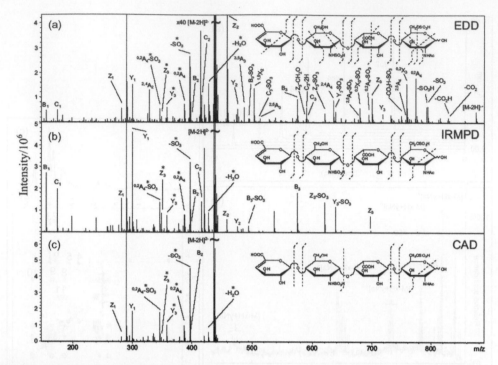

**FIGURE 3.7** Tandem mass spectra of the [M-2H]$^{2-}$ precursor ion of heparan sulfate tetrasaccharide, obtained by using (a) electron detachment dissociation (EDD), (b) infrared multiphoton dissociation (IRMPD), and (c) collisionally activated dissociation (CAD). Doubly-charged product ions are indicated with an asterisk.

in intact GAG chain mixture [123]. Bikunin is a serine protease inhibitor found in human amniotic fluid, plasma, and urine. Bikunin is posttranslationally modified with a CS chain, O-linked to a serine residue of the core protein. The CS chain of bikunin plays an important role in the physiological and pathological functions of this PG. While no PG or GAG has yet been sequenced, bikunin, the least complex PG, offered a compelling target. ESI-FTICR-MS permitted the identification of several major components in the GAG mixture having molecular masses in a range of 5505–7102 Da. This is the first report of a mass spectrum of an intact GAG component of a PG. FTICR-MS analysis of a size-uniform fraction of bikunin GAG mixture obtained by preparative polyacrylamide gel electrophoresis, allowed the determination of chain length and number of sulfo groups in the intact GAGs.

## Conformational Analysis

Nuclear magnetic resonance (NMR) is useful to characterize intact GAG chains. The composition of a chemoenzymatically synthesized HS GAG was characterized by $^1$H-NMR [124]. The sequence, linkage, and conformation were elucidated. The three-dimensional structure of GAGs can also be simulated based on these NMR data (Figure 3.8) [124,125].

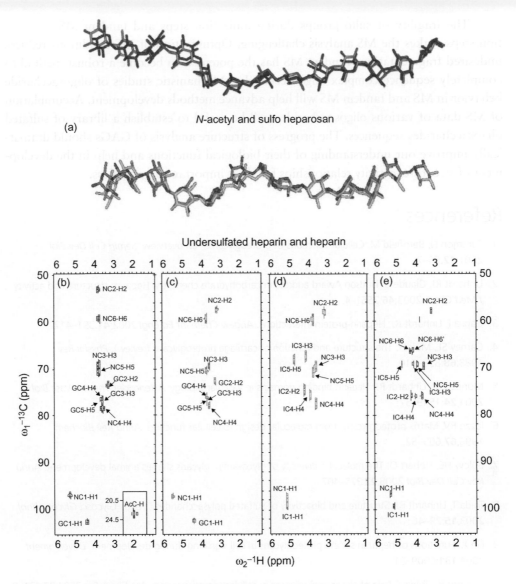

**FIGURE 3.8**  High-field NMR spectra and three-dimensional structures of chemo-enzymatic synthesized heparin (HP) and its precursors. (a) Three-dimensional structures of chemo-enzymatic synthesized HP and its precursors; (b–e) heteronuclear multiple quantum coherence of chemo-enzymatic synthesized HP and its precursors. (See color plate 8.)

## Conclusion

The progress of structure analysis of GAGs relies on highly efficient isolation/purification techniques, and high-sensitivity, information-rich analytical instruments. With modern MS techniques, structural information, such as molecular weight, monosaccharide composition, number and position of sulfo groups, composition of disaccharide blocks, and sequence of highly charged sulfated carbohydrates can be obtained. High sensitivity of MS is now available for the microanalysis of carbohydrates derived from biological samples.

The fragility of sulfo groups during ionization steps and tandem MS dissociation steps makes the MS analysis challenging. Optimization of MS conditions reduces undesired fragmentation. Tandem MS has the potential to become a robust method to completely sequence complex oligosaccharides. Mechanistic studies of oligosaccharide behavior in MS and tandem MS will help advance methods development. Accumulation of MS data of various oligosaccharides will be useful to establish a library of sulfated oligosaccharides sequences. The progress of structure analysis of GAGs should dramatically improve our understanding of their biological functions and help in the development of structure–activity relationships for these important biopolymers.

# References

1. Perrimon N, Bernfield M. Cellular functions of proteoglycans—an overview. *Semin Cell Dev Biol* 2001;**12**:65–7.

2. Linhardt RJ, Claude S. Hudson Award address in carbohydrate chemistry. Heparin: structure and activity. *J Med Chem* 2003;**46**:2551–4.

3. Capila I, Linhardt RJ. Heparin-protein interactions. *Angew Chem Int Ed Engl* 2002;**41**:391–412.

4. Carney SL, Muir H. The structure and function of cartilage proteoglycans (review). *Physiol Rev* 1988;**68**:858–910.

5. Munoz EM, Linhardt RJ. Heparin binding domains in vascular biology. *Arterioscler Thromb Vasc Biol* 2004;**24**:1549–57.

6. Iozzo RV. Matrix proteoglycans: from molecular design to cellular function. *Annu Rev Biochem* 1998;**67**:609–52.

7. Bülow HE, Hobert O. The molecular diversity of glycosaminoglycans shapes animal development. *Annu Rev Cell Dev Biol* 2006;**22**:375–407.

8. Toida T, Linhardt RJ. Structure and bioactivity of sulfated polysaccharides. *Trends Glycosci Glycotechnol* 2003;**15**:29–46.

9. Lin X. Functions of heparan sulfate proteoglycans in cell signaling during development. *Development* 2004;**131**:6009–21.

10. Linhardt RJ, Toida T. Role of glycosaminoglycans in cellular communication. *Acc Chem Res* 2004;**37**:431–8.

11. Sugahara K, Mikami T, Uyama T, Mizuguchi S, Nomura K, Kitagawa H. Recent advances in the structural biology of chondroitin sulfate and dermatan sulfate. *Curr Opin Struct Biol* 2003;**13**:612–20.

12. Hardingham TE, Fosang AJ. Proteoglycans: many forms and many functions. *FASEB J* 1992;**6**:861–70.

13. Linhardt RJ, Hileman RE. Dermatan sulfate as a potential therapeutic agent. *Gen Pharmacol* 1995;**26**:443–51.

14. Akiyama H, Sakai S, Linhardt RJ, Goda Y, Toida T, Maitani T. Chondroitin sulfate structure impacts its immunological effects on murine splenocytes sensitized with ovalbumin. *Biochem J* 2004;**382**:269–78.

15. Faissner A, Clement A, Lochter A, Streit C, Schachner M. Isolation of a neural chondroitin sulfate proteoglycan with neurite outgrowth promoting properties. *J Cell Biol* 1994;**126**:783–99.

16. Greiling H, Scott JE. *Keratan sulphate-chemistry, biology, chemical pathology*. London: The Biochemical Society; 1989.

17. Funderburgh JL. Mini review: Keratan sulfate: structure, biosynthesis, and function. *Glycobiology* 2000;**10**:951–8.

18. Krusius T, Finne J, Margolis RK, Margolis RU. Identification of an O-glycosidic mannose-linked sialylated tetrasaccharide and keratan sulfate oligosaccharides in the chondroitin sulfate proteoglycan of brain. *J Biol Chem* 1986;**261**:8237–42.

19. Tai GH, Nieduszynski IA, Fullwood NJ, Huckerby TN. Human corneal keratan sulfates. *J Bio Chem* 1997;**272**:28227–31.

20. Toole BP. Hyaluronan: from extracellular glue to pericellular cue. *Nat Rev Cancer* 2004;**4**:528–39.

21. Toole BP, Slomiany MG. Hyaluronan: a constitutive regulator of chemoresistance and malignancy in cancer cells. *Semin Cancer Biol* 2008;**18**:244–50.

22. Liu J, Shriver Z, Pope RM, Thorp SC, Duncan MB, Copeland RJ, Raska CS, Yoshida K, Eisenberg RJ, Cohen G, Linhardt RJ, Sasisekharan R. Characterization of a heparan sulfate octasaccharide that binds to herpes simplex viral type 1 glycoprotein D. *J Biol Chem* 2002;**277**:33456–67.

23. Venkataraman G, Shriver Z, Raman R, Sasisekharan R. Sequencing Complex Polysaccharides. *Science* 1999;**286**:537–42.

24. Liu J, Desai UR, Han XJ, Toida T, Linhardt RJ. Strategy for the Sequence analysis of heparin. *Glycobiology* 1995;**5**:765–74.

25. Esko JD. Special considerations for proteoglycans and glycosaminoglycans and their purification. Current Protocols in Molecular Biology 1993;17.2.1–17.2.9.

26. Hascall VC, Kimura JH. Proteoglycans: Isolation and characterization. *Methods Enzymol* 1982;**82**:769–800.

27. Linhardt RJ, Gunay NS. Production and chemical processing of low molecular weight heparins, Semin. *Thrombosis Hemostas* 1999;**25**:5–16.

28. Toida T, Yoshida H, Toyoda H, Koshiishi I, Imanari T, Hileman RE, Fromm JR, Linhardt RJ. Structural differences and the presence of unsubstituted aminogroups in heparan sulphates from different tissues and species. *Biochem J* 1997;**322**:499–506.

29. Warda M, Gouda EM, Toida T, Chi L, Linhardt RJ. Isolation and characterization of raw heparin from dromedary intestine: evaluation of a new source of pharmaceutical heparin. *Comparative Biochem Physiol Part C* 2004;**136**:357–65.

30. Vongchan P, Warda M, Toyoda H, Toida T, Marks RM, Linhardt RJ. Structural characterization of human liver heparan sulfate. *Biochemica Biophysica Acta* 2005;**1721**:1–8.

31. Nelson SR, Lyon M, Gallagher JT, Johnson EA, Pepys MB. Isolation and characterization of the integral glycosaminoglycan constituents of human amyloid A and monoclonal light-chain amyloid fibrils. *Biochem J* 1991;**275**:67–73.

32. Singe M, Chanbrasekaran EV, Cherian R, Bachhawatb K. Isolation and characterisation of glycosaminoglycans in brain of different species. *J Neurochem* 1969;**16**:1157–62.

33. Sobue M, Takeuchi J, Yoshida K, Akao S, Fukatsu T, Nagasaka T, Nakashima N. Isolation and characterization of proteoglycans from human nonepithelial tumors. *Cancer Res* 1987;**47**:160–8.

34. Skandalis S, Theocharis A, Papageorgakopoulou N, Zagris N. Glycosaminoglycans in early chick embryo. *Inter J Develop Biol* 2003;**47**:311–14.

35. Warda M, Zhang Radwan FM, Zhang Z, Kim N, Kim YN, Linhardt RJ, Han J. Is human placenta proteoglycan remodeling involved in pre-eclampsia? *Glycoconjugate J* 2008;**25**:441–50.

36. Hitchcock AM, Yates KE, Shortkroff S, Costello CE, Zaia J. Optimized extraction of glycosaminoglycans from normal and osteoarthritic cartilage for glycomics profiling. *Glycobiology* 2007;**17**:25–35.

37. Warda M, Toida T, Zhang F, Sun P, Munoz E, Linhardt RJ. Isolation and characterization of heparan sulfate from various murine tissues. *Glycoconjugate J* 2006;**23**:553–61.

38. Nairn AV, Kinoshita-Toyoda A, Toyoda H, Xie J, Harris K, Dalton S, Kulik M, Pierce JM , Toida T, Moremen KW, Linhardt RJ. Glycomics of proteoglycan biosynthesis in murine embryonic stem cell differentiation. *J Proteome Res* 2007;**6**:4374–87.

39. Studelska DR, Giljum K, McDowell LM, Zhang L. Quantification of glycosaminoglycans by reversed-phase HPLC separation of fluorescent isoindole derivatives. *Glycobiology* 2006;**16**:65–72.

40. Zhang F, Sun P, Munoz E, Chi L, Sakai S, Toida T, Zhang H, Mousa S, Linhardt RJ. Microscale isolation and analysis of heparin from plasma using an anion exchange spin column. *Anal Biochem* 2006;**353**:284–6.

41. Sakai S, Onose J, Nakamura H, Toyoda H, Toida T, Imanari T, Linhardt RJ. Pretreatment procedure for the microdetermination of chondroitin sulfate in plasma and urine. *Anal Biochem* 2002;**302**:169–74.

42. de Agostini AI, Dong J-C, de Vantéry Arrighi C, Ramus M-A, Dentand-Quadri I, Thalmann S, Ventura P, Ibecheole V, Monge F, Fischer A-M, HajMohammadi S, Shworak NW, Zhang L, Zhang Z, Linhardt RJ. Human follicular fluid heparan sulfate contain abundant 3-*O*-sulfated chains with anticoagulant activity. *J Biol Chem* 2008;**283**;28115–24.

43. Bellin ME, Ax RL. Purification of glycosaminoglycans from bovine follicular fluid. *J Dairy Sci* 1987;**70**:1913–19.

44. Toyoda H, Kinoshita-Toyoda A, Selleck S. Structural analysis of glycosaminoglycans in *Drosophila* and *Caenorhabditis elegans* and demonstration that tout-velu, a Drosophila gene related to EXT tumor suppressors, affects heparan sulfate in Vivo. *J Biol Chem* 2000;**275**:2269–75.

45. Sinnis P, Coppi A, Toida T, Toyoda H, Kinoshita-Toyoda A, Xie J, Kemp MM, Linhardt RJ. Mosquito heparan sulfate and its potential role in malaria infection and transmission. *J Biol Chem* 2007;**282**:25376–84.

46. Zhang F, Zhang Z, Thistle R, McKeen L, Hosoyama S, Toida T, Linhardt RJ, Page-McCaw P. Structural characterization of glycosaminoglycans from zebrafish in different ages. *Glycoconjugate J* 2009;**26**; 211–8.

47. Luppi E, Cesaretti M, Volpi N. Purification and characterization of heparin from the Italian clam *Callista chione*. *Biomacromolecules* 2005;**6**:1672–8.

48. Sakai S, Kim WS, Lee IS, Kim YS, Nakamura A, Toida T, Imanari T. Purification and characterization of dermatan sulfate from the skin of the eel, Anguilla japonica. *Carbohydrate Res* 2003;**338**:263–9.

49. Kim YS, Jo YT, Chang IM, Toida TE, Park Y, Linhardt RJ. A new glycosaminoglycan from the giant African snail *Achatina fulica. J Biol Chem* 1996;**271**:11750–5.

50. Linhardt RJ. Analysis of glycosaminoglycans with polysaccharide lyases. Current Protocols in Molecular Biology 1999;17.13B.1–17.13B.16

51. Conrad HE. Nitrous acid degradation of glycosaminoglycans. Current Protocols in Molecular Biology 1995;17.22.1–17.22.5.

52. Mulloy B, Gee C, Wheeler SF, Wait R, Gray E, Barrowcliffe TW. Mocular weight measurements of low molecular weight heparins by gel permeation chromatography. *Thromb Haemost* 1997;**77**:668–74.

53. Al-Hakim A, Linhardt RJ. Capillary electrophoresis for the analysis of chondroitin sulfate- and dermatan sulfate-derived disaccharides. *Anal Biochem* 1991;**195**:68–73.

54. Honda S, Ueno T, Kakehi K. High-performance capillary electrophoresis of unsaturated oligosaccharides derived from glycosaminoglycans by digestion with chondroitinase ABC as 1-phenyl-3-methyl-5-pyrazolone derivatives. *J Chromatogr* 1992;**608**:289–95.

55. Payan E, Presle N, Lapicque F, Jouzeau JY, Bordji K, Oerther S, Miralles G, Mainard D, Netter P. Separation and quantification by ion-association capillary zone electrophoresis of unsaturated disaccharide units of chondroitin sulfates and oligosaccharides derived from hyaluronan. *Anal Chem* 1998;**70**:4780–6.

56. Carney SL, Osborne DJ. The separation of chondroitin sulfate disaccharides and hyaluronan oligosaccharides by capillary zone electrophoresis. *Anal Biochem* 1991;**195**:132–40.

57. Hayase S, Oda Y, Honda S, Kakehi K. High-performance capillary electrophoresis of hyaluronic acid: determination of its amount and molecular mass. *J Chromatogr A* 1997;**768**:295–305.

58. Michaelsen S, Schrøder MB, Sørensen H. Separation and determination of glycosaminoglycan disaccharides by micellar electrokinetic capillary chromatography for studies of pelt glycosaminoglycans. *J Chromatogr A* 1993;**652**:503–15.

59. Scapol L, Marchi E, Viscomi GC. Capillary electrophoresis of heparin and dermatan sulfate unsaturated disaccharides with triethylamine and acetonitrile as electrolyte additives. *J Chromatogr A* 1996;**735**:367–74.

60. Desai UR, Wang H, Ampofo SA, Linhardt RJ. Oligosaccharide composition of heparin and low-molecular-weight heparins by capillary electrophoresis. *Anal Biochem* 1993;**213**:120–7.

61. Kerns RJ, Vlahov IR, Linhardt RJ. Capillary electrophoresis for monitoring chemical reactions: sulfation and synthetic manipulation of sulfated carbohydrates. *Carbohydr Res* 1995;**267**:143–52.

62. Ampofo SA, Wang HM, Linhardt RJ. Disaccharide compositional analysis of heparin and heparan sulfate using capillary zone electrophoresis. *Anal Biochem* 1991;**199**:249–55.

63. El Rassi Z. Recent developments in capillary electrophoresis of carbohydrate species. *Electrophoresis* 1997;**18**:2400–7.

64. Kitagawa H, Kinoshita A, Sugahara K. Microanalysis of glycosaminoglycan-derived disaccharides labeled with the fluorophore 2-aminoacridone by capillary electrophoresis and high-performance liquid chromatography. *Anal Biochem* 1995;**232**:114–21.

65. Pervin A, al-Hakim A, Linhardt RJ. Separation of glycosaminoglycan-derived oligosaccharides by capillary electrophoresis using reverse polarity. *Anal Biochem* 1994;**221**:182–8.

66. Karamanos NK, Axelsson S, Vanky P, Tzanakakis GN, Hjerpe A. Determination of hyaluronan and galactosaminoglycan disaccharides by high-performance capillary electrophoresis at the attomole level. Applications to analyses of tissue and cell culture proteoglycans. *J Chromatogr A* 1995;**696**:295–305.

67. Lamari F, Theocharis A, Hjerpe A, Karamanos NK. Ultrasensitive capillary electrophoresis of sulfated disaccharides in chondroitin/dermatan sulfates by laser-induced fluorescence after derivatization with 2-aminoacridone. *J Chromatogr B Biomed Sci Appl* 1999;**730**:129–33.

68. Mitropoulou TN, Lamari F, Syrokou A, Hjerpe A, Karamanos NK. Identification of oligomeric domains within dermatan sulfate chains using differential enzymic treatments, derivatization with 2-aminoacridone and capillary electrophoresis. *Electrophoresis* 2001;**22**:2458–63.

69. Pervin A, Gallo C, Jandik KA, Han XJ, Linhardt RJ. Preparation and structural characterization of large heparin-derived oligosaccharides. *Glycobiology* 1995;**5**:83–95.

70. Karamanos NK, Vanky P, Tzanakakis GN, Hjerpe A. High performance capillary electrophoresis method to characterize heparin and heparan sulfate disaccharides. *Electrophoresis* 1996;**17**:391–5.

71. Hileman RE, Smith AE, Toida T, Linhardt RJ. Preparation and structure of heparin lyase-derived heparan sulfate oligosaccharides. *Glycobiology* 1997;**7**:231–9.

72. Toida T, Yoshida H, Toyoda H, Koshiishi I, Imanari T, Hileman RE, Fromm JR, Linhardt RJ. Structural differences and the presence of unsubstituted amino groups in heparan sulphates from different tissues and species. *Biochem J* 1997;**322**(Pt 2):499–506.

73. Ruiz-Calero V, Puignou L, Galceran MT. Use of reversed polarity and a pressure gradient in the analysis of disaccharide composition of heparin by capillary electrophoresis. *J Chromatogr A* 1998;**828**:497–508.

74. Mastrogianni O, Lamari F, Syrokou A, Militsopoulou M, Hjerpe A, Karamanos NK. Microemulsion electrokinetic capillary chromatography of sulfated disaccharides derived from glycosaminoglycans. *Electrophoresis* 2001;**22**:2743–5.

75. Lamari FN, Militsopoulou M, Mitropoulou TN, Hjerpe A, Karamanos NK. Analysis of glycosaminoglycan-derived disaccharides in biologic samples by capillary electrophoresis and protocol for sequencing glycosaminoglycans. *Biomed Chromatogr* 2002;**16**:95–102.

76. Mao W, Thanawiroon C, Linhardt RJ. Capillary electrophoresis for the analysis of glycosaminoglycans and glycosaminoglycan-derived oligosaccharides. *Biomed Chromatogr* 2002;**16**:77–94.

77. Qiu G, Toyoda H, Toida T, Koshiishi I, Imanari T. Compositional analysis of hyaluronan, chondroitin sulfate and dermatan sulfate: HPLC of disaccharides produced from the glycosaminoglycans by solvolysis. *Chem Pharm Bull (Tokyo)* 1996;**44**:1017–20.

78. Sim JS, Jun G, Toida T, Cho SY, Choi DW, Chang SY, Linhardt RJ, Kim YS. Quantitative analysis of chondroitin sulfate in raw materials, ophthalmic solutions, soft capsules and liquid preparations. *J Chromatogr B Analyt Technol Biomed Life Sci* 2005;**818**:133–9.

79. Deakin JA, Lyon M. A simplified and sensitive fluorescent method for disaccharide analysis of both heparan sulfate and chondroitin/dermatan sulfates from biological samples. *Glycobiology* 2008;**18**:483–91.

80. Cramer JA, Bailey LC. A reversed-phase ion-pair high-performance liquid chromatography method for bovine testicular hyaluronidase digests using postcolumn derivatization with 2-cyanoacetamide and ultraviolet detection. *Anal Biochem* 1991;**196**:183–91.

81. Karamanos NK, Vanky P, Tzanakakis GN, Tsegenidis T, Hjerpe A. Ion-pair high-performance liquid chromatography for determining disaccharide composition in heparin and heparan sulphate. *J Chromatogr A* 1997;**765**:169–79.

82. Kariya Y, Herrmann J, Suzuki K, Isomura T, Ishihara M. Disaccharide analysis of heparin and heparan sulfate using deaminative cleavage with nitrous acid and subsequent labeling with paranitrophenyl hydrazine. *J Biochem* 1998;**123**:240–6.

83. Kuberan B, Lech M, Zhang L, Wu ZL, Beeler DL, Rosenberg RD. Analysis of heparan sulfate oligosaccharides with ion pair-reverse phase capillary high performance liquid chromatography-microelectrospray ionization time-of-flight mass spectrometry. *J Am Chem Soc* 2002;**124**:8707–18.

84. Thanawiroon C, Linhardt RJ. Separation of a complex mixture of heparin-derived oligosaccharides using reversed-phase high-performance liquid chromatography. *J Chromatogr A* 2003;**1014**:215–23.

85. Thanawiroon C, Rice KG, Toida T, Linhardt RJ. Liquid chromatography/mass spectrometry sequencing approach for highly sulfated heparin-derived oligosaccharides. *J Biol Chem* 2004;**279**:2608–15.

86. Ha YW, Jeon BT, Moon SH, Toyoda H, Toida T, Linhardt RJ, Kim YS. Characterization of heparan sulfate from the unossified antler of *Cervus elaphus. Carbohydr Res* 2005;**340**:411–16.

87. Vongchan P, Warda M, Toyoda H, Toida T, Marks RM, Linhardt RJ. Structural characterization of human liver heparan sulfate. *Biochim Biophys Acta* 2005;**1721**:1–8.

88. Warda M, Toida T, Zhang F, Sun P, Munoz E, Xie J, Linhardt RJ. Isolation and characterization of heparan sulfate from various murine tissues. *Glycoconj J* 2006;**23**:555–63.

89. Henriksen J, Roepstorff P, Ringborg LH. Ion-pairing reversed-phased chromatography/mass spectrometry of heparin. *Carbohydr Res* 2006;**341**:382–7.

90. Warda M, Zhang F, Radwan M, Zhang Z, Kim N, Kim YN, Linhardt RJ, Han J. Is human placenta proteoglycan remodeling involved in pre-eclampsia?. *Glycoconj J* 2008;**25**:441–50.

91. Zhang F, Zhang Z, Thistle R, McKeen L, Hosoyama S, Toida T, Linhardt RJ, Page-McCaw P. Structural characterization of glycosaminoglycans from zebrafish in different ages. *Glycoconj J* 2009;**26**:211–18.

92. Zhang Z, Xie J, Liu J, Linhardt RJ. Tandem MS can distinguish hyaluronic acid from *N*-acetylheparosan. *J Am Soc Mass Spectrom* 2008;**19**:82–90.

93. Zhang Z, Park Y, Kemp MM, Zhao W, Im AR, Shaya D, Cygler M, Kim YS, Linhardt RJ. Liquid chromatography-mass spectrometry to study chondroitin lyase action pattern. *Anal Biochem* 2009;**385**:57–64.

94. Volpi N, Zhang Z, Linhardt RJ. Mass spectrometry for the characterization of unsulfated chondroitin oligosaccharides from 2-mers to 16-mers. Comparison with hyaluronic acid oligomers. *Rapid Commun Mass Spectrom* 2008;**22**:3526–30.

95. Korir AK, Limtiaco JF, Gutierrez SM, Larive CK. Ultraperformance ion-pair liquid chromatography coupled to electrospray time-of-flight mass spectrometry for compositional profiling and quantification of heparin and heparan sulfate. *Anal Chem* 2008;**80**:1297–306.

96. Saad OM, Leary JA. Compositional analysis and quantification of heparin and heparan sulfate by electrospray ionization ion trap mass spectrometry. *Anal Chem* 2003;**75**:2985–95.

97. Desaire H, Leary JA. Detection and quantification of the sulfated disaccharides in chondroitin sulfate by electrospray tandem mass spectrometry. *J Am Soc Mass Spectrom* 2000;**11**:916–20.

98. Zhang Y, Conrad AH, Tasheva ES, An K, Corpuz LM, Kariya Y, Suzuki K, Conrad GW. Detection and quantification of sulfated disaccharides from keratan sulfate and chondroitin/dermatan sulfate during chick corneal development by ESI-MS/MS. *Invest Ophthalmol Vis Sci* 2005;**46**:1604–14.

99. Linhardt RJ, Rice KG, Kim YS, Lohse DL, Wang HM, Loganathan D. Mapping and quantification of the major oligosaccharide components of heparin. *Biochem J* 1988;**254**:781–7.

100. Loganathan D, Wang HM, Mallis LM, Linhardt RJ. Structural variation in the antithrombin III binding site region and its occurrence in heparin from different sources. *Biochemistry* 1990;**29**:4362–8.

101. Linhardt RJ, al-Hakim A, Liu SY, Kim YS, Fareed J. Molecular profile and mapping of dermatan sulfates from different origins. *Semin Thromb Hemost* 1991;**17**:15–22.

102. Linhardt RJ, Ampofo SA, Fareed J, Hoppensteadt D, Mulliken JB, Folkman J. Isolation and characterization of human heparin. *Biochemistry* 1992;**31**:12441–5.

103. Chi L, Munoz EM, Choi HS, Ha YW, Kim YS, Toida T, Linhardt RJ. Preparation and structural determination of large oligosaccharides derived from acharan sulfate. *Carbohydr Res* 2006;**341**:864–9.

104. Yang HO, Gunay NS, Toida T, Kuberan B, Yu G, Kim YS, Linhardt RJ. Preparation and structural determination of dermatan sulfate-derived oligosaccharides. *Glycobiology* 2000;**10**:1033–9.

105. Zaia J. Mass spectrometry of oligosaccharides. *Mass Spectrom Rev* 2004;**23**:161–227.

106. Kühn AV, Raith K, Sauerland V, Neubert RH. Quantification of hyaluronic acid fragments in pharmaceutical formulations using LC-ESI-MS. *J Pharm Biomed Anal* 2003;**30**:1531–7.

107. Kühn AV, Rüttinger HH, Neubert RH, Raith K. Identification of hyaluronic acid oligosaccharides by direct coupling of capillary electrophoresis with electrospray ion trap mass spectrometry. *Rapid Commun Mass Spectrom* 2003;**17**:576–82.

108. Mahoney DJ, Aplin RT, Calabro A, Hascall VC, Day AJ. Novel methods for the preparation and characterization of hyaluronan oligosaccharides of defined length. *Glycobiology* 2001;**11**:1025–33.

109. Tawada A, Masa T, Oonuki Y, Watanabe A, Matsuzaki Y, Asari A. Large-scale preparation, purification, and characterization of hyaluronan oligosaccharides from 4-mers to 52-mers. *Glycobiology* 2002;**12**:421–6.

110. Minamisawa T, Suzuki K, Hirabayashi J. Multistage mass spectrometric sequencing of keratan sulfate-related oligosaccharides. *Anal Chem* 2006;**78**:891–900.

111. Oguma T, Toyoda H, Toida T, Imanari T. Analytical method for keratan sulfates by high-performance liquid chromatography/turbo-ionspray tandem mass spectrometry. *Anal Biochem* 2001;**290**:68–73.

112. Zhang Y, Kariya Y, Conrad AH, Tasheva ES, Conrad GW. Analysis of keratan sulfate oligosaccharides by electrospray ionization tandem mass spectrometry. *Anal Chem* 2005;**77**:902–10.

113. Desaire H, Sirich TL, Leary JA. Evidence of block and randomly sequenced chondroitin polysaccharides: sequential enzymatic digestion and quantification using ion trap tandem mass spectrometry. *Anal Chem* 2001;**73**:3513–20.

114. McClellan JE, Costello CE, O'Connor PB, Zaia J. Influence of charge state on product ion mass spectra and the determination of 4S/6S sulfation sequence of chondroitin sulfate oligosaccharides. *Anal Chem* 2002;**74**:3760–71.

115. Zaia J, Costello CE. Compositional analysis of glycosaminoglycans by electrospray mass spectrometry. *Anal Chem* 2001;**73**:233–9.

116. Zaia J, McClellan JE, Costello CE. Tandem mass spectrometric determination of the 4S/6S sulfation sequence in chondroitin sulfate oligosaccharides. *Anal Chem* 2001;**73**:6030–9.

117. Chai W, Luo J, Lim CK, Lawson AM. Characterization of heparin oligosaccharide mixtures as ammonium salts using electrospray mass spectrometry. *Anal Chem* 1998;**70**:2060–6.

118. Pope RM, Raska CS, Thorp SC, Liu J. Analysis of heparan sulfate oligosaccharides by nano-electrospray ionization mass spectrometry. *Glycobiology* 2001;**11**:505–13.

119. Wolff JJ, Amster IJ, Chi L, Linhardt RJ. Electron detachment dissociation of glycosaminoglycan tetrasaccharides. *J Am Soc Mass Spectrom* 2007;**18**:234–44.

120. Wolff JJ, Chi L, Linhardt RJ, Amster IJ. Distinguishing glucuronic from iduronic acid in glycosaminoglycan tetrasaccharides by using electron detachment dissociation. *Anal Chem* 2007;**79**:2015–22.

121. Wolff JJ, Laremore TN, Busch AM, Linhardt RJ, Amster IJ. Electron detachment dissociation of dermatan sulfate oligosaccharides. *J Am Soc Mass Spectrom* 2008;**19**:294–304.

122. Wolff JJ, Laremore TN, Busch AM, Linhardt RJ, Amster IJ. Influence of charge state and sodium cationization on the electron detachment dissociation and infrared multiphoton dissociation of glycosaminoglycan oligosaccharides. *J Am Soc Mass Spectrom* 2008;**19**:790–8.

123. Chi L, Wolff JJ, Laremore TN, Restaino OF, Xie J, Schiraldi C, Toida T, Amster IJ, Linhardt RJ. Structural analysis of bikunin glycosaminoglycan. *J Am Chem Soc* 2008;**130**:2617–25.

124. Zhang Z, McCallum SA, Xie J, Nieto L, Corzana F, Jiménez-Barbero J, Chen M, Liu J, Linhardt RJ. Solution structures of chemoenzymatically synthesized heparin and its precursors. *J Am Chem Soc* 2008;**130**:12998–3007.

125. Silipo A, Zhang Z, Cañada FJ, Molinaro A, Linhardt RJ, Jiménez-Barbero J. Conformational analysis of a dermatan sulfate-derived tetrasaccharide by NMR, molecular modeling, and residual dipolar couplings. *Chembiochemistry* 2008;**9**:240–52.

# Glycotranscriptomics

# 4

# Isotopic Labeling of Glycans for Quantitative Glycomics

Richard D. Cummings and David F. Smith
*DEPARTMENT OF BIOCHEMISTRY, EMORY UNIVERSITY SCHOOL OF MEDICINE, ATLANTA, GA, USA*

## Introduction

The repertoire of glycans on glycoproteins, glycolipids, and proteoglycans and free glycans will vary qualitatively and quantitatively with every cell type and with different bodily fluids, such as urine and milk. A major goal of glycomics is the characterization of such glycosylation differences. This is often difficult, because samples are normally analyzed individually by comparison of data from mass spectrometry (MS), chromatography, or electrophoresis, where subtle qualitative and/or quantitative differences can easily be missed. A more modern approach for characterizing differences in glycosylation between two samples takes its cue from proteomic approaches using isotope-coded affinity tags (ICAT) [1] and MS. In such approaches polypeptides from one sample are isotopically tagged and then mixed with another polypeptide sample, which is also isotopically tagged, but with a different isotope. Thus, if two samples are identical, mass spectrometric analysis will reveal "doublets" of common structures with mass differences equal to the difference in the two isotopes used. However, if even a single component of one sample differs quantitatively or qualitatively, these differences will be reflected in an increased intensity of one of the doublets or a mass difference beyond the expected isotope tag size. Such differential isotopic labeling of proteins using ICAT has revolutionized proteomic studies [2]. Similar isotope labeling approaches have been used for cells in culture (SILAC) [3] and other approaches [4]. Such isotope labeling approaches have only recently been developed for glycomic analyses. While this area of research is still relatively new, there have been significant developments in creating isotope-based approaches for comparative analyses of glycan samples. This chapter will highlight the different approaches to stable isotope labeling of glycans for comparative and/or

**Handbook of Glycomics**
Copyright © 2009 by Elsevier Inc. All rights of reproduction in any form reserved.

quantitative glycomics and indicate how they can be used to explore major questions in the glycomics field. The use of isotopic labeling for glycan identification is also discussed in Chapters 1 and 2.

## Overall Strategy of Isotope Labeling of Glycans for Quantitative Glycomics

The basis of differential isotope labeling of glycans is the quantitative derivatization of free, reducing glycans by one of several approaches, including reductive amination of the reducing end of the glycans with heavy or light isotope labeled reagents, or derivatization of the glycan backbone by permethylation with reagents containing heavy or light isotopes. An illustration of this approach using reductive amination with $^{12}C_6$-aniline and $^{13}C_6$-aniline is shown in Figure 4.1. In this example, N-glycans are enzymatically released from two different samples, which could be serum glycoproteins, cell or tissue-derived glycoproteins, or even purified glycoproteins from different sources. The free glycans could also be O-glycans derived from non-reductive β-elimination, glycans derived from glycosphingolipids through chemical or enzymatic release, or any set of free glycans in animal fluids. In this illustration, free glycans are quantitatively derivatized by reductive amination with either $^{12}C_6$-aniline and $^{13}C_6$-aniline, which introduces a 6 Da mass difference ($\Delta m = 6$ Da) at the reducing end of each glycan [5–7]. The introduction of the aniline also provides a way of quantifying glycans by simply measuring absorbance at 260 nm. While the aromatic derivatives, such as aniline, 2-aminopyridine, or 2-anthranilic acid are not thought of as affinity tags, they do endow free glycans with a unique property, allowing the separation of underivatized glycans from derivatized ones. For example, aniline-tagged glycans are bound by C18 residues, whereas free glycans are not.

When equal amounts of differentially labeled glycans from the two samples are mixed in equimolar amounts and analyzed by matrix-assisted laser desorption/ionization-time-of-flight-mass spectroscopy (MALDI-TOF-MS), quantitative and qualitative differences in glycan expression and composition between the two samples may be apparent from the spectra. For example, it can be concluded that at least one glycan (numbers 1 and 4 in Figure 4.1) is found in both samples and appears to be present in equal amounts between the samples A and B, whereas all other glycans are different between the two samples. Alternatively, such studies could be conducted using a defined set of glycans labeled with the heavy isotope, in place of the unknown sample B. Such heavy-tagged standard glycans at known amounts could then be mixed with the unknown light-labeled sample A. The MS data then provide quantitative measurements of common structures based on the ration of peak heights and identify unique compounds not present in the standard mixture [5]. This alternative approach is typical of most of the recently published applications of stable isotope labeling to glycomic analysis (summarized in Table 4.1). In most cases, the unknown sample is labeled with a "light isotope" and is mixed with another sample or standard labeled with the "heavy isotope." Each approach shown in Table 4.1 offers distinct advantages and disadvantages, which will be discussed below.

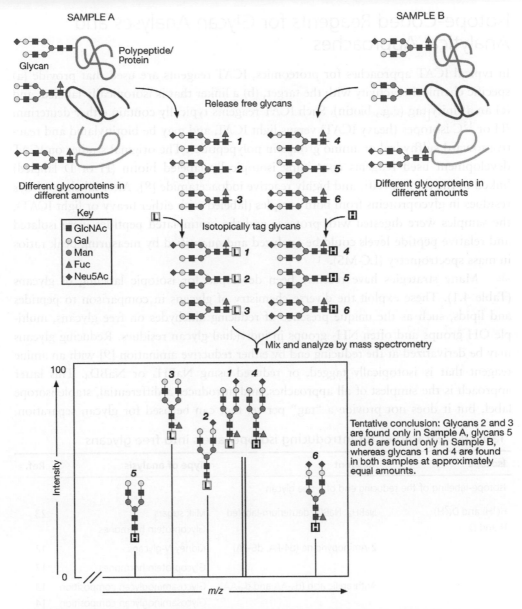

**FIGURE 4.1**   General strategy for isotope tagging of glycans for quantitative and comparative glycomics. In this example, free glycans either occurring naturally or released from glycoproteins or glycolipids, for example, may be derivatized with any of the reagents shown in Table 4.1, including $^{12}$C- or $^{13}$C-aniline, which are shown here as the tags for labeling by reductive amination [5–7]. The separately tagged glycan samples are mixed and then analyzed by mass spectrometry. The example shown here is a matrix-assisted laser desorption/ionization-time-of-flight (MALDI-TOF) profile of *N*-glycans showing that the only common glycan in the two samples being compared are glycans 1 and 4. Glycans with identical structures in the two mixed samples are identified as a doublet with a mass difference of 6 Da. Quantitative analyses of such mixed samples can be performed when one sample is a tagged library of known structures in known amounts [5].

# Isotope-Coded Reagents for Glycan Analyses and Analytical Approaches

In typical ICAT approaches for proteomics, ICAT reagents are used that provide (a) specific chemical reactivity with the target, (b) a linker that is isotopically labeled, and (c) an affinity tag (e.g., biotin). Such ICAT reagents typically contain either deuterium $^2$H or $^{13}$C isotopes (heavy ICAT) versus light ICAT, and may be biotinylated and reactive toward sulfhydryl or amino groups in polypeptides. The original studies on ICAT development used reagents containing isotopically labeled biotin (H or D labeled) linked to the thiol-specific and highly reactive iodoacetamide [8]. After labeling of Cys residues in glycoproteins from both samples (labeled with either heavy or light ICAT), the samples were digested with protease and the biotinylated peptides were isolated and relative peptide levels could be analyzed and quantified by measuring peak ratios in mass spectrometry (LC-MS).

Many strategies have recently been developed for isotopic labeling of glycans (Table 4.1). These exploit the diverse chemistry of glycans in comparison to peptides and lipids, such as the unique presence of reducing aldehydes on free glycans, multiple OH groups and often $NH_2$ groups in individual glycan residues. Reducing glycans may be derivatized at the reducing end by either reductive amination [9] with an amine reagent that is isotopically tagged, or reduced using $NaBH_4$ or $NaBD_4$. This latter approach is the simplest of all approaches, and introduces a differential, stable isotope label, but it does not provide a "tag" per se, that can be used for glycan separation,

Table 4.1  Strategies for introducing isotope labels into free glycans

| Isotope | Reagent | Type of analysis | Ref. |
|---|---|---|---|
| Isotope-labeling of the reducing end of a free glycan | | | |
| H($^1$H) and D($^2$H) H and D | $NaBH_4$, $NaBD_4$ deuterium-labeled | Milk sugars Glycoprotein hormones | 23 |
| | 2-Aminopyridine (d4-PA, d6-PA) | Kidney N-glycans | 12 |
| | | Glycoprotein hormones | 17 |
| H and D | Anthranilic acid ($d_0$-AA and $d_4$-AA) | Glycosaminoglycan composition | 13 |
| | | Glycosaminoglycan composition | 14 |
| $^{12}$C and $^{13}$C | $^{12}$C- or $^{13}$C-Aniline | Plant polysaccharide | 6 |
| | | Glycosaminoglycan composition | 5 |
| | | Serum glycoproteins | 7 |
| Permethylation of a glycan with isotope reagents | | | |
| $^{12}$C and $^{13}$C | $^{12}$C-$H_3$I or $^{13}CH_3$I methyl iodide | Commercial glycoproteins | 18 |
| H and D | $CH_3$I or $CD_3$I methyl iodide | Serum glycoproteins Cancer cells | 20 |
| $^{12}$C and $^{13}$C and H and D | $^{12}$C-$H_2$DI or $^{13}CH_3$I | Murine embryonic stem cells | 19 |
| Labeling of a probe to which glycans are linked | | | |
| H and D | $D_0$ and $D_8$-labeled biotin | Glycohydrolase assays | 22 |

such as biotin. Furthermore, the introduction of deuterium at the reducing end of a glycan provides a small change in mass ($\Delta m = 1\,Da$), and the analyses must contend with the natural isotope abundance of H and C, which must be calculated with underivatized samples [10] making analysis with a single mass unit difference difficult to interpret. The introduction of H- or D-labeled 2-aminopyridine ($d4$-PA and $d6$-PA) [11,12] and anthranilic acid ($d_0$-AA and $d_4$-AA) [13,14], in addition to providing a linked "tag," provides much more robust mass differences. Although it is possible that isotopes H ($^1$H) and D ($^2$H) could differentially affect the separation properties of peptides [15] and glycans in high-performance liquid chromatography (HPLC), at least for the 2-PA derivatives, such isotope effects appear to be negligible for H- or D-derivatized glycans on zwitterionic-type hydrophilic interaction liquid chromatography (ZIC-HILIC) [16].

Deuterium-labeled 2-aminopyridine ($d4$-PA, $d6$-PA) was first used in monosaccharide composition analyses by MS where monosaccharides from an unknown were labeled with $d4$-PA and then mixed with known amounts of standard monosaccharides derivatized with $d6$-PA, and then all samples were separated and analyzed by LC/MS [17]. This method was successfully applied to the monosaccharide composition analysis of model glycoproteins, fetuin, and erythropoietin. For quantitative glycan profiling, a similar approach tagged released glycans and those from a standard glycoprotein with $d0$- and $d4$-AP, respectively, and an equal amount of $d0$- and $d4$-PA glycans were co-injected into GCC-LC/MS [17]. The paired ions ($\Delta m = 4\,Da$) were directly detected and unique glycans present in only one sample were easily identified. This approach was successfully applied to studies of the glycan differences between recombinant human chorionic gonadotropin (rhCG) and human chorionic gonadotropin (rhCG) [17]. This pioneering study illustrated the principle of labeling glycans with isotopically distinguishable tags. In either situation where an isotope tag is introduced at the reducing end of unknown glycans or onto the OH groups of unknown glycans by methylation, it is possible to quantify and compare the "heavy" or "light" isotope labeled material to an alternatively labeled "light" or "heavy" set of standard, know glycans, respectively, that may be added to the unknowns.

In all cases the introduction of a light and heavy isotope label into glycans raises a question as to the linearity of detection of sensitivity of measurements of isotope ratios for one sample compared to the other. For most of the studies on isotope labeled glycans, the range of detection and linearity is approximately 10:1 for heavy: light or vice versa [7,17,18] or even better [19]. In terms of sensitivity, analyses of isotope-labeled glycans approach the sensitivity of the instrumentation available, and in terms of practical limits are typically in the low nanomoles of total glycoprotein (1–200 nmol). Thus, for typical samples that may mean 0.1–5 mg quantities of sample depending upon the complexity of the glycan repertoire and whether detailed structural analyses or only comparative analyses are to be performed.

A major advantage of isotope tags for glycan analysis is that multiple samples can be analyzed simultaneously and compared with an internal standard. The addition of the "tag" itself provides a method of separating derivatized from underivatized glycans, which is the advantage of ICAT approaches in proteomics. Reductive labeling approaches have the advantage of limited sample manipulation, easy and quantitative introduction of the tag, such as labeled 2-aminopyridine, anthranilic acid, or aniline, and the use of tags provide a means to detect and quantify the glycans using HPLC techniques for identification and separation when coupled with MS analyses, as in LC/MS [5–7,13]. In addition, tags such as isotope-labeled aniline and 2-aminopyridine are relatively inexpensive. This reductive labeling strategy, termed GRIL (glycan reductive isotope labeling) [7], is advantageous for many types of glycans, especially those derived from glycosaminoglycans [5], where permethylation analysis is not applicable due to the high content of sulfate and carboxyl groups. Also, reductively labeled glycans are water-soluble and allow a variety of separation and analytical techniques including capillary electrophoresis and conventional HPLC.

An example of GRIL using $^{12}$C- or $^{13}$C-aniline as tags is shown in Figure 4.2. In this example, mouse and human serum glycoproteins were treated with N-glycanase (PNGase F) to release N-glycans. The resulting N-linked glycans from each source were separated into two portions and then quantitatively derivatized in separate reactions with $^{12}$C- or $^{13}$C-aniline. The tagged glycans were mixed at equal amounts and analyzed by MALDI-TOF [7]. As can be seen in Figure 4.2, most of the glycans are identified based on the most probable structure corresponding to the compositions of the molecular ion from MALDI analysis. When the heavy and light isotope labels are from the same sample (mouse serum vs. mouse serum) the ratio (R) of each glycan to itself is approximately 1.0. Interestingly, when the $^{12}$C-analine-labeled mouse serum N-glycans were mixed with equivalent $^{13}$C-aniline-labeled N-glycans from human serum glycoproteins, there was little in common between the two samples. The exceptions were for the glycans at 1700.55/1706.60 Da that correspond to an asialo-biantennary structure and at 1743.66/1749.68 that correspond to a bisected asialo-agalactobiantennary structure. But even those compounds are apparently present in different amounts based on the ratio (0.20 and 0.13, respectively). Also, the sialylated mouse serum glycans differed from the human serum glycans in their sialic acid, which was Neu5Gc in the mouse sample and Neu5Ac in the human sample. This figure generally illustrates the power of comparative glycomic strategies using isotope-labeling techniques in that the results show the large and small differences in glycan composition and expression between the two serum samples. Thus, this approach provides a simple and direct comparative analysis of glycans in these two samples with relatively simple sample preparation.

Reductively tagged glycans can also be permethylated, but permethylation of glycan derivatives and subsequent MS analysis is sometime difficult to interpret due to partial methylation of the glycan or the aglycone component, along with ionization problems due to the aglycone moiety. Thus, a limitation of the GRIL approach is the limited structural analyses that can be performed on the non-permethylated glycans. However, this reductive labeling strategy without permethylation allows the rapid acquisition of very simple and easily interpretable data for comparing samples and quantifying amounts of glycans, rather than detailed sequencing of unknown glycans. Although the GRIL approach without permethylation is less sensitive than MS analyses with permethylated glycans, it is capable of detecting sulfated and phosphorylated glycans, which are lost in classical permethylation reactions.

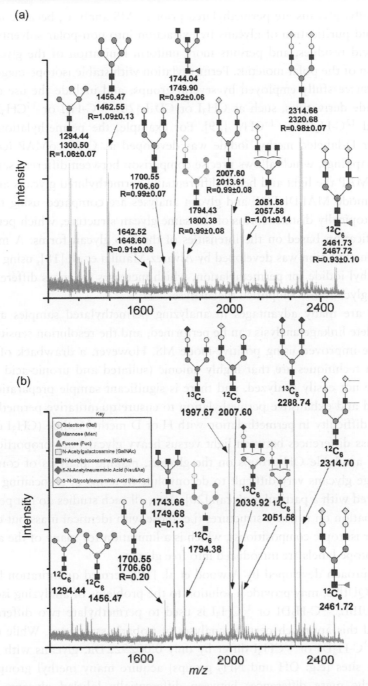

**FIGURE 4.2** Comparative glycomics on *N*-glycans from serum glycoproteins from mouse and human using isotope-labeled glycans. *N*-Glycans were released from human and mouse serums by PNGase F and separately labeled as in Figure 4.1 with either $^{12}$C- or $^{13}$C-aniline [7]. (a) Matrix-assisted laser desorption/ionization-time-of-flight (MALDI-TOF) profile of $^{12}$C$_6$-aniline-glycans and $^{13}$C$_6$-aniline-glycans from a single mouse serum sample where they were mixed in an equimolar amount and analyzed. (b) MALDI-TOF profile of an equimolar mixture of $^{12}$C$_6$-aniline-glycans from mouse serum glycoproteins and $^{13}$C$_6$-aniline-glycans from human serum glycoproteins. The masses of the individual glycans are shown associated with the most plausible structures represented with the symbols shown in the inset. In the case of the doublets found in the equimolar mixtures, the masses are shown below the structure and the ratio of the peak areas of each isotope-labeled glycans is indicated.

Typically, glycans are permethylated prior to MS analysis, because methylation permits rapid purification of glycans by extraction into non-polar solvents, stabilizes the sialic acid residues, and permits more uniform ionization of the glycans due to modification of the polar moieties. Permethylation with stable isotope-tagged reagents have been successfully employed by several groups, and include the use of different methyl iodide derivatives, such as $CH_3I$ or $CD_3I$ [20], $^{12}C$-$H_3I$ or $^{13}CH_3I$ [18], and dual-labeled $^{12}C$-$H_2DI$ or $^{13}CH_3I$ [19]. For example, the permethylation approach using H- or D-labeled methyl iodide was developed as C-GlycoMAP (comparative glycomic mapping), which allows precise comparison between different samples [20]. In C-GlycoMAP the light and heavy differentially permethylated glycans are analyzed in positive-mode MALDI-TOF and glycan analyses are compared using the ratio of the two isotopically distinct forms of the same glycan structure, which permits absolute quantification based on the intensities of the two glycan forms. A more robust type of permethylation was developed by Alvarez-Manilla et al. [18], using $^{12}C$-$H_3I$ or $^{13}CH_3I$ methyl iodide for permethylation, which increases the mass difference for the derivatized glycans and simplified the analyses.

There are many advantages in analyzing permethylated samples approaches, since complete linkage analysis can be performed, and the resolution sensitivity of the methods are improved using positive-mode MS. However, a drawback of these permethylation techniques are that highly anionic (sulfated and uronic-acid containing glycans) are not easily analyzed, and there is significant sample preparation, including repeated and exhaustive permethylation to ensure quantitative permethylation of samples. A difficulty in permethylation with H or D methyl iodide ($CH_3I$ or $CD_3I$) is that the mass differences between light versus heavy glycans are proportional to the numbers of available OH groups on the glycans, making analyses of complex mixtures of large glycans very difficult to deconvolute in terms of associating each mass peak observed with a particular glycan [20]. Also, all such studies do not permit structural information on isobaric structures (i.e. those with identical mass but differing in elemental or isotopic compositions), which is a limitation for many of the approaches in which isotope labels are introduced into free glycans.

An approach developed by Atwood et al. [19], termed quantitation by isobaric labeling or QUIBL, may provide a solution to the problems of analyzing isobaric glycans. In QUIBL, $^{12}C$-$H_2DI$ or $^{13}CH_3I$ is used to permethylate two different sets of glycans, and this approach permits analyses of isobaric structures. While the precise masses of $^{12}C$-$H_2DI$ or $^{13}CH_3I$ differ by only 0.002922 Da, glycans with many permethylation sites (e.g. OH and $NH_2$ groups) acquire many methyl groups, thereby increasing the mass differences between differentially labeled glycans. Following differential permethylation samples are mixed, and then analyzed by hybrid tandem mass spectrometry using a Fourier transform ion cyclotron resonance (FTICR) mass spectrometer or an ion trap–orbitrap. The masses are determined at a low resolution in the ion trap and the quasimolecular ions are analyzed in an FTICR or orbitrap at high resolution to identify ions arising from either $^{12}C$-$H_2DI$ or $^{13}CH_3I$. Direct comparisons of such differentially labeled glycans can provide a relative, but not absolute, quantitative measurement. Fragment ions may then be analyzed at high resolution to complete the structural identifications. If standard glycans are used in such an approach and they are labeled with $^{12}C$-$H_2DI$ for example, and a defined amount of them is then mixed with the unknown sample labeled with $^{13}CH_3I$, then it would be possible to provide near quantitative analyses of glycan amounts in samples.

The recent development of multiple isotope tags for glycans will make it possible in the future to analyze a large mixture of samples labeled separately with different isotope reagents so that each tag is clearly distinguished by several mass units [21]. A novel use of an ICAT-approach for glycomics was the development of biotinylated, cleavable, activity-based (AB)-ICAT reagents (light H8 and heavy D8) for the analysis of glycohydrolases. This technique was developed by Hekmat et al. [22] where two pairs of biotinylated and thiol-cleavable, AB-ICAT reagents (light H8 and heavy D8) were synthesized, where one incorporates a recognition determinant for cellulases and the other incorporates a recognition determinant for xylanases. Thus, both xylanases and cellulases can be quantified simultaneously in a sample with both reagents present.

## Concluding Remarks

The importance of glycans in many biological systems is well-established, but tremendous challenges are ahead. One of those challenges is defining changes in the glycome of a cell or organism in health and disease and as a function of development. Modern MS approaches utilizing isotope-based tags has been invaluable in proteomics, lipidomics, and metabolomics. Within the past few years, new developments of isotope-labeling strategies for comparative and quantitative glycomics will help to revolutionize the field in terms of high-throughput screening for comparative glycomics. As described in this chapter, there are many methods available for quantitative glycomics depending on the type of sample, amounts, the question asked, and the rigor needed. There are challenges ahead with these approaches, however, and this aspect of glycomics using stable isotope-labeling methods is still in its infancy. Most of the strategies available are ideal for comparative glycomics of relatively similar samples (i.e., serum glycoproteins and cultured cells). As glycomic techniques become more sophisticated and quantitative there is a need to balance development so that simple labeling technologies, inexpensive reagents, and simple MS approaches will still be available and used for analyses, so that these technologies can be applicable to many biological studies in laboratories that are not focused on glycobiology but interested in changes in glycan structures and expression. There is also the challenge of improving the chemical design and availability of isotope labeling reagents for glycomics, where in the future it is hoped that affinity tags, including biotin, can be incorporated into the isotope labeling strategies.

## Acknowledgments

The work of the authors was supported by NIH Grant P01 HL085607 to R.D.C., NIH Grant GM085448 to D.F.S., and in part by the Consortium for Functional Glycomics under NIH Grant GM62116.

# References

1. Gygi SP, Rist B, Gerber SA, Turecek F, Gelb MH, Aebersold R. Quantitative analysis of complex protein mixtures using isotope-coded affinity tags. *Nat Biotechnol* 1999;**17**:994–9.

2. Gevaert K, Impens F, Ghesquiere B, Van Damme P, Lambrechts A, Vandekerckhove J. Stable isotopic labeling in proteomics. *Proteomics* 2008;**8**:4873–85.

3. Ong SE, Blagoev B, Kratchmarova I, Kristensen DB, Steen H, Pandey A, Mann M. Stable isotope labeling by amino acids in cell culture, SILAC, as a simple and accurate approach to expression proteomics. *Mol Cell Proteomics* 2002;**1**:376–86.

4. Tao WA, Aebersold R. Advances in quantitative proteomics via stable isotope tagging and mass spectrometry. *Curr Opin Biotechnol* 2003;**14**:110–18.

5. Lawrence R, Olson SK, Steele RE, Wang L, Warrior R, Cummings RD, Esko JD. Evolutionary differences in glycosaminoglycan fine structure detected by quantitative glycan reductive isotope labeling. *J Biol Chem* 2008;**283**:33674–84.

6. Ridlova G, Mortimer JC, Maslen SL, Dupree P, Stephens E. Oligosaccharide relative quantitation using isotope tagging and normal-phase liquid chromatography/mass spectrometry. *Rapid Commun Mass Spectrom* 2008; **22**:2723–30.

7. Xia B, Feasley CL, Sachdev GP, Smith DF, Cummings RD. Glycan reductive isotope labeling for quantitative glycomics. *Anal Biochem* 2009;**387**:162–70.

8. Gygi SP, Rist B, Gerber SA, Turecek F, Gelb MH, Aebersold R. Quantitative analysis of complex protein mixtures using isotope-coded affinity tags. *Nat Biotechnol* 1999;**17**:994–9.

9. Bigge JC, Patel TP, Bruce JA, Goulding PN, Charles SM, Parekh RB. Nonselective and efficient fluorescent labeling of glycans using 2-amino benzamide and anthranilic acid. *Anal Biochem* 1995;**230**:229–38.

10. Xie Y, Liu J, Zhang J, Hedrick JL, Lebrilla CB. Method for the comparative glycomic analyses of O-linked, mucin-type oligosaccharides. *Anal Chem* 2004;**76**:5186–97.

11. Hashii N, Kawasaki N, Itoh S, Hyuga M, Kawanishi T, Hayakawa T. Glycomic/glycoproteomic analysis by liquid chromatography/mass spectrometry: analysis of glycan structural alteration in cells. *Proteomics* 2005;**5**:4665–72.

12. Hashii N, Kawasaki N, Itoh S, Nakajima Y, Kawanishi T, Yamaguchi T. Alteration of N-glycosylation in the kidney in a mouse model of systemic lupus erythematosus: relative quantification of N-glycans using an isotope-tagging method. *Immunology* 2008;**126**:336–45.

13. Hitchcock AM, Costello CE, Zaia J. Glycoform quantification of chondroitin/dermatan sulfate using a liquid chromatography-tandem mass spectrometry platform. *Biochemistry* 2006;**45**:2350–61.

14. Hitchcock AM, Yates KE, Costello CE, Zaia J. Comparative glycomics of connective tissue glycosaminoglycans. *Proteomics* 2008;**8**:1384–97.

15. Zhang R, Sioma CS, Thompson RA, Xiong L, Regnier FE. Controlling deuterium isotope effects in comparative proteomics. *Anal Chem* 2002;**74**:3662–9.

16. Takegawa Y, Hato M, Deguchi K, Nakagawa H, Nishimura S. Chromatographic deuterium isotope effects of derivatized N-glycans and N-glycopeptides in a zwitterionic type of hydrophilic interaction chromatography. *J Sep Sci* 2008;**31**:1594–7.

17. Yuan J, Hashii N, Kawasaki N, Itoh S, Kawanishi T, Hayakawa T. Isotope tag method for quantitative analysis of carbohydrates by liquid chromatography-mass spectrometry. *J Chromatogr A* 2005;**1067**:145–52.

18. Alvarez-Manilla, G, Warren NL, Abney T, Atwood 3rd J, Azadi P, York WS, Pierce M, Orlando R. Tools for glycomics: relative quantitation of glycans by isotopic permethylation using 13CH3I. *Glycobiology* 2007;**17**:677–87.

19. Atwood 3rd JA, Cheng L, Alvarez-Manilla G, Warren NL, York WS, Orlando R. Quantitation by isobaric labeling: applications to glycomics. *J Proteome Res* 2008;**7**:367–74.

20. Kang P, Mechref Y, Kyselova Z, Goetz JA, Novotny MV. Comparative glycomic mapping through quantitative permethylation and stable-isotope labeling. *Anal Chem* 2007;**79**:6064–73.

21. Bowman MJ, Zaia J. Tags for the stable isotopic labeling of carbohydrates and quantitative analysis by mass spectrometry. *Anal Chem* 2007;**79**:5777–84.

22. Hekmat O, He S, Warren RA, Withers SG. A mechanism-based ICAT strategy for comparing relative expression and activity levels of glycosidases in biological systems. *J Proteome Res* 2008;**7**:3282–92.

23. Ninonuevo MR, Ward RE, LoCascio RG, German JB, Freeman SL, Barboza M, Mills DA, Lebrilla CB. Methods for the quantitation of human milk oligosaccharides in bacterial fermentation by mass spectrometry. *Anal Biochem* 2007;**361**:15–23.

# 5

# Glycotranscriptomics

Alison Nairn and Kelley Moremen

*COMPLEX CARBOHYDRATE RESEARCH CENTER, THE UNIVERSITY OF GEORGIA, ATHENS, GA, USA*

## Introduction

Carbohydrate structures attached to glycoproteins, glycolipids, and proteoglycans play key roles in a variety of biological recognition events [1]. At the cellular level, N-linked, O-linked, and glycolipid glycan structures contribute to several essential aspects of biological recognition, including immune surveillance, inflammatory reactions, hormone action, viral infection, arthritis, and metastasis of oncogenically transformed cells [2–5]. Changes in glycoprotein and glycolipid structures are also believed to influence cell migration during embryogenesis and normal mammalian physiology [1,6]. Most of our understanding on the roles of cellular glycosylation in physiology and pathology comes from a combination of glycan structural analysis on specific glycoproteins, cell surfaces, or total tissue extracts in combination with years of study on the biochemistry and enzymology of glycan biosynthetic and degradative enzymes [7–10]. While the regulation of expression for many glycan-related genes have been examined in a gene-by-gene approach [11], broad-based transcriptional profiling of glycan-related genes has recently begun to provide insights into the global regulation of glycan synthesis and degradation in animal systems [12].

A major goal in the field of "glycobiology" is to understand how glycan structures are globally regulated in abundance and the impact that these changes have on the physiology and pathology of an organism [1,4]. Since glycan biosynthesis is not a directly template-driven process like the synthesis of polypeptide structures from genome-derived transcripts, there are often difficulties in attempting to examine the regulation of glycan structures in complex biological systems. Factors that impact the efficiency and penetrance of individual glycosylation steps on protein and lipid acceptors include enzyme accessibility to glycan modification sites, the abundance of the respective protein or lipid acceptors, the availability of sugar-nucleotide precursors, and the relative enzyme levels or relative localizations of biosynthetic enzymes that can compete for the same glycan substrates [12]. Despite these complexities in glycan biosynthesis, several lines of evidence indicate that one of the major modes of regulating cellular glycosylation is transcriptional regulation of the enzymes involved in

**Handbook of Glycomics**
Copyright © 2009 by Elsevier Inc. All rights of reproduction in any form reserved.

glycan synthesis and catabolism [12–14]. One method for testing whether the elaboration of glycan structures is controlled at the transcriptional level is the comparison of glycan structural data with transcript abundance measurements in multiple biological samples, where differences in glycan structures are known to occur [12,14]. Broad-based profiling of glycan structures in complex biological systems has been accomplished over the past decade and has provided detailed snapshots into the diversity of glycosylation that can occur in a given cell or tissue source [15–17]. While these analyses have revealed critical changes in glycan structures during development or between biological samples, it is only recently that they have been paired with broad-based transcript analysis methods to determine if transcriptional regulation is the major mechanism driving the structural alterations [12–14,18–24].

Transcript profiling of glycan-related genes has its own set of complexities. It has been estimated that a total of over 700 proteins are involved in glycan-related processes in animal systems, including sugar-nucleotide biosynthesis and interconversion, sugar and sugar-nucleotide transport, glycan extension, modification, recognition, catabolism, and numerous glycosylated core proteins [12]. Many of the proteins involved in glycan synthesis, modification, and degradation have been collated into multigene families based on sequence and structural similarities. In mammalian cells, glycosyltransferases number ~200 members and are subdivided into 40 families (CAZy database [25,26]), but in many cases the acceptor specificity of individual family members are not well defined or potential enzymatic redundancy may exist between multiple members of the same enzyme family. Thus, one-to-one mapping of individual gene products to steps in glycan biosynthetic pathways is difficult to achieve or may have ambiguity among multiple family members. Existing web-based resources (CAZy [25,26], KEGG [27–29], Consortium for Functional Glycomics [13,18,21], SOURCE [30]) have collated and annotated many of the genes related to glycan biosynthesis and catabolism, but more comprehensive resources for mapping enzymes to complex glycan biosynthetic pathways for glycoprotein, glycolipid, and proteoglycan biosynthesis and catabolism are still in their early stages [12].

An additional complexity for the study of glycan-related gene expression is the relatively low abundance of transcripts encoding many of the critical enzymes involved in glycan modifications. These low transcript levels make it difficult to employ broad-based survey methods, such as microarray approaches [13,18–22,24], for global transcriptome analyses. More focused approaches employing quantitative real-time polymerase chain reaction (qRT-PCR) analysis have also been employed effectively [23,31,32], but most commonly the technique is employed with a relatively small number of target genes. Broad-based transcript profiling by qRT-PCR has been developed recently for medium-throughput analysis of glycan-related transcripts, but challenges still exist for the correlation of transcript levels with corresponding glycan structures [12].

This chapter will discuss the range of proteins and activities that are relevant to the broad-based profiling of glycan-related gene expression in animal systems. An overview of the relevant pathways will be briefly described to provide a frame of reference for the scope of target genes that need to be profiled in order to fully understand the global regulation of glycan structures in animal systems. These pathways include enzyme systems required for activated sugar precursor biosynthesis and transport, glycan biosynthesis, modification, and catabolism, and glycan recognition by lectins. This will be followed by a brief discussion of cloning strategies employed to identify the respective glycan-related genes. The discussion will then review the methods that have been used for transcript analysis of glycan-related genes and the challenges that exist in the correlation of transcript levels with glycan structures. The use of biochemical pathway information as a bridge between the two disparate data sets will then be discussed.

## Overview of Glycan Biosynthesis in Animal Systems

### Uptake, Activation, and Transport of Sugar Donors

#### *Uptake of Sugars*

In animal systems several of the sugars that are needed for glycan biosynthesis come from dietary sources (e.g., Glc, Fru, Man, and Fuc) and others must be generated either by de novo biosynthesis or by interconversion from other sugar sources [33]. Uptake of monosaccharides from the intestine and transport into tissues generally occurs by the combined use of energy-dependent transporters that are found in intestinal epithelial cells and kidney tubules (e.g., SGLTs) [34] along with energy-independent facilitated diffusion transporters that are common to most cell types (e.g., GLUT transporters) [35]. In the intestine, the sodium-dependent glucose transporters (SGLT family) use the energy of a downhill sodium gradient to transport glucose across the apical membrane against an uphill glucose gradient [36]. Further transport into the bloodstream is accomplished by the GLUT uniporters on the basolateral membrane to deliver glucose to the peritubular capillaries [36]. The energy-independent facilitated sugar transporters similar to GLUT1 then allow the uptake of glucose from the bloodstream into cells of peripheral tissues [37]. The GLUT transporters (SLC2A gene family) contain 12 transmembrane amphipathic helices that presumably provide a pore structure for selective hexose transport [37,38]. The 13 members of the GLUT protein family have been divided into three families with varied expression patterns, affinities, and selectivities for monosaccharide transport [39,40]. Among the most characterized are the class I isoforms, including GLUT1–4 [35,37]. GLUT1 is broadly expressed in fetal tissues, while expression is highest in erythrocytes and endothelial cells of the blood–brain barrier in adult tissues. GLUT2 is expressed in renal tubular cells, liver, and pancreatic beta cells as a high-capacity, low-affinity isoform. GLUT3 is expressed in placenta and neurons as a high affinity isoform. GLUT4 is found in adipose tissue and skeletal muscle and is the major insulin-regulated glucose transporter, where insulin signaling triggers its transport from intracellular vesicles to the cell surface [41]. The other transporter families include the class II transporters (GLUT5, GLUT7, GLUT9, and GLUT11) [35], at least one of which transports fructose, and the less characterized Class III transporters (GLUT6, GLUT8, GLUT10, GLUT12, and HMIT). Some of the GLUT transporters have the capacity to transport Man or Gal, but their low affinities and high circulating Glc and Fru levels indicate that the latter monosaccharides are the most relevant substrates [33].

In the intestinal brush border and kidney tubule epithelial cells an energy-dependent SLGT-like transporter is also responsible for entry of mannose into the bloodstream, similar to the energy-dependent glucose uptake of digested macromolecules [33]. Transport of Man into cells appears to be accomplished through a Man-selective facilitated transporter that is poorly inhibited by Glc, suggesting that Man uptake may supply biosynthetic sources for glycoconjugate biosynthesis [42].

Selective transport of fucose into cells has also been reported [43], but the corresponding coding region has not yet been cloned or expressed and minimal characterization of the transport activity has been obtained. Transporters for galactose, amino sugars, and sialic acids have not been reported and these sugars are presumably only obtained by salvage of degraded glycoconjugates or interconversions from other sugars or sugar precursors [33].

## Activation and Interconversion of Sugar-Nucleotides

The synthesis of glycoconjugates in animal systems involves the ordered assembly of monosaccharides into linear and branched glycan structures by the action of glycosyltransferases that employ activated sugar precursors as substrates (usually as NDP-sugars or NMP-sugars for sialic acid extension) [7]. In general, monosaccharides taken up into cells via cell surface transporters or salvaged following the degradation of pre-existing glycoconjugate structures are subsequently trapped in the cytoplasm following phosphorylation (as sugar-6-phosphates for Glc, Man, GlcN [44] or as sugar-1-phosphates for Gal, Fuc, GalNAc, GlcNAc) [33]. Sugar-6-phosphates are converted to sugar-1-phosphates by phospho-sugar isomerases and activated nucleotide-diphospho-sugars are generated using nucleoside triphosphate (NTPs) as co-substrates by pyrophosphorylases or sugar-1-phosphate nucleotidyl transferases [33]. An exception is the generation of CMP-sialic acids from NeuAc and NeuGc, which are activated directly from free monosaccharides and CTP by CMP-NANA synthase (*Cmas*) [45]. Activated NDP-sugar precursors are generated in the cytosol of mammalian cells [46] (or in the nucleus for the synthesis of CMP-sialic acids [47]) and many sugars can be interconverted either at the level of monosaccharide, sugar-phosphate, or nucleotide-sugar intermediates [12]. Thus, sugar precursors for all glycan extension reactions are generated and available in the cytosol of mammalian cells.

## Transfer of Activated Sugar Donors to Compartments where They are Used for Glycan Extension

A topological problem exists in the elaboration of many classes of glycan structures in animal systems. Activated sugar-nucleotide precursors synthesized in the cytosol generally must gain access to luminal compartments of the secretory pathway for use in glycan extension reactions [46]. Exceptions exist for the elaboration of several classes of glycan polymers. The synthesis of the first seven sugar linkages on the lipid-linked precursor for N-glycan biosynthesis occurs on the cytosolic face of the endoplasmic reticulum (ER) membrane using cytosolic UDP-GlcNAc and GDP-Man as substrates [48]. Similarly, the first glycosylation step in the synthesis of the glyco-sylphosphatidylinositol (GPI) anchor precursor structure, the addition of GlcNAc to phosphatidylinositol, occurs on the cytosolic face of the ER using UDP-GlcNAc as a substrate [49]. Generation of *O*-GlcNAc linkages on cytosolic and nuclear proteins also employs the direct use of a cytosolic sugar-nucleotide donor (UDP-GlcNAc) as a substrate [50]. The elaboration of extracellular hyaluronic acid co-polymers appears to employ the direct use of cytosolic UDP-GlcA and UDP-GlcNAc for the transmembrane extrusion of the extended alternating polymer at the plasma membrane [51]. Finally, the synthesis of glycogen polymers in the cytosol employs the direct use of UDP-Glc from cytosolic pools of the sugar-nucleotide.

Other families of glycoconjugates, including *N*-linked and *O*-linked glycoproteins, proteoglycans, glycolipids, and GPI precursors are elaborated in the lumen of the secretory pathway and require transport of activated sugar donors into the respective compartments for extension reactions [46]. In the lumen of the ER, the extension and completion of the lipid-linked precursor for *N*-glycan biosynthesis is accomplished by the shuttling of activated sugar-nucleotides (GDP-Man and UDP-Glc) to dolichol intermediates (Dol-P-Man and Dol-P-Glc) on the cytoplasmic face of the ER membrane [52]. The Dol-P-sugar intermediates are then flipped to the luminal leaflet of the membrane and subsequently used as donors for extension of the glycan precursor on the luminal side of the membrane [53]. Dol-P-Man is also used as a donor for extension of GPI anchor glycan structures after the flipping of the phosphatidylinositol-GlcN intermediate to the luminal face of the ER membrane [52]. Other classes of glycan extension reactions directly employ sugar-nucleotides within the lumen of the secretory pathway and use antiporters for the transport of sugar nucleotides into the lumen of these membrane compartments [46]. Import of nucleotide-sugars corresponds with an antiport of nucleotide-monophosphate products of glycosyltransferase reactions back out to the cytoplasm. Presently 23 human genes have been identified with sequence similarity to known sugar-nucleotide antiporters (Slc35 family of proteins) and they have been divided into five subfamilies [54]. Less than half of these have been characterized in regard to transporter activity, but the subset that has been characterized is capable of antiporting CMP-Sia (Slc35a1), UDP-Gal (Slc35a2, Slc35a4, and Slc35b1), UDP-GlcNAc (Slc35a3), PAPS (Slc35b2 and Slc35b3), bifunctional UDP-GlcNAc/UDP-Xyl (Slc35b4), GDP-Fuc (Slc35c1), and bifunctional UDP-GlcA/UDP-GalNAc (Slc35d1). Thus, genes and corresponding antiport activities for many of the key nucleotide-sugars involved in glycan extension reactions have been identified and additional members of the Slc35 family may account for additional classes of sugar donors in the secretory pathway.

## Biosynthesis and Modification of Major Glycan Classes

### N-*Linked Glycan Biosynthesis*

Synthesis of *N*-glycan structures on cellular and secreted glycoproteins in animal systems is initiated by the synthesis of a $Glc_3Man_9GlcNAc_2$ precursor attached to Dol-pyrophosphate in the ER membrane [7]. Initial stages of glycan extension up to

$Man_5GlcNAc_2$ occurs on the cytoplasmic leaflet and employs at least seven distinct integral membrane enzymes that use sugar nucleotide donors for glycan extension. Recently, many of these enzymes have been shown to work as multi-enzyme complexes to allow the efficient tunneling of substrates through the biosynthetic pathway [55]. Following the flipping of the glycan to the luminal leaflet of the ER membrane, extension of the final seven sugars to the lipid-linked precursor employs a collection of integral membrane proteins that use Dol-P-sugars as donors [52,53]. Transfer of the glycan structure to nascent polypeptide chains occurs co-translationally through the action of a multiprotein oligosaccharyltransferase (OST) complex [48]. Trimming of the glycan starts immediately after transfer to the polypeptide and involves removal of both glucose and mannose residues in the ER and further mannose trimming in the Golgi complex [56,57]. Critical roles for trimmed $GlcMan_9GlcNAc_2$ glycan intermediates on newly synthesized glycoproteins are found in the ER, where they engage the lectin-mediated chaperones, Cnx and/or Crt, during conformational maturation of the polypeptide [48]. Mannose trimming of glycan structures also defines the residence time for newly synthesized glycoproteins in the ER lumen prior to disposal of misfolded proteins through a quality control mechanism termed ER-associated degradation (ERAD) [58].

Further elaboration of N-glycan structures occurs in the lumen of the Golgi complex, where mannose trimming overlaps with glycan branching and extension [7]. Cleavage to $Man_5GlcNAc_2$ occurs via a collection of Golgi α-mannosidases, followed by extension with a single GlcNAc and further mannose trimming to the $GlcNAcMan_3GlcNAc_2$ intermediate. This trimmed structure is then further branched by a collection of Golgi GlcNAc transferases [59]. The GlcNAc termini of complex type N-glycans are further extended and capped with a huge diversity of glycan structures that can include a variety of monosaccharides (e.g., Gal, GalNAc, GlcNAc, Fuc, Sia, and GlcUA) in numerous linkages as well as several positions where sulfation may occur [12]. The ensemble of glycan structures that are generated within a given cell type or on a single glycoprotein substrate largely depends upon the array of glycosyltransferases that are expressed in the given cell and the potential competition between enzymes for extension and modification on a given glycan structure.

## O-Linked Glycan Biosynthesis

In contrast to N-glycans that are transferred to nascent polypeptides as a large glycan precursor, O-glycan structures are appended to protein Ser and Thr hydroxyl side chains by a stepwise sequential addition of monosaccharides to the polypeptide backbone. Several different classes of O-glycan structures can be generated in mammalian cells based on the type of monosaccharide that is initially added to the polypeptide Ser/Thr side chains [60]. The simplest of the O-glycan structures includes the addition of a single GlcNAc residue to a wide variety of nuclear and cytoplasmic proteins [50]. This modification has been shown to play a role in the regulation of many molecular processes in the cytosol, in some cases by competing with protein phosphorylation of the same or nearby Ser/Thr sites in a given protein substrate.

In contrast to cytosolic *O*-GlcNAc addition, *O*-glycans in the secretory pathway are extended beyond the transfer of a single monosaccharide to generate complex and highly branched glycan structures, or highly modified linear polymers. The type of initiating monosaccharide attached to the polypeptide core can distinguish each of the *O*-glycan classes. Examples include the initiation of mucin-type structures by GalNAc addition, initiation of proteoglycan structures by Xyl addition, and the initiation of several recently discovered glycan structural families by *O*-linked Glc, Man, or Fuc addition. Mucin-type *O*-glycan structures are highly abundant on secreted and cell surface glycoproteins, where they can be synthesized on highly-glycosylated mucin peptide backbones that have a high content of Ser, Thr, Pro, and Gly, or as more sparsely glycosylated forms on a wide array of intracellular, cell surface, and secreted glycoproteins [61,62]. Initiation of mammalian *O*-linked mucin-type GalNAc linkages occurs through the action of a family of 20 polypeptide-GalNAc transferases that have unique expression patterns and acceptor specificities [62]. The GalNAc core linkage can subsequently be branched (i.e., with GlcNAc, Gal, GalNAc, Sia), extended (i.e., with GlcNAc, Gal, GalNAc, Fuc, Sia, GlcA), and modified (i.e., with $SO_4$ addition) with a collection of sugars and linkages with comparable diversity to the structures that are found on the termini of complex *N*-glycans [61].

Proteoglycan structures are also abundant on mammalian cell surfaces as well as contributing to the extracellular matrix in a variety of mammalian tissues [63]. In contrast to polypeptide-GalNAc linkages, initiation of the core region of proteoglycan structures is accomplished by the action of only two polypeptide-Xyl transferase isoforms [64] followed by monosaccharide extensions to generate a proteoglycan tetrasaccharide core linker region [63]. The latter extension steps are each accomplished by single enzyme isoforms. Proteoglycans are also distinguished from the highly branched and diverse *O*-GalNAc mucin structures in that they are extended as linear co-polymers of HexNAc-HexA that can then be further modified by epimerization of the uronic acid and multiple sulfations of both sugars in the heteropolymer [65]. Complex multiprotein assemblies of multidomain co-polymerases are used to extend the heteropolymer disaccharide repeats [66,67]. Several co-polymerase enzyme isoforms have been identified for the extension of heparan sulfate (GlcNAcα1,4GlcAβ1,4-) polymer backbones [68] and a distinctive set of enzyme isoforms extend chondroitin/dermatan sulfate co-polymers (GalNAcβ1,4GlcAβ1,3-) [69]. The structural diversity in proteoglycan structures is largely achieved through the sulfation of the polymer backbone on as many as four positions in the heparan sulfate repeat unit and three positions in the chondroitin/dermatan repeat. Clusters of highly sulfated regions of the extended linear polymers are commonly interspersed by less modified regions of the polymer. Thus, GlcNAc residues in heparan sulfate disaccharide repeat unit can be deacetylated and *N*-sulfated and additionally sulfated in the 3′ and 6′ hydroxyl positions, in addition to the 2′ sulfation of the GlcA/IdoA residues [68]. Each of the sulfation reactions is catalyzed by multiple enzyme isoforms. De-*N*-acetylation and *N*-sulfation does not occur in dermatan/chondroitin sulfate, but sulfation of the 4′ and 6′ positions of GalNAc and 2′ position of GlcA/IdoA does occur, again through the action of multiple enzyme isoforms [69]. Both heparan and chondroitin polymers can be epimerized at the GlcA residue to IdoA by single enzyme isoforms. The clustered patterns of sulfation and epimerization of the extended co-polymers can then act as ligand binding sites and co-receptors for a large number of extracellular binding proteins, including extracellular matrix components and cell surface signaling receptors and hormones. Glycosaminoglycan repeat polymers can also be directly extended at the plasma membrane through the co-polymerization and extrusion of (-4GlcAβ1,3GlcNAcβ-) disaccharide repeats by the action of the bifunctional hyaluroran synthase (Has) isoforms [51]. Three Has isoforms are found in mammalian systems that differ in the length of polymers synthesized and their regulation in response to different control mechanisms [70].

An emerging field of protein O-glycosylation has focused on the selective modification of distinctive sets of protein acceptors through short O-linked glycans with unique characteristics. The first to be discovered were O-linked Fuc modifications of EGF domains on secreted forms of plasma glycoproteins, including uPAR, tPA, and blood clotting factors VII, IX, and XII [71]. Similar O-Fuc structures were subsequently found on EGF repeats associated with the Notch receptor involved in *Drosophila* wing development catalyzed by a single enzyme isoform, *Pofut1* [72]. Modifications to the O-Fuc core glycan on Notch influence its interaction with cognate ligands, Delta or Serrate, on adjacent cell surfaces and trigger subsequent downstream signaling events [73]. In *Drosophila*, extension of O-Fuc with a GlcNAc residue by the GlcNAc transferase, Fringe, has been shown to modulate the responsiveness of Notch by inhibiting interaction and subsequent signaling through Serrate, but promoting interaction and signaling through Delta [74]. Similar modulation of Notch signaling in the mouse, through modification of O-Fuc glycans, has also been shown to influence vertebrate development [75]. In mice, the GlcNAcβ1,3Fuc disaccharide is further extended to a tetrasaccharide structure by the addition of β1,4Gal and α2,3Sia residues [76]. O-Fuc glycans have also been found on EGF repeats of Cripto, a membrane-bound co-receptor for Nodal signaling [77]. Notch EGF repeats are also modified by the addition of O-Glc residues [78] that can be extended to a linear trisaccharide by the addition of two additional α1,3Xyl residues [79]. Similar O-Glc structures have been found in thrombospondin, bovine factor VII, human factor IX and protein Z [71]. The functional roles of these glycan structures are still unknown. Protein O-Fuc addition has also been found in association with distinctive small cysteine-knot motifs, termed thrombospondin type 1 repeat units (TSR), and this latter modification is catalyzed by a distinctive O-fucosyltransferase, *Pofut2* [80–82]. In contrast to extensions of O-Fuc structures on EGF repeats to tetrasaccharides of Siaα2,3Galβ1,4GlcNAcβ1,3Fuc in mammalian systems, mammalian TSR repeats are extended to Glcβ1,3Fuc disaccharides [83]. Finally, a distinctive glycan structure has been found associated with the extended central core structure of α-dystroglycan (α-DG) [84], a highly glycosylated extracellular component of the dystrophin-associated glycoprotein complex that links the actin cytoskeleton to the extracellular matrix in neural and skeletal tissues [85]. Glycans associated with α-DG include more common mucin-type GalNAc-Ser/Thr linkages as well as novel O-Man structures that can be extended to linear Siaα2,3Galβ1,4GlcNAcβ1,2Man glycans [84] and potentially branched structures containing a GlcNAcβ1,2(GlcNAcβ1,6)Man core. Initiation of O-Man structures is catalyzed by two protein-O-Man transferase isoforms [86] and extended with a GlcNAc residue catalyzed by single enzyme isoform [87]. Defects in O-Man glycan initiation and extension lead to congenital muscular dystrophies, termed dystroglycanopathies, where severity of the disease correlates with the degree of the truncation of the O-Man glycan structures on α-DG [88,89].

## Glycolipid Biosynthesis

Glycan extensions to ceramide cores comprise another class of diverse glycoconjugate structures (glycosphingolipids) at the plasma membrane and luminal compartments of the secretory pathway [90]. Initiation of glycosphingolipid structures occur through addition of either Gal (galactosylceramide) or Glc (glucosylceramide). Galactosylceramides show limited diversity, by capping with α2,3Sia, 3'-sulfation, or extension with α1,4Gal and potential 3'-sulfation. In contrast, glucosylceramide is generally extended by the addition of a β1,4Gal residue that can be further extended and branched with an incredible diversity of structures that are classified based on the linkage position and sugar attached to the Gal and subsequent residues [91]. Among the common classes of mammalian glycolipid core families are the ganglio- series (Galβ1,3GalNAcβ1,4Galβ1,4Glc-) [92], the lacto- series (Galβ1,3GlcNAcβ1,3Galβ1,4Glc-) [93], the neolacto- series (Galβ1, 4GlcNAcβ1,3Galβ1,4Glc-), the globo- series (GalNAcβ1,3Galα1,4Galβ1,4Glc-), and the isoglobo- series (GalNAcβ1,3Galα1,3Galβ1,4Glc-) [33]. Highly extended and branched glycosphingolipid structures in each glycolipid series can be generated by addition of GlcNAc, GalNAc, Gal, Fuc, and Sia with a structural diversity that is similar to the capping reactions on N-glycan and mucin-type O-glycan structures. Many of the enzyme isoforms that elaborate glycolipid structures are distinct from similar enzymes that work on glycoprotein acceptors, in that they prefer glycolipid substrates [11]. In addition, tissue-specific and developmental expression patterns for the enzymes that elaborate glycosphingolipids are regulated quite differently from enzymes that elaborate glycoprotein structures [12].

## Biosynthesis of Glycosylphosphatidylinositol-Anchored Proteins

The biosynthesis of the precursor for GPI addition to a protein backbone is similar in overall features to the biosynthesis of the N-glycan lipid-linked precursor [94]. A lipid core, in this case phosphatidylinositol, is modified by the addition of GlcNAc on the cytoplasmic leaflet of the ER membrane and subsequent de-acetylation to GlcN by a multiprotein complex containing as many as seven separate enzyme components [49]. Additional fatty acylation of the inositol occurs through the transfer of palmitate from palmitoyl-CoA. The glycoconjugate is then flipped into the luminal leaflet of the ER membrane and further extended by Man and phosphoethanolamine addition generally using single enzyme isoforms for each step [94]. In mammals, addition of phosphoethanolamine to the 2'-hydroxyl position of the terminal Man residue on the

Manα1,4GlcNα1,6PI intermediate is required prior to the extension to a final Manα1, 2Manα1,6(2′PE)Manα1,4GlcNα1,6PI structure. Further addition of ethanolamine to 6′-OH positions on Man residues completes the synthesis of the precursor structure. Transfer of the GPI precursor to nascent polypeptides occurs via a transaminase complex that results in the cleavage of C-terminal polypeptide membrane anchor regions and transfer to the amino group of the terminal or subterminal phosphoethanolamine residue. Thus, a total of at least 24 protein products are required for the synthesis and transfer of the GPI anchor precursor onto nascent polypeptides [94].

### Glycogen Biosynthesis and Catabolism

The synthesis of glycogen in animal systems is initiated by the self-glycosylation of glycogenin, a glycosyltransferase in the cytosol that acts as a primer for the extension of the highly branched glycogen dendrimer [95]. The protein exists as a dimer and employs UDP-Glc as a substrate to initiate glycan extension to a protein-associated Tyr residue and subsequently extend the structure with α1,4Glc residues [96]. Following extension to at least eight Glc units, further extension is accomplished by two isozymes of glycogen synthase [97,98] and branched by a single branching enzyme isoform [99]. Three phosphorylase isoforms and a single debranching enzyme accomplish depolymerization under conditions of low glucose concentration. Additional catabolism of glycogen presumably occurs in lysosomes, since genetic defects in lysosomal acid α-glucosidase leads to glycogen accumulation in glycogen storage disease, type II (Pompe disease) [100]. The complex control that balances the extension and trimming of the glycogen dendrimer is predominately coordinated by allosteric control, hormonally linked control by phosphorylation of glycogen synthase and phosphorylase, and transcriptional control of the multiple isoforms of glycogen synthase and phosphorylase. Synthesis of the glycogen polymer in liver and skeletal muscle can reach >60 000 Glc units with each core remaining attached to the glycogenin priming protein. Thus a relatively small number of protein-coding regions (~9 genes) are directly involved in glycogen metabolism in animal systems.

# Identification of Genes Involved in Glycan Biosynthesis

## Overview of Methods for Glycan-Related Gene Identification in Animal Systems

Our understanding of the biochemistry of glycan biosynthesis and catabolism is generally derived from over seven decades of studies on the glycan structures and enzymology of glycan metabolism. As the relevant enzymes were being purified throughout the mid-1980s, efforts were initiated to isolate their corresponding cDNAs and genes in order to provide a better understanding of glycan-related gene regulation and the molecular defects responsible for human glycoconjugate-related diseases. Numerous approaches were taken to clone the respective protein-coding regions that reflected the evolution of cloning technologies at the time. In the early 1980s mammalian cDNA cloning was in its infancy, and the major approaches for identification of coding regions were based either on peptide sequence data available from the purified proteins or on cross-reactivities of antibodies raised to the proteins. The discussion below attempts to trace some of the major trends in the cloning of mammalian glycan-related cDNAs over the last two decades, focusing initially on the early methods used to identify cDNA clones and how these efforts provided a foundation for the subsequent generation of comprehensive glycan-related gene lists based on mammalian genome sequencing.

### Early cDNA Cloning of Mammalian Glycan-Related cDNAs by Oligonucleotide Hybridization and Antibody Screening of Bacteriophage Expression Libraries

Glycan-related enzyme purification in the mid-1980s often resulted in the isolation of milligram quantities of purified enzymes for biochemical and enzymatic characterization. With the availability of the purified proteins, monoclonal or polyclonal antibodies were commonly generated for use in determining the subcellular localization of the proteins in the mammalian secretory pathway [101]. In addition, the purified proteins were also fragmented by proteolysis and partial sequence maps were generated following Edman sequencing of the isolated peptide fragments. Two strategies for cDNA cloning were available at the time based on the availability of the high-affinity antibody probes and protein sequence data. A common cDNA cloning approach in the mid-1980s was the use of antibodies to probe cDNA expression libraries by detecting a cross-reactive polypeptide fusion protein product linked to β-galactosidase expressed in the λgt11 bacteriophage vector [102]. Several mammalian glycosyltransferases were cloned by this method, including bovine β1,4Gal transferase [103], α2,6Sia transferase [104], and bovine α1,3Gal transferase [105]. Protein sequence data was also used to design degenerate oligonucleotide probes based on protein sequence data and the radioactive probes were used to screen cDNA libraries. Mammalian glycosyltransferases that were identified by this approach include bovine [106] and human [107] β1,4Gal transferase.

### Cloning of Mammalian cDNAs by Degenerate Oligonucleotide PCR Amplification Based on Protein Sequence Data

In 1988, a significant advance in the efficiency of cDNA cloning was introduced with the introduction of a PCR-based method for generating segments within a protein coding region by the design of degenerate primers based on peptide sequence data. The corresponding cDNA fragment within the coding region was then either used as a high-stringency probe for the isolation of full-length cDNAs by hybridization or it was used as a template for generating full length coding regions by 5' and 3' RACE [108]. The PCR cloning method, termed mixed oligonucleotide primed amplification of cDNA (MOPAC) [109], was originally employed for the cloning of porcine urate oxidase, but the method was rapidly applied to the cloning of glycan-related genes (the first was Golgi α-mannosidase II [110]) using protein sequence data from purified enzymes. Over the next decade the method was applied in numerous variations for the cloning of glycan-related genes where protein sequence data were available (e.g., blood group A transferase [111], GlcNAc transferase I [112], α2,3Sia transferase [113], GlcNAc transferase V [114], polypeptide-GalNAc transferase [115,116], GlcNAc transferase II [117], Golgi α-mannosidase I [118], and core 1 β1,3Gal transferase [119]) or when conserved motifs were available for enzymes with multiple sequence isoforms (e.g. Golgi mannosidase I isoforms [120], ER mannosidase I [121], and sialyltransferase isoforms [122–125]).

## Expression Cloning in Mammalian Cells

An alternative approach for identifying cDNAs encoding glycan-related enzymes was by expression cloning using gene transfer into mammalian cells followed by detection of the enzymatic products of the transfected gene on the recipient host cell surface. The approach generally involved the transfection of a recipient mammalian host cell line that does not express the glycan enzymatic product. Introduction of a cDNA library in an episomal expression plasmid vector into the host cells would be followed by the cell surface probing of the transfected cells with antibodies or lectins that recognize the glycan enzymatic product. The antibody or lectin was usually immobilized on the surface of a plastic culture dish in order to isolate transformants by "panning" on the antibody- or lectin-coated dish. The expression plasmids would then be re-isolated from the adherent transfected cells and subjected to additional rounds of enrichment by repeated transfection and panning. This approach was used to clone α1,3Gal transferase [126], several Fuc transferases [127,128], and β1,4GalNAc transferase [129].

## Identification of Enzyme Homologs Based on Cross-hybridization and Sequence Searches of EST and Genomic Sequence Data

As cDNAs were being isolated in the late 1980s and early 1990s it rapidly became clear that many glycan-related enzyme activities were encoded by multigene families of enzymes that could be isolated by cross-hybridization. Low stringency hybridization approaches were initially used to identify novel sequence homologs in the same species (e.g., Golgi mannosidase IIx [130], Fuc transferase isoforms [131–135]) or across species (e.g., mouse β1,4Gal transferase [136], mouse Fuc transferases [137,138]). As high-throughput DNA sequence databases were becoming available to the general scientific community in the late 1990s, these databases also rapidly became a resource for identification of novel related enzyme isoforms (e.g., β1,3Gal transferases [139,140] and a lysosomal α1,6-mannosidase [141]).

## Identification of Genes by Complementation in Lower Eukaryotes or in Mutant Mammalian Cell Lines

Another approach for identification of glycan-related genes has been the cloning and characterization of enzymes in other organisms that carry out similar biochemical pathways followed by the subsequent identification of mammalian homologs by sequence similarity searches. A prominent example in mammalian glycobiology was the initial identification of genes involved in the lipid-linked precursor biosynthetic pathway for N-glycan biosynthesis, which was first characterized by *Saccharomyces cerevisiae*, prior to identification of mammalian homologs by sequence similarity. Yeast genes involved in several steps of lipid-linked precursor biosynthesis were first cloned by complementation of mutants defective in various steps in Asn-linked glycosylation (*Alg* mutants) [142–145]. In a similar approach, mutant murine lymphoma cell lines defective in GPI anchor biosynthesis were used in genetic complementation studies to identify many of the proteins involved in GPI anchor biosynthesis [146–149].

### Filling Out the Sequence Databases: Use of Genomic Sequence Mining to Identify the Remaining Sequence Homologs

As individual glycan-related enzymes and proteins were purified and cloned based on protein sequence data or genetic complementation, separate efforts were being made to complete the sequencing of several animal genomes. As genome sequences were being assembled, independent efforts were also being made to annotate both whole genome and EST sequence records and to develop sequence-searching tools that allow the identification of novel genes based on search algorithms that could detect more complicated patterns and motifs of sequence similarity. As a result of these developments, sequence homologs of numerous glycan-related proteins were rapidly identified within the same species or across species over the period of several years. Novel enzyme catalytic activities continue to be identified, especially for less abundant glycoconjugates, as a result of ongoing efforts to purify the respective proteins or identify genes based on genetic approaches. While numerous protein-coding regions have been identified by sequence similarity to enzymes of known function or based on their involvement in glycan-related diseases, the catalytic specificities for many of the novel sequences remain to be identified.

## Organization of Mammalian Glycan-Related Genes into Databases

The complexity of glycan structures in animal systems and the multigene families of enzymes involved in their biosynthesis and catabolism presents a significant challenge for the organization of comprehensive databases that can: (a) convey sequence relationships among families of related enzymes, (b) provide functional annotations for multiple enzyme isoforms, and (c) assign individual isoforms to specific steps in biochemical pathways. Existing databases have generally attempted to address the individual goals above (e.g., assembly and cataloging of protein sequence families without providing functional or pathway annotations), but no existing database presently provides a comprehensive assembly of data on sequence relationships paired with detailed functional annotations.

Assembly of protein and gene sequence data on glycan-related enzymes was first initiated with the creation of the database of Carbohydrate Active Enzymes (CAZy, www.cazy.org). The CAZy database, developed by Dr. Bernard Henrissat and coworkers [26] as of 2008 covers collections of enzyme families for glycosidases (113 families), glycosyltransferases (91 families), polysaccharide lyases (19 families), carbohydrate lyases (15 families), and carbohydrate-binding modules (52 families) in bacteria (662 genomes), archaea (52 genomes), eukaryotes (28 genomes), and viruses (55 genomes). This organized sequence assembly is a tremendous resource for the glycobiology community and provides additional links and functional annotations at the family level. Additional annotated sequence databases of glycan-related genes classified into protein families include the genomic resource for animal lectins (www.imperial. ac.uk/research/animallectins/ [150]) organized by Dr. Kurt Drikamer and the transporter classification database (TCDB, www.tcdb.org [54]) organized by Dr. Milton Saier. Organization of databases based on enzyme function in glycan biosynthetic and catabolic pathways includes the databases of the Kyoto Encyclopedia of Genes and Genomes (www.genome.jp/kegg/ [27–29,151]), initial efforts within the Complex Carbohydrate Research Center at the University of Georgia (glycomics.ccrc.uga.edu [12]), and the enzyme database of the Consortium for Functional Genomics (CFG, www.functionalglycomics.org). More comprehensive databases containing sequence records and associated annotation information can also be found at the National Center for Biotechnology Information (NCBI, www.ncbi.nlm.nih.gov [152]) and the SOURCE database (source.stanford.edu [30]).

Recent efforts to profile glycan-related transcripts in animal systems have employed glycan-specific oligonucleotide or cDNA microarrays or medium-throughput transcript profiling by qRT-PCR. In each instance gene lists were assembled from multiple database sources, including those listed above, as well as the primary literature and suggestions from specialists in the glycobiology field. Thus, glycan-specific gene lists were employed for the assembly of oligonucleotide microarrays by the CFG [13,18,21,153]) and for the creation of cDNA arrays by the Glyco-Chain Expression Laboratory [20,22]. A more comprehensive gene list of >700 glycan-related genes was also compiled for qRT-PCR-based transcript profiling at the University of Georgia (glycomics.ccrc.uga.edu [12]).

# Overview of Transcript Analysis and Gene Expression Methods

Since the advent of Northern blotting in 1977 for the detection of transcript abundance in biological samples [154], many other techniques have been developed for the detection and quantitation of transcript levels. These methods can be categorized as either hybridization, sequence, or amplification-based methods. An overview of each method along with their benefits and drawbacks is presented below. Examples of information gained by the application of these techniques within the field of glycobiology are also discussed.

## Hybridization-Based Transcript Analysis Methods

### Northern Blots

Northern blotting is a hybridization-based technique where isolated RNA is separated by gel electrophoresis, transferred to a membrane, and detected by hybridization with a DNA or RNA probe. The first detection methods involved radioactive probes. Subsequently, methods were developed using chemiluminescent detection, which provided a safer technique with a greater range of detection than radioactive methods. This technique allows the determination of several transcript characteristics; mRNA transcript size by comparison with RNA size standards, alternatively spliced transcripts, RNA half-life, and amount of RNA degradation. The predominant use of Northern blots has been to determine the expression level of a detected transcript via comparison with the signal intensity of a housekeeping gene. Compared with more recently developed sequence-based methods, Northern blotting is a time-intensive and much less sensitive technique for the quantitation of transcripts due to the limited dynamic range of the detection method.

The tissue-specific expression patterns for many glycan-related genes were first determined by Northern blotting [155]. Some studies were done on an individual gene basis. For example, Northern blots of *B4galt1* in mouse tissues revealed transcripts that were detected in most of the tissues assayed, but barely detectable in brain and testis, which led to a further investigation of how this gene is regulated during spermatogenesis using a male germ cell-specific transcriptional start site [156]. The identification of a new member of the GT29 sialyltransferase family, *ST8Sia6*, was accompanied by Northern blot analysis which determined that the gene was ubiquitously expressed with the highest expression levels found in kidney [157].

There are also examples of more broadly based Northern blot studies. An investigation of tissue-specific expression of all six members of the β4Gal transferase family using Northern blots revealed that *B4galt2* and *B4galt6* are expressed in only a restricted set of tissues and that only *B4galt1* transcripts were detected in lactating mammary glands [158]. Comparison of the expression the three GT11 family α1,2Fuc transferases using Northern blotting demonstrated restricted expression for *Fut1* and *Fut3* in epididymis and testis, respectively, while *Fut2* was expressed more broadly, with particularly high expression in tissues of the digestive tract [159].

## In Situ Hybridization

This technique uses a small nucleotide sequence as a probe to detect specific mRNA sequences within preserved tissue sections or cells. No initial isolation of RNA is required for this technique. Tissue sections are frozen or embedded in paraffin before sectioning for in situ hybridization. Cells in suspension can also be used if they are cryospun onto glass slides and fixed. Four types of probes can be used: oligonucleotide probes, single-stranded DNA probes, double-stranded DNA probes and RNA probes. As with Northern blotting, the probes are either radiolabeled or labeled with biotin, FITC, rhodamine, or DIG. Prior to hybridizing the probe with the sample, a permeabilization step is required so that the probe can reach the target and various prehybridization steps may be needed to reduce background staining. Following hybridization, probes are detected with photographic film or emulsion. Although highly sensitive (10–20 copies per cell), this technique is time-intensive due to the need for extensive control and optimization experiments. Information on the location of a specific mRNA within a cell or tissue can be gained from this technique, but no information on the amount or size of a transcript, alternative splicing, or half-life are determined.

Expression of β4Gal transferases were investigated during brain development using in situ hybridization, which revealed that while both *B4galt2* and *B4galt5* could be detected in the hippocampus, they were expressed in different regions of that structure [160]. In situ hybridization of β1,6GlcNAc transferase-V (*Mgat5*) and the core 2 GlcNAc transferase (*Gcnt1*) in developing mouse embryos revealed distinct temporal and spatial expression patterns for these two genes [161]. Recently, in situ hybridization was used to investigate the localization of *B4galt1* and *B4galt5* expression in rat muscles before and after sciatic nerve injury [162]. The differential expression pattern for Golgi class I α-mannosidases was also examined during mouse development by in situ hybridization [163].

## Ribonuclease Protection Assay

A ribonuclease protection assay (RPA) is another hybridization technique that requires RNA isolation. Unlike, Northern blots and in situ hybridization, the actual hybridization occurs in solution. The probe used in RPA is a labeled antisense probe (isotopic or non-isotopic) which is hybridized with total RNA then subjected to RNase (S1 Nuclease) digestion to remove any unhybridized material. The resulting material is precipitated and resolved on a denaturing polyacrylamide gel and then detected by autoradiography for isotopic probes, or transferred to a membrane and detected for non-isotopic probes. The RPA technique can be used to determine relative or absolute transcript abundance and to map mRNA termini and intron/exon boundaries. It is more sensitive than Northern blot analysis (4000–50 000 copies/sample) and can be used for multi-probe analyses without the necessity of stripping and re-probing as with a Northern blot analysis.

The presence of two distinct mRNA transcripts encoding murine β1,4Gal transferase-I was determined using S1 protection assays and was confirmed by Northern blots [136]. As part of a study to determine the genomic organization of human *MGAT2*, RPA was used to identify multiple transcription initiation sites [164]. RPA has also been used to determine the presence of tissue-specific alternative splicing of α2,6Sia transferase in rat kidney [165] and later used to determine different transcriptional start sites for the kidney and liver specific transcripts [166].

## Microarray

Compared with the previously described hybridization methods, microarray approaches are much higher throughput techniques for profiling mRNA expression. Microarrays can be used to determine expression levels of a few hundred genes up to thousands of genes (low-density arrays) or even entire genomes (high-density arrays). Gene expression studies using microarrays involve the isolation of RNA from cells or tissues to be analyzed, followed by labeling via reverse transcription with fluorescently tagged nucleotides. The labeled cDNA is then incubated with an array containing immobilized gene-specific sequences allowing for hybridization of complementary sequences. The degree of hybridization of fluorescent molecules is then detected and quantified. Several different types of microarray platforms are available that differ by the type of probes that are immobilized and scale of analysis.

Affymetrix GeneChip™ arrays (http://www.affymetrix.com) are high-density arrays which employ short oligonucleotides (~25mers) that are lithographically synthesized directly onto the chip surface [167]. Multiple different oligonucleotide sequences (11–16 pairs) are selected for each gene of interest which are either "exact matches" to the known sequence for expression determination or "mis-matched" for the purpose of determining non-specific background hybridization. A cDNA sample is hybridized to a single chip followed by detection of signal intensity and data analysis. The incorporation of multiple oligonucleotide sequences for each gene increases the statistical power of the data obtained from these arrays and is a definite advantage for this approach. Some disadvantages of the array platform include the expense and the inflexibility for designing chips for specific experiments.

An array platform that allows more flexibility for individual experimental design is the NimbleGen array (http://www.nimblegen.com). These arrays contain oligonucleotide probes synthesized directly on the chip surface similar to the Affymetrix arrays, but the selection of assay genes is more flexible and they use longer oligonucleotide sequences which increases the sensitivity of the platform [168]. Athough this platform provides the advantage of a more flexible and custom analysis, the increased cost is a disadvantage.

In contrast to the Affymetrix and Nimblegen array platforms, spotted arrays are created using spotting robots that place synthesized oligonucleotide sequences (60–70 bp), cDNA clones, or PCR products directly on the surface of a glass slide. The spotting on these arrays is less dense and tends to be less uniform due to the technical differences between this platform and the platforms discussed above [169]. To circumvent problems with inconsistent spotting from one array to the next, two-color experiments are used, where two samples are labeled with different fluorescent dyes followed by hybridization to the same array. Spotted arrays have the advantage of low cost and flexibility of target gene selection.

Analysis of the data from microarray experiments includes normalization [170,171], calculation of expression values, and statistical analysis, among other types of analyses [172–175]. Several software packages are available either as commercially produced analysis packages or freely available programs, including Statistical Analysis of Microarray (SAM) [176,177] and GeneCAT, which incorporate new co-expression analysis and data mining tools [178].

The CFG has generated custom designed "Glyco-gene chips" in conjunction with Affymetrix that focus on glycosyltransferases, carbohydrate-binding proteins, and proteoglycans, among other related genes [153]. Since the creation of the first Glyco-gene chip, GLYCOv1, three updated versions have been produced with the latest being the GLYCOv4 chip, which was produced in 2008. The latest version has the ability to determine the expression of ~2000 human and mouse transcripts related to mammalian glycobiology.

The GLYCOv1 chip was used to investigate glycan-related transcript profiles in nine mouse tissues [13]. The study revealed a distinct difference in overall expression profiles for immune-related tissues as compared to non-immune tissues. It was also noted that genes involved in the synthesis of N-linked and O-linked glycan core structures were constitutively expressed, whereas those involved in the addition of terminal sugars exhibited tissue-specific expression patterns. Other examples where the GLYCOv1 chip was employed include studies on changes in N-linked glycan structures that occur during T lymphocyte differentiation and activation [179] and an analysis of glycan-related transcripts in cell populations derived from mouse thymus with a specific focus on Galectin-8 expression [180]. The GLYCOv2 and GLYCOv3 chips have been used for expression profiling of both mouse [12,21,181,182] and human [18,183–186] samples. The application of these custom microarrays to investigate changes in glyco-related gene expression during cellular tissue development [21], immune system development [18], bacterial infection [181,185], and disease states [182–184,186] highlights the usefulness of this platform as a survey method in the field of glycobiology.

In addition to glycan-related Affymetrix-type microarray studies, several groups employed spotted microarrays to perform focused glyco-gene expression studies in various biological systems. Kroes et al. used a "human glycobiology microarray" containing 45mer oligomers for 359 genes assembled from fully curated gene sequences consisting of genes encoding glycosyltransferases, glycosylhydrolases, polysaccharide lyases, and carbohydrate esterases [187] to investigate changes in gene expression in both normal and malignant brain tissue (gliomas). These studies detected elevated levels of Golgi α-mannosidase II and α2,6Sia transferase-V (*ST6GALNAC5*) expression in normal tissue compared to glioma tissue, and high levels of *POFUT1* and *CHI3L1* in the glioma tissue compared with normal tissue. A 60mer spotted "GlycoProfiler" microarray (Scienion, Berlin, Germany) containing ~220 probes for human glycosyltransferases and lectins was used to investigate the effects of tumor supressor p16$^{INK4a}$ expression on the abundance of glycosyltransferase transcripts [188]. This study revealed changes in the expression of several β1,4Gal transferases, the α2,3Sia transferases, and α2,6Sia transferases resulting from transfection of human pancreatic carcinoma cells with p16$^{INK4a}$.

The RIKEN Frontier Human Glyco-gene cDNA microarray (version 2) is probably the largest spotted glycan-related gene array developed to date, containing probes for 888 genes, including those encoding glycosyltransferases and genes involved in sugar metabolism, glycan remodeling and recognition, and lipid metabolism [20]. This array was used to investigate the expression of sialyltransferase genes in several human B-cell lines. In order to compare several samples to each other, a commercially available universal reference RNA was competitively hybridized to each array to normalize each data set. As an extension of their microarray analysis, the same group also developed a Correlation Index-Based Responsible-Enzyme Gene Screening (CIRES) method to correlate changes in gene expression with changes in cell surface glycan formation detected by lectin binding visualized by flow cytometry [22].

## Sequence-Based Transcript Analysis Methods

### *Expressed Sequence Tags (ESTs)*

The development of commercially available automated technologies for DNA sequencing in the mid-1980s provided the foundation for genome scale sequencing efforts [189]. Using this technology to generate short, single-pass sequence reads (400–600 bp) from the 5′ or 3′ ends of cDNA produced sequences known as expressed sequence tags or ESTs [190]. The abundance of these sequences, which were deposited with NCBI, prompted the formation of dbEST in 1992 to provide a searchable and annotated database of the deposited sequences [191]. To reduce the problem of redundancy of deposited ESTs that comprise portions of the same transcript, the UniGene database (www.ncbi.nlm.nih.gov/unigene) was created by NCBI to arrange the ESTs into sets of gene-oriented clusters. Within the UniGene database, expression profiles are generated from the analysis of deposited EST sequences. In silico Nothern blots are created by generating spots with varying intensity based on transcript counts (expressed as transcripts per million, TPM). For a given organism and gene, data are classified by tissue source, health state, or developmental stage. Information from the EST databases is helpful for gene identification and can provide qualitative inferences about transcript abundance in profiled tissues. However, since the isolation of mRNA is not consistent across all cell types or developmental stages, the data may not provide directly comparable quantification across the data sets.

### *Serial Analysis of Gene Expression*

Serial Analysis of Gene Expression (SAGE) requires the isolation of mRNA and the generation of cDNA from which unique small sequences (~14 bp), or tags, are generated using restriction enzyme digestion. The tags are then concatenated by ligation with other tags, amplified in a bacterial host and then sequenced. The number of times that a specific sequence tag is found determines the relative abundance of the transcript in that sample. Thousands of transcripts within a cell can be analyzed simultaneously with this technique [192]. SAGE is very useful for discovering new transcripts for which the gene sequences were not previously known. However, generation of a small sequence tag may not provide enough information to clearly determine the identity of a new sequence [193,194]. A modified version of this technique that uses longer tags has been shown to be more useful [195]. A variation of SAGE was developed as a tool to provide information about transcript abundance as well as transcriptional regulation. Cap analysis gene expression (CAGE) involves the isolation of short sequence tags at the 5′ end of full-length mRNAs, which contrasts with SAGE where tags originate at the 3′ end of the mRNA [196,197]. The 5′ end tags are sequenced and can be used to determine the relative transcript abundance and can also be used to determine the location of the transcription start site for promoter analyses.

Since SAGE is generally used for gene discovery between two populations of cells or tissues, there are few, if any, glycobiology-targeted studies employing SAGE as the predominant analytical method. However, there are examples of SAGE analyses where glycan-related genes have been identified as abundantly expressed genes. For example, a SAGE analysis of human marrow mesenchymal stem cells found a high level of expression for genes involved in extracellular matrix (ECM) formation (e.g., collagen and matrix metalloproteinase-2) and cell adhesion (galectin-1) [198]. A recent study of the transcriptome of vascular endothelial cells subjected to static and shear stress conditions using SAGE revealed a decrease in expression for several ECM genes resulting from shear stress [199].

## Cyclic Array Sequencing

Several new platforms for DNA sequencing have emerged that use multiple sequencing reactions conducted in a cyclic manner. This process is referred to as massively parallel sequencing because it has the potential to generate megabases of sequence information from templates immobilized on a solid substrate. The reduced cost of the new systems compared to previous Sanger sequencing methodologies, make these techniques more attractive for performing transcript analyses. The 454 (Roche) and the Massively Parallel Signature Sequencing systems are briefly discussed below. For a detailed discussion of the respective platforms, refer to the review by Shendure et al. [200].

Transcript analysis using the 454 platform involves nebulization of cDNA to produce fragments for ligation to adaptor fragments followed by annealing to DNA capture beads. The sequences on the beads are then subjected to clonal amplification in an emulsion and then loaded into wells of a fiber-optic slide for sequencing [201]. The templates are sequenced using a "sequencing by synthesis" method called pyrosequencing. This method involves sequencing the DNA strand as it is synthesized by monitoring the addition of nucleotides via detection of the luciferase-based release of pyrophosphate [202]. The data are acquired in real-time and do not require the separation of synthesized fragments by gel electrophoresis as used in Sanger-type sequencing approaches. The main advantage of this platform is the ability to obtain read lengths >100 base pairs. A disadvantage of the platform is that the processes of emulsion, amplification, and bead recovery are more difficult than other cyclic array methods.

Massively Parallel Signature Sequencing (MPSS) combines gene cloning with the signature sequencing characteristics of SAGE, and a high-throughput cyclic sequencing methodology developed by Lynx Theraputics, Inc. (Heidelberg, Germany). This technique utilizes in vitro cloning of tagged templates from cDNA libraries on microbeads [203] with sequencing of tags, or "signatures," using a ligation-based sequencing technique [204]. The microbeads, with their amplified sequences, are arranged in a planar array within a flow cell and the 20 base signature sequences of each bead are determined, potentially identifying as many as a million transcripts in a single run. The high sensitivity of this technique is particularly useful for detecting low abundance transcripts.

Due to the high-throughput nature of these methods, these studies are not primarily targeted toward collecting data purely relevant to the glycobiology community. However, the huge amount of data generated from these techniques may be available for mining through database searching. The establishment of publicly available databases, like the Gene Expression Omnibus (GEO), www.ncbi.clm.nih.gov/geo, which is maintained by NCBI [205], allows users to search for information on their genes of interest from a repository of mRNA expression data. The GEO database is designed to accept data from microarray platforms and high-throughput sequencing platforms (e.g., SAGE). Numerous databases are currently available and have been published annually in the January issue of *Nucleic Acids Research* since 1996. With the introduction of less expensive high-throughput sequencing technologies, the amount of sequencing data should increase exponentially. The development of annotated databases will allow more of these data to be accessed and utilized by the glycobiology community.

## Amplification-Based Transcript Analyses

### Reverse-Transcriptase PCR

The use of PCR to analyze DNA target abundance following reverse transcription of an mRNA template source (RT-PCR) is a technique that can be used to analyze transcript abundance from small amounts of starting material [206–208]. Early applications of RT-PCR for determining transcript abundance involved the comparison of final amplimer quantities following gel electrophoresis and densitometry and the methods were generally qualitative in nature. Modifications to this method include the co-amplification of a housekeeping gene within the PCR reaction for normalization ("Primer-Dropping RT-PCR" [209]) and cDNA synthesis of the RNA sample in the presence of a synthetic competitor RNA followed by amplification ("Competitive RT-PCR" [210,211]). These techniques are considered semi-quantitative RT-PCR methods because they detect only the final end-product at the plateau phase of the amplification, where reaction components can be limiting [212]. The requirement for the detection of end-products by densitometry results in limited sensitivity and dynamic range for this approach [213].

Initial studies of glycan-related transcripts employed semi-quantitative RT-PCR analysis to examine transcript abundance for several glycosyltransferases related to L-selectin-mediated signaling in infected mice [214]. Densitometry was used to ensure that products were analyzed in the exponential phase of the reaction instead of the plateau phase. Primers were designed for Core1 and Core2 GlcNAc transferases, *FucTVII*, and *GlcNAc6ST-2*, and were used to determine transcript abundances in normal and *Helicobacter felis*-infected mouse gastric mucosa. Amplimer band intensity for each gene was normalized to GAPDH and an increase in transcripts for Core2 GlcNAc transferase-1 was observed. Another study using semi-quantitative RT-PCR investigated changes in transcript abundance for three isoforms of *UGT1A*, from the glycosyltransferase 1 (GT1) family, in human hepatocytes following treatment with rifampicin (Rif) and 3-methylcholanthrene (3-MC) [215]. Expression of *UGT1A8*, *-1A9*, and *-1A10* differed between six different donors as detected by densitometry. Treatment of hepatocytes with Rif or 3-MC resulted in increases in transcript abundance for *UGT1A8* and *UGT1A10* isoforms. A fairly large-scale semi-quantitative RT-PCR analysis was performed analyzing 68 glycosyltransferase genes in 27 different human tissues [216]. Hierarchical clustering was used to visualize differences in expression across multiple tissues.

### Real-Time RT-PCR

Advances in sensitivity and dynamic range for determining transcript abundance were introduced with the advent of fluorescent dyes or probes to monitor the progress of the PCR reaction by continuous collection of fluorescent signals [217–219]. This new technology was termed "real-time PCR" which is sometimes confused with reverse-transcriptase PCR. For clarity, in this chapter, this technique will be referred to as quantitative RT-PCR, or qRT-PCR, to differentiate it from the non-real-time methods. The advantage of detecting the appearance of amplimer products throughout the amplification reaction is that amplimers are detected during the exponential phase of the reaction. It is during the exponential phase where the amount of amplified target is directly proportional to the amount of input target allowing for quantification of the starting material [220]. Two types of quantitation, either absolute or relative, are used for qRT-PCR transcript analyses. Absolute quantification entails determining the exact copy within an amplified sample by generation of a calibration curve from quantitative standards [221–223]. Relative quantitation involves comparison of changes in expression to an internal or external standard or a reference sample [222–226].

Several different detection chemistries are available for qRT-PCR. The least specific and least expensive is the use of intercalating dyes like SYBR® green I (Molecular Probes, Invitrogen) [227,228], EvaGreen™ (Biotium, Hayward, CA, USA) [229], and BOXTO™ (TATAA Biocenter, Gothenburg, Sweden) [230]. These dyes are non-specifically incorporated into the double-stranded amplimer which produces a fluorescent signal proportional to the amount of product. The disadvantage of this method of detection is that these dyes will be equally incorporated into any double-stranded molecule in the PCR reaction. However, with rigorous primer design, appropriate quality control, and post-run melt-curve analysis, this method has proven to be a viable and inexpensive option for qRT-PCR analysis [231]. Other chemistries are available which use custom-designed fluorescent probes. This adds both increased specificity and cost to the qRT-PCR analysis. Detection probes are generally classified as: hydrolysis probes (i.e., TaqMan Probes) [217], hybridization probes (i.e., HybProbes/LightCycler) [232], hairpin probes (i.e., Molecular Beacons, LUX) [233,234] and stem-loop probes (i.e., Scorpions) [235]. An in-depth discussion of each of these types of probes can be found in [236]. qRT-PCR is extremely sensitive, has a dynamic range of 7–8 log orders of magnitude, and has the ability to generate highly quantitative data [227]. Historically, this technique was conducted on a relatively small scale (i.e., <100 genes), and was primarily used as a validation method following microarray analyses [237,238]. However, with the introduction of 384-well reaction blocks for performing qRT-PCR analysis in standard real-time PCR machines in low microliter reaction volumes [225] and new microfluidic technologies that perform reactions in nanoliter reaction volumes [239,240], experiments can be now conducted at a scale to analyze hundreds to thousands of target transcripts at a time.

TaqMan qRT-PCR was used to analyze 16 isoforms of UDP-GalNAc:polypeptide *N*-acetylgalactosaminyltransferase (pp-GALNAC-T) in mouse tissues, which revealed strong tissue-specific expression of the T13 isoform in brain, and a putative isoform (Tb) in testis [23]. The expression of pp-GALNAc-T12 was similarly analyzed in normal and human cancer cell lines, which revealed a downregulation of the gene in cancer tissues [241]. Other examples of the use of TaqMan qRT-PCR analysis in glycan-related studies include: determination of the expression of mucin core 2 β1,6GlcNAc transferase-M in human tissues and airway epithelial cells [242], analysis of *EXT1* and *EXT2* gene expression in human osteochondromas and chondrosarcomas [243], and expression of α2,8Sia transferases (*ST8SIA1–5*) in immune cells [244].

Using a SYBR green qRT-PCR approach, human sialidase gene expression was analyzed during human monocyte differentiation [245]. Transcripts for *NEU1* and *NEU3* increased during differentiation of monocytes into macrophages, which correlated with increases in specific protein levels detected by immunoprecipitation. Garcia-Vallejo et al. also used a SYBR Green qRT-PCR approach to investigate the effects of tumor necrosis factor-α (TNFα) on β1,4Gal transferase-1 expression in human endothelial cells [246]. This study revealed an increase in transcript abundance following TNFα treatment in a time- and concentration-dependent fashion. Continuing their studies on TNFα, the same group performed a larger-scale study of 74 glycan-related genes [31,32] and compared the expression of glycosyltransferases, mannosidases, and sulfotransferases in both human umbilical vein endothelial cells (HUVEC) and human foreskin microvascular endothelial cells (FMVEC) in untreated and TNFα-treated cells. The genes involved in proteoglycan biosynthesis in mouse embryonic stem cells were also analyzed using a SYBR Green qRT-PCR approach [14]. In this study a parallel structural analysis was conducted to compare changes in transcript abundance with changes in abundance of heparan, chondroitin/dermatan and the relative amounts of sulfation in undifferentiated and differentiated mouse stem cells. A larger scale qRT-PCR study of >700 glycan-related genes using SYBR Green qRT-PCR was recently published, which investigated the differences in relative transcript abundance in four mouse tissues [12]. A comparison between microarray and qRT-PCR analysis with paired samples revealed an increase in dynamic range by several orders of magnitude for qRT-PCR, as well as the ability to detect more low abundance transcripts than microarray approaches.

## Analysis of Gene Expression Data

Analysis of simple data sets of transcript expression data, where only a single gene or several genes are investigated, is fairly straightforward. However, newer high-throughput analyses like microarray, sequencing based platforms, and high-throughput qRT-PCR approaches require the aid of bioinformatic tools to derive trends and correlations from the data.

One way to organize and display large data sets is to group them by function using gene ontology information. The gene ontology project (GO) website (www.geneontology.org) classifies genes according to cellular component, biological process and molecular function. Gene ontology information can be used to analyze the expression data with the use of several web-based applications like DAVID [247] or Onto-Express [248].

Another way to analyze the data sets is by grouping them by regulatory or metabolic pathways to discover functional interactions. Several commercial software packages are available to correlate gene expression data with biochemical pathways and proteomic data. Analysis tools like PathwayAssist™ from Ariadne Genomics, Inc. (www.arianegenomics.com) and Ingenuity Pathways Analysis (IPA) from Ingenuity Systems (www.ingenuity.com), incorporate proprietary databases of molecular interactions to generate pathway diagrams and interpret microarray data and proteomic data, among other functions. The databases provided with these software packages are constantly updated to incorporate recent additions to the literature.

The KEGG database is a web-accessible tool that contains pathway diagrams with associated genes from multiple organisms [28]. In 2005, microarray-based pathway analysis tools added additional functionality to the KEGG website [249]. Additional tools like KOBAS [250] and DGEM [251] were also added more recently to provide statistical analysis methods that allow investigators to determine correlations between microarray data sets and metabolic pathways or disease states. Recently, a standalone JAVA-based program, called Eu.Gene, was also introduced that uses pathway information from KEGG and other databases to correlate genes and pathways and provide statistical significance values [252]. A comprehensive discussion of these applications and other tools for pathway analysis can be found in a recent article by Werner [253].

Pathways for the synthesis and catabolism of many glycan-related structures are included in the KEGG database. We recently utilized these pathways and other literature sources to assemble more completely annotated sets of glycan biosynthetic pathways in order to display data generated from qRT-PCR analysis of glycan-related genes [12,14]. By displaying the data in order of synthetic steps, we were able to make predictions about structural changes and then look for correlations between transcript abundance and corresponding glycan structures.

## Examples of Correlation of Transcript Analysis Data with Protein and Structural Data

Determination of quantitative values for transcript abundance, regardless of the analytical methods employed, is just one piece of a larger puzzle of understanding the regulation of biochemical processes in complex biological systems. In many instances the mRNA transcript levels in a given cell or tissue may, or may not, correlate with the amount of active, properly folded protein in the cell due to transcript stability, efficiency of translation, or posttranslational regulatory events. Due to this limitation, other methods are commonly used to determine if regulatory mechanisms are predominately controlled at the transcript level or by downstream events.

Changes in transcript abundance for a specific gene may correlate with changes in the amount of protein encoded by that gene. Western blots, although more qualitative than quantitative, can be used to investigate correlations between protein and transcript levels. For example, a study on sialidase expression in human monocytes revealed that *NEU1* and *NEU3* transcripts increased as monocytes differentiated into macrophages [245]. Western blots utlizing antibodies specific for each sialidase revealed increases in protein amounts for both enzymes following differentiation. Lectin binding has also been used to correlate changes in specific glycosylation events with changes in corresponding enzyme transcript levels. Lectin binding was used in conjunction with flow cytometry to correlate changes in cell-surface glycan formation with changes in transcript abundance detected by microarray [22]. A separate study employed qRT-PCR to analyze 74 genes involved in *N*-, *O*-, and glycolipid-core formation in HUVEC cells before and after treatment with TNF$\alpha$ and included parallel cell surface staining with lectins and monoclonal antibodies to detect changes in glycan structures [32]. Treatment with TNF$\alpha$ resulted in changes in expression of 19 glycosyltransferases, which was correlated to changes in epithelial cell suface glycosylation.

Changes in proteoglycan biosynthesis in mouse stem cell differentiation were analyzed by combining GAG compositional analysis with qRT-PCR transcript analysis [14]. GAG disaccharides were produced by enzymatic digestion and then separated by reversed-phase ion-pairing HPLC to determine changes in composition between cell stages. Disaccharide composition was then compared with transcript abundance data from paired samples. Stem cell differentiation resulted in increases in hyaluronan (HA) content which correlated with increased transcript levels for *Has1* and *Has2* and increased *GalNAc4S6ST* transcript abundance correlated with increased levels of the disaccharides, 4S6S (SE) and 2S4S (SB), in chondroitin and dermatan sulfate. Transcripts involved in the generation of GAG biosynthetic precursors (i.e., nucleotide-sugars) did not reveal correlations with changes in GAG structures, suggesting that these transcripts were present in excess and were not regulating the resulting GAG content in these cells.

Glycan profiling by mass spectrometry has been used to correlate changes in glycan structures with changes in transcript abundance [12,13,18,179]. Microarray transcript abundance data were compared with MALDI-TOF profiles of *N*-glycan structures and revealed that *Fut9* transcript levels correlated with the synthesis of Lewis x structures in mouse kidney [13].

Recently, an analysis comparing glycan structure data with qRT-PCR transcript analysis in mouse tissues was published by our group [12]. Comparison of transcript abundance data with glycan structural data revealed some instances of correlations between the two data sets, including reduced levels of bisecting GlcNAc and core $\alpha$1,6Fuc residues in liver *N*-glycans corresponding with reduced transcript abundance for *Mgat3* and *Fut8* in these tissues, respectively. In other cases, the comparisons were not as straightforward. For example, reduced sialylation of *N*-glycans in kidney was accompanied by only minor reductions in some of the sialyltransferase transcripts and elevation of some neuraminidase transcripts in this tissue.

## Future Directions

The development of tools to determine the presence and abundance of transcripts encoding the full spectrum of glycan-related proteins in cells and tissues is a first step in the broader goal of determining the global regulation of glycan structures in animal systems. There are several major hurdles that remain that hinder our ability to fully understand the underlying mechanisms that control glycan abundance in animal systems and their roles in physiology and pathology. The first problem is that, while many glycan-related enzyme isoforms have been identified, in many cases their substrate specificities and relative capacities for catalysis have not yet been examined. The second, and even greater concern, is the inability to easily quantify the levels for many glycan-related proteins or catalytic activities in complex biological samples. Broad-based transcript profiles can capture the steady state abundance of glycan-related mRNAs in a biological sample and recent advances in glycan structural analysis can now provide surprising levels of breadth, depth, and sensitivity to effectively profile many classes of glycan structures. Correlations between transcript levels and corresponding glycan structures have already revealed many cases where glycan structural diversity is controlled at the transcriptional level [12]. However, these two types of data sets are the two extremes in the informatic and metabolomic pipeline that must eventually bridge between gene structure, transcript expression levels, protein abundance, enzyme specificity and catalytic capacity, substrate abundance, abundance of competing enzymes, relative localization of the glycosylation enzymes, and finally, the structures of the corresponding glycan enzymatic products. The tools of genetics and molecular biology can provide tremendous insights by perturbing transcript expression by RNA interference or gene knockouts and the corresponding changes in glycan structures can be profiled to look at compensation and competition among enzymes in the complex biosynthetic systems. In the long term, additional approaches must eventually be developed in order to provide a greater capability to measure individual enzyme levels in complex biological systems that will begin to provide the missing pieces in the metabolomic pipeline. Initial attempts to create mathematical models for even the best understood glycosylation pathways in complex biological systems are still in their early stages [254]. Concerted developments must be made in the coming years to identify the specificities of the full range of glycan-related protein isoforms and characterize their ability to compete for glycan substrates in complex biological systems. Sensitive analytical methods must also be developed to quantify protein and enzyme levels for the low abundance glycosylation machinery in complex biological systems. These are daunting challenges for biochemists and analytical chemists in the glycobiology field, but the rewards of being able to model the metabolomic profile of complex glycosylation systems will eventually provide significant benefits in our understanding of human disease and allow the selective engineering of glycosylated therapeutics in the future.

## Acknowledgments

Work in the Moremen lab described in this chapter was supported by NIH Grant RR0118502.

# References

1. Varki A. Biological roles of oligosaccharides: all of the theories are correct. *Glycobiology* 1993;**3**:97–130.

2. Brockhausen I, Schutzbach J, Kuhns W. Glycoproteins and their relationship to human disease. *Acta Anat (Basel)* 1998;**161**:36–78.

3. Dennis JW, Granovsky M, Warren CE. Protein glycosylation in development and disease. *Bioessays* 1999;**21**:412–21.

4. Haltiwanger RS, Lowe JB. Role of glycosylation in development. *Annu Rev Biochem* 2004;**73**:491–537.

5. Taniguchi N, Miyoshi E, Gu J, Honke K, Matsumoto A. Decoding sugar functions by identifying target glycoproteins. *Curr Opin Struct Biol* 2006;**16**:561–6.

6. Marth JD. Will the transgenic mouse serve as a Rosetta Stone to glycoconjugate function? *Glycoconj J* 1994;**11**:3–8.

7. Kornfeld R, Kornfeld S. Assembly of asparagine-linked oligosaccharides. *Annu Rev Biochem* 1985;**54**:631–64.

8. Schachter H. Glycoproteins: their structure, biosynthesis and possible clinical implications. *Clin Biochem* 1984;**17**:3–14.

9. Schachter H. Coordination between enzyme specificity and intracellular compartmentation in the control of protein-bound oligosaccharide biosynthesis. *Biol Cell* 1984;**51**:133–45.

10. Schachter H. The "yellow brick road" to branched complex N-glycans. *Glycobiology* 1991;**1**:453–61.

11. Taniguchi N, Honke K, Fukuda M, editors. *Handbook of Glycosyltransferases and Related Genes*. Tokyo: Springer-Verlag; 2002:670.

12. Nairn AV, York WS, Harris K, Hall EM, Pierce JM, Moremen KW. Regulation of glycan structures in animal tissues: Transcript profiling of glycan-related genes. *J Biol Chem* 2008;**283**:17298–313.

13. Comelli EM, Head SR, Gilmartin T, Whisenant T, Haslam SM, North SJ, Wong N-K, Kudo T, Narimatsu H, Esko JD, Drickamer K, Dell A, Paulson JC. A focused microarray approach to functional glycomics: transcriptional regulation of the glycome. *Glycobiology* 2006;**16**:117–31.

14. Nairn AV, Kinoshita-Toyoda A, Toyoda H, Xie J, Harris K, Dalton S, Kulik M, Pierce JM, Toida T, Moremen KW, Linhardt RJ. Glycomics of proteoglycan biosynthesis in murine embryonic stem cell differentiation. *J Proteome Res* 2007;**6**:4374–87.

15. Aoki K, Perlman M, Lim JM, Cantu R, Wells L, Tiemeyer M. Dynamic developmental elaboration of N-linked glycan complexity in the *Drosophila melanogaster* embryo. *J Biol Chem* 2007;**282**:9127–42.

16. Kui Wong N, Easton RL, Panico M, Sutton-Smith M, Morrison JC, Lattanzio FA, Morris HR, Clark GF, Dell A, Patankar MS. Characterization of the oligosaccharides associated with the human ovarian tumor marker CA125. *J Biol Chem* 2003;**278**:28619–34.

17. Sutton-Smith M, Morris HR, Grewal PK, Hewitt JE, Bittner RE, Goldin E, Schiffmann R, Dell A. MS screening strategies: investigating the glycomes of knockout and myodystrophic mice and leukodystrophic human brains. *Biochem Soc Symp* 2002;**69**:105–15.

18. Bax M, Garcia-Vallejo JJ, Jang-Lee J, North SJ, Gilmartin TJ, Hernandez G, Crocker PR, Leffler H, Head SR, Haslam SM, Dell A, van Kooyk Y. Dendritic cell maturation results in pronounced changes in glycan expression affecting recognition by siglecs and galectins. *J Immunol* 2007;**179**:8216–24.

19. Hemmoranta H, Satomaa T, Blomqvist M, Heiskanen A, Aitio O, Saarinen J, Natunen J, Partanen J, Laine J, Jaatinen T. N-glycan structures and associated gene expression reflect the characteristic N-glycosylation pattern of human hematopoietic stem and progenitor cells. *Exp Hematol* 2007;**35**:1279–92.

20. Naito Y, Takematsu H, Koyama S, Miyake S, Yamamoto H, Fujinawa R, Sugai M, Okuno Y, Tsujimoto G, Yamaji T, Hashimoto Y, Itohara S, Kawasaki T, Suzuki A, Kozutsumi Y. Germinal center marker GL7 probes activation-dependent repression of N-glycolylneuraminic acid, a sialic acid species involved in the negative modulation of B-cell activation. *Mol Cell Biol* 2007;**27**:3008–22.

21. Smith FI, Qu Q, Hong SJ, Kim KS, Gilmartin TJ, Head SR. Gene expression profiling of mouse postnatal cerebellar development using oligonucleotide microarrays designed to detect differences in glycoconjugate expression. *Gene Expr Patterns* 2005;**5**:740–9.

22. Yamamoto H, Takematsu H, Fujinawa R, Naito Y, Okuno Y, Tsujimoto G, Suzuki A, Kozutsumi Y. Correlation Index-Based Responsible-Enzyme Gene Screening (CIRES), a novel DNA microarray-based method for enzyme gene involved in glycan biosynthesis. *PLoS ONE* 2007;**2**:e1232.

23. Young Jr WW, Holcomb DR, Ten Hagen KG, Tabak LA. Expression of UDP-GalNAc:polypeptide N-acetylgalactosaminyltransferase isoforms in murine tissues determined by real-time PCR: a new view of a large family. *Glycobiology* 2003;**13**:549–57.

24. Ishii A, Ikeda T, Hitoshi S, Fujimoto I, Torii T, Sakuma K, Nakakita S, Hase S, Ikenaka K. Developmental changes in the expression of glycogenes and the content of N-glycans in the mouse cerebral cortex. *Glycobiology* 2007;**17**:261–76.

25. Coutinho PM, Deleury E, Davies GJ, Henrissat B. An evolving hierarchical family classification for glycosyltransferases. *J Mol Biol* 2003;**328**:307–17.

26. Coutinho PM, Henrissat B. Carbohydrate-active enzymes: an integrated database approach. In: Davies G, Gilbert HJ, Henrissat B, Svensson B, editors. *Recent advances in carbohydrate bioengineering*. Cambridge: The Royal Society of Chemistry; 1999, p. 3–12.

27. Kanehisa M, Araki M, Goto S, Hattori M, Hirakawa M, Itoh M, Katayama T, Kawashima S, Okuda S, Tokimatsu T, Yamanishi Y. KEGG for linking genomes to life and the environment. *Nucleic Acids Res* 2007;**36**:D480–4.

28. Kanehisa M, Goto S. KEGG: kyoto encyclopedia of genes and genomes. *Nucleic Acids Res* 2000;**28**:27–30.

29. Kanehisa M, Goto S, Hattori M, Aoki-Kinoshita KF, Itoh M, Kawashima S, Katayama T, Araki M, Hirakawa M. From genomics to chemical genomics: new developments in KEGG. *Nucleic Acids Res* 2006;**34**:D354–7.

30. Diehn M, Sherlock G, Binkley G, Jin H, Matese JC, Hernandez-Boussard T, Rees CA, Cherry JM, Botstein D, Brown PO, Alizadeh AA. SOURCE: a unified genomic resource of functional annotations, ontologies, and gene expression data. *Nucleic Acids Res* 2003;**31**:219–23.

31. Garcia-Vallejo JJ, Gringhuis SI, van Dijk W, van Die I. Gene expression analysis of glycosylation-related genes by real-time polymerase chain reaction. *Methods Mol Biol* 2006;**347**:187–209.

32. Garcia-Vallejo JJ, Van Dijk W, Van Het Hof B, Van Die I, Engelse MA, Van Hinsbergh VW, Gringhuis SI. Activation of human endothelial cells by tumor necrosis factor-alpha results in profound changes in the expression of glycosylation-related genes. *J Cell Physiol* 2006;**206**:203–10.

33. Varki A, Cummings RD, Esko JD, Freeze HH, Stanley P, Bertozzi CR, Hart GW, Etzler ME, editors. *Essentials of Glycobiology*. 2nd edn. NY: Cold Spring Harbor Laboratory Press Cold Spring Harbor; 2008.

34. Hediger MA, Rhoads DB. Molecular physiology of sodium-glucose cotransporters. *Physiol Rev* 1994;**74**:993–1026.

35. Wood IS, Trayhurn P. Glucose transporters (GLUT and SGLT): expanded families of sugar transport proteins. *Br J Nutr* 2003;**89**:3–9.

36. Thorens B. Glucose transporters in the regulation of intestinal, renal, and liver glucose fluxes. *Am J Physiol* 1996;**270**:G541–53.

37. Bell GI, Burant CF, Takeda J, Gould GW. Structure and function of mammalian facilitative sugar transporters. *J Biol Chem* 1993;**268**:19161–4.

38. Mueckler M, Caruso C, Baldwin SA, Panico M, Blench I, Morris HR, Allard WJ, Lienhard GE, Lodish HF. Sequence and structure of a human glucose transporter. *Science* 1985;**229**:941–5.

39. Henderson PJ. The 12-transmembrane helix transporters. *Curr Opin Cell Biol* 1993;**5**:708–21.

40. Joost HG, Thorens B. The extended GLUT-family of sugar/polyol transport facilitators: nomenclature, sequence characteristics, and potential function of its novel members (review). *Mol Membr Biol* 2001;**18**:247–56.

41. Watson RT, Pessin JE. Subcellular compartmentalization and trafficking of the insulin-responsive glucose transporter, GLUT4. *Exp Cell Res* 2001;**271**:75–83.

42. Panneerselvam K, Freeze HH. Mannose enters mammalian cells using a specific transporter that is insensitive to glucose. *J Biol Chem* 1996;**271**:9417–21.

43. Leck JR, Wiese TJ. Purification and characterization of the L-fucose transporter. *Protein Expr Purif* 2004;**37**:288–93.

44. Cardenas ML, Cornish-Bowden A, Ureta T. Evolution and regulatory role of the hexokinases. *Biochim Biophys Acta* 1998;**1401**:242–64.

45. Munster AK, Eckhardt M, Potvin B, Muhlenhoff M, Stanley P, Gerardy-Schahn R. Mammalian cytidine 5′-monophosphate N-acetylneuraminic acid synthetase: a nuclear protein with evolutionarily conserved structural motifs. *Proc Natl Acad Sci USA* 1998;**95**:9140–5.

46. Hirschberg CB, Robbins PW, Abeijon C. Transporters of nucleotide sugars, ATP, and nucleotide sulfate in the endoplasmic reticulum and Golgi apparatus. *Annu Rev Biochem* 1998;**67**:49–69.

47. Munster AK, Weinhold B, Gotza B, Muhlenhoff M, Frosch M, Gerardy-Schahn R. Nuclear localization signal of murine CMP-Neu5Ac synthetase includes residues required for both nuclear targeting and enzymatic activity. *J Biol Chem* 2002;**277**:19688–96.

48. Helenius A, Aebi M. Roles of N-linked glycans in the endoplasmic reticulum. *Annu Rev Biochem* 2004;**73**:1019–49.

49. Tiede A, Bastisch I, Schubert J, Orlean P, Schmidt RE. Biosynthesis of glycosylphosphatidylinositols in mammals and unicellular microbes. *Biol Chem* 1999;**380**:503–23.

50. Vosseller K, Wells L, Hart GW. Nucleocytoplasmic O-glycosylation: O-GlcNAc and functional proteomics. *Biochimie* 2001;**83**:575–81.

51. Weigel PH, DeAngelis PL. Hyaluronan synthases: a decade-plus of novel glycosyltransferases. *J Biol Chem* 2007;**282**:36777–81.

52. Orlean P. Enzymes that recognize dolichols participate in three glycosylation pathways and are required for protein secretion. *Biochem Cell Biol* 1992;**70**:438–47.

53. Herscovics A, Orlean P. Glycoprotein biosynthesis in yeast. *FASEB J* 1993;**7**:540–50.

54. Saier Jr MH, Tran CV, Barabote RD. TCDB: the Transporter Classification Database for membrane transport protein analyses and information. *Nucleic Acids Res* 2006;**34**:D181–6.

55. Gao XD, Nishikawa A, Dean N. Physical interactions between the Alg1, Alg2, and Alg11 mannosyltransferases of the endoplasmic reticulum. *Glycobiology* 2004;**14**:559–70.

56. Moremen K. a-Mannosidases in Asparagine-linked oligosaccharide processing and catabolism. In: Ernst B, Hart G, Sinay P, editors. *Oligosaccharides in Chemistry and Biology: A Comprehensive Handbook*, vol. II: Biology of Saccharides, Part 1: Biosynthesis of Glycoconjugates. New York, NY: Wiley and Sons, Inc.; 2000, p. 81–117.

57. Moremen KW, Trimble RB, Herscovics A. Glycosidases of the asparagine-linked oligosaccharide processing pathway. *Glycobiology* 1994;**4**:113–25.

58. Wu Y, Swulius MT, Moremen KW, Sifers RN. Elucidation of the molecular logic by which misfolded alpha 1-antitrypsin is preferentially selected for degradation. *Proc Natl Acad Sci USA* 2003;**100**:8229–34.

59. Schachter H. The joys of HexNAc. The synthesis and function of *N*- and *O*-glycan branches. *Glycoconj J* 2000;**17**:465–83.

60. Spiro RG. Protein glycosylation: nature, distribution, enzymatic formation, and disease implications of glycopeptide bonds. *Glycobiology* 2002;**12**:43R–56R.

61. Hanisch FG. O-glycosylation of the mucin type. *Biol Chem* 2001;**382**:143–9.

62. Ten Hagen KG, Fritz TA, Tabak LA. All in the family: the UDP-GalNAc:polypeptide *N*-acetylgalactosaminyltransferases. *Glycobiology* 2003;**13**:1R–16R.

63. Esko JD, Selleck SB. Order out of chaos: assembly of ligand binding sites in heparan sulfate. *Annu Rev Biochem* 2002;**71**:435–71.

64. Wilson IB. The never-ending story of peptide *O*-xylosyltransferase. *Cell Mol Life Sci* 2004;**61**:794–809.

65. Taylor KR, Gallo RL. Glycosaminoglycans and their proteoglycans: host-associated molecular patterns for initiation and modulation of inflammation. *FASEB J* 2006;**20**:9–22.

66. Busse M, Feta A, Presto J, Wilen M, Gronning M, Kjellen L, Kusche-Gullberg M. Contribution of EXT1, EXT2, and EXTL3 to heparan sulfate chain elongation. *J Biol Chem* 2007;**282**:32802–10.

67. Izumikawa T, Koike T, Shiozawa S, Sugahara K, Tamura J, Kitagawa H. Identification of chondroitin sulfate glucuronyltransferase as chondroitin synthase-3 involved in chondroitin polymerization: chondroitin polymerization is achieved by multiple enzyme complexes consisting of chondroitin synthase family members. *J Biol Chem* 2008;**283**:11396–406.

68. Sugahara K, Kitagawa H. Heparin and heparan sulfate biosynthesis. *IUBMB Life* 2002;**54**:163–75.

69. Silbert JE, Sugumaran G. Biosynthesis of chondroitin/dermatan sulfate. *IUBMB Life* 2002;**54**:177–86.

70. Itano N, Kimata K. Mammalian hyaluronan synthases. *IUBMB Life* 2002;**54**:195–9.

71. Okajima T, Matsuura A, Matsuda T. Biological functions of glycosyltransferase genes involved in *O*-fucose glycan synthesis. *J Biochem* 2008;**144**:1–6.

72. Panin VM, Shao L, Lei L, Moloney DJ, Irvine KD, Haltiwanger RS. Notch ligands are substrates for protein *O*-fucosyltransferase-1 and Fringe. *J Biol Chem* 2002;**277**:29945–52.

73. Haines N, Irvine KD. Glycosylation regulates Notch signalling. *Nat Rev Mol Cell Biol* 2003;**4**:786–97.

74. Shao L, Moloney DJ, Haltiwanger R. Fringe modifies *O*-fucose on mouse Notch1 at epidermal growth factor-like repeats within the ligand-binding site and the Abruptex region. *J Biol Chem* 2003;**278**:7775–82.

75. Shao L, Haltiwanger RS. *O*-fucose modifications of epidermal growth factor-like repeats and thrombospondin type 1 repeats: unusual modifications in unusual places. *Cell Mol Life Sci* 2003;**60**:241–50.

76. Stanley P. Regulation of Notch signaling by glycosylation. *Curr Opin Struct Biol* 2007;**17**:530–5.

77. Shi S, Ge C, Luo Y, Hou X, Haltiwanger RS, Stanley P. The threonine that carries fucose, but not fucose, is required for Cripto to facilitate Nodal signaling. *J Biol Chem* 2007;**282**:20133–41.

78. Stanley P. Glucose: a novel regulator of notch signaling. *ACS Chem Biol* 2008;**3**:210–13.

79. Nishimura H, Yamashita S, Zeng Z, Walz DA, Iwanaga S. Evidence for the existence of O-linked sugar chains consisting of glucose and xylose in bovine thrombospondin. *J Biochem* 1992;**111**:460–4.

80. Loriol C, Dupuy F, Rampal R, Dlugosz MA, Haltiwanger RS, Maftah A, Germot A. Molecular evolution of protein O-fucosyltransferase genes and splice variants. *Glycobiology* 2006;**16**:736–47.

81. Luo Y, Koles K, Vorndam W, Haltiwanger RS, Panin VM. Protein O-fucosyltransferase 2 adds O-fucose to thrombospondin type 1 repeats. *J Biol Chem* 2006;**281**:9393–9.

82. Luo Y, Nita-Lazar A, Haltiwanger RS. Two distinct pathways for O-fucosylation of epidermal growth factor-like or thrombospondin type 1 repeats. *J Biol Chem* 2006;**281**:9385–92.

83. Sato T, Sato M, Kiyohara K, Sogabe M, Shikanai T, Kikuchi N, Togayachi A, Ishida H, Ito H, Kameyama A, Gotoh M, Narimatsu H. Molecular cloning and characterization of a novel human beta1,3-glucosyltransferase, which is localized at the endoplasmic reticulum and glucosylates O-linked fucosylglycan on thrombospondin type 1 repeat domain. *Glycobiology* 2006;**16**:1194–206.

84. Chiba A, Matsumura K, Yamada H, Inazu T, Shimizu T, Kusunoki S, Kanazawa I, Kobata A, Endo T. Structures of sialylated O-linked oligosaccharides of bovine peripheral nerve alpha-dystroglycan. The role of a novel O-mannosyl-type oligosaccharide in the binding of alpha-dystroglycan with laminin. *J Biol Chem* 1997;**272**:2156–62.

85. Barresi R, Campbell KP. Dystroglycan: from biosynthesis to pathogenesis of human disease. *J Cell Sci* 2006;**119**:199–207.

86. Manya H, Chiba A, Yoshida A, Wang X, Chiba Y, Jigami Y, Margolis RU, Endo T. Demonstration of mammalian protein O-mannosyltransferase activity: coexpression of POMT1 and POMT2 required for enzymatic activity. *Proc Natl Acad Sci USA* 2004;**101**:500–5.

87. Manya H, Sakai K, Kobayashi K, Taniguchi K, Kawakita M, Toda T, Endo T. Loss-of-function of an N-acetylglucosaminyltransferase, POMGnT1, in muscle-eye-brain disease. *Biochem Biophys Res Commun* 2003;**306**:93–7.

88. Jimenez-Mallebrera C, Torelli S, Feng L, Kim J, Godfrey C, Clement E, Mein R, Abbs S, Brown SC, Campbell KP, Kroger S, Talim B, Topaloglu H, Quinlivan R, Roper H, Childs AM, Kinali M, Sewry CA, Muntoni F. A comparative study of alpha-dystroglycan glycosylation in dystroglycanopathies suggests that the hypoglycosylation of alpha-dystroglycan does not consistently correlate with clinical severity. *Brain Pathol* 2008.

89. Godfrey C, Clement E, Mein R, Brockington M, Smith J, Talim B, Straub V, Robb S, Quinlivan R, Feng L, Jimenez-Mallebrera C, Mercuri E, Manzur AY, Kinali M, Torelli S, Brown SC, Sewry CA, Bushby K, Topaloglu H, North K, Abbs S, Muntoni F. Refining genotype phenotype correlations in muscular dystrophies with defective glycosylation of dystroglycan. *Brain* 2007;**130**:2725–35.

90. Lahiri S, Futerman AH. The metabolism and function of sphingolipids and glycosphingolipids. *Cell Mol Life Sci* 2007;**64**:2270–84.

91. Sandhoff K, Kolter T. Biosynthesis and degradation of mammalian glycosphingolipids. *Philos Trans R Soc Lond B Biol Sci* 2003;**358**:847–61.

92. Furukawa K. Recent progress in the analysis of gangliosides biosynthesis. *Nagoya J Med Sci* 1998;**61**:27–35.

93. Chatterjee S, Pandey A. The Yin and Yang of lactosylceramide metabolism: implications in cell function. *Biochim Biophys Acta* 2008;**1780**:370–82.

94. Orlean P, Menon AK. Thematic review series: lipid posttranslational modifications. GPI anchoring of protein in yeast and mammalian cells, or: how we learned to stop worrying and love glycophospholipids. *J Lipid Res* 2007;**48**:993–1011.

95. Lomako J, Lomako WM, Whelan WJ. Glycogenin: the primer for mammalian and yeast glycogen synthesis. *Biochim Biophys Acta* 2004;**1673**:45–55.

96. Hurley TD, Stout S, Miner E, Zhou J, Roach PJ. Requirements for catalysis in mammalian glycogenin. *J Biol Chem* 2005;**280**:23892–9.

97. Browner MF, Nakano K, Bang AG, Fletterick RJ. Human muscle glycogen synthase cDNA sequence: a negatively charged protein with an asymmetric charge distribution. *Proc Natl Acad Sci USA* 1989;**86**:1443–7.

98. Westphal SA, Nuttall FQ. Comparative characterization of human and rat liver glycogen synthase. *Arch Biochem Biophys* 1992;**292**:479–86.

99. Greenberg CC, Jurczak MJ, Danos AM, Brady MJ. Glycogen branches out: new perspectives on the role of glycogen metabolism in the integration of metabolic pathways. *Am J Physiol Endocrinol Metab* 2006;**291**:E1–8.

100. Matsuishi T, Yoshino M, Terasawa K, Nonaka I. Childhood acid maltase deficiency. A clinical, biochemical, and morphologic study of three patients. *Arch Neurol* 1984;**41**:47–52.

101. Roth J. Protein glycosylation in the endoplasmic reticulum and the Golgi apparatus and cell type-specificity of cell surface glycoconjugate expression: analysis by the protein A-gold and lectin-gold techniques. *Histochem Cell Biol* 1996;**106**:79–92.

102. Young RA, Davis RW. Efficient isolation of genes by using antibody probes. *Proc Natl Acad Sci USA* 1983;**80**:1194–8.

103. Shaper NL, Shaper JH, Meuth JL, Fox JL, Chang H, Kirsch IR, Hollis GF. Bovine galactosyltransferase: identification of a clone by direct immunological screening of a cDNA expression library. *Proc Natl Acad Sci USA* 1986;**83**:1573–7.

104. Weinstein J, Lee EU, McEntee K, Lai PH, Paulson JC. Primary structure of beta-galactoside alpha 2,6-sialyltransferase. Conversion of membrane-bound enzyme to soluble forms by cleavage of the NH$_2$-terminal signal anchor. *J Biol Chem* 1987;**262**:17735–43.

105. Joziasse DH, Shaper JH, Van den Eijnden DH, Van Tunen AJ, Shaper NL. Bovine alpha 1→3-galactosyltransferase: isolation and characterization of a cDNA clone. Identification of homologous sequences in human genomic DNA. *J Biol Chem* 1989;**264**:14290–7.

106. Narimatsu H, Sinha S, Brew K, Okayama H, Qasba PK. Cloning and sequencing of cDNA of bovine *N*-acetylglucosamine (beta 1–4)galactosyltransferase. *Proc Natl Acad Sci USA* 1986;**83**:4720–4.

107. Appert HE, Rutherford TJ, Tarr GE, Wiest JS, Thomford NR, McCorquodale DJ. Isolation of a cDNA coding for human galactosyltransferase. *Biochem Biophys Res Commun* 1986;**139**:163–8.

108. Frohman MA, Dush MK, Martin GR. Rapid production of full-length cDNAs from rare transcripts: amplification using a single gene-specific oligonucleotide primer. *Proc Natl Acad Sci USA* 1988;**85**:8998–9002.

109. Lee CC, Wu XW, Gibbs RA, Cook RG, Muzny DM, Caskey CT. Generation of cDNA probes directed by amino acid sequence: cloning of urate oxidase. *Science* 1988;**239**:1288–91.

110. Moremen KW. Isolation of a rat liver Golgi mannosidase II clone by mixed oligonucleotide-primed amplification of cDNA. *Proc Natl Acad Sci USA* 1989;**86**:5276–80.

111. Yamamoto F, Marken J, Tsuji T, White T, Clausen H, Hakomori S. Cloning and characterization of DNA complementary to human UDP-GalNAc: Fuc alpha 1–2Gal alpha 1→3GalNAc transferase (histo-blood group A transferase) mRNA. *J Biol Chem* 1990;**265**:1146–51.

112. Sarkar M, Hull E, Nishikawa Y, Simpson RJ, Moritz RL, Dunn R, Schachter H. Molecular cloning and expression of cDNA encoding the enzyme that controls conversion of high-mannose to hybrid and complex *N*-glycans: UDP-*N*-acetylglucosamine: alpha-3-D-mannoside beta-1,2-*N*-acetylglucosaminyltransferase I. *Proc Natl Acad Sci USA* 1991;**88**:234–8.

113. Wen DX, Livingston BD, Medzihradszky KF, Kelm S, Burlingame AL, Paulson JC. Primary structure of Gal beta 1,3(4)GlcNAc alpha 2,3-sialyltransferase determined by mass spectrometry sequence analysis and molecular cloning. Evidence for a protein motif in the sialyltransferase gene family. *J Biol Chem* 1992;**267**:21011–19.

114. Shoreibah M, Perng GS, Adler B, Weinstein J, Basu R, Cupples R, Wen D, Browne JK, Buckhaults P, Fregien N, Pierce M. Isolation, characterization, and expression of a cDNA encoding *N*-acetylglucosaminyltransferase V. *J Biol Chem* 1993;**268**:15381–5.

115. Homa FL, Hollander T, Lehman DJ, Thomsen DR, Elhammer AP. Isolation and expression of a cDNA clone encoding a bovine UDP-GalNAc:polypeptide *N*-acetylgalactosaminyltransferase. *J Biol Chem* 1993;**268**:12609–16.

116. White T, Bennett EP, Takio K, Sorensen T, Bonding N, Clausen H. Purification and cDNA cloning of a human UDP-*N*-acetyl-alpha-*D*-galactosamine:polypeptide *N*-acetylgalactosaminyltransferase. *J Biol Chem* 1995;**270**:24156–65.

117. D'Agostaro GA, Zingoni A, Moritz RL, Simpson RJ, Schachter H, Bendiak B. Molecular cloning and expression of cDNA encoding the rat UDP-*N*-acetylglucosamine:alpha-6-*D*-mannoside beta-1,2-*N*-acetylglucosaminyltransferase II. *J Biol Chem* 1995;**270**:15211–21.

118. Lal A, Schutzbach JS, Forsee WT, Neame PJ, Moremen KW. Isolation and expression of murine and rabbit cDNAs encoding an alpha 1,2-mannosidase involved in the processing of asparagine-linked oligosaccharides. *J Biol Chem* 1994;**269**:9872–81.

119. Ju T, Brewer K, D'Souza A, Cummings RD, Canfield WM. Cloning and expression of human core 1 beta1,3-galactosyltransferase. *J Biol Chem* 2002;**277**:178–86.

120. Herscovics A, Schneikert J, Athanassiadis A, Moremen KW. Isolation of a mouse Golgi mannosidase cDNA, a member of a gene family conserved from yeast to mammals. *J Biol Chem* 1994;**269**:9864–71.

121. Gonzalez DS, Karaveg K, Vandersall-Nairn AS, Lal A, Moremen KW. Identification, expression, and characterization of a cDNA encoding human endoplasmic reticulum mannosidase I, the enzyme that catalyzes the first mannose trimming step in mammalian Asn-linked oligosaccharide biosynthesis. *J Biol Chem* 1999;**274**:21375–86.

122. Livingston BD, Paulson JC. Polymerase chain reaction cloning of a developmentally regulated member of the sialyltransferase gene family. *J Biol Chem* 1993;**268**:11504–7.

123. Kitagawa H, Paulson JC. Cloning and expression of human Gal beta 1,3(4)GlcNAc alpha 2,3-sialyltransferase. *Biochem Biophys Res Commun* 1993;**194**:375–82.

124. Kitagawa H, Paulson JC. Differential expression of five sialyltransferase genes in human tissues. *J Biol Chem* 1994;**269**:17872–8.

125. Kitagawa H, Paulson JC. Cloning of a novel alpha 2,3-sialyltransferase that sialylates glycoprotein and glycolipid carbohydrate groups. *J Biol Chem* 1994;**269**:1394–401.

126. Larsen RD, Rajan VP, Ruff MM, Kukowska-Latallo J, Cummings RD, Lowe JB. Isolation of a cDNA encoding a murine UDPgalactose:beta-*D*-galactosyl- 1,4-*N*-acetyl-*D*-glucosaminide alpha-1,3-galactosyltransferase: expression cloning by gene transfer. *Proc Natl Acad Sci USA* 1989;**86**:8227–31.

127. Kukowska-Latallo JF, Larsen RD, Nair RP, Lowe JB. A cloned human cDNA determines expression of a mouse stage-specific embryonic antigen and the Lewis blood group alpha(1,3/1,4)fucosyltransferase. *Genes Dev* 1990;**4**:1288–303.

128. Smith PL, Lowe JB. Molecular cloning of a murine *N*-acetylgalactosamine transferase cDNA that determines expression of the T lymphocyte-specific CT oligosaccharide differentiation antigen. *J Biol Chem* 1994;**269**:15162–71.

129. Nagata Y, Yamashiro S, Yodoi J, Lloyd KO, Shiku H, Furukawa K. Expression cloning of beta 1,4 *N*-acetylgalactosaminyltransferase cDNAs that determine the expression of GM2 and GD2 gangliosides. *J Biol Chem* 1992;**267**:12082–9.

130. Misago M, Liao YF, Kudo S, Eto S, Mattei MG, Moremen KW, Fukuda MN. Molecular cloning and expression of cDNAs encoding human alpha-mannosidase II and a previously unrecognized alpha-mannosidase IIx isozyme. *Proc Natl Acad Sci USA* 1995;**92**:11766–70.

131. Lowe JB, Kukowska-Latallo JF, Nair RP, Larsen RD, Marks RM, Macher BA, Kelly RJ, Ernst LK. Molecular cloning of a human fucosyltransferase gene that determines expression of the Lewis x and VIM-2 epitopes but not ELAM-1-dependent cell adhesion. *J Biol Chem* 1991;**266**:17467–77.

132. Weston BW, Smith PL, Kelly RJ, Lowe JB. Molecular cloning of a fourth member of a human alpha (1,3)fucosyltransferase gene family. Multiple homologous sequences that determine expression of the Lewis x, sialyl Lewis x, and difucosyl sialyl Lewis x epitopes. *J Biol Chem* 1992;**267**:24575–84.

133. Weston BW, Nair RP, Larsen RD, Lowe JB. Isolation of a novel human alpha (1,3)fucosyltransferase gene and molecular comparison to the human Lewis blood group alpha (1,3/1,4)fucosyltransferase gene. Syntenic, homologous, nonallelic genes encoding enzymes with distinct acceptor substrate specificities. *J Biol Chem* 1992;**267**:4152–60.

134. Natsuka S, Gersten KM, Zenita K, Kannagi R, Lowe JB. Molecular cloning of a cDNA encoding a novel human leukocyte alpha-1,3-fucosyltransferase capable of synthesizing the sialyl Lewis x determinant. *J Biol Chem* 1994;**269**:16789–94.

135. Rouquier S, Lowe JB, Kelly RJ, Fertitta AL, Lennon GG, Giorgi D. Molecular cloning of a human genomic region containing the H blood group alpha(1,2)fucosyltransferase gene and two H locus-related DNA restriction fragments. Isolation of a candidate for the human Secretor blood group locus. *J Biol Chem* 1995;**270**:4632–9.

136. Shaper NL, Hollis GF, Douglas JG, Kirsch IR, Shaper JH. Characterization of the full length cDNA for murine beta-1,4-galactosyltransferase, Novel features at the, 5' -end predict two translational start sites at two in-frame AUGs. *J Biol Chem* 1988;**263**:10420–8.

137. Gersten KM, Natsuka S, Trinchera M, Petryniak B, Kelly RJ, Hiraiwa N, Jenkins NA, Gilbert DJ, Copeland NG, Lowe JB. Molecular cloning, expression, chromosomal assignment, and tissue-specific expression of a murine alpha-(1,3)-fucosyltransferase locus corresponding to the human ELAM-1 ligand fucosyl transferase. *J Biol Chem* 1995;**270**:25047–56.

138. Domino SE, Zhang L, Lowe JB. Molecular cloning, genomic mapping, and expression of two secretor blood group alpha (1,2)fucosyltransferase genes differentially regulated in mouse uterine epithelium and gastrointestinal tract. *J Biol Chem* 2001;**276**:23748–56.

139. Hennet T, Dinter A, Kuhnert P, Mattu TS, Rudd PM, Berger EG. Genomic cloning and expression of three murine UDP-galactose: beta-*N*-acetylglucosamine beta1,3-galactosyltransferase genes. *J Biol Chem* 1998;**273**:58–65.

140. Amado M, Almeida R, Carneiro F, Levery SB, Holmes EH, Nomoto M, Hollingsworth MA, Hassan H, Schwientek T, Nielsen PA, Bennett EP, Clausen H. A family of human beta3-galactosyltransferases. Characterization of four members of a UDP-galactose:beta-*N*-acetyl-glucosamine/beta-nacetyl-galactosamine beta-1,3-galactosyltransferase family. *J Biol Chem* 1998;**273**:12770–8.

141. Park C, Meng L, Stanton LH, Collins RE, Mast SW, Yi X, Strachan H, Moremen KW. Characterization of a human core-specific lysosomal {alpha}1,6-mannosidase involved in *N*-glycan catabolism. *J Biol Chem* 2005;**280**:37204–16.

142. Huffaker TC, Robbins PW. Temperature-sensitive yeast mutants deficient in asparagine-linked glycosylation. *J Biol Chem* 1982;**257**:3203–10.

143. Huffaker TC, Robbins PW. Yeast mutants deficient in protein glycosylation. *Proc Natl Acad Sci USA* 1983;**80**:7466–70.

144. Runge KW, Huffaker TC, Robbins PW. Two yeast mutations in glucosylation steps of the asparagine glycosylation pathway. *J Biol Chem* 1984;**259**:412–17.

145. Runge KW, Robbins PW. A new yeast mutation in the glucosylation steps of the asparagine-linked glycosylation pathway. Formation of a novel asparagine-linked oligosaccharide containing two glucose residues. *J Biol Chem* 1986;**261**:15582–90.

146. Hirose S, Mohney RP, Mutka SC, Ravi L, Singleton DR, Perry G, Tartakoff AM, Medof ME. Derivation and characterization of glycoinositol-phospholipid anchor-defective human K562 cell clones. *J Biol Chem* 1992;**267**:5272–8.

147. Mohney RP, Knez JJ, Ravi L, Sevlever D, Rosenberry TL, Hirose S, Medof ME. Glycoinositol phospholipid anchor-defective K562 mutants with biochemical lesions distinct from those in Thy-1- murine lymphoma mutants. *J Biol Chem* 1994;**269**:6536–42.

148. Takahashi M, Inoue N, Ohishi K, Maeda Y, Nakamura N, Endo Y, Fujita T, Takeda J, Kinoshita T. PIG-B, a membrane protein of the endoplasmic reticulum with a large lumenal domain, is involved in transferring the third mannose of the GPI anchor. *EMBO J* 1996;**15**:4254–61.

149. Takeda J, Kinoshita T. GPI-anchor biosynthesis. *Trends Biochem Sci* 1995;**20**:367–71.

150. Taylor ME, Drickamer K. *Introduction to Glycobiology*. Second edn. Oxford: Oxford University Press; 2006.

151. Kanehisa M, Goto S, Kawashima S, Nakaya A. The KEGG databases at GenomeNet. *Nucleic Acids Res* 2002;**30**:42–6.

152. Wheeler DL, Barrett T, Benson DA, Bryant SH, Canese K, Chetvernin V, Church DM, DiCuccio M, Edgar R, Federhen S, Geer LY, Kapustin Y, Khovayko O, Landsman D, Lipman DJ, Madden TL, Maglott DR, Ostell J, Miller V, Pruitt KD, Schuler GD, Sequeira E, Sherry ST, Sirotkin K, Souvorov A, Starchenko G, Tatusov RL, Tatusova TA, Wagner L, Yaschenko E. Database resources of the National Center for Biotechnology Information. *Nucleic Acids Res* 2007;**35**:D5–D12.

153. Comelli EM, Amado M, Head SR, Paulson JC. Custom microarray for glycobiologists: considerations for glycosyltransferase gene expression profiling. *Biochem Soc Symp* 2002;**69**:135–42.

154. Alwine JC, Kemp DJ, Stark GR. Method for detection of specific RNAs in agarose gels by transfer to diazobenzyloxymethyl-paper and hybridization with DNA probes. *Proc Natl Acad Sci USA* 1977;**74**:5350–4.

155. Sato T, Furukawa K, Bakker H, Van den Eijnden DH, Van Die I. Molecular cloning of a human cDNA encoding beta-1,4-galactosyltransferase with 37% identity to mammalian UDP-Gal:GlcNAc beta-1,4-galactosyltransferase. *Proc Natl Acad Sci USA* 1998;**95**:472–7.

156. Shaper NL, Wright WW, Shaper JH. Murine beta 1,4-galactosyltransferase: both the amounts and structure of the mRNA are regulated during spermatogenesis. *Proc Natl Acad Sci USA* 1990;**87**:791–5.

157. Takashima S, Ishida HK, Inazu T, Ando T, Ishida H, Kiso M, Tsuji S, Tsujimoto M. Molecular cloning and expression of a sixth type of alpha 2,8-sialyltransferase (ST8Sia VI) that sialylates *O*-glycans. *J Biol Chem* 2002;**277**:24030–8.

158. Lo NW, Shaper JH, Pevsner J, Shaper NL. The expanding beta 4-galactosyltransferase gene family: messages from the databanks. *Glycobiology* 1998;**8**:517–26.

159. Lin B, Saito M, Sakakibara Y, Hayashi Y, Yanagisawa M, Iwamori M. Characterization of three members of murine alpha 1,2-fucosyltransferases: Change in the expression of the Se gene in the intestine of mice after administration of microbes. *Arch Biochem Biophys* 2001;**388**:207–15.

160. Nakamura N, Yamakawa N, Sato T, Tojo H, Tachi C, Furukawa K. Differential gene expression of beta-1,4-galactosyltransferases I, II and V during mouse brain development. *J Neurochem* 2001;**76**:29–38.

161. Granovsky M, Fode C, Warren CE, Campbell RM, Marth JD, Pierce M, Fregien N, Dennis JW. GlcNAc-transferase V and core 2 GlcNAc-transferase expression in the developing mouse embryo. *Glycobiology* 1995;**5**:797–806.

162. Yan M, Cheng C, Ding F, Jiang J, Gao L, Xia C, Shen A. The expression patterns of beta1,4 galactosyltransferase I and V mRNAs, and Galbeta1–4GlcNAc group in rat gastrocnemius muscles post sciatic nerve injury. *Glycoconj J* 2008;**25**:685–701.

163. Tremblay LO, Nagy Kovacs E, Daniels E, Wong NK, Sutton-Smith M, Morris HR, Dell A, Marcinkiewicz E, Seidah NG, McKerlie C, Herscovics A. Respiratory distress and neonatal lethality in mice lacking Golgi alpha1,2-mannosidase IB involved in N-glycan maturation. *J Biol Chem* 2007;**282**:2558–66.

164. Chen SH, Zhou S, Tan J, Schachter H. Transcriptional regulation of the human UDP-GlcNAc:alpha-6-D-mannoside beta-1–2-*N*-acetylglucosaminyltransferase II gene (MGAT2) which controls complex *N*-glycan synthesis. *Glycoconj J* 1998;**15**:301–8.

165. O'Hanlon TP, Lau KM, Wang XC, Lau JT. Tissue-specific expression of beta-galactoside alpha-2,6-sialyltransferase. Transcript heterogeneity predicts a divergent polypeptide. *J Biol Chem* 1989;**264**:17389–94.

166. Wang X, O'Hanlon TP, Young RF, Lau JT. Rat beta-galactoside alpha 2,6-sialyltransferase genomic organization: alternate promoters direct the synthesis of liver and kidney transcripts. *Glycobiology* 1990;**1**:25–31.

167. Lockhart DJ, Dong H, Byrne MC, Follettie MT, Gallo MV, Chee MS, Mittmann M, Wang C, Kobayashi M, Horton H, Brown EL. Expression monitoring by hybridization to high-density oligonucleotide arrays. *Nat Biotechnol* 1996;**14**:1675–80.

168. Nuwaysir EF, Huang W, Albert TJ, Singh J, Nuwaysir K, Pitas A, Richmond T, Gorski T, Berg JP, Ballin J, McCormick M, Norton J, Pollock T, Sumwalt T, Butcher L, Porter D, Molla M, Hall C, Blattner F, Sussman MR, Wallace RL, Cerrina F, Green RD. Gene expression analysis using oligonucleotide arrays produced by maskless photolithography. *Genome Res* 2002;**12**:1749–55.

169. Katagiri F, Glazebrook J. Overview of mRNA expression profiling using microarrays. *Curr Protoc Mol Biol* 2004. Chapter 22: Unit 22 24.

170. Bolstad BM, Irizarry RA, Astrand M, Speed TP. A comparison of normalization methods for high density oligonucleotide array data based on variance and bias. *Bioinformatics* 2003;**19**:185–93.

171. Quackenbush J. Microarray data normalization and transformation. *Nat Genet* 2002;**32**(Suppl): 496–501.

172. Page GP, Zakharkin SO, Kim K, Mehta T, Chen L, Zhang K. Microarray analysis. *Methods Mol Biol* 2007;**404**:409–30.

173. Simon R. Challenges of microarray data and the evaluation of gene expression profile signatures. *Cancer Invest* 2008;**26**:327–32.

174. Simon R. Microarray-based expression profiling and informatics. *Curr Opin Biotechnol* 2008;**19**:26–9.

175. Yang JY. Microarrays—planning your experiment. *Methods Mol Med* 2008;**141**:71–85.

176. Tusher VG, Tibshirani R, Chu G. Significance analysis of microarrays applied to the ionizing radiation response. *Proc Natl Acad Sci USA* 2001;**98**:5116–21.

177. Gautier L, Cope L, Bolstad BM, Irizarry RA. Affy—analysis of Affymetrix GeneChip data at the probe level. *Bioinformatics* 2004;**20**:307–15.

178. Mutwil M, Obro J, Willats WG, Persson S. GeneCAT—novel webtools that combine BLAST and co-expression analyses. *Nucleic Acids Res* 2008;**36**:W320–6.

179. Comelli EM, Sutton-Smith M, Yan Q, Amado M, Panico M, Gilmartin T, Whisenant T, Lanigan CM, Head SR, Goldberg D, Morris HR, Dell A, Paulson JC. Activation of murine CD4+ and CD8+ T lymphocytes leads to dramatic remodeling of *N*-linked glycans. *J Immunol* 2006;**177**:2431–40.

180. Tribulatti MV, Mucci J, Cattaneo V, Aguero F, Gilmartin T, Head SR, Campetella O. Galectin-8 induces apoptosis in the CD4(high)CD8(high) thymocyte subpopulation. *Glycobiology* 2007;**17**:1404–12.

181. Kobayashi M, Lee H, Schaffer L, Gilmartin TJ, Head SR, Takaishi S, Wang TC, Nakayama J, Fukuda M. A distinctive set of genes is upregulated during the inflammation-carcinoma sequence in mouse stomach infected by Helicobacter felis. *J Histochem Cytochem* 2007;**55**:263–74.

182. Desplats PA, Denny CA, Kass KE, Gilmartin T, Head SR, Sutcliffe JG, Seyfried TN, Thomas EA. Glycolipid and ganglioside metabolism imbalances in Huntington's disease. *Neurobiol Dis* 2007;**27**:265–77.

183. Sampathkumar SG, Jones MB, Meledeo MA, Campbell CT, Choi SS, Hida K, Gomutputra P, Sheh A, Gilmartin T, Head SR, Yarema KJ. Targeting glycosylation pathways and the cell cycle: sugar-dependent activity of butyrate-carbohydrate cancer prodrugs. *Chem Biol* 2006;**13**:1265–75.

184. Narayan S, Head SR, Gilmartin TJ, Dean B, Thomas EA. Evidence for disruption of sphingolipid metabolism in schizophrenia. *J Neurosci Res* 2008;**87**:278–88.

185. Marcos NT, Magalhaes A, Ferreira B, Oliveira MJ, Carvalho AS, Mendes N, Gilmartin T, Head SR, Figueiredo C, David L, Santos-Silva F, Reis CA. Helicobacter pylori induces beta3GnT5 in human gastric cell lines, modulating expression of the SabA ligand sialyl-Lewis x. *J Clin Invest* 2008;**118**:2325–36.

186. Diskin S, Kumar J, Cao Z, Schuman JS, Gilmartin T, Head SR, Panjwani N. Detection of differentially expressed glycogenes in trabecular meshwork of eyes with primary open-angle glaucoma. *Invest Ophthalmol Vis Sci* 2006;**47**:1491–9.

187. Kroes RA, Dawson G, Moskal JR. Focused microarray analysis of glyco-gene expression in human glioblastomas. *J Neurochem* 2007;**103**(Suppl 1):14–24.

188. Andre S, Sanchez-Ruderisch H, Nakagawa H, Buchholz M, Kopitz J, Forberich P, Kemmner W, Bock C, Deguchi K, Detjen KM, Wiedenmann B, von Knebel Doeberitz M, Gress TM, Nishimura S, Rosewicz S, Gabius HJ. Tumor suppressor p16INK4a—modulator of glycomic profile and galectin-1 expression to increase susceptibility to carbohydrate-dependent induction of anoikis in pancreatic carcinoma cells. *FEBS J* 2007;**274**:3233–56.

189. Hood LE, Hunkapiller MW, Smith LM. Automated DNA sequencing and analysis of the human genome. *Genomics* 1987;**1**:201–12.

190. Adams MD, Kelley JM, Gocayne JD, Dubnick M, Polymeropoulos MH, Xiao H, Merril CR, Wu A, Olde B, Moreno RF, et al. Complementary DNA sequencing: expressed sequence tags and human genome project. *Science* 1991;**252**:1651–6.

191. Boguski MS, Lowe TM, Tolstoshev CM. dbEST—database for "expressed sequence tags." *Nat Genet* 1993;**4**:332–3.

192. Velculescu VE, Zhang L, Vogelstein B, Kinzler KW. Serial analysis of gene expression. *Science* 1995;**270**:484–7.

193. Polyak K, Riggins GJ. Gene discovery using the serial analysis of gene expression technique: implications for cancer research. *J Clin Oncol* 2001;**19**:2948–58.

194. Yamamoto M, Wakatsuki T, Hada A, Ryo A. Use of serial analysis of gene expression (SAGE) technology. *J Immunol Methods* 2001;**250**:45–66.

195. Ryo A, Kondoh N, Wakatsuki T, Hada A, Yamamoto N, Yamamoto M. A modified serial analysis of gene expression that generates longer sequence tags by nonpalindromic cohesive linker ligation. *Anal Biochem* 2000;**277**:160–2.

196. Shiraki T, Kondo S, Katayama S, Waki K, Kasukawa T, Kawaji H, Kodzius R, Watahiki A, Nakamura M, Arakawa T, Fukuda S, Sasaki D, Podhajska A, Harbers M, Kawai J, Carninci P, Hayashizaki Y. Cap analysis gene expression for high-throughput analysis of transcriptional starting point and identification of promoter usage. *Proc Natl Acad Sci USA* 2003;**100**:15776–81.

197. de Hoon M, Hayashizaki Y. Deep cap analysis gene expression (CAGE): genome-wide identification of promoters, quantification of their expression, and network inference. *Biotechniques* 2008;**44**:627–32.

198. Silva Jr. WA, Covas DT, Panepucci RA, Proto-Siqueira R, Siufi JL, Zanette DL, Santos AR, Zago MA. The profile of gene expression of human marrow mesenchymal stem cells. *Stem Cells* 2003;**21**:661–9.

199. Chu TJ, Peters DG. Serial analysis of the vascular endothelial transcriptome under static and shear stress conditions. *Physiol Genomics* 2008;**34**:185–92.

200. Shendure JA, Porreca GJ, Church GM. Overview of DNA sequencing strategies. *Curr Protoc Mol Biol* 2008. Chapter 7: Unit 7 1.

201. Margulies M, Egholm M, Altman WE, Attiya S, Bader JS, Bemben LA, Berka J, Braverman MS, Chen YJ, Chen Z, Dewell SB, Du L, Fierro JM, Gomes XV, Godwin BC, He W, Helgesen S, Ho CH, Irzyk GP, Jando SC, Alenquer ML, Jarvie TP, Jirage KB, Kim JB, Knight JR, Lanza JR, Leamon JH, Lefkowitz SM, Lei M, Li J, Lohman KL, Lu H, Makhijani VB, McDade KE, McKenna MP, Myers EW, Nickerson E, Nobile JR, Plant R, Puc BP, Ronan MT, Roth GT, Sarkis GJ, Simons JF, Simpson JW, Srinivasan M, Tartaro KR, Tomasz A, Vogt KA, Volkmer GA, Wang SH, Wang Y, Weiner MP, Yu P, Begley RF, Rothberg JM. Genome sequencing in microfabricated high-density picolitre reactors. *Nature* 2005;**437**:376–80.

202. Ronaghi M, Uhlen M, Nyren P. A sequencing method based on real-time pyrophosphate. *Science* 1998;**281**:363–5.

203. Brenner S, Williams SR, Vermaas EH, Storck T, Moon K, McCollum C, Mao JI, Luo S, Kirchner JJ, Eletr S, DuBridge RB, Burcham T, Albrecht G. In vitro cloning of complex mixtures of DNA on microbeads: physical separation of differentially expressed cDNAs. *Proc Natl Acad Sci USA* 2000;**97**:1665–70.

204. Brenner S, Johnson M, Bridgham J, Golda G, Lloyd DH, Johnson D, Luo S, McCurdy S, Foy M, Ewan M, Roth R, George D, Eletr S, Albrecht G, Vermaas E, Williams SR, Moon K, Burcham T, Pallas M, DuBridge RB, Kirchner J, Fearon K, Mao J, Corcoran K. Gene expression analysis by massively parallel signature sequencing (MPSS) on microbead arrays. *Nat Biotechnol* 2000;**18**:630–4.

205. Barrett T, Suzek TO, Troup DB, Wilhite SE, Ngau WC, Ledoux P, Rudnev D, Lash AE, Fujibuchi W, Edgar R. NCBI GEO: mining millions of expression profiles—database and tools. *Nucleic Acids Res* 2005;**33**:D562–6.

206. Mullis KB. Target amplification for DNA analysis by the polymerase chain reaction. *Ann Biol Clin (Paris)* 1990;**48**:579–82.

207. Saiki RK, Scharf S, Faloona F, Mullis KB, Horn GT, Erlich HA, Arnheim N. Enzymatic amplification of beta-globin genomic sequences and restriction site analysis for diagnosis of sickle cell anemia. *Science* 1985;**230**:1350–4.

208. Powell LM, Wallis SC, Pease RJ, Edwards YH, Knott TJ, Scott J. A novel form of tissue-specific RNA processing produces apolipoprotein-B48 in intestine. *Cell* 1987;**50**:831–40.

209. Wong H, Muzik H, Groft LL, Lafleur MA, Matouk C, Forsyth PA, Schultz GA, Wall SJ, Edwards DR. Monitoring MMP and TIMP mRNA expression by RT-PCR. *Methods Mol Biol* 2001;**151**:305–20.

210. Wells GM, Catlin G, Cossins JA, Mangan M, Ward GA, Miller KM, Clements JM. Quantitation of matrix metalloproteinases in cultured rat astrocytes using the polymerase chain reaction with a multi-competitor cDNA standard. *Glia* 1996;**18**:332–40.

211. Vu HL, Troubetzkoy S, Nguyen HH, Russell MW, Mestecky J. A method for quantification of absolute amounts of nucleic acids by (RT)-PCR and a new mathematical model for data analysis. *Nucleic Acids Res* 2000;**28**:E18.

212. Cha RS, Thilly WG. Specificity, efficiency, and fidelity of PCR. *PCR Methods Appl* 1993;**3**:S18–29.

213. Zhao S, Consoli U, Arceci R, Pfeifer J, Dalton WS, Andreeff M. Semi-automated PCR method for quantitating MDR1 expression. *Biotechniques* 1996;**21**:726–31.

214. Kobayashi M. Expression profiling of glycosyltransferases using RT-PCR. *Methods Enzymol* 2006;**416**:129–40.

215. Li X, Bratton S, Radominska-Pandya A. Human UGT1A8 and UGT1A10 mRNA are expressed in primary human hepatocytes. *Drug Metab Pharmacokinet* 2007;**22**:152–61.

216. Yamamoto M, Yamamoto F, Luong TT, Williams T, Kominato Y. Expression profiling of 68 glycosyltransferase genes in 27 different human tissues by the systematic multiplex reverse transcription-polymerase chain reaction method revealed clustering of sexually related tissues in hierarchical clustering algorithm analysis. *Electrophoresis* 2003;**24**:2295–307.

217. Heid CA, Stevens J, Livak KJ, Williams PM. Real time quantitative PCR. *Genome Res* 1996;**6**:986–94.

218. Steuerwald N, Cohen J, Herrera RJ, Brenner CA. Analysis of gene expression in single oocytes and embryos by real-time rapid cycle fluorescence monitored RT-PCR. *Mol Hum Reprod* 1999;**5**:1034–9.

219. Gibson UE, Heid CA, Williams PM. A novel method for real time quantitative RT-PCR. *Genome Res* 1996;**6**:995–1001.

220. Schmittgen TD, Zakrajsek BA, Mills AG, Gorn V, Singer MJ, Reed MW. Quantitative reverse transcription-polymerase chain reaction to study mRNA decay: comparison of endpoint and real-time methods. *Anal Biochem* 2000;**285**:194–204.

221. Bustin SA. Absolute quantification of mRNA using real-time reverse transcription polymerase chain reaction assays. *J Mol Endocrinol* 2000;**25**:169–93.

222. Pfaffl MW. A new mathematical model for relative quantification in real-time RT-PCR. *Nucleic Acids Res* 2001;**29**:e45.

223. Pfaffl MW, Hageleit M. Validities of mRNA quantification using recombinant RNA and recombinant DNA external calibration curves in real-time RT-PCR. *Biotechnol Lett* 2001;**23**:275–82.

224. Livak KJ, Schmittgen TD. Analysis of relative gene expression data using real-time quantitative PCR and the 2(-Delta Delta C(T)) Method. *Methods* 2001;**25**:402–8.

225. Schmittgen TD, Lee EJ, Jiang J. High-throughput real-time PCR. *Methods Mol Biol* 2008;**429**:89–98.

226. Schmittgen TD, Livak KJ. Analyzing real-time PCR data by the comparative C(T) method. *Nat Protoc* 2008;**3**:1101–8.

227. Morrison TB, Weis JJ, Wittwer CT. Quantification of low-copy transcripts by continuous SYBR (R) green I monitoring during amplification. *Biotechniques* 1998;**24**:954–8.

228. Karlsen F, Steen HB, Nesland JM. SYBR green I DNA staining increases the detection sensitivity of viruses by polymerase chain reaction. *J Virol Methods* 1995;**55**:153–6.

229. Mao F, Leung WY, Xin X. Characterization of EvaGreen and the implication of its physicochemical properties for qPCR applications. *BMC Biotechnol* 2007;**7**:76.

230. Ahmad AI. BOXTO as a real-time thermal cycling reporter dye. *J Biosci* 2007;**32**:229–39.

231. Ponchel F. Real-time PCR using SYBR Green. In: Dorak MT, editor. *Real-time PCR*. New York: Taylor & Francis Group; 2006, p. 139–54.

232. Bernard PS, Ajioka RS, Kushner JP, Wittwer CT. Homogeneous multiplex genotyping of hemochromatosis mutations with fluorescent hybridization probes. *Am J Pathol* 1998;**153**:1055–61.

233. Tyagi S, Kramer FR. Molecular beacons: probes that fluoresce upon hybridization. *Nat Biotechnol* 1996;**14**:303–8.

234. Nazarenko I, Lowe B, Darfler M, Ikonomi P, Schuster D, Rashtchian A. Multiplex quantitative PCR using self-quenched primers labeled with a single fluorophore. *Nucleic Acids Res* 2002;**30**:e37.

235. Whitcombe D, Theaker J, Guy SP, Brown T, Little S. Detection of PCR products using self-probing amplicons and fluorescence. *Nat Biotechnol* 1999;**17**:804–7.

236. Bustin SA, editor. A-Z of Quantitative PCR. In: Tsigelny IF, editor. IUL Biotechnology Series. La Jolla: International University Line; 2004, p. 882.

237. Canales RD, Luo Y, Willey JC, Austermiller B, Barbacioru CC, Boysen C, Hunkapiller K, Jensen RV, Knight CR, Lee KY, Ma Y, Maqsodi B, Papallo A, Peters EH, Poulter K, Ruppel PL, Samaha RR, Shi L, Yang W, Zhang L, Goodsaid FM. Evaluation of DNA microarray results with quantitative gene expression platforms. *Nat Biotechnol* 2006;**24**:1115–22.

238. Provenzano M, Mocellin S. Complementary techniques: validation of gene expression data by quantitative real time PCR. *Adv Exp Med Biol* 2007;**593**:66–73.

239. Morrison T, Hurley J, Garcia J, Yoder K, Katz A, Roberts D, Cho J, Kanigan T, Ilyin SE, Horowitz D, Dixon JM, Brenan CJ. Nanoliter high throughput quantitative PCR. *Nucleic Acids Res* 2006;**34**:e123.

240. King KR, Wang S, Irimia D, Jayaraman A, Toner M, Yarmush ML. A high-throughput microfluidic real-time gene expression living cell array. *Lab Chip* 2007;**7**:77–85.

241. Guo JM, Chen HL, Wang GM, Zhang YK, Narimatsu H. Expression of UDP-GalNAc:polypeptide *N*-acetylgalactosaminyltransferase-12 in gastric and colonic cancer cell lines and in human colorectal cancer. *Oncology* 2004;**67**:271–6.

242. Tan S, Cheng PW. Mucin biosynthesis: identification of the cis-regulatory elements of human C2GnT-M gene. *Am J Respir Cell Mol Biol* 2007;**36**:737–45.

243. Hameetman L, David G, Yavas A, White SJ, Taminiau AH, Cleton-Jansen AM, Hogendoorn PC, Bovee JV. Decreased EXT expression and intracellular accumulation of heparan sulphate proteoglycan in osteochondromas and peripheral chondrosarcomas. *J Pathol* 2007;**211**:399–409.

244. Avril T, North SJ, Haslam SM, Willison HJ, Crocker PR. Probing the cis interactions of the inhibitory receptor Siglec-7 with alpha2,8-disialylated ligands on natural killer cells and other leukocytes using glycan-specific antibodies and by analysis of alpha2,8-sialyltransferase gene expression. *J Leukoc Biol* 2006;**80**:787–96.

245. Stamatos NM, Liang F, Nan X, Landry K, Cross AS, Wang LX, Pshezhetsky AV. Differential expression of endogenous sialidases of human monocytes during cellular differentiation into macrophages. *FEBS J* 2005;**272**:2545–56.

246. Garcia-Vallejo JJ, van Dijk W, van Die I, Gringhuis SI. Tumor necrosis factor-alpha up-regulates the expression of beta1,4-galactosyltransferase I in primary human endothelial cells by mRNA stabilization. *J Biol Chem* 2005;**280**:12676–82.

247. Dennis Jr. G, Sherman BT, Hosack DA, Yang J, Gao W, Lane HC, Lempicki RA. DAVID: Database for Annotation, Visualization, and Integrated Discovery. *Genome Biol* 2003;**4**:P3.

248. Draghici S, Khatri P, Martins RP, Ostermeier GC, Krawetz SA. Global functional profiling of gene expression. *Genomics* 2003;**81**:98–104.

249. Arakawa K, Kono N, Yamada Y, Mori H, Tomita M. KEGG-based pathway visualization tool for complex omics data. *In Silico Biol* 2005;**5**:419–23.

250. Wu J, Mao X, Cai T, Luo J, Wei L. KOBAS server: a web-based platform for automated annotation and pathway identification. *Nucleic Acids Res* 2006;**34**:W720–4.

251. Xia Y, Campen A, Rigsby D, Guo Y, Feng X, Su EW, Palakal M, Li S. DGEM—a microarray gene expression database for primary human disease tissues. *Mol Diagn Ther* 2007;**11**:145–9.

252. Cavalieri D, Castagnini C, Toti S, Maciag K, Kelder T, Gambineri L, Angioli S, Dolara P. Eu. Gene Analyzer: a tool for integrating gene expression data with pathway databases. *Bioinformatics* 2007;**23**:2631–2.

253. Werner T. Bioinformatics applications for pathway analysis of microarray data. *Curr Opin Biotechnol* 2008;**19**:50–4.

254. Krambeck FJ, Betenbaugh MJ. A mathematical model of *N*-linked glycosylation. *Biotechnol Bioeng* 2005;**92**:711–28.

# Protein–Glycan Interactions

# 6

# Glycan-Binding Proteins and Glycan Microarrays

David F. Smith and Richard D. Cummings

*DEPARTMENT OF BIOCHEMISTRY, EMORY UNIVERSITY SCHOOL OF MEDICINE, ATLANTA, GA, USA*

## Introduction

The binding interactions between proteins and carbohydrates have been subjects of research since the discovery that all cells interact with their environment through their surface sugars or glycoconjugates within their glycocalyx and that many secreted proteins are glycosylated [1]. By the 1960s it was also clear that human blood group antigens were carbohydrates, and part of that development was based on using plant lectins, which have specific carbohydrate-binding properties. Plant lectins were also shown to activate lymphocytes and to differentiate between normal and transformed cells, indicating altered expression of carbohydrates during malignant transformation. Thus, plant lectins and anti-carbohydrate antibodies (collectively referred to here as glycan-binding proteins or GBPs) became important tools for detecting specific oligosaccharide (glycan) structures. Accordingly, methods were developed for measuring the interactions of GBPs with glycans as a way of also defining the specificity of the protein–carbohydrate interaction. One of the most widely used methods to defining glycan-binding specificity of GBPs was to inhibit their precipitation of a glycoprotein or agglutination of a particular cell type with available structurally defined glycans. The inhibition of precipitin formation, which was measured by quantifying precipitated protein or visible agglutination with known glycan structures, provided information about the relative binding affinity and overall specificity. This was a labor-intensive process, even after the introduction of the 96-well microtiter plate, which could be used to monitor many agglutination reactions. The major limitations, however, were the lack of structurally defined complex glycans that could be introduced into the inhibition analyses and the vast amounts of materials needed. This limitation has now largely been overcome by the development of solid-phase binding assays, including printed glycan microarrays, which permit the rapid analysis of potentially hundreds of defined glycans in a single analysis.

**Handbook of Glycomics**
Copyright © 2009 by Elsevier Inc. All rights of reproduction in any form reserved.

One of the most successful solid-phase assays in the historical context involved direct binding of proteins to glycolipids that were separated by thin-layer chromatography in a glycolipid overlay technique. This approach was originally used to explore binding of GBPs and bacterial toxins (which are also GBPs), and even viruses and bacteria which express GBPs [2–4]. The glycolipid overlay technique was instrumental in identifying a specific glycolipid ligand for a particular GBP within complex mixtures of glycolipids. However, due to the limited availability of defined, naturally occurring glycolipid structures, this approach were not used for extensive definition of GBP specificities. The TLC overlay technique was later adapted to defined glycan structures by synthesizing neoglycolipids by reductive amination of defined glycan structures to dipalmitoylphosphatidylethanolamine (DPPE) to produce neoglycolipids (NGLs) that were separated by thin-layer chromatography [5,6]. More recently NGLs of glycans reductively aminated to 1,2-dihexadecyl-*sn*-glycero-3-phosphoethanolamine (DHPE) have been printed on nitrocellulose to prepare glycan microarrays for use in defining GBP specificities [7]. In such NGL microarrays, the NGLs are printed on nitrocellulose or polyvinylidene fluoride (PVDF) membranes at relatively high concentrations to afford non-covalent, but virtually irreversible association of the NGL with the slide surface. One of the most widely used approaches for exploring glycan binding is the printed covalent microarray, in which glycans are coupled covalently to *N*-hydroxysuccinimide-activated (NHS) glass slides [8]. This approach and the exploration of GBP binding to such glycan microarrays will be the major focus of this chapter, since the experiments and results of such assays with hundreds of samples are publicly available in the database of the NIH-funded Consortium for Functional Glycomics (CFG) at http://www.functionalglycomics.org/.

## Construction of Glycan Microarrays

The most common format for glycan microarrays is the 1×3 inch microscope slide, which evolved due to the availability of fluorescence readers that were developed for the analysis of microarrays of nucleic acids and proteins on microscope slides in genomic and proteomic applications. Immobilization methods are generally grouped into two approaches, covalent or non-covalent processes. Park et al. [9] have recently reviewed the chemistry of glycan immobilization on derivatized slides and a wide variety of functionalized slides are commercially available. The selection of immobilization strategy depends on the properties of the glycans being printed and the linker or linkers being used. The general strategy for construction of a glycan microarray is shown in Figure 6.1.

The limiting factor for glycan microarray production is the availability of glycans for printing on a microarray. This factor should not be underestimated in importance, since there are very few derivatized and activated glycans commercially available. The CFG, which is sponsored by the National Institute of General Medical Sciences of the National Institutes of Health in the United States, supported the development of the first robot-produced, publicly available array of 167 defined glycans

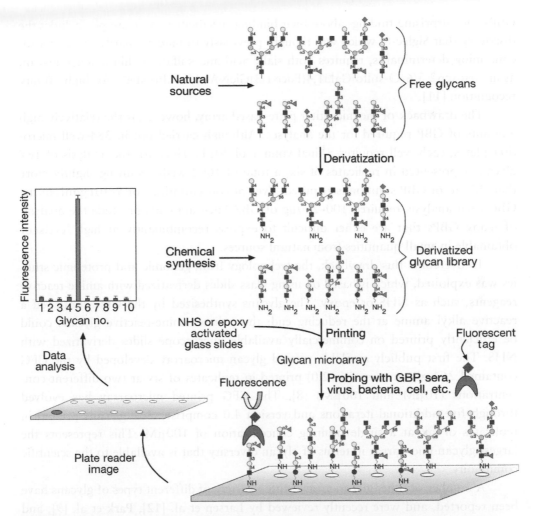

**FIGURE 6.1**   The general strategy for preparing covalent glycan microarrays printed on *N*-hydroxysuccinimide (NHS)-activated glass slides. Free glycans obtained from natural sources or chemically synthesized to contain reactive amine space groups at the reducing termini, are used for covalent printing. The immobilized glycans are deposited on slides in replicate (usually 3–6) by robotic printing as small spots (50–150 μm) and stored in desiccated conditions. The slides may then be hydrated and interrogated or incubated with compounds or cells that may bind glycans, including purified natural or recombinant GBPs, antibodies, toxins, serum, cells, viruses, and bacteria. Binding is detected by direct or indirect fluorescence approaches, which provide an image of fluorescent spots that may be identified to coincide with deposited glycans. The data intensity of fluorescence is measured and then graphically represented as a bar graph where fluorescence intensity is graphed versus glycan number on the microarray.

that were biotinylated and coupled to streptavidin coated 384-well microtiter plates [10]. The glycans were synthesized chemically or made by a combination of chemical and enzymatic synthesis to represent the terminal sequences of mammalian glycans. The microtiter plate-based array was made available to the public by allowing investigators to submit samples to the Protein-Carbohydrate Interactions Core component of the CFG. Data from requests approved by a steering committee were collected and made available to the public on the CFG website. The microtiter plate-based array was highly successful and resulted in the determination of glycan specificities of many

GBPs and surprising findings about their binding specificities. An example includes the discovery that Siglec-8, which was thought previously to bind primarily to sialic acid-containing determinants, requires both sialic acid and sulfate within a single glycan, as in NeuAcα2,3(6-O-sulfo)Galβ1,4(Fucα1,3)GlcNAc (6'-sulfo-sLe$^x$), for high-affinity recognition [11].

The drawback of the microtiter plate-based array, however, is the relatively high amounts of GBP required for the analyses. Although carried out in 384-well microtiter plates, each well requires a final volume of 20 μL. Thus, for the analysis of 167 glycans represented in replicates of six, a total of 1002 wells requiring slightly more than 20 mL of GBP solution is required. Even at concentrations of 5–10 μg/mL of the GBP, each analysis required 100–200 μg of GBP. Such amounts preclude the analysis of many GBPs that are either difficult to express recombinantly at high levels or obtainable in small quantities from natural sources.

To overcome this drawback, the technology from genomic and proteomic studies was exploited, which is based on using glass slides derivatized with amine-reactive reagents, such as NHS or epoxy. The glycans synthesized by the CFG contained a reactive alkyl amine at the reducing end; thus, these amine-reactive glycans could be efficiently printed on commercially available microscope slides derivatized with NHS. The first publicly available printed glycan microarray developed by the CFG contained 264 glycans (version 2.0) printed in replicates of six at two different concentrations (10 μM and 100 μM) [8]. The CFG printed microarray has evolved through five additional iterations and version 4.0 comprises 442 glycans printed in replicates of six at a single printing concentration of 100 μM. This represents the largest glycan microarray in terms of glycan diversity that is available to the scientific community.

A number of other glycan arrays with a variety of different types of glycans have been reported, and were recently reviewed by Larsen et al. [12], Park et al. [9], and Smith and Cummings [13]. Many different strategies for creating glycan microarrays are available and are obviously useful [14–18]. Most of these microarrays are not publicly available and the raw data are generally private, as opposed to the CFG databases, which are available during and after the published studies. In most cases the private microarrays are limited in respect to the diversity of glycans, and in addition they involve diverse chemistries and glycan derivatization strategies. Given the limited variety of glycans available, it is important that different assay methods be compared to facilitate the interpretation of data. The CFG microarray is printed on a single slide, whereas many private glycan microarrays are printed as multiple subarrays on a single microscope slide. For subarray printing, commercially available silicone gaskets are used to form up to 16 "wells" on each slide where each "well" contains a subarray. With spot diameters of 150–250 μm and center-to-center spacing of 200–300 μm, this configuration permits a high-throughput processing of multiple arrays containing 400–900 spots in each subarray.

Printing functionalized glycans on derivatized slides is limited by the technology of microarray printers, which are available from several vendors. The printing technology continues to evolve, but the most common instruments that are available are contact printers, which deposit the glycans solutions upon physical contact with a surface, and non-contact printers, which may either eject a defined sample by inkjet (Linomat IV, Camag, Switzerland) or deposit a droplet of sample from the constriction of a filled capillary with a piezo collar (Piezorray, PerkinElmer, Shelton, CT, USA). The number of replicates that are printed should be a function of the precision of the printing process. For example, contact printed glycans typically require a larger number of replicates than non-contact printed glycans, since the latter print with significantly greater precision. Our group at the Core H of the CFG have observed that %CV (coefficient of variation) (standard deviation/mean $\times$ 100) <20% can be obtained using $n=6$ with contact printers (PerkinElmer Spotarray) and $n=3$ with a Piezorray printer (PerkinElmer).

The concentration of glycan derivative used during the printing process is easily determined empirically by printing at multiple concentrations and evaluating the effectiveness of the printed microarray. Unless the glycan itself is fluorescent, it is virtually impossible to accurately determine the density or "surface concentration" of a glycan on the microarray. Selecting a glycan concentration for printing is generally accomplished by using a defined glycan structure and a corresponding GBP that is detected on the microarray using a fluorescent label. It is very important to understand the range and maximum signals of the fluorescent reader being used so that data are collected in the linear range of the instrument. There may be some debate on what is considered an optimal concentration for printing, but it can be empirically defined as that concentration of glycan, which when printed, generates a signal close to saturation of the instrument's detector at a relatively high concentration of GBP. Thus, printing at increasing concentrations of glycan and monitoring the signal of increasing concentrations of the GBP being bound will provide the information to define a useful printing concentration. In cases where the glycan derivative is fluorescent, as in cases using fluorescently labeled glycolipids [19] or fluorescent-labeled linkers for glycans [20,21], determining a saturating concentration of glycan derivative is relatively straightforward.

Evaluating the utility of glycan microarrays, which is often referred to as validation of a microarray, is generally accomplished using commercially available plant lectins. This is possible due to the large amount of data that are available on the glycan specificity of these proteins, which are commercially available as biotinylated forms. In addition to evaluating the utility of a glycan array, it is imperative that the quality of the array be monitored from printing to printing. The absolute values of fluorescence signals, which are generally reported as relative fluorescence units or RFU, from GBPs binding to glycans on arrays can be affected by a number of potential variables. These include the concentration of the glycan printed on the array, the precision of the printer used for printing the array, the efficiency of the coupling reaction and the quality of the derivatized surface, the concentration of the GBP applied to the final array product, the quality of the secondary or tertiary labeled components of the detection reaction(s), and the laser power settings and photomultiplier gain settings on the fluorescence scanner.

Most of these can be controlled by carefully monitoring the quality of reagents and reproducibility of the protocols, but quality testing is an important issue in understanding and interpreting data from different versions of the same array. Comparison and interpretation of data between different glycan arrays and data from the same array obtained under different conditions must be done with caution. For example, in earlier glycan microarray experiments, a fixed and relatively high concentration of GBP was used, such as for the CFG studies of GBPs at 200 μg/mL. For example, at high concentrations of galectin-8 (200 μg/mL), the protein bound to a wide variety of glycans containing terminal or penultimate galactose residues [22,23] (Figure 6.2a). However, at lower concentrations (0.2 μg/mL), the lectin bound to a small subset of glycans [23], and primarily only those containing blood group antigens and α2,3-sialylated and sulfated sequences (Figure 6.2b). Moreover, it was shown that at these low concentrations, the N-terminal domain recognized a few α2,3-sialylated and sulfated glycans, whereas the C-terminal bound to glycans expressing blood group A and B antigens [23]. Thus, it has become routine now for the Core H of the CFG to explore GBP interactions at various concentrations and focus on those glycans bound by the GBP at the lowest concentrations.

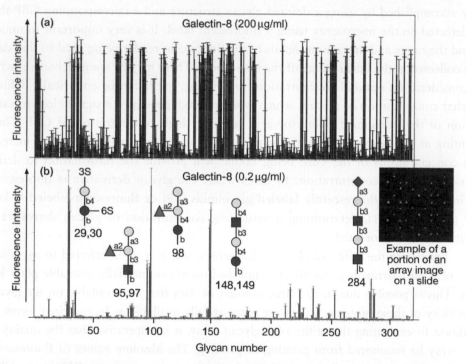

FIGURE 6.2   Example of the binding of human galectin-8, a tandem-repeat galectin, to the printed glycan microarray. Recombinant form of human galectin-8 was biotinylated and applied to the glycan microarray at high concentrations (a) 200 μg/mL and low concentration (b) 0.2 μg/mL [23]. After washing, the bound galectin was detected with glycan interactions following incubation with fluorescein isothiocyanate (FITC)–streptavidin. The slide was washed, dried, and an image of bound fluorescence was obtained using a microarray scanner (Scan Array Express, PerkinElmer Life Sciences). The integrated spot intensities were determined using IMAGENE image analysis software (BioDiscovery, El Segundo, CA, USA) and data were plotted using Microsoft Excel software for evaluation. Raw data results files, generated in Excel format, were uploaded to the Consortium database as ".dat" files and IMAGENE data were uploaded to the database as ".txt" files.

## Overall Strategy of Glycan Arrays: the Big Picture

The rationale for developing a glycan array is to interrogate it with GBPs or potential GBPs and quantify the specific fluorescent signals that indicate the glycan structures to which the GBP in question binds (Figure 6.1). By comparing relative binding of the GBP to related glycan structures, it is possible to define a glycan motif or motifs that the GBP recognizes with higher binding compared to other related structures. Thus, it is easy to see that the value of an array of structurally defined glycans is in some ways proportional to the number and variety of the glycans that are available for printing on the array. It is difficult to define a minimum number of glycans necessary for a useful array because the size and complexity of any particular glycome is not known. A glycome is the "list" of all of the glycan structures produced by a cell line, tissue, organ or organism. The perfect array would be one that was comprised of a complete glycome, but that is not possible at the present time.

Since most protein interactions with glycans occur at the non-reducing termini of glycans coupled to glycoconjugates, one strategy has been to develop an array of defined structures that represent terminal glycan sequences found in animal glycolipids and glycoproteins. The CFG has made a concerted effort to generate such a collection of glycans and maintain a publicly available array for defining the specificities of GBPs and viruses and other organisms. The CFG array is largely based on amine functionalized glycans obtained by chemical synthesis or a combination of chemical and enzymatic synthesis and has avoided putting any mixtures or undefined glycans on the array to keep interpretation of data as simple as possible. The glycans are printed by contact printing on glass slides derivatized with *N*-hydroxysuccinimide (NHS) and the current version (v4.0) contains 442 glycan targets coupled through a variety of linkers or spacers, which all terminate in a primary amino group to keep the coupling reaction as consistent as possible. In some cases identical glycans are coupled via different linkers or spacers, which may vary from a few carbon atoms to a peptide containing a few amino acids (see Figure 6.2b, where several identical glycans are shown with different numbers due to the different spacer used). Since the glycans are practically invisible once they are printed, the effects of the linker on the coupling reaction are impossible to define, thus differential binding of a GBP to the identical glycan with a different linker may represent a difference in presentation of the glycan, but could also represent a difference in the concentration of the glycan on the surface.

Currently, GBPs, protein complexes, virus particles, viruses, bacteria, or cells are detected on the array by a fluorescent-based approach. Fluorescent readers can be equipped with multiple lasers to permit the detection of any commercially available fluorescent label. The obvious drawbacks of fluorescence are the difficulties in quantifying fluorescence, background issues, and possible quenching and bleaching effects that are difficult to control. Some of these variables can be overcome by using well-behaved, and well-characterized fluorescent labels such as Cyanine dyes and Alexa dyes that have been developed by Molecular Probes (Eugene, OR, USA). The fluorescence readers are designed to monitor background fluorescence and correct for background using sophisticated software provided by the instrument manufacturer. An important parameter that must be monitored is the saturation of the reader when high levels of fluorescence are produced in the assay. When the fluorescence is beyond the linear range of the detector, it is impossible to differentiate binding among the different glycans and the GBP in question may appear to be relatively non-specific in binding (see Figure 6.2a). A simple approach to overcome this problem is to analyze GBPs or organisms at lower concentrations (see Figure 6.2b). Defining relative binding strengths of various glycans and thus permitting reliable interpretation of specificity is possible by evaluating the binding profiles at a variety of GBP concentrations. In addition, for a single GBP it is expected, and commonly found, that if it interacts with several different glycans, then those glycans would have a shared structural motif. If no such motif is apparent, then more detailed studies are required to validate both the purity of the GBP and the reproducibility of the results.

The chemistries and the physical properties of glycan array surface are highly debated within the field. This issue does not seem to be as highly debated among investigators in genomics and proteomics, which may be due to the generally accepted belief that, unlike protein and nucleic acids, protein–carbohydrate interactions are weak (in the range of $K_d = 10^{-2}$–$10^{-5}$ M) and that multivalent presentation of glycan is required for proteins to bind. In fact, most GBPs appear to bind with relatively high affinity (in the range of $K_d = 10^{-5}$–$10^{-8}$ M) toward complex glycan determinants, and exhibit low-affinity binding to relatively simple glycans. Nevertheless, the concept that multivalency is required for high-affinity binding by GBPs has raised debates about the degree to which measurable binding is dependent upon glycan density. A simpler way to look at this issue is probably to accept the fact that, in spite of the generally accepted belief regarding the requirement for multivalency, not all protein–carbohydrate interactions are weak and that for any given array the physical properties of the surface will be the same from one analysis to another. Thus, the solid phase presentation of the glycan can be assumed to simply represent a constant concentration of the glycan being analyzed, thereby eliminating one variable in the analytic approach. Thus, measuring binding of GBP as a function of its concentration can be used to determine relative binding strengths to constant glycan presentations.

## Methods for Detecting Binding of GBPs to Glycan Arrays

Fluorescence readers for microarrays on microscope slides were originally developed for glycomic and proteomic applications and were adapted for production of glycan microarrays. The availability of this instrumentation has made fluorescence the method of choice for glycan microarray analysis. This is a convenient choice due to the availability of a large spectrum of fluorescence-labeled reagents that can be used for the immunochemical detection of molecules on the array. Fluorescence suffers many disadvantages relative to quantification and background issues, but as a detection system it is convenient, and sufficiently reproducible and sensitive to generate reliable data. It is also difficult to determine the sensitivity of detection using this technology since it is not possible to relate fluorescence measurements on the glycan array with a primary standard of a known amount of labeled protein on the array. For this reason the data are reported as relative fluorescence units or RFUs. As mentioned above, a major problem with fluorescence is the limited dynamic range in the context of glycan microarrays. Although photomultiplier technology is capable of monitoring signals over many logs by automatic gain changes, the dynamic range of most instruments is between 0 and about 50 000 RFU for a dynamic range of approximately 4 logs. Increasing the dynamic range to 5 logs would be a major improvement for glycan arrays, where the evaluation of relative binding is useful. Thus, it is currently necessary to evaluate multiple concentrations of GBP due to the limited dynamic range.

The simplest detection system for evaluating GBPs is to directly label the protein with fluorescent dyes, which are commercially available in a number of different activated forms for coupling to a variety of different functional groups on GBPs, viruses, bacteria, and other organisms. Most manufacturers provide protocols for labeling proteins with their activated dyes and these can be modified for particles, viruses, or other cells and organisms to be directly labeled. Of course, a major drawback of this approach is the possibility that the functional group(s) derivatized by the fluorescent label could be in the carbohydrate-binding domain (CRD) of the GBP. Thus, an independent assay that confirms the retention of activity of the labeled molecule or organism is imperative for generating conclusive results when no binding to the array is detected. As a general precaution to avoid the possibility of inactivating a GBP during a labeling procedure, it is possible to include a high concentration (10–50 mM) of a monosaccharide cocktail in the labeling reaction. This works well if a monosaccharide inhibitor for the GBP is known (i.e., 50 mM lactose during the labeling of galectins). In cases where the GBP specificity is unknown, a cocktail of a variety of different glycans can be employed with the idea that at high concentrations even a low-affinity ligand can compete for an activated dye in the glycan-binding region of the GBP. Alexa 488-NHS (Molecular Probes) has been a popular labeling reagent for many proteins including lectins and antibodies. This reagent has been successfully used for directly labeling viruses and virus particles [24,25]. Other approaches for direct labeling include preparing fusion proteins with various fluorescent proteins, such as green fluorescent protein (GFP) and Phycoerythrin. It is conceivable that microorganisms or cells could be used directly on the glycan microarray after labeling with lipophilic dyes or nucleic acid-binding dyes that do not alter the binding properties of the relevant GBP on their surfaces. Alternatively, cells expressing recombinant fluorescent protein such as GFP could be useful directly for binding to the glycan microarray.

A successful example of binding cells to the microarray was analysis of *Candida glabrata* adhesin specificity by Zupancic et al. [26]. The epithelial adhesins (Epas) of *Candida glabrata,* an agent of invasive candidiasis, are thought to be responsible for adhesion of the yeast to host epithelium. These investigators expressed the N-terminal ligand-binding domain of Epa1, Epa6, and Epa7 on the surface of *Saccharomyces cerevisiae* and detected fluorescent whole yeast preparations that had been stained with $DiOC_6$. They demonstrated that all three adhesions bound to ligands containing a terminal galactose residue with varying specificity among the three Epa classes.

The application of samples to the array can be accomplished by a number of different methods that allow solutions of labeled protein or suspensions of labeled particles to come into contact with the surface of the glycan array. The preferred method will depend on the nature and amount of the sample. In all cases the printed slides, which are generally stored desiccated, are rehydrated for 5 min in a buffer that is compatible with the buffer used for binding the GBP to the array contained in a Copeland jar designed for processing slides for histology. After hydration, the slide is removed and the buffer is allowed to drain to the bottom and is gently blotted on a tissue to remove excess buffer, but not dry the slide. With small amounts of GBP (i.e., difficult to express recombinant proteins or rare preparations of particles or organisms) as little as 50 μL and as much as 70 μL can be applied to a glycan array under a cover slip to avoid evaporation during the incubation period. The slide is then placed in a dark, humidified chamber at the temperature appropriate for the experiment and incubated for an appropriate time period. A stainless steel immuno-slide staining tray is a convenient apparatus for this procedure as it protects the experiment and eliminates any light that could affect the fluorescent dye. Although any buffer system (even those containing polar organic solvents such as DMSO and DMF) can be used on the array, the protein–carbohydrate interaction component of the CFG generally initiates an analysis of a GBP with unknown glycan binding properties at a final concentration of 200 μg/mL in a Tris-buffer saline, pH 7.4, containing 2 mM $CaCl_2$, 2 mM $MgCl_2$, 1% bovine serum albumin (BSA), and 0.05% Tween-20 (binding buffer). The GBP can be analyzed at any concentration, but 200 μg/mL is selected since it has generally been observed that GBPs that do not bind at this concentration do not bind at higher concentrations; however, it is common to see strong binding of GBPs in the 0.1–10 μg/mL range. Ideally experiments should demonstrate concentration-dependent binding of the GBP and binding at the lower concentrations may represent more physiologically relevant conditions.

For samples that are available in large quantities it is possible to perform the incubation in larger volumes by either drawing a barrier with a hydrophobic barrier pen around the array or subarrays on a slide and applying a large volume, generally up to 1 mL, inside the barrier. It is difficult to agitate or rotate such a configuration due to the instability of the large volume, which can spill over the barrier with motion. If movement is desired with the incubation, it is possible to perform the binding assays in appropriate incubation chambers that permit complete submersion of the slide in 2–5 mL of sample with agitation or rotation.

After primary incubation of the sample on the array in binding buffer, the slides are gently washed by dipping them in a series of buffers contained in Copeland jars. If cover slips are used, they must be carefully removed to avoid damaging the array. This can be accomplished by allowing the cover slip to gently slide off into the first washing buffer, which is binding buffer containing no BSA. Washing simply involves dipping the slide into and removing from the buffer 3–4 times. The second wash is in binding buffer without BSA or Tween-20. If the primary incubation was with a labeled GBP, the final wash is in water to remove salts, which interfere with the fluorescence signal. Excess water is removed from the slide using a slide centrifuge (15 s) or with a gentle stream of nitrogen. The slide is allowed to dry for 5–10 min at room temperature prior to reading in a fluorescent slide reader.

When it is not possible to directly label a GBP, particle, cell, or organism, detection may be effected using indirect immunochemical reagents. This can be accomplished using a fluorescent-labeled secondary detecting reagent or an unlabeled secondary reagent and a fluorescent-labeled tertiary detecting reagent. Secondary and tertiary reagents can be introduced to the array for detection of the bound GBP using any of the procedures mentioned above, but applying reagents under a cover slip is a convenient method, which conserves reagents. It is necessary to wash slides between application of reagents and this is accomplished as described above with the water wash being reserved for the last wash just prior to drying and reading the slide. It should also be noted that there are recent developments in glycan microarray technology in which evanescent field fluorescence is used [27,28]. In this technology the fluorescently labeled GBP is applied to slides of glycan microarrays and fluorescence measurements are made in real time without washing using the principle of applying an electromagnetic wave, termed an evanescent field, which is propagated within a few hundred nanometers from the sensor surface, which represents a wavelength distance, in a low refractive index sample medium. While this approach has the advantage of eliminating a washing step and allowing measurements in near equilibrium conditions without washing, it has the disadvantage of needed specialized equipment that is not widely available at this time.

Biotinylated GBPs are conveniently detected with fluorescent-labeled streptavidin as a secondary reagent. Commercially available antibodies, many of which are fluorescent-labeled, can detect recombinant protein fused with a variety of tags. These include anti-IgG for detecting Fc fusions, and the corresponding antibodies for detecting His-tagged, GST-tagged, and other fusion tags. Successful detection conditions can be approximated using the assay parameters used for detecting the fusion protein in Western blots.

GBPs can be detected with specific antibodies, but these are generally not available commercially as labeled reagents. The cleanest results are obtained with monoclonal antibodies or monospecific polyclonal antibodies that are affinity purified on the antigen. After the unlabeled secondary reagent is applied and excess is washed away, detection can be carried out with the appropriate anti-antibody, most of which are commercially available as fluorescent-labeled reagents. Polyclonal antiserum should be avoided since all animal serum contains anti-glycan antibodies, and control experiments analyzing the detection system without GBP must be performed. Such experiments are recommended even with monoclonal or monospecific antibodies, but with polyclonal antiserum, these assays are sometimes difficult if not impossible to interpret due to the complexity of the polyclonal antibody profile.

Anti-carbohydrate monoclonal antibodies are easily analyzed using the appropriate fluorescent anti-antibody reagent (i.e., anti-mouse IgG or IgM) because they are directed against a single epitope. On the other hand, analyses of polyclonal anti-bodies present significant difficulties due to the large number of anti-glycan antibodies that are normal components of animal sera. In monitoring the immune response of an animal to carbohydrate immunogens using the glycan array, it is necessary to have samples of preimmune serum for immunized individuals to be able to evaluate individual responses. Mammalian sera including humans present very complex binding patterns to glycans on large arrays such as the CFG glycan array. Not only is the pattern complex, it is extremely variable from individual to individual presumably due to the great variety of potential exposures to environmental agents and pathogens that present carbohydrate immunogens to the immune system. In general the glycan array has not been useful in identifying specific anti-glycan antibody responses to any particular condition or disease. This may be due to the large variation of naturally occurring antibodies among individuals, but it could also be due to the absence of the appropriate glycan on the array being interrogated.

## Collection and Presentation of Data from a Large Glycan Microarray

Fluorescence scanners like the PerkinElmer ProScanArray can be equipped with appropriate lasers and filters so that they can be used for monitoring most commercial fluorescent tags. Current instruments, however, are not designed for use with excitation wavelengths in the near ultraviolet range, which are required for excitation of some of the newer bifunctional, fluorescent glycan derivatives. This permits detection of fluorescent-labeled GBPs, which are detected at longer wavelengths without interference from the fluorescent glycans on the array, but precludes quantitative analysis of the printed glycans on the same reader.

The fluorescence scanner generates an image of the fluorescence of GBP binding to the array, and a portion of the slide image is shown in Figure 6.2b. Such images are analyzed using software that was originally developed for analysis of microarrays of nucleic acids (Gene Arrays), but software for analyzing these images is now available from the instrument manufacturers, who are interested in the expanding market of proteomics and glycomics. The image is well populated with fluorescent spots and a grid corresponding to the entire array can be layered over the spots so that the fluorescent signals can be quantified. When the number of bound glycans is low as in the case of a GBP with a restricted specificity and only a few glycans with the corresponding structure are on the array, positioning the grid for the analysis is difficult. For this reason, biotin is printed in strategic, known locations on the array and fluorescent-labeled streptavidin can be included in the assay to permit grid location. In version 4.0 of the CFG array, the glycans were printed at $100\mu M$ in replicates of $n=6$ using an 18-pin contact printer that distributes the individual spots over the array in a defined pattern, which is recorded by the software. The software then sorts the data into appropriate groups of six, determines the high and low values within each set of replicates, and generates a table of glycans numbers corresponding to defined structures, the average fluorescence reading of the four remaining values after removing the high and low values with the standard deviation, standard error of the mean, and %CV (100 × standard deviation/the mean). The data are then converted to an Excel spreadsheet that displays a table of glycan numbers, glycan structures in an IUPAC nomenclature, average relative fluorescence units (RFU), and the statistical parameters.

An example of such an experiment using *Aleuria aurantia* lectin (AAL) is shown in Figure 6.3. Such tables of data (example Figure 6.3b) can be viewed as a downloaded data or "raw file" for a wide assortment of experiments on the glycan microarray at Core H of the CFG at http://www.functionalglycomics.org/static/consortium/resources/resourcecoreh.shtml. The list of glycans and the definition of the spacers or linkers (i.e., -Sp8, -Sp9, -Sp0, etc.) for all versions of the glycan array can also be found on the CFG website. The data are also presented as a histogram of RFU vs. glycan number, as shown for the analysis of *Aleuria aurantia* lectin (AAL) at 0.1 μg/mL in Figure 6.3a, which can be compared with the histogram data of RFU vs. glycan number for other GBPs, such as for galectin-8 at high and low concentrations (Figure 6.2). Another table showing the list of glycans sorted in descending order of RFU is also generated, and a portion of this table from this analysis in the case of binding of AAL is shown in Figure 6.3c. Interestingly, it was observed in the experiments shown in Figure 6.3 that AAL bound to all glycans containing α-linked fucose residues, independently of lectin concentration, which is consistent with the extreme specificity of the lectin for only those fucose-containing glycans [29–31]. Thus, the behavior of AAL on the glycan microarray differed significantly from that of galectin-8, which showed specific binding to selected glycans at low lectin concentrations and less specific binding to a variety of glycans at very high concentrations (Figure 6.2).

For investigators who send samples to the protein–carbohydrate interaction component of the CFG for analysis, this large spreadsheet is provided for interpretation of the binding assay. These data are uploaded on the CFG website as an Excel file (data or "raw file"), and are also uploaded into a CFG database that provides an interactive histogram, which defines candidate glycans, provides general information on each glycan, several structural representations, links to other analyses in the CFG database where the glycan was detected as a high-affinity glycan for other GBP. In addition, researchers are also able to search the entire website for GBPs of interest. Results obtained from the protein–carbohydrate interaction Core H of the CFG are made available to CFG members in confidence for six weeks prior to being made public as mandated by the NIGMS. The availability of public databases of glycan-binding data is of paramount importance to bioinformatic approaches to linking glycan structure to function and expression among different cellular and organismal glycomes.

## Use of Glycan Arrays to Investigate Specificity of Glycan-Binding Proteins

Specificities of anti-carbohydrate antibodies are conveniently analyzed using a defined glycan microarray, and an interesting example is a mouse monoclonal antibody that was produced against lacto-*N*-nor-hexaosylceramide (Galβ1,4GlcNAcβ1,3Galβ1, 4GlcNAcβ1,3Galβ1,4Glcβ1,1-ceramide) that was non-covalently adsorbed to naked

(a)

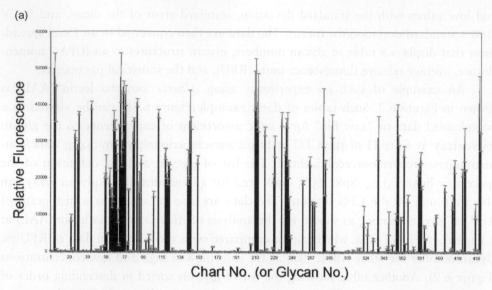

(b)

| Chart Number | Structure | RFU | STDEV | SEM | %CV |
|---|---|---|---|---|---|
| 1 | Gala-Sp8 | 107 | 20 | 10 | 19 |
| 2 | Glca-Sp8 | 78 | 30 | 15 | 38 |
| 3 | Mana-Sp8 | 68 | 25 | 12 | 37 |
| 4 | GalNAca-Sp8 | 36 | 22 | 11 | 61 |
| 5 | Fuca-Sp8 | 49784 | 8544 | 4272 | 17 |
| 6 | Fuca-Sp9 | 40121 | 9822 | 4911 | 24 |
| 7 | Rha-Sp8 | 646 | 68 | 34 | 10 |
| 8 | Neu5Aca-Sp8 | 45 | 31 | 15 | 68 |
| 9 | Neu5Aca-Sp11 | 21 | 14 | 7 | 64 |
| 10 | Neu5Acb-Sp8 | 44 | 12 | 6 | 27 |
| 11 | Galb-Sp8 | 124 | 45 | 22 | 36 |
| 12 | Glcb-Sp8 | 75 | 14 | 7 | 18 |
| 13 | Manb-Sp8 | 139 | 62 | 31 | 44 |

(c)

| Chart Number | Structure | RFU | STDEV | SEM | %CV |
|---|---|---|---|---|---|
| 5 | Fuca-Sp8 | 49784 | 8544 | 4272 | 17 |
| 73 | Fuca1-2Galb1-4GlcNAcb-Sp8 | 47230 | 13504 | 6797 | 29 |
| 211 | Fuca1-2[9OSO3]Galb1-4GlcNAc-Sp0 | 45991 | 13243 | 6622 | 29 |
| 76 | Fuca1-3GlcNAcb-Sp8 | 44514 | 3650 | 1825 | 8 |
| 210 | [3OSO3]Galb1-4(Fuca1-3)[6OSO3]GlcNAc-Sp8 | 43751 | 10849 | 5424 | 25 |
| 78 | Fucb1-3GlcNAcb-Sp8 | 42341 | 5422 | 2711 | 13 |
| 213 | Fuca1-2[9OSO3]Galb1-4[6OSO3]Glc-Sp0 | 40553 | 15943 | 7971 | 39 |
| 441 | Fuca1-2Galb1-4GlcNAcb1-2(Fuca1-2Galb1-4 GlcNAcb1-4)Mana1-3(Fuca1-2Galb1-4 GlcNAcb1-2Mana1-6)Manb1-4GlcNAcb1-4GlcNAcb-N | 40403 | 15153 | 7576 | 38 |
| 90 | GalNAca1-4(Fuca1-2)Galb1-4GlcNAcb-Sp8 | 40380 | 2152 | 1076 | 5 |
| 6 | Fuca-Sp9 | 40121 | 9822 | 4911 | 24 |
| 43 | Neu5Aca2-3[6OSO3]Galb1-4GlcNAcb-Sp8 | 39762 | 2153 | 1076 | 5 |
| 357 | Fuca1-2Galb1-4GlcNAcb-2Mana1-3(Fuca1-2Galb1-4GlcNAcb1-2Mana1-6)Manb1-4GlcNAcb1-4GlcNAcb-Sp20 | 38483 | 3286 | 1643 | 9 |

**FIGURE 6.3** Example of the binding of *Aleuria aurantia* lectin (AAL) to a glycan microarray. The biotinylated AAL was applied to the version 4.0 of the CFG glycan microarray at 0.1 μg/ml. (a) Histogram of the binding of AAL detected as in Figure 6.2, using Alexa 488-labeled streptavidin. (b) An example of a portion of the spreadsheet data is shown for AAL binding, where the glycan or chart number is indicated, along with glycan structure, relative fluorescent units (RFU), standard deviation (STDEV), standard error of the mean (SEM), and % coefficient of variation (CV). (c) The glycans are reordered in this example of the spreadsheet showing the descending order of binding of glycans from the top starting with those having the highest RFUs of AAL binding.

*Salmonella minnesota* [32]. This monoclonal IgM antibody, 1B2, was shown to recognize terminal *N*-acetyl lactosamine, based on its binding to a variety of glycosphingolipids and has been used to identify terminal *N*-acetyl lactosamine in cells and tissues [33,34]. This antibody was applied to version 3.2 of the CFG glycan microarray in response to request cfg_rRequest_1326. The sample was a hybridoma supernatant, which was assayed under a cover slip at a 1:5 dilution and neat in binding buffer, and binding to glycans on the array was detected using fluorescent-labeled anti-mouse IgM (Figure 6.4a,b). At the lower concentration (Figure 6.4a) the antibody binds

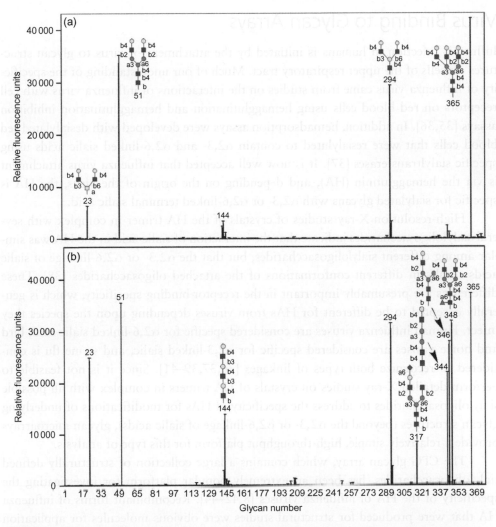

**FIGURE 6.4** Example of the binding of a mouse monoclonal antibody that was produced against lacto-*N*-nor-hexaosylceramide (Galβ1,4GlcNAcβ1,3Galβ1,4GlcNAcβ1,3Galβ1,4Glcβ1,1-ceramide) to the microarray. The IgM antibody, 1B2, was applied as a hybridoma supernatant (b) and a 1:5 dilution of the hybridoma supernatant (a) to version 3.2 of the Consortium for Functional Glycomics glycan microarray in response to request cfg_rRequest_1326 (used by permission of Dr. Gerald A. Schwarting, University of Massachusetts Medical School, Worcester, MA, USA). After washing the slide with the bound antibody, it was incubated with Alexa 488-labeled anti-mouse IgM, washed, dried, and an image of bound fluorescence was obtained using a microarray scanner (Scan Array Express, PerkinElmer Life Sciences) as described in the legend to Figure 6.2.

only biantennary and triantennary presentations of the terminal Galβ1–4GlcNAc within each glycan; however, at a fivefold higher concentration, specificity for the linear presentation of Galβ1–4GlcNAc is observed (Figure 6.4b). GBP specificities can appear to be different depending on the protein concentration, and this concentration dependence, which is presumably related to binding strength, is a simple approach to defining the GBP's specificity based on the strongest binding glycan on the array. Striking examples of this property were observed in studies on the carbohydrate specificity of recombinant human galectins, as shown in Figure 6.2.

## Virus Binding to Glycan Arrays

Influenza infection in humans is initiated by the attachment of virus to glycan structures on cells of the upper respiratory tract. Much of our understanding of the specificity of influenza virus came from studies on the interactions of influenza virus with cell receptors on red blood cells using hemagglutination and hemagglutination inhibition assays [35,36]. In addition, hemadsorption assays were developed with desialylated red blood cells that were resialylated to contain α2,3- and α2,6-linked sialic acids using specific sialyltransferases [37]. It is now well accepted that influenza virus attachment is via the hemagglutinin (HA), and depending on the origin of the virus, the HA is specific for sialylated glycans with α2,3- or α2,6-linked terminal sialic acid.

High-resolution X-ray studies of crystals of the HA trimer in complex with several sialyloligosaccharides indicate that the interaction of sialic acid in the HA was similar among different sialyloligosaccharides, but that the α2,3- or α2,6-linkage of sialic acids results in different conformations of the attached oligosaccharides [38]. These differences are presumably important in the receptor-binding specificity, which is generally accepted to be different for HAs from viruses depending upon the species they infect. Human influenza viruses are considered specific for α2,6-linked sialic acid, bird and horse viruses are considered specific for α2,3-linked sialic, and swine flu is considered to recognize both types of linkages [36,37,39–41]. Since it is not feasible to perform detailed X-ray studies on crystals of HA trimers in complex with all possible sialyloligosaccharides to address the specificity of HAs for modifications of underlying glycan structures (beyond the α2,3- or α2,6-linkage of sialic acids), glycan microarrays provide a relatively simple, high-throughput platform for this type of analysis.

The CFG glycan array, which contains a large collection of structurally defined sialyloligosaccharides, has been an extremely popular platform for investigating the specificity of the HA of influenza viruses [8,42–44]. Recombinant forms of influenza HA that were produced for structural studies were obvious molecules for application to the array. Because the isolated HA trimers have a relatively low affinity ($K_d$ ~2 mM) for their simple sialylated binding partners [36], a multivalent presentation of the hemagglutinin is typically required to detect them binding on the array. Hemagglutinin multivalency was accomplished by pre-complexing recombinant C-terminal, His-tagged hemagglutinin trimers with fluorescent-labeled, mouse anti-penta-His IgG and a fluorescent-labeled anti-mouse IgG in a molar ratio of 4:2:1 [42,43]. This method has been used successfully for the analysis of recombinantly expressed influenza HAs and has the advantage of requiring only small amounts of protein (1–10 μg per assay). The CFG glycan array has also been used with whole influenza virus, where the HAs are presented in their natural multivalent arrangement. The differences in multivalency generated by an artificial "pre-complex" of HA and the presentation of HA in a whole virus are not known; however, recent comparative analyses [45] suggest that the specificity of whole virus binding is broader than that of the corresponding recombinant, His-tagged HA indicating the intact virus surface amplifies weak interactions, which may or may not be physiologically important.

Whole viruses are detected on glycan arrays by direct chemical labeling with fluorescent dyes [24,25] or by indirect immunochemical methods using specific anti-virus antibodies and labeled secondary antibodies [46]. In all cases the virus suspensions are evaluated for an optimal binding concentration based on "plaque-forming units" (PFU) or hemagglutinating activity expressed in hemagglutination units or titer and applied to the array in a buffered saline solution containing protein and detergent (i.e., 1–3% BSA and 0.05% Tween) to prevent non-specific binding. As is the case with GBPs and HA pre-complexes, the concentration of virus will affect the pattern of binding presumably by revealing higher affinity glycan ligands at lower concentrations. The virus suspension can be applied as a relatively large volume (2–3 mL) with the slide totally submerged in the suspension covering the slide, 0.5–1 mL with the suspension retained within a hydrophobic boundary, or in small volumes (50–70 μL) under a cover slip. After incubation for appropriate times, the slides are washed by dipping in washing solutions as described for analyses of GBPs and detected by directly assaying for fluorescence of labeled virus using a scanner or by further processing with primary and labeled secondary antibodies prior to scanning.

These methods have been successfully used to evaluate influenza binding to the diverse sialylated glycans of the CFG array [42] or to a microarray containing only sialyloligosaccharides from the CFG library prepared exclusively for the Centers for Disease Control [45]. The resulting studies not only confirmed the clear differentiation of receptor preference for the α2,3- and α2,6-linked sialic by bird and human influenza-derived HA, but also detected preferences in their specificities for underlying sulfates and fucose [24,25,46] and for the presentation of the sialylated terminal sequences on long versus short lactosamine chains [44].

Binding of virus to glycan arrays is not limited to influenza. The glycan binding specificities of many viruses, including adeno-associated viruses (AAVs) and SV40 virus have been analyzed on the CGF glycan array. For example, results show that AAV1 primarily binds a single glycan on the microarray, NeuAcα2–3GalNAcβ1–4GlcNAcβ-R [47], and SV40 is specific for GM1 [48]. When interpreting data obtained from the glycan array, it is very tempting to conclude that the glycan bound at the lowest concentration of GBP or virus represents the physiological coupling partner. In the case of SV40 virus, the only glycan bound on the CFG array was GM1 [48]; however, using an array of defined glycolipids that contained GM1 with an *N*-glycolyl-5-neuraminic acid (Neu5Gc) rather than *N*-acetyl-5-neuraminic acid

(Neu5Ac), it was shown that the increase of SV40 binding to Neu5Gc-GM1 compared with Neu5Ac-GM1 was approximately tenfold [49]. These results suggest that the natural glycan ligand for SV40 may be Neu5Gc-GM1, which is consistent with its propensity to infect primate cells that have Neu5Gc [50].

## Limitations of Glycan Arrays

While it is clear that glycan microarrays offer tremendous potential to gain insight into the glycan recognition by GBPs and even cells and microorganisms, the use of such microarrays is still in the very early stages and the technology has significant limitations. A tremendously important limiting factor is the diversity and number of glycans on the microarray. While the size and complexity of the human glycome is unknown, it is likely to contain thousands of different glycan determinants or sequences that are recognizable by some type of GBP. Thus, the current size of the glycan microarray from the CFG, which is the largest to date and contains nearly 500 glycans, is probably far short of representing the total glycome complexity. In addition, few microarrays incorporate any types of glycosaminoglycans, although there has been some progress on that front recently with microarrays containing a few dozen glycosaminoglycan-related sequences [51–55]. However, the development of such glycosaminoglycan microarrays is still very early in that they are far short of representing the complex determinants of glycosaminoglycans, since there are likely to be many thousands of hexa- and octa-saccharide sequences in typical heparan sulfate, for example. Thus, there are great challenges ahead to expand the diversity and number of glycans on microarrays. Chemical synthesis is expensive, slow and practically limited to relatively small glycan structures. Enzymatic syntheses of glycans is achievable, but the enzymes are difficult to prepare and the sugar nucleotides and glycan acceptors are also not readily available. A general solution to the limitation of available glycans for microarray preparation may be to directly utilize naturally occurring glycans and derivatize them for inclusion on printed glycan microarrays. This may be possible, since labeling strategies have been introduced that allow fluorescent tagging of glycans released from natural sources and thereby facilitates the glycan purification and characterization [19–21]. However, those approaches are also limited by the large complexity of the glycome, the difficulty in purifying glycans, and the small quantities of glycans available even from natural sources. Thus, there is a great need for improved technologies to promote the saturation of glycan microarrays with all possible glycans from the human, animal, and microbial/pathogen, glycomes.

It should be self-evident that the exploration of glycan recognition by GBPs on microarrays is a limited approach for defining the "physiological" ligands for the GBPs. Microarray analyses may provide clues and insights into potential glycan structures of biological importance, but further experiments with biological samples are required to validate the interpretations and predictions arising from glycan microarray analyses. In addition, microarray analyses of GBP binding only provide an indirect assessment of GBP affinity, and more detailed biophysical approaches, including surface plasmon resonance, microthermal calorimetry, and other equilibrium and direct binding approaches are required to define association/dissociation constants. The use of frontal affinity chromatography to define the binding affinity of glycans to glycan-binding proteins is discussed in Chapter 7.

## Concluding Remarks

Glycan microarrays afford an unprecedented opportunity to explore a wide range of glycan recognition by GBPs, cells, and even organisms. Such microarrays, though still relatively limited in complexity and diversity of glycans, have made profound impacts on a number of fields of study, including immunology, development, cancer, and infectious diseases. Without doubt the glycomic era we are entering is unthinkable without such glycan microarray technologies. Nevertheless, the public availability of such microarrays is under threat from lack of funding, difficulty in procuring glycans, distribution of glycans, costs of preparing microarrays, etc. It is imperative that the National Institutes of Health continue to provide strong support for glycan microarray generation and availability, since effective commercialization of such microarrays with a wide diversity of glycans is still years ahead. The support of glycan microarray development by the Consortium of Functional Glycomics funded by the NIGMS/NIH has been instrumental in bringing such microarray technology to the forefront and making it available to non-experts in glycobiology. Also, the public access of the glycan microarrays and the bioinformatic storage and access of the results has opened up new vistas for data analysis and interpretation. Thus, if all goes well in the future and funding is available for the continuance of glycan microarray access, this technology will continue to enrich the development of glycosciences and provide broad access to biologists interested in exploring glycan functions and recognition.

### Acknowledgments

The work of the authors was supported by NIH Grant P01 HL085607 to R.D.C., NIH Grant GM085448 to D.F.S., and in part by the Consortium for Functional Glycomics under NIH Grant GM62116. We also thank Dr. Gerald A. Schwarting, University of Massachusetts Medical School, Worcester, MA, for permission to use binding data of 1B2 on the glycan microarray.

## References

1. Varki A, Cummings RD, Esko JD, Freeze HH, Stanley P, Bertozzi CR, Hart GW, Etzler ME. *Essentials of glycobiology*. Cold Spring Harbor, New York: Cold Spring Harbor Laboratory Press; 2009.

2. Magnani JL, Smith DF, Ginsburg V. Detection of gangliosides that bind cholera toxin: direct binding of 125I-labeled toxin to thin-layer chromatograms. *Anal Biochem* 1980;**109**:399–402.

3. Magnani JL, Brockhaus M, Smith DF, Ginsburg V. Detection of glycolipid ligands by direct binding of carbohydrate-binding proteins to thin-layer chromatograms. *Methods Enzymol* 1982;**83**:235–41.

4. Karlsson KA, Stromberg N. Overlay and solid-phase analysis of glycolipid receptors for bacteria and viruses. *Methods Enzymol* 1987;**138**:220–32.

5. Tang PW, Gool HC, Hardy M, Lee YC, Feizi T. Novel approach to the study of the antigenicities and receptor functions of carbohydrate chains of glycoproteins. *Biochem Biophys Res Commun* 1985;**132**:474–80.

6. Feizi T, Childs RA. Neoglycolipids: probes in structure/function assignments to oligosaccharides. *Methods Enzymol* 1994;**242**:205–17.

7. Fukui S, Feizi T, Galustian C, Lawson AM, Chai W. Oligosaccharide microarrays for high-throughput detection and specificity assignments of carbohydrate-protein interactions. *Nat Biotechnol* 2002;**20**:1011–17.

8. Blixt O, Head S, Mondala T, Scanlan C, Huflejt ME, Alvarez R, Bryan MC, Fazio F, Calarese D, Stevens J, Razi N, Stevens DJ, Skehel JJ, van Die I, Burton DR, Wilson IA, Cummings R, Bovin N, Wong CH, Paulson JC. Printed covalent glycan array for ligand profiling of diverse glycan binding proteins. *Proc Natl Acad Sci USA* 2004;**101**:17033–8.

9. Park S, Lee MR, Shin I. Chemical tools for functional studies of glycans. *Chem Soc Rev* 2008;**37**: 1579–91.

10. Alvarez RA, Blixt O. Identification of ligand specificities for glycan-binding proteins using glycan arrays. *Methods Enzymol* 2006;**415**:292–310.

11. Bochner BS, Alvarez RA, Mehta P, Bovin NV, Blixt O, White JR, Schnaar RL. Glycan array screening reveals a candidate ligand for Siglec-8. *J Biol Chem* 2005;**280**:4307–12.

12. Larsen K, Thygesen MB, Guillaumie F, Willats WG, Jensen KJ. Solid-phase chemical tools for glycobiology. *Carbohydr Res* 2006;**341**:1209–34.

13. Smith DF, Cummings RD. Deciphering lectin ligands through glycan arrays. In: Vasta GR, Ahmed H, editors. *Animal Lectins: A Functional View*. New York: CRC Press; 2008, p. 49–62.

14. Wang D. Carbohydrate microarrays. *Proteomics* 2003;**3**:2167–75.

15. de Paz JL, Seeberger PH. Deciphering the glycosaminoglycan code with the help of microarrays. *Mol Biosyst* 2008;**4**:707–11.

16. Liang PH, Wu CY, Greenberg WA, Wong CH. Glycan arrays: biological and medical applications. *Curr Opin Chem Biol* 2008;**12**:86–92.

17. Timmer MS, Stocker BL, Seeberger PH. Probing glycomics. *Curr Opin Chem Biol* 2007;**11**:59–65.

18. Culf AS, Cuperlovic-Culf M, Ouellette RJ. Carbohydrate microarrays: survey of fabrication techniques. *Omics* 2006;**10**:289–310.

19. Stoll MS, Feizi T, Loveless RW, Chai W, Lawson AM, Yuen CT. Fluorescent neoglycolipids. Improved probes for oligosaccharide ligand discovery. *Eur J Biochem* 2000;**267**:1795–804.

20. Song X, Xia B, Lasanajak Y, Smith DF, Cummings RD. Quantifiable fluorescent glycan microarrays. *Glycoconj J* 2008;**25**:15–25.

21. Song X, Xia B, Stowell SR, Lasanajak Y, Smith DF, Cummings RD. Novel fluorescent glycan microarray strategy reveals ligands for galectins. *Chem Biol* 2009;**16**:36–47.

22. Carlsson S, Oberg CT, Carlsson MC, Sundin A, Nilsson UJ, Smith D, Cummings RD, Almkvist J, Karlsson A, Leffler H. Affinity of galectin-8 and its carbohydrate recognition domains for ligands in solution and at the cell surface. *Glycobiology* 2007;**17**:663–76.

23. Stowell SR, Arthur CM, Slanina KA, Horton JR, Smith DF, Cummings RD. Dimeric Galectin-8 induces phosphatidylserine exposure in leukocytes through polylactosamine recognition by the C-terminal domain. *J Biol Chem* 2008;**283**:20547–59.

24. Amonsen M, Smith DF, Cummings RD, Air GM. Human parainfluenza viruses hPIV1 and hPIV3 bind oligosaccharides with alpha2–3-linked sialic acids that are distinct from those bound by H5 avian influenza virus hemagglutinin. *J Virol* 2007;**81**:8341–5.

25. Kumari K, Gulati S, Smith DF, Gulati U, Cummings RD, Air GM. Receptor binding specificity of recent human H3N2 influenza viruses. *Virol J* 2007;**4**:42.

26. Zupancic ML, Frieman M, Smith D, Alvarez RA, Cummings RD, Cormack BP. Glycan microarray analysis of *Candida glabrata* adhesin ligand specificity. *Mol Microbiol* 2008;**68**:547–59.

27. Kuno A, Uchiyama N, Koseki-Kuno S, Ebe Y, Takashima S, Yamada M, Hirabayashi J. Evanescent-field fluorescence-assisted lectin microarray: a new strategy for glycan profiling. *Nat Methods* 2005;**2**:851–6.

28. Tateno H, Mori A, Uchiyama N, Yabe R, Iwaki J, Shikanai T, Angata T, Narimatsu H, Hirabayashi J. Glycoconjugate microarray based on an evanescent-field fluorescence-assisted detection principle for investigation of glycan-binding proteins. *Glycobiology* 2008;**18**:789–98.

29. Debray H, Montreuil J. Aleuria aurantia agglutinin. A new isolation procedure and further study of its specificity towards various glycopeptides and oligosaccharides. *Carbohydr Res* 1989;**185**:15–26.

30. Kochibe N, Furukawa K. Purification and properties of a novel fucose-specific hemagglutinin of Aleuria aurantia. *Biochemistry* 1980;**19**:2841–6.

31. Yamashita K, Kochibe N, Ohkura T, Ueda I, Kobata A. Fractionation of L-fucose-containing oligosaccharides on immobilized Aleuria aurantia lectin. *J Biol Chem* 1985;**260**:4688–93.

32. Young Jr. WW, Portoukalian J, Hakomori S Two monoclonal anticarbohydrate antibodies directed to glycosphingolipids with a lacto-*N*-glycosyl type II chain. *J Biol Chem* 1981;**256**:10967–72.

33. Campos L, Portoukalian J, Bonnier S, Shi ZH, Calmard-Oriol P, Treille D, Guyotat D. Specific binding of anti-*N*-acetyllactosamine monoclonal antibody 1B2 to acute myeloid leukaemia cells. *Eur J Cancer* 1992;**28**:37–41.

34. Henion TR, Raitcheva D, Grosholz R, Biellmann F, Skarnes WC, Hennet T, Schwarting GA. Beta1,3-*N*-acetylglucosaminyltransferase 1 glycosylation is required for axon pathfinding by olfactory sensory neurons. *J Neurosci* 2005;**25**:1894–903.

35. Levinson B, Pepper D, Belyavin G. Substiuted sialic acid prosthetic groups as determinants of viral hemagglutination. *J Virol* 1969;**3**:477–83.

36. Gambaryan AS, Tuzikov AB, Pazynina GV, Webster RG, Matrosovich MN, Bovin NV. H5N1 chicken influenza viruses display a high binding affinity for Neu5Acalpha2–3Galbeta1–4(6-HS03)GlcNAc-containing receptors. *Virology* 2004;**326**:310–16.

37. Rogers GN, D'Souza BL. Receptor binding properties of human and animal H1 influenza virus isolates. *Virology* 1989;**173**:317–22.

38. Skehel JJ, Wiley DC. Receptor binding and membrane fusion in virus entry: the influenza hemagglutinin. *Annu Rev Biochem* 2000;**69**:531–69.

39. Matrosovich MN, Gambaryan AS, Teneberg S, Piskarev VE, Yamnikova SS, Lvov DK, Robertson JS, Karlsson KA. Avian influenza A viruses differ from human viruses by recognition of sialyloligosaccharides and gangliosides and by a higher conservation of the HA receptor-binding site. *Virology* 1997;**233**:224–34.

40. Connor RJ, Kawaoka Y, Webster RG, Paulson JC. Receptor specificity in human, avian, and equine H2 and H3 influenza virus isolates. *Virology* 1994;**205**:17–23.

41. Ito T, Couceiro JN, Kelm S, Baum LG, Krauss S, Castrucci MR, Donatelli I, Kida H, Paulson JC, Webster RG, Kawaoka Y. Molecular basis for the generation in pigs of influenza A viruses with pandemic potential. *J Virol* 1998;**72**:7367–73.

42. Stevens J, Blixt O, Glaser L, Taubenberger JK, Palese P, Paulson JC, Wilson IA. Glycan microarray analysis of the hemagglutinins from modern and pandemic influenza viruses reveals different receptor specificities. *J Mol Biol* 2006;**355**:1143–55.

43. Stevens J, Blixt O, Paulson JC, Wilson IA. Glycan microarray technologies: tools to survey host specificity of influenza viruses. *Nat Rev Microbiol* 2006;**4**:857–64.

44. Chandrasekaran A, Srinivasan A, Raman R, Viswanathan K, Raguram S, Tumpey TM, Sasisekharan V, Sasisekharan R. Glycan topology determines human adaptation of avian H5N1 virus hemagglutinin. *Nat Biotechnol* 2008;**26**:107–13.

45. Stevens J, Blixt O, Chen LM, Donis RO, Paulson JC, Wilson IA. Recent avian H5N1 viruses exhibit increased propensity for acquiring human receptor specificity. *J Mol Biol* 2008;**381**:1382–94.

46. Belser JA, Blixt O, Chen LM, Pappas C, Maines TR, Van Hoeven N, Donis R, Busch J, McBride R, Paulson JC, Katz JM, Tumpey TM. Contemporary North American influenza H7 viruses possess human receptor specificity: Implications for virus transmissibility. *Proc Natl Acad Sci USA* 2008;**105**:7558–63.

47. Wu Z, Miller E, Agbandje-McKenna M, Samulski RJ. Alpha2,3 and alpha2,6 *N*-linked sialic acids facilitate efficient binding and transduction by adeno-associated virus types 1 and 6. *J Virol* 2006;**80**:9093–103.

48. Neu U, Woellner K, Gauglitz G, Stehle T. Structural basis of GM1 ganglioside recognition by simian virus 40. *Proc Natl Acad Sci USA* 2008;**105**:5219–24.

49. Campanero-Rhodes MA, Smith A, Chai W, Sonnino S, Mauri L, Childs RA, Zhang Y, Ewers H, Helenius A, Imberty A, Feizi T. *N*-glycolyl GM1 ganglioside as a receptor for simian virus 40. *J Virol* 2007;**81**:12846–58.

50. Varki A. Loss of *N*-glycolylneuraminic acid in humans: Mechanisms, consequences, and implications for hominid evolution. *Am J Phys Anthropol* 2001;**33**(suppl):54–69.

51. de Paz JL, Noti C, Seeberger PH. Microarrays of synthetic heparin oligosaccharides. *J Am Chem Soc* 2006;**128**:2766–7.

52. de Paz JL, Spillmann D, Seeberger PH. Microarrays of heparin oligosaccharides obtained by nitrous acid depolymerization of isolated heparin. *Chem Commun (Camb)* 2006;**29**:3116–18.

53. Wakao M, Saito A, Ohishi K, Kishimoto Y, Nishimura T, Sobel M, Suda Y. Sugar Chips immobilized with synthetic sulfated disaccharides of heparin/heparan sulfate partial structure. *Bioorg Med Chem Lett* 2008;**18**:2499–504.

54. Zhi ZL, Powell AK, Turnbull JE. Fabrication of carbohydrate microarrays on gold surfaces: direct attachment of nonderivatized oligosaccharides to hydrazide-modified self-assembled monolayers. *Anal Chem* 2006;**78**:4786–93.

55. Zhang F, Ronca F, Linhardt RJ, Margolis RU. Structural determinants of heparan sulfate interactions with Slit proteins. *Biochem Biophys Res Commun* 2004;**317**:352–7.

# 7

# Chromatographic and Mass Spectrometric Techniques

Jun Hirabayashi

*RESEARCH CENTER FOR MEDICAL GLYCOSCIENCE, NATIONAL INSTITUTE OF ADVANCED INDUSTRIAL SCIENCE AND TECHNOLOGY, TSUKUBA, IBARAKI, JAPAN*

## Current Status of Understanding about Glycans

"Glycomics" is an emerging field in the post-genome/proteome era, which is directed toward not simply structural identification, but also the elucidation of glycan functions in particular biological phenomena, including cell–cell interaction events, such as differentiation, development, morphogenesis, embryogenesis, immunity, infection, and tumorigenesis, including metastasis [1–5]. This emergence is a quite reasonable consequence considering the fact that all living organisms comprise cells that are, with no known exceptions, covered abundantly with diverse forms of glycoconjugates [6]. It is an unavoidable fact in proteomics that >50% of proteins expressed in eukaryotes are glycosylated [7], and thus it is logical to include the study of glycosylation in individual fields of biology.

In the light of recent research, glycans should no longer be considered less important than nucleic acids and proteins, which have been favored traditionally. In many instances the ultimate biological recognition processes have been shown to be mediated by glycans. Similarly, glycomics should be considered to be just as important as genomics and proteomics.

## Basic Approaches to Structural Glycomics

The most fundamental issue in glycomics is resolving the structural approaches to adopt for these highly heterogeneous and complex molecules with multiple branches and linkage isomers. Moreover, glycans are usually attached to either proteins or lipids in the form of glycoconjugates. This explains why there is still no established procedure for automated glycan synthesis and glycan sequencing. Currently, glycans are probably best characterized in a "matching manner" with respect to authentic standards. This type of structural analysis is referred to as "glycan profiling." In general, glycan profiling requires speed, sensitivity, and high throughput, rather than time-consuming laborious structure determination. Although there is a conventional method for de novo determination of complete covalent structures (e.g., permethylation analysis), much faster and highly sensitive alternatives for purposes of glycan profiling are eagerly awaited [8–12].

There are four key technologies available for glycan profiling (Figure 7.1). The first method introduced in this field is multi-dimensional liquid chromatography (LC), which is usually coupled with prior labeling with a fluorescence reagent, typically 2-aminopyridine [13]. The method was first established for neutral N-linked glycans as two-dimensional mapping [14] and later for those including acidic glycans as three-dimensional mapping [15]. The separation principle is based on differences in the chemical structures of glycans: in normal-phase (hydrophilic interaction) chromatography the separation substantially depends on the size of each glycan (roughly, the number of hydroxyl groups), while in reversed-phase chromatography the separation is based on the hydrophobicity of each glycan, though the actual mechanism is much more complex. For acidic glycans, anion-exchange chromatography is used as the third mode of separation. The method is basically applicable to all kinds of glycans once they are pyridylaminated, and has proved to be able to discriminate >500 glycans. It should also be mentioned that the method is indispensable for preparative purposes, and that coupling with conventional glycosidase treatments is highly effective. As a significant defect, the method requires a relatively long time for each chromatographic analysis (>30 min), and thus has low throughput.

The second approach is mass spectrometry (MS). This powerful technique is an aspect of advanced biotechnology. Its applicability to complex glycans was

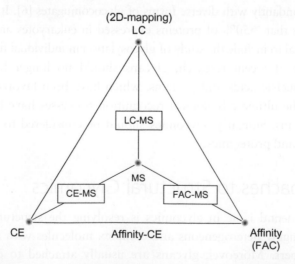

**FIGURE 7.1**  The four major separation technologies for structural glycomics: MS (mass spectrometry), LC (liquid chromatography), CE (capillary electrophoresis) and affinity techniques (FAC, frontal affinity chromatography). Appropriate combinations of these techniques have also been developed, such as LC–MS, FAC-MS, affinity-CE and CE–MS.

first achieved by the development of a tandem MS technique, generally called $MS^n$ [10–12,16–19]. Since individual glycans tend to show a characteristic degradation pattern, they are effectively differentiated usually in the process of up to $MS^3$ or $MS^4$. Essentially, each MS experiment provides only a set of *m/z* values, which explain theoretically possible compositions in terms of elementary saccharides (e.g., hexose (Hex, *m/z*=162), N-acetylhexosamine (HexNAc, *m/z*=203), deoxyhexose (dHex, *m/z*=142) and N-acetylneuraminic acid (NeuAc, *m/z*=291). Thus MS analysis only affords possible combinations of Hex/HexNAc/dHex/NeuAc, and therefore cannot alone reveal component saccharide identities. Nevertheless, the method is highly useful for both confirmation and estimation of glycan structures, including glycosaminoglycans [20–22]. The method is superior in its accuracy (resolution). However, its practice needs, at least at the present level, pre-treatments (i.e., liberation of glycans from protein and lipid and subsequent modification with an appropriate labeling reagent, such as 2-aminopyridine). Direct application to clinical samples will require great improvements. It should also be noted that MS is utilized only for detection, not for preparative purposes.

The third approach, capillary electrophoresis (CE) is based on a different separation principle from the ones described previously [23–25]. Compared with multi-dimensional LC, CE has advantages in speed and sensitivity, while the method is used, together with MS, predominantly for analytical purposes. Considering the extremely high number of theoretical plates (e.g., >100 000) and speed of CE, it seems likely that multi-dimensional LC could well be substituted by multi-dimensional CE. From a practical viewpoint, however, coupling of multiple separation modes is of practical difficulty, because the method usually deals with an extremely small volume of analyte solution (i.e., nanoliter level) though recent advances in CE technology enables combination of a CE device with an MS detector system (i.e., CE–MS [26]).

The final approach is to use lectins for structural analysis of glycans [27]. This approach is rather exceptional considering that the other approaches are based on established separation sciences, and have already been utilized in various fields of life science. In this context, these approaches are simple applications to structural glycomics. On the other hand, this final approach, described in most detail in this mini-review, utilizes special tools (i.e., lectins). Lectins have long been used for detection and purification of glycans in various research areas [28]. Attempts to separate various types of glycans have been elaborated in several groups [29–31]. However, practical use of lectins had never been achieved before a highly quantitative, sensitive, and rapid analytical method emerged in the form of reinforced frontal affinity chromatography (FAC) [32–37], of which the principle and theory were originally derived by Kasai [38]. FAC provides values for the strength of binding between analyte (glycan) and ligand (lectin) [the latter being immobilized on a matrix gel (e.g., agarose)] in terms of a dissociation constant ($K_d$) or an affinity constant ($K_a$), where $K_d=1/K_a$. Since the recently developed system for automated FAC [39–42] enables 100 analyses per day, systematic data on lectin sugar-binding specificity may be accumulated rapidly [36,37,39,43–52] and lectin specificity can now be discussed in terms of $K_d$ in a more systematic manner than ever.

## The Use of Lectins in Structural Glycomics

Consistent with such a trend to utilize a variety of lectins, an increase in the number of papers dealing with lectins is a feature in the field of proteomics in the last few years (Figure 7.2). As of January 27, 2008, PubMed (http://www.ncbi.nlm.nih.gov/sites/entrez?db=pubmed) hits with the query "glycomics OR glycome" total 214, among which those with "lectin" terminology account for as many as 55 (26%). As can be seen from Figure 7.2, the increase in "Lectin/Glycomics" papers is obvious in the last four years. This trend runs parallel to the development of an automated FAC machine (designated FAC-1) [27], which appeared immediately after the first statements using the term "glycome" around the year 2000 [53–56]. It is noteworthy that the concept "proteome", which was first presented by Wasinger [57], was accompanied with the supporting technologies to put forward the practical approach (i.e., separation of proteins by 2D polyacrylamide gel electrophoresis followed by structural identification by MS). The same situation should be applied to glycomics, which is performed with the four basic technologies described previously—LC, CE, MS, and lectin affinity. However, as lectin affinity is an emerging technology, rapid advances can be expected in its sensitivity, resolution, reproducibility, and applications (e.g., combinations with other methodologies).

So far, a number of lectins derived from diverse origins have been identified, which show a variety of carbohydrate specificities in their biological contexts. Lectin

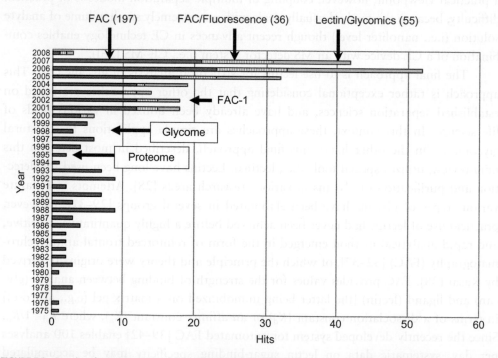

**FIGURE 7.2** Results of a PubMed search for papers dealing with lectin, glycomics and frontal affinity chromatography (FAC). Note that after the glycome concept was presented in 1998 the number of lectin/glycomics papers published shows a rapid increase in association with the increase in FAC/fluorescence papers.

is a technical term for "select" in Latin, and in its generally accepted definition—"a protein which shows affinity to a group of carbohydrates." Any such protein, which had previously been categorized as some other functional protein (such as enzyme, receptor, or cytokine) is entitled to become a new lectin if its sugar-binding ability can be demonstrated.

## Methods for Interaction Analysis of Lectins and Glycans: A Basis of Lectin Study

Since the use of lectins assumes their specific recognition of some glycans, separation technologies based on lectin affinity should be properly evaluated, considering both their merits and demerits (e.g., sensitivity, resolution, reproducibility, speed, through-put, feasibility, etc.). Major analytical methods so far reported are listed in Table 7.1. First, the method requires accuracy in the determination of affinity strength between lectins and glycans with satisfactory throughput and speed. For the latter two requirements, equilibrium dialysis and isothermal calorimetry (ITC) may not be adequate, though these methods may be used to determine reliable affinity constants ($K_a$) or dissociation constants ($K_d$). Among the remaining methods in Table 7.1, FAC has a range of methods for detection: radioisotope (RI) [59], MS [60], and fluorescence (FD). For FAC-RI, however, $N$-glycans must be radiolabeled (e.g., with $NaB[^3H]_4$). Similarly, for FAC-MS, modification of glycans with an appropriate alkyl reagent is

**Table 7.1** Quantitative methods to determine $K_d$ for lectin–carbohydrate interactions

| Methods | Examples of analyses | Ref |
| --- | --- | --- |
| Equilibrium dialysis | ConA and PA-oligosaccharides | 58 |
| FAC-RI[a] | ConA and ovalbumin $N$-glycans | 59 |
| FAC-FD[b] | Galectins and PA-oligosaccharides | 36 |
| | mJRLs and PA-oligosaccharides | 48 |
| FAC-MS[c] | Chorela toxin and ganglioside series derivatives | 60 |
| | Galectin-3 and lacto-$N$-biose libraries | 61 |
| ITC[d] | Galectins and synthetic glycans | 62 |
| | Artocarpin and deoxytrimannoside derivatives | 63 |
| SPR[e] | C-type macrophage lectin and glycopeptides | 64 |
| | Plant lectins and asialofetuin glycopeptide | 65 |
| | Galectin-4 and glycosphingolipids | 66 |
| FP[f] | Galectins and their inhibitors | 67 |

[a] Frontal affinity chromatography radioisotope detection method.
[b] Frontal affinity chromatography fluorescence detection method.
[c] Frontal affinity chromatography mass spectrometry detection method.
[d] Isothermal calorimetry.
[e] Surface plasmon resonance.
[f] Fluorescence polarization.

necessary to increase ionization efficiency in MS. On the other hand, FAC-FD with pyridylaminated (PA)-glycans enables both efficient separation and high-sensitivity detection in LC. In this context, PA-glycans are particularly suited for FAC-FD in that they afford not only sufficient sensitivity (<1 pmol/analysis) but also high reliability and reproducibility, because the labeled glycans show no detectable non-specific adsorption on the resin on which lectins are immobilized.

FAC-FD enables precise determination of interactions in terms of $K_d$ between lectins and glycans in a highly systematic manner. Thus, the method provides us with extensive data regarding lectin specificity. Since $K_d$ (or $K_a$) values are inherent to individual lectin–glycan combinations at any given temperature, they are fundamental knowledge essential for development of a lectin-based glycan profiling system (to be described subsequently) as well as for understanding lectins from a global viewpoint. At present, we are convinced that the use of lectins is advantageous over antibodies for the purpose of glycan profiling for the following reason: glycan structures are extremely diverse. Though it is difficult to determine an actual size of the glycome for each biological species, it would certainly exceed the order of $10^3$–$10^4$. Therefore, it is almost impossible to prepare a set of antibodies strong and specific enough to detect the glycome from a practical viewpoint. Anti-carbohydrate antibodies, if any, do not seem to cover a full spectrum of the glycome. On the other hand, legume lectins, for example, are well-known to have a wide range of binding specificities for $N$-glycans (both high-mannose-type and complex-types with and without sialic acid), $O$-glycans, and glycolipid-type glycans. In contrast to antibodies, lectins show generally much weaker binding affinities (i.e., in terms of $K_d$, $10^{-3}$ to $10^{-7}$M) compared with the known affinities of antibodies ($10^{-6}$ to $10^{-9}$M). This is a consistent feature of bio-molecules (Figure 7.3): the higher the specificity of a bio-molecule, the higher its affinity. On the other hand, lower specificity is suitable for the sake of glycan profiling, because it requires for the first place "coverage." Because most lectins show cross-affinity to structurally related glycans to different extents, a set of lectins would be expected to cover a much wider range of structures than a set of antibodies (Figure 7.4). It is not necessarily true that some specific lectins discriminate closely related structures (e.g., Lewis x (Galβ1–4 (Fucα1–3) GlcNAc) and Lewis a antigens (Galβ1–3 (Fucα1–4) GlcNAc)). In fact, these structural isomers were successfully differentiated by the combination of *Bauhinia purpurea* lectin known as BPA and *Ulex europaeus* isolectin-1 (UEA-1), neither of which is a specific probe for these Lewis antigens [68].

This unexpected result typifies the characteristics of lectin-based glycan profiling.

In the following two sections, the author describes two profiling methodologies which have been investigated and improved in our laboratory.

## Frontal Affinity Chromatography

The first method to be addressed is FAC-FD. Since the original theory of FAC [38] as well as recent improvements using PA-glycans [32–37] and an automated machine

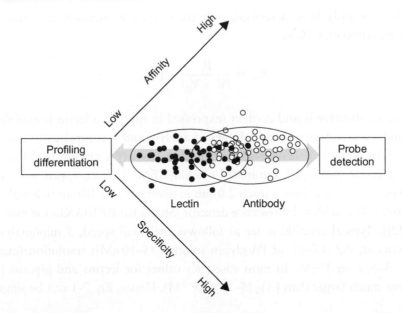

**FIGURE 7.3** A schematic diagram to demonstrate lectin and antibody performance. In general, antibodies show a high specificity and affinity toward their antigen, while lectins show much less specificity and affinity toward their counterpart (carbohydrate) ligand. The former provides the best characteristics for specific detection and probing in different biochemical scenarios, whereas the latter is more suitable for glycan profiling and differentiation purposes in glycomics.

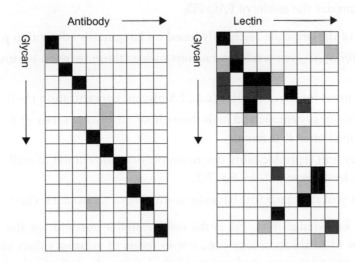

**FIGURE 7.4** Schematic diagram to show the concept of glycan profiling; in particular, comparison with specific detection means using antibodies. In general, lectins show a range of affinities to a wide variety of glycans, and thus, have wide epitope coverage to glycans. This unique property, however, is superior for the sake of glycan profiling aiming at comprehensive characterization of glycosylation features of glycoproteins, cells, and tissue extracts, as well as body fluids. Also note that lectins are considered to function in vivo in various biological processes to decipher glycan structures.

[40–42] have already been described, only the essence is mentioned in this chapter. The basic equation of FAC is:

$$K_d = \frac{B_t}{(V - V_0)} - [A]_0 \qquad (7\text{-}1)$$

where $B_t$ is an effective ligand content (expressed in mol) of a lectin-immobilized column, $V$ and $V_0$ are elution front volumes of analyte and a control substance, respectively, and $[A]_0$ is the initial concentration of the analyte (PA-glycans in this case).

The recently developed automated machine (FAC-1) is equipped with a pair of capsule-type miniature columns (each 2.0 mm in diameter and 10 mm in length, 31.4 μL bed volume) in line with a fluorescence detector (Shimadzu RF10AXL: for more details, see [40–42]). Typical conditions are as follows: analytical speed, 5 min/analysis; sample requirement, 0.3–1.0 mL of PA-glycan solution (5–10 nM); resolution (experimental error), 3–5 μL in $V-V_0$. In most cases, $K_d$ values for lectins and glycans ($10^{-3}$ to $10^{-7}$ M) are much larger than $[A]_0$ ($5-10 \times 10^{-9}$ M). Hence, Eq. 7-1 can be simplified to Eq. 7-2:

$$K_d = \frac{B_t}{(V - V_0)} \qquad (7\text{-}2)$$

As Eq. 7-2 does not include $[A]_0$, interaction experiments avoid experimental errors attributed to PA-glycan concentration. This is a basic reason why FAC is able to provide precise data with sufficient reproducibility and reliability.

To summarize the merits of FAC-FD:

- Clarity of the principle essentially based on Langmuir's adsorption principle.
- Applicability to even weak interactions such as those between lectins and glycans.
- Requirement for a low analyte (e.g., PA-glycan) concentration (5–10 nM).
- Suitability to systematic and high-throughput analysis (a series of $K_d$ values can be obtained once $B_t$ is fixed).
- Accuracy and reproducibility due to simple isocratic elution as well as independence from $[A]_0$ by Eq. 7-2.
- Simplicity of the apparatus basically consisting of an isocratic elution system.

To our knowledge, FAC-FD is the only available method for the quantitative determination of lectin specificity (i.e., not in terms of relative values such as $I_{50}$ in the hemagglutination assay and enzyme-linked assay, but of absolute values, i.e., $K_d$ or $K_a$) with a full panel (>100) of oligosaccharides. Lectin–glycan interactions determined so far number as many as 20000. Approximately 200 lectins have been analyzed for interaction with 100 glycans, among which 12000 interactions were determined in terms of $K_d$. A global view of the interaction analyses so far achieved is given in Figure 7.5.

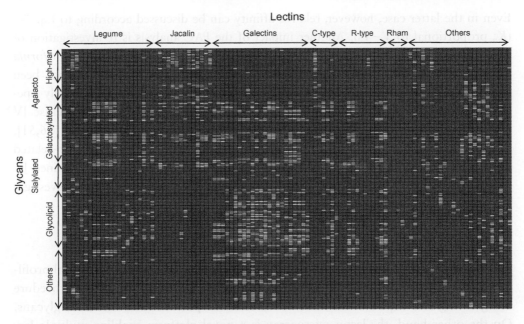

**FIGURE 7.5**   A matrix summarizing the present state of comprehensive interaction analysis by FAC, designated the "Hect-by-Hect" Project. Here, only interactions of 120 lectins, for which $K_d$ (or $K_a$) values have already been determined, are listed. The strength of each interaction is expressed according to six different colors; i.e., *red* ("strongest" to show >30 μL in terms of $V-V_0$ (see text)), *orange* ("second strongest" to show 25–30 μL of $V-V_0$), *dark yellow* ("third strongest" to show 20–25 μL of $V-V_0$), *yellow* ("medium" to show 15–20 μL of $V-V_0$), *green* ("second weakest" to show 10–15 μL of $V-V_0$), *sky blue* ("weakest but significant" to show 5–10 μL of $V-V_0$), and *blue* ("no interaction" showing <5 μL of $V-V_0$). (See color plate 9.)

On the other hand, the method has the following potential drawbacks:

- Immobilization of lectins to the matrix (gel) might result in modification (reduction) of their binding properties. This is particularly the case for sialic acid-binding lectins which have lysine residue(s) in their active site. Appropriate immobilization methods other than the amino-coupling method usually used for immobilization can improve this problem.

- Even using a miniature column (2×10 mm, 31.4 μL) a relatively large amount of lectins (e.g., approximately 500 μg) is required for completion of the analysis.

- Crude samples (glycans) cannot be analyzed. Unlike FAC-MS, mixed samples cannot directly be analyzed.

- For the determination of $B_t$, only a limited repertoire of saccharide derivatives having a UV-sensitive group (e.g., *p*-nitrophenyl, *p*-methoxyphenyl, methotrexate) is available.

As a result of comprehensive interaction analysis using approximately 200 lectins and 100 glycans, systematic elucidation is made for most lectins in terms of $K_d$, whereas it is not possible, at the moment, to determine the $K_d$ of some lectins, because $B_t$ values cannot easily be obtained due to the lack of appropriate sugar derivatives (e.g., *p*-nitrophenyl derivative) required for concentration-dependency analysis to calculate $B_t$.

Even in the latter case, however, relative affinity can be discussed according to Eq. 7-2 ($K_a$ proportional to $V - V_0$). Another impact of the FAC analysis is re-investigation of lectin specificity: this is represented by a GlcNAc-binding plant lectin from *Grifornia simplicifonia*, GSL-II, for which the detailed sugar-binding specificity has not been elucidated. Recent FAC analysis revealed that GSL-II recognizes, in a highly specific manner, a GlcNAc residue transferred by the action of GlcNAc transferase IV [47]. Other examples of re-investigation of lectin specificity include galectins [36,51], *Agaricus bisporus* lectin (known as ABA) [48], mannose-binding-type Jacalin-related lectins [49], and animal C-type lectins [50]. Thus, new information can be gained by re-examination of older work using FAC. FAC is also a powerful tool for the investigation of the sugar-binding specificity of novel lectins.

## Lectin Microarray

The second approach to glycan profiling in the particular context of differential profiling is lectin microarray. Basically, FAC is performed by a chromatographic procedure using a single or a pair of columns and a series of purified (i.e., standard) glycans. On the other hand, the lectin microarray is a novel platform enabling multiple lectin interaction analyses simultaneously. Moreover, all kinds of glycans (even a mixture and crude samples) can be applied to the lectin array. The method is therefore expected to provide an extremely high-throughput means for glycan profiling. To realize this, however, there is the basic issue that usual microarray techniques (e.g., for DNA and antibody) require extensive washing procedures before scanning of the bound probes on the microarray. As emphasized previously, lectin–glycan interactions are weak ($K_d = 10^{-3}$ to $10^{-7}$M) compared with those of antigen–antibody and DNA–RNA hybrids. Such washing procedures would eliminate weakly bound glycans on the lectin microarray, resulting in loss of significant information on glycan (probe) affinity, which provides important clues to glycan structures.

Recently, we developed a unique infrastructure for the lectin microarray, which is based on an evanescent-field activated fluorescence detection principle [69,70]. In the procedure, about 40 lectins, for which specificity has been elucidated by FAC, are immobilized on a glass slide. After probe (e.g., Cy3-labeled glycoprotein) incubation, excitation light is injected from both sides of the slide glass at a certain angle to make a total reflection (reaction chamber formed by a rubber sheet). Here, an evanescent wave is generated within a limited space from the surface (so-called "near optic field"), which is approximately a half distance of the wavelength used for excitation. When Cy3-labeled glycoprotein is used as a probe, the near optic field should be within ~250 nm, for an excitation wavelength of 488 nm. Since the intensity of the evanescent wave is exponentially reduced from the surface, it is considered that a substantially effective near optic field is 200 nm or shorter. The system enables specific detection of fluorescently labeled (e.g., Cy3) glycans without any washing process. The evanescent wave fluorescence detection system thus enables liquid-phase observation in an equilibrium state, unlike the confocal detection principle, which most of other microarray

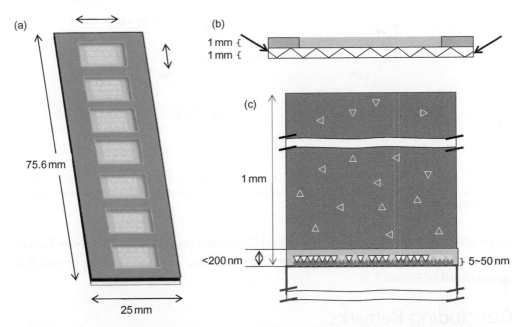

**FIGURE 7.6** Diagram showing total dimensions of lectin microarray, which is based on an evanescent-field activated fluorescence detection principle. (a) Overall view of the commercial lectin microarray with a rubber sheet to make seven reaction chambers. (b) Front view of the array with a rubber sheet. To generate an evanescent wave, an excitation light is injected from both sides of the slide glass (shown by *arrows*). (c) Rough estimates of the inside dimensions of each reaction chamber. Triangles represent fluorescently labeled glyco-materials (Cy3-labeled glycoprotein). Note that they are concentrated in the extremely restricted space almost within the generated evanescent field.

techniques adopt. Figure 7.6 illustrates the geometry of the lectin microarray platform, and shows the total dimensions of this unique detection system, and how the evanescent principle works. In our standard protocol, 0.1 mL of probing solution (Cy3-labeled glycan solution) of 1 mm depth is used in each reaction chamber. Since the evanescent field is generated within 200 nm, in theory the effective activation field occupies only 0.02% of the bottom of the reaction chamber. This means 0.02% of the applied probe solution is exposed to the evanescent wave activation. Thus the background level is extremely low. In fact, the evanescent-field activated fluorescence detection system shows the highest sensitivity among the lectin microarray systems reported so far: i.e., limit of detection (LOD) values are 100 pg/mL of glycoprotein (asialofetuin) and 100 pM glycan (asialo-biantennary *N*-glycan) probes, respectively [68].

Unlike other profiling methods, the lectin microarray has the extremely useful property of direct applicability to crude samples. These are typically clinical samples such as sera and tissue extracts. Moreover, combination target glycoprotein is highly feasible for high-throughput analysis of clinical samples for biomarker investigation, because individual labeling (e.g., with Cy3-SE reagent) and removal procedures can be omitted with the aid of Cy3-labeled antibody (or appropriate second antibody). Alternatively, the biotin–avidin system can be used. Such applications of the lectin microarray will be described elsewhere, as well as its practical application to glycoprotein–biomarker

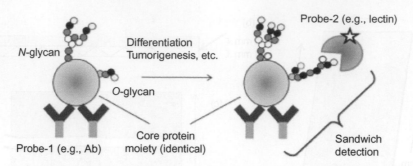

**FIGURE 7.7** Developmental principle of glycoprotein biomarker. The scheme shows that significant glycosylation change occurs in a manner associated with various changes in cell states (e.g., in development, differentiation, and tumorigenesis).

investigations in the context of the Medical Glycomics project of NEDO (New Energy and Industrial Technology Development Organization), the basic concept of which is summarized in Figure 7.7.

## Concluding Remarks

In this chapter, recent advances in lectin-based glycotechnologies are described. The concept of glycan profiling is realized by means of the lectin microarray, but this emerging technique is extensively supported by fundamental information provided by FAC. Probably, both lectin microarray and FAC are necessary tools for structural studies of glycans (i.e., for structural glycomics) as well as other analytical methods. All of these are necessary for elucidation of glycan biological functions (i.e., functional glycomics). With these tools, we have been able to approach even evolutionary issues involving glycans (i.e., comparative glycomics). Overall, why do we study glycans? The author would answer: because they are old (for comparative glycomics), important (for functional glycomics), and difficult (for structural glycomics). By any measure, glycobiology occupies a most fundamental position in life science, because we now know that "all scientific roads lead to the glycome."

## References

1. Mahal LK. Glycomics: towards bioinformatic approaches to understanding glycosylation. *Anticancer Agents Med Chem* 2008;**8**:37–51.

2. Packer NH, von der Lieth CW, Aoki-Kinoshita KF, Lebrilla CB, Paulson JC, Raman R, Rudd P, Sasisekharan R, Taniguchi N, York WS. Frontiers in glycomics: bioinformatics and biomarkers in disease. An NIH White Paper prepared from discussions by the focus groups at a workshop on the NIH campus, Bethesda MD (September 11–13, 2006). *Proteomics* 2006;**8**:8–20.

3. Seeberger PH, Werz DB. Synthesis and medical applications of oligosaccharides. *Nature* 2007;**446**:1046–51.

4. Turnbull JE, Field RA. Emerging glycomics technologies. *Nat Chem Biol* 2007;**3**:74–7.

5. Wada Y, Azadi P, Costello CE, Dell A, Dwek RA, Geyer H, Geyer R, Kakehi K, Karlsson NG, Kato K, Kawasaki N, Khoo KH, Kim S, Kondo A, Lattova E, Mechref Y, Miyoshi E, Nakamura K, Narimatsu H,

Novotny MV, Packer NH, Perreault H, Peter-Katalinic J, Pohlentz G, Reinhold VN, Rudd PM, Suzuki A, Taniguchi N. Comparison of the methods for profiling glycoprotein glycans–HUPO human disease Glycomics/Proteome Initiative multi-institutional study. *Glycobiology* 2007;**17**:411–22.

6. Valid A, Cummings R, Esko J, Freeze H, Hart G, Marth J, editors. *Essentials of glycobiology*. New York: Cold Springer Harbor Laboratory Press; 1999.

7. Apweiler R, Hermjakob H, Sharon N. On the frequency of protein glycosylation, as deduced from analysis of the SWISS-PROT database. *Biochim Biophys Acta* 1999;**1473**:4–8.

8. Kang P, Mechref Y, Kyselova Z, Goetz JA, Novotny MV. Comparative glycomic mapping through quantitative permethylation and stable-isotope labeling. *Anal Chem* 2007;**79**:6064–73.

9. Liu X, McNally DJ, Nothaft H, Szymanski CM, Brisson JR, Li JM. Spectrometry-based glycomics strategy for exploring N-linked glycosylation in eukaryotes and bacteria. *Anal Chem* 2006;**78**:6081–7.

10. Lapadula AJ, Hatcher PJ, Hanneman AJ, Ashline DJ, Zhang H, Reinhold VN. Congruent strategies for carbohydrate sequencing. 3. OSCAR: an algorithm for assigning oligosaccharide topology from $MS^n$ data. *Anal Chem* 2000;**77**:6271–9.

11. Zhang H, Singh S, Reinhold VN. Congruent strategies for carbohydrate sequencing. 2. FragLib: an $MS^n$ spectral library. *Anal Chem* 2005;**77**:6263–70.

12. Ashline D, Singh S, Hanneman A, Reinhold V. Congruent strategies for carbohydrate sequencing. 1. Mining structural details by $MS^n$. *Anal Chem* 2005;**77**:6250–62.

13. Hase S, Ikenaka T, Matsushima Y. Structure analyses of oligosaccharides by tagging of the reducing end sugars with a fluorescent compound. *Biochem Biophys Res Commun* 1978;**85**:257–63.

14. Tomiya N, Awaya J, Kurono M, Endo S, Arata Y, Takahashi N. Analyses of N-linked oligosaccharides using a two-dimensional mapping technique. *Anal Biochem* 1988;**171**:73–90.

15. Takahashi N, Nakagawa H, Fujikawa K, Kawamura Y, Tomiya N. Three-dimensional elution mapping of pyridylaminated N-linked neutral and sialyl oligosaccharides. *Anal Biochem* 1995;**226**:139–46.

16. Kameyama A, Kikuchi N, Nakaya S, Ito H, Sato T, Shikanai T, Takahashi Y, Takahashi K, Narimatsu H. A strategy for identification of oligosaccharide structures using observational multistage mass spectral library. *Anal Chem* 2005;**77**:4719–25.

17. Kameyama A, Nakaya S, Ito H, Kikuchi N, Angata T, Nakamura M, Ishida HK, Narimatsu H. Strategy for simulation of CID spectra of N-linked oligosaccharides toward glycomics. *J Proteome Res* 2006;**5**:808–14.

18. Sekiya S, Yamaguchi Y, Kato K, Tanaka K. Mechanistic elucidation of the formation of reduced 2-aminopyridine-derivatized oligosaccharides and their application in matrix-assisted laser desorption/ionization mass spectrometry. *Rapid Commun Mass Spectrom* 2005;**19**:3607–11.

19. Haslam SM, North SJ, Dell A. Mass spectrometric analysis of N- and O-glycosylation of tissues and cells. *Curr Opin Struct Biol* 2006;**16**:584–91.

20. Tissot B, Gasiunas N, Powell AK, Ahmed Y, Zhi ZL, Haslam SM, Morris HR, Turnbull JE, Gallagher JT, Dell A. Towards GAG glycomics: analysis of highly sulfated heparins by MALDI-TOF mass spectrometry. *Glycobiology* 2007;**17**:972–82.

21. Hitchcock AM, Costello CE, Zaia J. Glycoform quantification of chondroitin/dermatan sulfate using a liquid chromatography-tandem mass spectrometry platform. *Biochemistry* 2006;**45**:2350–61.

22. Minamisawa T, Suzuki K, Hirabayashi J. Multistage mass spectrometric sequencing of keratan sulfate-related oligosaccharides. *Anal Chem* **78**:891–900.

23. Kamoda S, Nakanishi Y, Kinoshita M, Ishikawa R, Kakehi K. Analysis of glycoprotein-derived oligosaccharides in glycoproteins detected on two-dimensional gel by capillary electrophoresis using online concentration method. *J Chromatogr A* 2006;**1106**:67–74.

24. Dang F, Kakehi K, Nakajima K, Shinohara Y, Ishikawa M, Kaji N, Tokeshi M, Baba Y. Rapid analysis of oligosaccharides derived from glycoproteins by microchip electrophoresis. *J Chromatogr A* 2006;**1109**:138–43.

25. Mechref Y, Novotny MV. Miniaturized separation techniques in glycomic investigations. *J Chromatogr B Analyt Technol Biomed Life Sci* 2006;**841**:65–78.

26. Balaguer E, Neusüss C. Glycoprotein characterization combining intact protein and glycan analysis by capillary electrophoresis-electrospray ionization-mass spectrometry. *Anal Chem* 2006;**78**:5384–93.

27. Hirabayashi J. Lectin-based structural glycomics: glycoproteomics and glycan profiling. *Glycoconj J* 2004;**21**:35–40.

28. Sharon N. Lectins: carbohydrate-specific reagents and biological recognition molecules. *J Biol Chem* 2007;**282**:2753–64.

29. Cummings RD, Kornfeld S. Fractionation of asparagine-linked oligosaccharides by serial lectin-Agarose affinity chromatography. A rapid, sensitive, and specific technique. *J Biol Chem* 1982;**257**:11235–40.

30. Harada H, Kamei M, Tokumoto Y, Yui S, Koyama F, Kochibe N, Endo T, Kobata A. Systematic fractionation of oligosaccharides of human immunoglobulin G by serial affinity chromatography on immobilized lectin columns. *Anal Biochem* 1987;**164**:374–81.

31. Bierhuizen MF, Hansson M, Odin P, Debray H, Obrink B, van Dijk W. Structural assessment of the N-linked oligosaccharides of cell-CAM 105 by lectin-agarose affinity chromatography. *Glycoconj J* 1989;**6**:195–208.

32. Hirabayashi J, Arata Y, Kasai K. Reinforcement of frontal affinity chromatography for effective analysis of lectin-oligosaccharide interactions. *J Chromatogr A* 2000;**890**:261–71.

33. Hirabayashi J, Kasai K. Separation technologies for glycomics. *J Chromatogr B Analyt Technol Biomed Life Sci* 2002;**771**:67–87.

34. Hirabayashi J, Arata Y, Kasai K. Frontal affinity chromatography as a tool for elucidation of sugar recognition properties of lectins. *Methods Enzymol* 2003;**362**:353–68.

35. Arata Y, Hirabayashi J, Kasai K. Application of reinforced frontal affinity chromatography and advanced processing procedure to the study of the binding property of a *Caenorhahditis elegans* galectin. *J Chromatogr A* 2001;**905**:337–43.

36. Hirabayashi J, Hashidate T, Arata Y, Nishi N, Nakamura T, Hirashima M, Urashima T, Oka T, Futai M, Muller WE, Yagi F, Kasai K. Oligosaccharide specificity of galectins: a search by frontal affinity chromatography. *Biochim Biophys Acta* **1572**:232–54.

37. Arata Y, Hirabayashi J, Kasai K. Sugar binding properties of the two lectin domains of the tandem repeat-type galectin LEC-1 (N32) of caenorhabditis elegans. Detailed analysis by an improved frontal affinity chromatography method. *J Biol Chem* 2001;**276**:3068–77.

38. Kasai K, Ishii S. Quantitative analysis of affinity chromatography of trypsin. A new technique for investigation of protein-ligand interaction. *J Biochem* 1975;**77**:261–4.

39. Nakamura S, Yagi F, Totani K, Ito Y, Hirabayashi J. Comparative analysis of carbohydrate-binding properties of two tandem repeat-type Jacalin-related lectins, *Castanea crenata* agglutinin and *Cycas revoluta* leaf lectin. *FEBS J* 2005;**272**:2784–99.

40. Nakamura-Tsuruta S, Uchiyama N, Hirabayashi J. High-throughput analysis of lectin-oligosaccharide interactions by automated frontal affinity chromatography. *Methods Enzymol* 2006;**415**:311–25.

41. Tateno H, Nakamura-Tsuruta S, Hirabayashi J. Frontal affinity chromatography: sugar-protein interactions. *Nat Protoc* 2007;**2**:2529–37.

42. Nakamura-Tsuruta S, Uchiyama N, Kominami J, Hirabayashi J. Frontal affinity chromatography: systematization for quantitative interaction analysis between lectins and glycans. In: Nilsson CL, editor. *Lectins: analytical technologies*. Amsterdam: Elsevier; 2007.

43. Itakura Y, Nakamura-Tsuruta S, Kominami J, Sharon N, Kasai K, Hirabayashi J. Systematic comparison of oligosaccharide specificity of *Ricinus communis* agglutinin I and *Erythrina* lectins: a search by frontal affinity chromatography. *J Biochem* **142**:459–69.

44. Van Damme EJ, Nakamura-Tsuruta S, Hirabayashi J, Rougé P, Peumans WJ. The *Sclerotinia sclerotiorum* agglutinin represents a novel family of fungal lectins remotely related to the Clostridium botulinum non-toxin haemagglutinin HA33/A. *Glycoconj J* 2007;**24**:143–456.

45. Van Damme EJ, Nakamura-Tsuruta S, Smith DF, Ongenaert M, Winter HC, Rougé P, Goldstein IJ, Mo H, Kominami J, Culerrier R, Barre A, Hirabayashi J, Peumans WJ. Phylogenetic and specificity studies of two-domain GNA-related lectins: generation of multispecificity through domain duplication and divergent evolution. *Biochem J* 2007;**404**:51–61.

46. Huang X, Tsuji N, Miyoshi T, Nakamura-Tsuruta S, Hirabayashi J, Fujisaki K. Molecular characterization and oligosaccharide-binding properties of a galectin from the argasid tick *Ornithodoros moubata*. *Glycobiology* 2007;**17**:313–23.

47. Nakamura-Tsuruta S, Kominami J, Kamei M, Koyama Y, Suzuki T, Isemura M, Hirabayashi J. Comparative analysis by frontal affinity chromatography of oligosaccharide specificity of GlcNAc-binding lectins, *Griffonia simplicifolia* lectin-II (GSL-II) and *Boletopsis leucomelas* lectin (BLL). *J Biochem* 2006;**140**:285–91.

48. Nakamura-Tsuruta S, Kominami J, Kuno A, Hirabayashi J. Evidence that Agaricus bisporus agglutinin (ABA) has dual sugar-binding specificity. *Biochem Biophys Res Commun* 2006;**347**:215–20.

49. Nakamura-Tsuruta S, Uchiyama N, Peumans WJ, Van Damme EJM, Totani K, Ito Y, Hirabayashi J. Analysis of the sugar-binding specificity of mannose-binding-type Jacalin-related lectins by frontal affinity chromatography: an approach to functional classification. *FEBS J* 2008;**275**:1227–39.

50. Oo-Puthinan S, Maenuma K, Sakakura M, Denda-Nagai K, Tsuiji M, Shimada I, Nakamura-Tsuruta S, Hirabayashi J, Bovin NV, Irimura T. The amino acids involved in the distinct carbohydrate specificities between macrophage galactose-type C-type lectins 1 and 2 (CD301a and b) of mice. *Biochim Biophys Acta* 2008;**1780**:89–100.

51. Nagae M, Nishi N, Nakamura-Tsuruta S, Hirabayashi J, Wakatsuki S, Kato R. Structural analysis of the human galectin-9 N-terminal carbohydrate recognition domain reveals unexpected properties that differ from the mouse orthologue. *J Mol Biol* 2007;**375**:119–35.

52. Watanabe Y, Shiina N, Shinozaki F, Yokoyama H, Kominami J, Nakamura-Tsuruta S, Hirabayashi J, Sugahara K, Kamiya H, Matsubara H, Ogawa T, Muramoto K. Isolation and characterization of l-rhamnose-binding lectin, which binds to microsporidian Glugea plecoglossi, from ayu (*Plecoglossus altivelis*) eggs. *Dev Comp Immunol* 2007;**32**:487–99.

53. Kobata A. A journey to the world of glycobiology. *Glycoconj J* 2000;**17**:443–64.

54. Feizi T. Progress in deciphering the information content of the "glycome"–a crescendo in the closing years of the millennium. *Glycoconj J* **17**:553–65.

55. Taniguchi N, Ekuni A, Ko JH, Miyoshi E, Ikeda Y, Ihara Y, Nishikawa A, Honke K, Takahashi M. A glycomic approach to the identification and characterization of glycoprotein function in cells transfected with glycosyltransferase genes. *Proteomics* 2001;**1**:239–47.

56. Hirabayashi J, Arata Y, Kasai K. Glycome project: concept, strategy and preliminary application to *Caenorhabditis elegans*. *Proteomics* **1**:295–303.

57. Wasinger VC, Cordwell SJ, Cerpa-Poljak A, Yan JX, Gooley AA, Wilkins MR, Duncan MW, Harris R, Williams KL, Humphery-Smith I. Progress with gene-product mapping of the mollicutes: mycoplasma genitalium. *Electrophoresis* 1995;**16**:1090–4.

58. Mega T, Hase S. Determination of lectin-sugar binding constants by microequilibrium dialysis coupled with high performance liquid chromatography. *J Biochem* 1991;**109**:600–3.

59. Ohyama Y, Kasai K, Nomoto H, Inoue Y. Frontal affinity chromatography of ovalbumin glycoasparagines on a concanavalin A-sepharose column. A quantitative study of the binding specificity of the lectin. *J Biol Chem* 1985;**260**:6882–7.

60. Schriemer DC, Bundle DR, Li L, Hindsgaul O. Micro-scale frontal affinity chromatography with mass spectrometric detection: a new method for the screening of compound libraries. *Angew Chem Int Ed* 1998;**37**:3383–7.

61. Fort S, Kim HS, Hindsgaul O. Screening for galectin-3 inhibitors from synthetic lacto-N-biose libraries using microscale affinity chromatography coupled to mass spectrometry. *J Org Chem* 2006;**71**:7146–54.

62. Dam TK, Brewer CF. Multivalent protein-carbohydrate interactions: isothermal titration microcalorimetry studies. *Methods Enzymol* 2004;**379**:107–28.

63. Rani PG, Bachhawat K, Reddy GB, Oscarson S, Surolia A. Isothermal titration calorimetric studies on the binding of deoxytrimannoside derivatives with artocarpin: implications for a deep-seated combining site in lectins. *Biochemistry* 2000;**39**:10755–60.

64. Yamamoto K, Ishida C, Shinohara Y, Hasegawa Y, Konami Y, Osawa T, Irimura T. Interaction of immobilized recombinant mouse C-type macrophage lectin with glycopeptides and oligosaccharides. *Biochemistry* 1994;**33**:8159–66.

65. Shinohara Y, Kim F, Shimizu M, Goto M, Tosu M, Hasegawa Y. Kinetic measurement of the interaction between an oligosaccharide and lectins by a biosensor based on surface plasmon resonance. *Eur J Biochem* 1994;**223**:189–94.

66. Ideo H, Seko A, Yamashita K. Galectin-4 binds to sulfated glycosphingolipids and carcinoembryonic antigen in patches on the cell surface of human colon adenocarcinoma cells. *J Biol Chem* 2005;**280**:4730–7.

67. Sörme P, Kahl-Knutsson B, Wellmar U, Magnusson BG, Leffler H, Nilsson UJ. Design and synthesis of galectin inhibitors. *Methods Enzymol* 2003;**363**:157–69.

68. Uchiyama N, Kuno A, Tateno H, Kubo Y, Mizuno M, Noguchi M, Hirabayashi J. Optimization of evanescent-field fluorescence-assisted lectin microarray for high-sensitive detection of monovalent oligosaccharides and glycoproteins. *Proteomics* 2008;**15**:3042–50.

69. Kuno A, Uchiyama N, Koseki-Kuno S, Ebe Y, Takashima S, Yamada M, Hirabayashi J. Evanescent-field fluorescence-assisted lectin microarray: a new strategy for glycan profiling. *Nat Methods* 2005;**2**:851–6.

70. Uchiyama N, Kuno A, Koseki-Kuno S, Ebe Y, Horio K, Yamada M, Hirabayashi J. Development of a lectin microarray based on an evanescent-field fluorescence principle. *Methods Enzymol* 2006;**415**:341–51.

# Glycobioinformatics

# Integration of Glycomics Knowledge and Data

## William S. York[a,b,c], Krys J. Kochut[c] and John A. Miller[c]

[a]COMPLEX CARBOHYDRATE RESEARCH CENTER,
THE UNIVERSITY OF GEORGIA, ATHENS, GA, USA
[b]DEPARTMENT OF BIOCHEMISTRY AND MOLECULAR BIOLOGY,
THE UNIVERSITY OF GEORGIA, ATHENS, GA, USA
[c]DEPARTMENT OF COMPUTER SCIENCE,
THE UNIVERSITY OF GEORGIA, ATHENS, GA, USA

## Background and Motivation

### The Complexity of Glycobiology as a Scientific Problem

During the last several decades, great progress has been made in understanding the biology and biochemistry of proteins and nucleic acids. However, it has been much more difficult to gain knowledge regarding the biology of glycans, which are often attached to proteins and lipids and make up a large proportion of the material forming the cell surface [1]. The biological mechanisms by which organisms synthesize glycans are completely different from those used to synthesize proteins and nucleic acids. The complexity of glycan biosynthesis has made it extremely difficult to analyze, limiting our knowledge of the mechanisms by which glycans carry out their biological functions.

The field of glycomics, which seeks to understand the structure and biosynthesis of glycans, has emerged only recently as a result of technical advances that allow high-throughput, high-sensitivity analysis of these structurally complex molecules [2]. These powerful new analytical techniques produce large amounts of information-rich raw data, making it necessary to develop equally powerful informatics techniques that process and mine this data in order to understand its biological implications in any detail.

### The Structural Complexity of Glycans

Glycans have several features that distinguish them from nucleic acids and proteins. All three classes of molecules are composed of monomeric subunits (residues) that can be released by hydrolysis of the polymer. However, unlike nucleic acids and proteins,

**Handbook of Glycomics**
Copyright © 2009 by Elsevier Inc. All rights of reproduction in any form reserved.

glycans are branched molecules. This is due to the fact that two residues in a glycan can be linked together in several different ways; that is, any of the hydroxyl groups of a monosaccharide residue can be glycosylated (i.e., bear another glycosyl residue). In addition, formation of a glycosidic bond almost always involves the formation of a cyclic form of the monosaccharide residue: a furanosyl residue has five atoms in the ring and a pyranosyl residue has six atoms in the ring. The glycosidic bond between two glycosyl residues can adopt one of two different configurations, labeled $\alpha$ or $\beta$, depending on the stereochemistry of the anomeric carbon. (The anomeric carbon is covalently attached to a ring oxygen and to the oxygen or nitrogen that spans the glycosidic linkage.) Together, these structural features provide a considerably greater potential for structural diversity than is possible for nucleic acids or proteins. This structural complexity and diversity makes glycan characterization a technically challenging endeavor [3]. Although several rapid and accurate methods for determining the sequences of nucleic acids and proteins exist, no single method for rigorously determining glycan structure has yet been developed.

A typical cell or tissue sample contains a highly complex mixture of glycoconjugates (i.e., glycoproteins and glycolipids). Even with the most advanced analytical techniques that are currently available, determining the overall glycosylation patterns present in such a complex mixture remains an extremely difficult endeavor. For example, the glycans linked to each N-glycosylation site of a glycoprotein usually exhibit structural "microheterogeneity," resulting in a potentially large number of distinct glycoforms for each of the many different glycoproteins present on the cell surface. Robust computational tools are necessary in order to process the large volumes of raw data that are becoming available as analytical glycoproteomics methodology advances. Other informatics tools are required to store, retrieve, and interpret the resulting structural information in the context of biological models.

## The Complexity of Glycan Biosynthesis

Unlike nucleic acids and proteins, glycans are not synthesized by a template-directed mechanism, but rather by a sequential acceptor substrate-recognition mechanism. Addition of each monosaccharide residue to the glycan is catalyzed by the presence of a glycosyltransferase that specifically recognizes the glycosyl donor substrate (often a sugar nucleotide) and the acceptor substrate (usually a partially formed nascent glycan). A single glycosyltransferase may recognize several distinct but structurally related acceptor substrates, each with a potentially different binding constant ($K_M$) and catalytic constant ($k_{cat}$). Moreover, most glycosyltransferases are targeted to specific cellular compartments (such as the endoplasmic reticulum or Golgi) in which nascent glycans are processed. The rate at which each glycosyltransfer reaction occurs depends on the local concentration of donor substrates within these compartments, which can also contain glycosylhydrolases that remove specific glycosyl residues from the nascent glycan during its biosynthetic maturation. Thus, glycan biosynthesis is a complex process in which diverse substrates compete for interaction with a limited pool of glycosylhydrolases and glycosyltransferases. Clearly, a deep understanding of

the mechanisms by which cells control the glycosylation patterns of individual glycoproteins and glycolipids and the physiological consequences of modulating these patterns will not be attained solely using reductionist approaches. The effects of altering
the expression of a glycan-processing enzyme depend on a myriad of subtle interacting factors, such as the sensitivity of local substrate concentrations to changes in local
glycosyltransferase activity. Robust informatics tools are required to reveal the
detailed information that is implicit in the large, complex data sets obtained by quantitative glycomics analyses and thereby provide insight into the mechanistic details of
glycan biosynthesis within a cell or tissue.

## The Complexity of Glycan Function

Glycans have diverse biological functions that are distinct from those of nucleic acids
and proteins. No examples of a glycan acting as a template for the biosynthesis of
another biopolymer are known. Glycans with catalytic activity are extremely rare [4].
Nevertheless, glycans do have critical roles in the maintenance of cell or tissue structure, molecular signal transduction, and cell recognition. The mechanisms by which
glycans perform these diverse functions typically involve the specific interaction of the
glycan with another biopolymer (e.g., lectin, receptor, enzyme, carbohydrate-binding
module, or another glycan). Thus, elucidation of the biological mechanism underlying a glycan's function usually depends on the identification of its specific binding
partner(s). Glycans are often characterized by a high degree of molecular flexibility
and no archetypal molecular interaction model (analogous to nucleotide base-pairing)
has been demonstrated for the interaction of glycans with their binding partners.

One approach to gain insight into the structural basis for the interactions of
glycans with other biomolecules is to systematically search for glycan motifs whose
presence can be correlated to binding specificity. The identification of such motifs and
their biological significance will be facilitated by robust informatics tools that make
it possible to integrate the results of diverse experimental and theoretical analyses
that provide specific data regarding biochemical phenomena (e.g., molecular binding, structure, and dynamics) and biological phenomena (e.g., genotype, phenotype,
development, and disease progression). Key to the success of this approach are methods to digitally represent glycan structure in a way that makes it possible to identify
and quantify structural homology within sets of glycans that share a common binding
specificity or biological activity. Due to the complex, branched nature of glycans, the
identification of structural homology is a fundamentally different problem for glycans
than it is for linear polymers, for which powerful sequence comparison methods (e.g.,
BLAST [5]) have been developed.

## The Explosion of Glycomics Data

Recent developments in analytical technology, especially mass spectrometry, have led
to a rapid increase in the amount of experimental glycomics data being generated.
This is paralleled by other technical developments that have brought about similar
increases in genomic, transcriptomic, and proteomic data. Due to the complexity of

the underlying chemistry, interpretation of glycomics data such as multiple mass spectrometry ($MS^n$) can be very time-consuming for the human analyst. This precludes manual interpretation of the vast amounts of glycomics data currently being generated. Although the development of automated methods for the interpretation, structural assignment, and quantification of high-throughput glycomics data is a high priority, much work remains to be done in this area.

## The Need for Context in Data Interpretation

Even when high-quality structural interpretation of glycomics data is available, assessing its biological relevance is extremely challenging, given the inherent complexity of glycan synthesis and function. The full significance of these data can be ascertained only if they are interpreted within the context of an immense amount of diverse background knowledge. A major goal of the Integrated Technology Resource for Biomedical Glycomics at the University of Georgia is to develop bioinformatics tools that automatically identify pertinent background knowledge based on the type and content of an experimental data set in order to graphically present the data in the context of this knowledge. Realization of this ambitious goal is in sight as a result of recent enabling technologies for knowledge representation, discovery, and presentation.

# Bioinformatic Approaches to Glycobiology

## Ontologies as Repositories of Shared Scientific Knowledge

As described above, methods to represent, store, and extract fundamental knowledge regarding the chemical and biological processes underlying glycan synthesis and function are required in order to make sense of glycomics data. This need can be partially fulfilled by "ontologies," which are formally defined descriptions of the terminology and relationships that comprise a conceptual domain of interest [6]. We have developed three highly expressive, interrelated ontologies (GlycO, ReactO, and EnzyO) to describe the glycomics domain [7,8] (http://glycomics.ccrc.uga.edu/ontologies/).

### Formal Specification of an Ontology

An ontology is often defined as a hierarchical structure of concepts that are organized by their relationships. Each concept is represented by a node in a graph and each relationship is represented by an edge connecting two nodes. Each node and edge is assigned a unique name. In the text that follows, **bold font** is used to indicate each ontological name. For example, the edge designated by the name is_a specifies the hierarchical relationship between a class and its superclass (as in **enzyme is_a catalyst**). Another way of saying this is that the class **catalyst** subsumes the class **enzyme**. Ontologies provide a consistent formal mechanism to categorize objects by specifying their membership in a specific class. The relationship between an object and its class is specified by an edge called is_an_instance_of (as in **mannosidase_1 is_an_instance_of** enzyme). This infrastructure defines a common vocabulary for the domain and a context within which a human or computer program can distinguish objects with similar names. For example, the ontological assertion that Kelvin is_an_instance_of temperature_unit would distinguish the concept specified by the name Kelvin from distinct concepts (such as "Lord Kelvin") with similar names. This semantic annotation provides a context that allows a computer program to retrieve descriptions of relevant digital objects in response to semantic queries (e.g., "find all instances of the class temperature_unit" or "identify the class of the object called Kelvin").

Several different computer languages are available for specifying ontologies. One of the most commonly used is the Web Ontology Language—OWL [9]. OWL builds upon many descriptive languages developed in the 1990s, especially Resource Description Framework (RDF, http://www.w3.org/RDF/), which defines objects and their relationships in terms of subject–predicate–object statements called "RDF triples." Each RDF triple corresponds to an ontological assertion (introduced in the previous paragraph), which itself corresponds to the connection of two nodes by an edge in the graphical representation of the ontology. In order for an ontology to be useful, its representation must be syntactically correct (i.e., conform to the rules of the computer language used to describe it) and logically consistent. This requirement for strict logical consistency places severe constraints on the ontology, distinguishing it from a human language description of the domain. However, human understanding of the ontology is facilitated by software for interpretation and/or graphical representation of the knowledge that it contains.

We have employed OWL to create the GlycO, ReactO, and EnzyO ontologies. Although each of these ontologies contains distinct knowledge, they have been designed to be highly integrated, allowing concepts defined in one ontology to be used to make assertions in another.

## *The GlycO Ontology*

The structural features of glycans and glycoconjugates are formally specified in the GlycO ontology (Figure 8.1a). Hierarchical relationships, especially taxonomy (also called class subsumption, as introduced above) and partonomy (described in the next paragraph), play a critical role in describing this knowledge. The GlycO taxonomy is rooted using two top-level superclasses (chemical_entity and chemical_feature). Subclasses subsumed by the chemical_entity superclass include molecule, residue, and atom while subclasses subsumed by the chemical_feature superclass include ring_form and anomeric_configuration. The is_a relationship is used to construct assertions that define progressively more restrictive subclasses (Figure 8.1b). For example, residue is_a direct subclass of polyatomic_molecular_fragment, which itself is an indirect subclass of chemical_entity. As a result of this hierarchical structure, one can infer, for example, that instances (such as *N*-glycan_core_β-D-Manp) of the subclass residue are also instances of the superclass chemical_entity. This so-called transitive relationship implies that each residue embodies all of the properties that have been assigned to the chemical_entity superclass when it was defined.

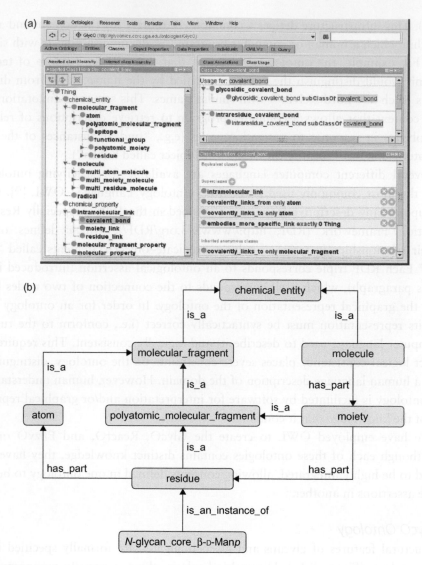

**FIGURE 8.1** (a) Class taxonomy of the GlycO ontology, as viewed in the graphical ontology editing tool Protégé (http://protege.stanford.edu). (b) Class subsumption and partonomy relationships specified in GlycO. Only a small portion of the complex graph corresponding to GlycO is shown. For example, Nglycan_core_β-D-Man$p$ is an indirect instance of residue, and there are several intervening subclasses.

Hierarchical partonomy relationships (e.g., has_part, Figure 8.1b) also play key roles in GlycO. These specify, for example, that a molecule has_part moiety, a moiety has_part residue, and a residue has_part atom. This supports the specification of structure at different levels of complexity in a way that is familiar to biochemists. These relationships are illustrated in the following assertions.

glycopeptide is_a molecule

carbohydrate_moiety is_a moiety

peptide_moiety is_a moiety

glycopeptide has_part carbohydrate_moiety

glycopeptide has_part peptide_moiety
carbohydrate_residue is_a residue
carbohydrate_moiety has_part carbohydrate_residue
carbohydrate_residue has_part atom

This hierarchical scheme, which we call PARCHMENT (PARtonomy of CHeMical ENTities) supports complete (atomistic) representation of structure while facilitating inference based on structural homology at several levels of granularity. Each glycan structure in the GlycO ontology is represented using a canonical representation scheme. As illustrated in Figure 8.2, a canonical tree structure is constructed by connecting all of the residues that are found in a collection of structurally and biosynthetically related glycans (e.g., the set of all *N*-glycans). Each glycan in this collection can then be unambiguously represented by selecting a subtree composed of interconnected nodes in this tree. Each of these canonical residues is a semantically defined object whose structure is explicitly described (e.g., as a β-D-Man*p* residue) and whose chemical context is defined by its position in the canonical tree. In addition to providing an explicit specification of chemical structure, this canonical representation constitutes a framework that allows a considerable amount of biological information to be inferred for each glycan. For example, the "rules" for biosynthesis of each collection of glycans are implied by a distinct canonical tree. The canonical tree constructed for the set of all eukaryotic *N*-glycans (Figure 8.2) includes a node corresponding to the β-D-Man*p* residue in the core structure common to all of these glycans. The presence of this canonical β-D-Man*p* residue in a glycan structure implies that the pathway leading to biosynthesis of the glycan includes a step catalyzed by a specific β-mannosyltransferase (i.e., EC 2.4.1.142). Furthermore, the presence of this canonical residue is also required for interaction with specific proteins (such as peptide *N*-glycosidase F (PNGase F)) that bind to or otherwise interact with the glycan. While these inferences based on the presence of a core β-D-Man*p* residue may be obvious to most glycobiologists, the presence of other combinations of canonical residues may imply the participation of the glycan in a multitude of more poorly understood biological interactions. That is, biological interactions depend on combinations of local chemical structure and chemical context, which is precisely the information that is encoded in the canonical representations. The structural (and perhaps functional) homology of two different glycans is inferred if they share a specific combination of canonical residues.

This semantic infrastructure facilitates the correlation of specific combinations of canonical residues to a specific biological activity or process. Once such correlations are identified, this information can be formally incorporated into the ontology itself. That is, a list corresponding to a specific combination of canonical residues that is associated with a particular biological process (e.g. binding to a receptor) can be generated and specified as an object in the ontology. Then, participation of a previously unknown glycan structure in the biological process is implied if it contains the canonical residues in the relevant list. The robustness of this model is readily testable by comparing such semantic inferences to experimental data.

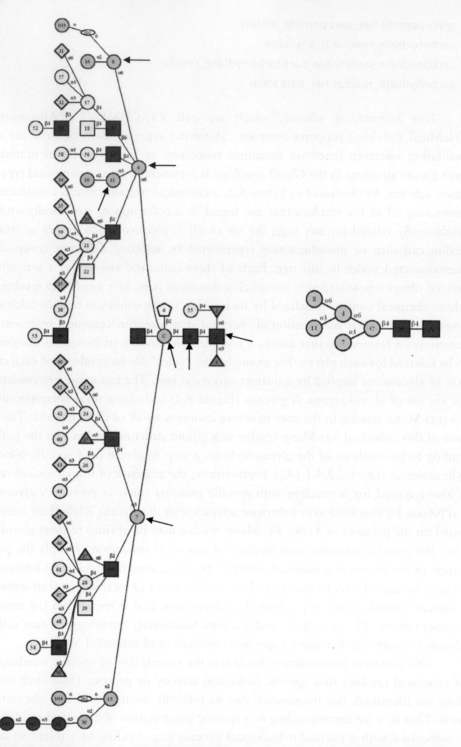

**FIGURE 8.2** Canonical tree representing all *N*-glycans in the GlycO ontology. An individual *N*-glycan is specified by selecting an interconnected subset of the nodes in this tree (i.e., a subtree). This is illustrated by specifying the *N*-glycan commonly known as Man5 by selecting seven nodes (indicated by arrows) in the canonical tree.

## The ReactO Ontology

By providing a means to unambiguously represent the semantics of glycan structure, the GlycO ontology makes it possible to generate semantic assertions that define biochemical reactions and pathways involved in glycan biosynthesis. As metabolism is a broad domain that encompasses far more than glycan biosynthesis, it would be shortsighted to develop an ontology that is limited to reactions involving glycans. Therefore, we developed a separate ontology, which we call ReactO, to describe the basic concepts in the domain of biochemical reactions.

The most important class defined in ReactO is reaction. Each instance of reaction is defined in part by assertions involving the properties consumes_reactant and generates_product (Figure 8.3). In the case of reactions involved in glycan biosynthesis, these assertions refer to structures defined in GlycO. It is well known that biochemical reactions are reversible and that a molecule specified as a reactant in one reaction is a product in the reverse reaction. Nevertheless, each reaction defined in ReactO has a specific direction. This facilitates the specification of kinetic and thermodynamic parameters (such as the equilibrium constant) whose formal definitions often depend on the direction of the reaction. Only the forward reaction is specified in ReactO, and the reverse reaction is merely implied.

Another fundamental class defined in ReactO is catalytic_system, which specifies a particular catalyst (such as an enzyme), along with the kinetic parameters (such as the catalytic constant $k_{cat}$ and the Michaelis constant $K_M$ for each reactant) that characterize the catalyzed reaction (Figure 8.3). As several different enzymes (each with its

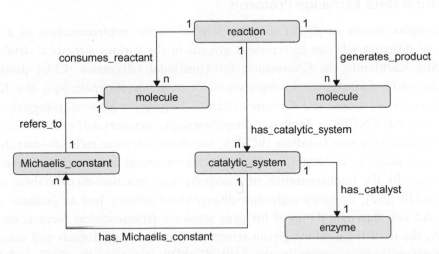

**FIGURE 8.3** Relationships in the ReactO ontology that define a reaction and its catalysis. The consumes_reactant and generates_product relationships define the chemistry of the reaction. A reaction can be associated with more than one catalytic_system, each of which corresponds to the catalysis of the reaction by a particular enzyme. Specification of the Michaelis_constant for the enzyme-catalyzed reaction is also illustrated. As the reaction may consume more than one reactant, more than one Michaelis_constant can be associated with a particular catalytic_system (i.e., a particular enzyme that catalyzes the reaction). Each Michaelis_constant refers_to a particular molecule (i.e., a substrate) that is consumed by the reaction. Note that a Michaelis_constant is not a simple number, but a full-fledged object in the ontology, allowing its relationships to other objects to be fully specified. The catalytic_constant $k_{cat}$ and equilibrium_constant $K_{eq}$ can also be specified, but are not illustrated to simplify the figure.

own set of catalytic constants) can catalyze the same reaction, there is a one-to-many relationship between each reaction and a unique set of relevant catalytic systems. Thus, the reaction itself and its catalysis are logically separated. The enzymes that catalyze the reactions specified in ReactO are defined as instances in a third ontology (described next), which is called EnzyO.

### The EnzyO Ontology

A large body of knowledge regarding the identity of enzymes that catalyze biochemical reactions is available from various sources such as BRENDA [10]. Significant effort has been expended in classifying these enzymes and the reactions that they catalyze. We developed EnzyO as a semantic repository of this information. EnzyO implements two separate but related enzyme taxonomies, which are based on two well-developed enzyme classification systems: the (more general) Enzyme Commission (EC) taxonomy, which classifies enzymes according to the reactions they catalyze, and the CAZy (Carbohydrate Active enZymes) system, which classifies enzymes according to their domain structures and catalytic mechanisms [11]. Furthermore, semantic assertions map each enzyme that is instantiated in EnzyO to the gene that encodes it. Thus, representation of enzymes as instances in EnzyO not only facilitates the specification of catalytic systems in the ReactO ontology but also provides a means of identifying related knowledge regarding the genetics, gene expression patterns and a broad range of other information that provides context required for understanding glycomics data.

## Structural Data Exchange Protocols

The complex nature of glycan structure has led to the implementation of a multitude of different schemes to represent glycans in the various structural databases, including CarbBank, the Consortium for Functional Glycomics (CFG) databases (http://www.functionalglycomics.org/glycomics/common/jsp/firstpage.jsp), the Kyoto Encyclopedia for Genes and Genomes (KEGG, http://www.genome.jp/kegg/glycan/), the GLYCOSCIENCES.de databases (http://www.glycosciences.de/) and the Bacterial Carbohydrate Structure DataBase (BCSDB, http://www.glyco.ac.ru/bcsdb/start.shtml). Significant progress in integrating these diverse structural information sources has been made by the implementation of a carbohydrate structure metadatabase called GlycomeDB ([12], http://www.glycome-db.org/About.action). Just as genomic databases did not eliminate the need for gene sequence representation formats such as FASTA, the need for a robust glycan structural data exchange format still exists. In collaboration with scientists in the EUROCarbDB initiative, the CFG and Soka University in Japan, we developed the GLYcan Data Exchange format GLYDE-II to meet this need (http://glycomics.ccrc.uga.edu/GLYDE-II/). GLYDE-II is an extension of XML that incorporates many of the features of the GlycoCT glycan representation format [13] along with the ability to represent glycans and glycoconjugates at different levels of granularity. The partonomy relationships implemented in GLYDE-II are analogous to those described above for the GlycO ontology. In addition, we have developed a glycan object model (GOM), which is implemented as a collection of Java classes for internal representation of glycan structural data. GOM provides a convenient method for conversion between the file-based data representations, such as GLYDE-II and glycan representations that are implemented within relational databases and ontologies such as GlycO.

## Workflows for Processing Glycomics Data

A large amount of raw data is generated by current analytical glycomics technology, especially mass spectrometry. This data must be processed in order to identify and quantify the glycans and glycoconjugates that give rise to it. Usually, this involves a combination of automated tasks (e.g., generation of mass spectral peak lists by "deisotoping" of raw data) and manual tasks (e.g., interpretation of $MS^n$ data). As algorithms for the interpretation of mass spectral data improve, the role of automated tasks in this process will increase. Automated workflows that control the execution of individual data-processing tasks and keep track of the results are indispensable for the processing of the many large data sets produced by glycomics analysis.

Metadata associate individual pieces of information (i.e., the identity and abundance of a specific glycan) obtained by processing glycomics data with the biological and analytical context in which it was generated. This information is required in order to determine the biological relevance of the glycomics data. It includes, for example, the identity and physiological state of the biological sample that was analyzed and the protocol used for preparing glycans for analysis. Ontologies provide a robust framework for organizing such information by specifying a controlled vocabulary and a uniform taxonomy of metadata concepts. Thus, annotation with semantic metadata is a key aspect of the workflows we design for processing glycomics data.

The technology used for analytical glycomics is advancing rapidly, and new instrumental and data-processing protocols are continually being developed. This volatility in experimental design presents a "moving target" for those charged with developing and maintaining robust workflows to implement these protocols. However, this problem can be successfully managed if it is broken down into reusable, well-defined subtasks, such as conversion to a common file format and extraction of individual spectra or spectral regions for detailed analysis. Our approach to workflow development is thus based on the establishment of a well-defined collection of data-processing modules, each of which performs a discrete task. Thorough descriptions of the data that each module processes (input), the operations performed by each module, and the data that each module generates (output) are necessary to facilitate their reuse as workflow components.

New algorithms that extract complex structural information from analytical glycomics data are continually being developed, often in response to or as part of innovations in analytical instrumentation. These algorithms are developed in many different laboratories, which have different data processing requirements and environments. Therefore, incorporation of the most robust of these algorithms into glycomics data processing workflows is a major bioinformatics challenge because they use different (often proprietary or poorly documented) data formats and are implemented on different computational platforms. Our goal is to work with the glycomics community to establish a Web portal where developers can deposit glycomics data-processing modules (including basic data manipulation modules and complex algorithms for information extraction) for use by the community. We are developing a prototype of this portal (http://glycomics.ccrc.uga.edu/portal/), which will provide a complete description of each module and a search interface to identify modules with appropriate input, output, and functionality, facilitating their incorporation into workflows. We expect that many of the modules will be implemented as Web Services (described below), and thus it will be necessary for the portal to support registries and tools (such as BioMOBY, http://biomoby.org/) that facilitate standardization of Web service implementation and tools for workflow implementation such as Taverna (http://taverna. sourceforge.net/) and jBPM (http://www.jboss.org/jbossjbpm/jbpm_overview/). In some cases, processing large volumes of high-throughput glycomics data would be more efficient if performed locally, so the portal must also support downloadable code for local execution on the client's computer. Developing this suite of highly interactive data processing modules will require considerable communication among the various analytical glycomics laboratories, so the portal should also support a forum for discussing technical issues and requesting new modules or modification of existing modules.

## Web Services

As described above, flexible and modular workflows are required for robust processing of experimental glycomics data. The various modules that comprise such workflows may be written in different computer languages and use different data formats for input and output. Integration of these diverse modules into workflows will be facilitated by implementation of Web services, defined by the World Wide Web Consortium (W3C) as "software system(s) designed to support interoperable machine-to-machine interaction over a network" (http://www.w3.org/TR/ws-arch/). Web services provide an infrastructure for flexible composition and efficient implementation of complex workflows. Each task in the workflow can be carried out on a computer that is best suited for the task (e.g., by virtue of having access to a particular database). This also allows work to be allocated to different computers based on availability of resources and the priority of each processing task. Such a distributed computing infrastructure can be implemented globally, by invoking Web services offered at geographically separated sites, or locally, within a group of computers that are hidden behind a firewall.

A major impediment to the use of Web services is difficulty in identifying specific Web services with the desired functionality and properly formatted input and output content that are compatible with the data to be processed. Several approaches have been implemented to circumvent this barrier, including Web service registries such as BioMOBY (http://biomoby.org/) and semantic annotation protocols such as WSDL-S (http://www.w3.0rg/Submission/WSDL-S/) and SAWSDL (http://www.w3.0rg/2002/

ws/sawsdl/) that aid in the discovery and reuse of Web services. A specialized Web portal (as described above) can also facilitate the use of a focused collection of Web services for use by relatively small groups of scientists (such as the glycomics community) by providing relevant information, a discussion forum, and access to software in the form of open source code or Web services themselves.

## Integrated Data Repositories

Semantic interpretation of large volumes of analytical glycomics data in the context of domain knowledge demands a well-organized computing environment that enables the interaction of experimental data in diverse 10 formats (spreadsheets, raw data files, relational tables, etc.) and ontological representations of knowledge. Our approach to this requirement is GlycoVault [14], a data repository that contains information that is encoded in many different formats but functions essentially as if it were a single information source. Data in GlycoVault are integrated and made available using an applications programming interface (API) that facilitates the querying of this diverse data in the context of concepts defined in the ontologies. Notably, GlycoVault supports the mapping of data to specific concepts and objects that are specified within our ontologies. This capability is absolutely essential in order to implement the graphical interfaces and knowledge discovery tools for complex glycomics data. We are currently exploring the implementation of a more flexible data management system within GlycoVault. One possibility is adapting the GlycoVault API so that it can access data that are stored in large-scale distributed data systems such as Tranche (https://proteomecommons.org/tranche/about.jsp).

## Intuitive Graphical User Interfaces

A glycoscientist seeking to form and test hypotheses on the basis of large volumes of glycomics data must have access to robust tools that facilitate the exploration and interpretation of experimental data in the context of diverse background knowledge. We are developing several such tools, which are implemented as modules in an intuitive graphical user interface (GUI) that obtains information by invoking the GlycoVault API (see above). These tools provide access to the data in the repository by facilitating data queries based on concepts defined in the ontologies, present the retrieved data in the context of the relevant concepts and provide a mechanism to expand, update and curate knowledge embedded in the ontologies.

### *GlycoBrowser*

The GlycO, ReactO, and EnzyO ontologies (described above) embody considerable knowledge regarding glycan structure and biosynthesis. GlycoBrowser (Figure 8.4) provides a graphical interface that exposes this knowledge and allows experimental glycomics data to be presented in the context of this knowledge [15]. The ontologies encode information in the form of a large, complex graph composed of nodes and edges. This includes, for example, all of the information needed to define a metabolic

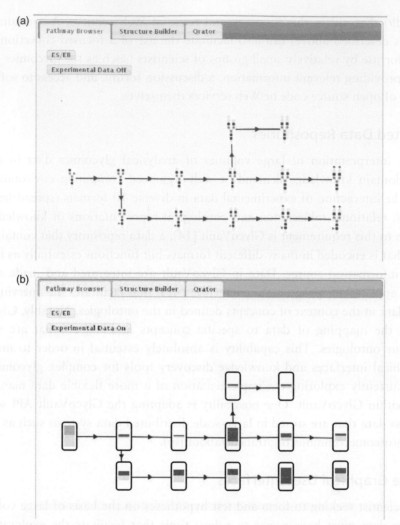

**FIGURE 8.4** Screenshots of the GlycoBrowser graphical interface. (a) GlycoBrowser rendition of the pathway for biosynthesis of biantennary N-glycans. The pathway was constructed by browsing the ontologies via a series of mouse clicks. A specific N-glycan structure (Man3GlcNAc3) was selected as an entry point. By clicking on this structure, the graph was expanded to include reactions in which this structure participates. A single reaction was then chosen. The product of that reaction was then clicked to expand the graph as before. This process of expanding and pruning the graph was continued until the pathway of interest was generated. (b) The same pathway as in (a), except the abundances of the glycans isolated from embryonic stem cells (ES) and embroyd body cells (EB) derived from those cells are illustrated. The ES (green) and EB (red) data are superimposed such that the the total height of the colored bar corresponds to the amount of the glycan in the sample in which it is most abundant. The region of the bar in which the two colors overlap is yellow, analogous to the combination of red and green fluorescence data in immunolocalization microscopy. These data show, for example, that although the initial intermediate (Man3GlcNAc3) is more abundant in ES cells, all the other glycans in this pathway are more abundant in EB cells. These data also suggest that fucosylation of the core GlcNAc promotes sialylation of these biantennary structures. Conceptual correlation of structure and abundance is facilitated by clicking on the "Experimental Data On/Off" button to toggle between the two views (a and b).

pathway responsible for the biosynthesis of a glycan. GlycoBrowser extracts such pathway information by querying the ontologies to identify a path through this complex graph that includes nodes corresponding to the reactants and products of each reaction and the enzymes that catalyze these reactions. The general methods used to accomplish this task can also be applied to identifying and retrieving the nodes and edges corresponding to paths in the ontological graph that correspond to other conceptual viewpoints of the knowledge, such as structural connectivity and partonomy relationships (see above).

The abundance of a glycan in a cell or tissue sample varies, in part, as a result of changes in the expression levels of gene products that catalyze individual steps in biosynthetic pathways. A deeper understanding of these relationships can be obtained by correlating glycomics data (i.e., glycan abundances) to qRT-PCR data (i.e., transcript abundances) or proteomics data. Although some knowledge may be gained by unsupervised statistical correlation analysis of these large data sets, the probability of identifying biologically relevant relationships between the two data sets increases if it is guided by knowledge of the metabolic relationships between glycans and the enzymes that catalyze their synthesis. GlycoBrowser provides the glycoscientist with a tool to identify and graphically represent a metabolic pathway and overlay this graph with quantitative glycomics, transcriptomics, and other experimental data. Thus, changes in the abundance of specific glycosyltransferases and specific glycan products of these enzymes can be simultaneously viewed within a specific biological context chosen by the user.

## Qrator

The usefulness of the GlycO ontology depends on the extent and quality of the knowledge that it contains. The initial versions of GlycO were populated by automated extraction of structural information from several sources, including the KEGG glycan and CarbBank databases. The KEGG glycan database is described further in Chapter 9 and an overview of carbohydrate structure databases is provided in Chapter 10. This process, which required mapping of these structural features to specific canonical objects in the ontology, revealed a large number of errors in the structural databases and highlighted the need for effective curational tools. We are developing a graphical tool called Qrator for this purpose. Qrator imports a glycan structure (in the form of a GLYDE-II file) and converts this to an internal tree representation using our glycan object model (GOM, see above) (Figure 8.5). The canonical tree representing all of the known glycans within the particular class (e.g., all N-glycans) is also imported and converted to GOM format. Then Qrator attempts to map the tree representing the new glycan to a specific subtree of the canonical tree. If an exact match is found, the new structure is imported into GlycO in the form of RDF code that specifies a list of canonical residues that map to the residues in the glycan. If no exact match is found, the best mappings are graphically presented, highlighting exactly where the match was not perfect. A mismatch can occur for two reasons: (1) the new structure is incorrect, and thus does not correspond to a well-established (canonical) structural motif, or (2) the canonical tree within GlycO is incorrect or incomplete. Qrator allows the curator to decide, by simple mouse clicks, whether to discard the incorrect structure, edit it so that it corresponds to a canonical subtree chosen from the graphical list of canonical subtrees, or extend GlycO to include all the canonical structures required to faithfully represent the new structure. Thus, Qrator not only provides a method for curating and importing new structures into GlycO, it allows the curator to generate entirely new canonical objects that can be used to define new classes of glycans.

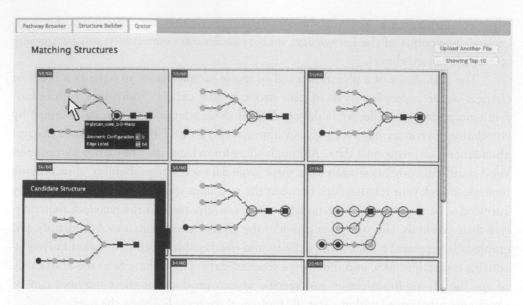

**FIGURE 8.5** Screenshot of the Qrator graphical interface for curating new glycan structures that are to be imported into the GlycO ontology. The new structure (i.e., the candidate) is shown in a "drawer" at the lower-left corner of the window. Mappings of this structure onto the canonical structure tree in GlycO are shown in the other panels. A score corresponding to the accuracy of each mapping is shown in the upper left corner of each panel. Each residue in the candidate that does not match the corresponding residue in the mapping is circled in red. Hovering over any residue with the cursor provides structural information about the residue. As illustrated here, this information includes the identity of the canonical residue and the structural features of the residue in the candidate structure and in its potential canonical representation. In the case illustrated here, the candidate structure has an α-D-Man*p* residue in the core, while the corresponding canonical residue is a β-D-Man*p* residue. Thus, Qrator rapidly detects and highlights structural differences between the candidate structure and its potential canonical representations. The user can choose to select one of the proposed canonical representations (thus correcting perceived errors in the target structure) or extend the canonical tree such that it is consistent with the candidate structure. In the case illustrated here, it is quite likely that the candidate structure is in error, as α-D-Man*p* residues are not usually present in the *N*-glycan core.

## Conclusions

The basic infrastructure for semantic integration of glycomics data is now in place. This includes knowledge repositories, data repositories and tools for glycoinformatics that are required for high-throughput data acquisition, integration of the data, and discovery of knowledge that is inferred by the data. Further development of these tools is required to create informatics systems that significantly advance our understanding of glycobiology. The highest priority at the present time is the extension and utilization of curational tools to populate ontologies with trusted knowledge in the glycomics domain. Ontological specification of this fundamental knowledge will serve as a springboard for the further development of powerful tools for interpreting glycomics data by providing a basis for the expressive annotation of experimental data, which will facilitate the realization of its biological relevance. In addition, cooperation of the global scientific community in the development of data exchange standards, Web services, and integrated ontologies will be critical for the advancement of glycobiology and its application to biomedicine and biotechnology.

## Acknowledgments

This work was funded by the NIH/NCRR-funded Integrated Technology Resource for Biomedical Glycomics (P41 RR018502).

# References

1. Ohtsubo K, Marth JD. Glycosylation in cellular mechanisms of health and disease. *Cell* 2006; **126**:855–67.

2. Zaia J. Mass spectrometry and the emerging field of glycomics. *Chemistry and Biology* 2008;**15**:881–92.

3. Laine RA. A calculation of all possible oligosaccharide isomers both branched and linear yields 1.05×1012 structures for a reducing hexasaccharide: the isomer barrier to development of single-method saccharide sequencing or synthesis systems. *Glycobiology* 1994;**4**:759–67.

4. de Figueiredo P, Terra B, Anand JK, Hikita T, Sadilek M, Monks DE, Lenskiy A, Hakomori S, Nester EW. A catalytic carbohydrate contributes to bacterial antibiotic resistance. *Extremophiles* 2007;**11**:133–43.

5. Altschul SF, Gish W, Miller W, Myers EW, Lipman DJ. Basic local alignment search tool. *J Mol Biol* 1990;**215**:403–10.

6. Gruber TR. A translation approach to portable ontology specifications. *Knowledge Acquisition* 1993;**5**:199–220.

7. Sahoo SS, Thomas C, Tartir S, York WS, Sheth A. Knowledge modeling and its applications in life sciences: A tale of two ontologies. The World Wide Web Conference 2006, Edinburgh, Scotland, May 2006;22–26.

8. Thomas CJ, Sheth AP, York WS. Modular ontology design using canonical building blocks in the biochemistry domain. In: Proceedings of the International Conference on Formal Ontology in Information Systems. IOS Press, Baltimore, Maryland (USA) November 2006;9–11.

9. Horrocks I, Patel-Schneider PF, van Harmelen F. From SHIQ and RDF to OWL: The making of a Web Ontology Language. *J Web Semantics* 2003;**1**:7–26.

10. Schomburg I, Chang A, Schomburg D. BRENDA, enzyme data and metabolic information. *Nucleic Acids Res* 2002;**30**:47–9.

11. Cantarel BL, Coutinho PM, Rancurel C, Bernard T, Lombard V, Henrissat B. The Carbohydrate-Active EnZymes database (CAZy): an expert resource for glycogenomics. *Nucleic Acids Res* 2009;**37**:D233–8.

12. Ranzinger R, Herget S, Wetter T, von der Lieth C-W. GlycomeDB–integration of open-access carbohydrate structure databases. *BMC Bioinformatics* 2008;**9**:384.

13. Herget S, Ranzinger R, Maass K, Lieth CW. GlycoCT-a unifying sequence format for carbohydrates. *Carbohydr Res* 2008;**343**:2162–71.

14. Nimmagadda S, Basu A, Eavenson M, Han J, Janik M, Narra R, Nimmagadda K, Sharma A, Kochut KJ, Miller JA, York WS. GlycoVault: A bioinformatics infrastructure for glycan pathway visualization, analysis, and modeling. In: Fifth International Conference on Information Technology: New Generation, p. 692–7.

15. Eavenson M, Janik M, Nimmagadda S, Miller JA, Kochut KJ, York WS. GlycoBrowser: A tool for contextual visualization of biological data and pathways using ontologies. In: Istrail S, Pevzner P, Waterman M, editors. *Bioinformatics Research and Applications*. Berlin/Heidelberg: Springer; 2008, p. 305–16.

# 9

# KEGG GLYCAN for Integrated Analysis of Pathways, Genes, and Glycan Structures

Kosuke Hashimoto and Minoru Kanehisa
*BIOINFORMATICS CENTER, INSTITUTE FOR CHEMICAL RESEARCH, KYOTO UNIVERSITY, KYOTO, JAPAN*

## Introduction

Bioinformatics approaches have recently been applied to various issues in the field of glycan research, such as phylogenetic analysis of genes and proteins, predictions of glycan structures from the output of mass spectrometry or gene microarrays, and development of algorithms for comparison and feature extraction of glycan structures. One aspect behind this trend is that the data obtained in experiments is rapidly increasing because of the improvement of experimental techniques. In other words, the sheer quantity of data requires the development of new computational methods. Another aspect is that the database resources for glycomics analysis have become more widely available. Chapter 8 provides an overview of the general problem of integrating glycomics information and structural databases and Chapter 10 provides an overview of carbohydrate structure databases. This chapter will specifically focus on the KEGG approach for integration of glycomics databases.

KEGG is an integrated knowledge base for understanding higher level functions of cellular processes and organism behaviors [1]. KEGG consists of genomic, chemical, and systemic functional information, and the KEGG GLYCAN resource for glycomics research is also organized according to these three perspectives [2]. Systemic functional information in glycobiology includes how glycans are synthesized and how they function, which are represented by pathway maps and structure maps. Genomic information includes glycogenes, such as for glycosyltransferases and lectins, as well as orthologous relationships across organisms. Chemical information includes not only known glycan structures but also possible structures that can be synthesized by a given set of genes. Here we describe how different types of data and knowledge are represented, provided, and related in KEGG GLYCAN (http://www.genome.jp/kegg/glycan/).

**Handbook of Glycomics**
Copyright © 2009 by Elsevier Inc. All rights of reproduction in any form reserved.

**Table 9.1  Major resources in KEGG GLYCAN**

| Systemic functional information | Pathway maps/Structure maps | http://www.genome.jp/kegg/pathway.html#glycan |
| | CSM (Composite Structure Map) | http://www.genome.jp/kegg-bin/draw_csm |
| Genomic information | Glycosyltransferases | http://www.genome.jp/kegg-bin/get_htext?k001003.keg |
| | Glycan-binding proteins | http://www.genome.jp/kegg-bin/get_htext?k004091.keg |
| | Glycosyltransferase reactions | http://www.genome.jp/kegg-bin/get_htext?br08203.keg |
| | GECS (Gene Expression to Chemical Structure) | http://www.genome.jp/tools/gecs/ |
| Chemical information | Glycan Structure Database | http://www.genome.jp/dbget-bin/www_bfind?glycan |
| | KCaM (searching for glycans) | http://www.genome.jp/tools/kcam/ |
| | KegDraw (drawing glycans) | http://www.genome.jp/download/ |

# Glycan Biosynthesis and Metabolism

## Metabolic Pathway Maps

The KEGG PATHWAY is a collection of manually drawn pathway maps representing molecular interaction networks, including metabolic pathways, regulatory pathways, molecular complexes, and structural relationships. In the maps of glycan biosynthesis and metabolism (see Table 9.1 for the major resources and tools in KEGG GLYCAN), alterations of glycan structures are illustrated with their catalytic enzymes. Figure 9.1 shows the overall relationships among the 15 currently available pathway maps. A connection between two maps indicates that metabolites synthesized in one map are used as substrates in the other. For example, glycans synthesized in N-glycan biosynthesis are degraded in the N-glycan degradation pathway or are elongated in the pathways of keratan sulfate biosynthesis and high-mannose type N-glycan biosynthesis.

To demonstrate how pathways are represented in the computer, we use the N-glycan biosynthesis pathway as an example. Figure 9.2a shows part of the N-glycan biosynthesis pathway in humans, preceding the transfer to an asparagine residue. Circles and arrows represent glycan structures and reactions, respectively. Enzymes that catalyze reactions are represented by gene names or EC numbers in boxes attached to the arrow. Each object has a hyperlink to detailed information; for example, the boxes link to the genes encoding the enzymes in the KEGG GENES database. Circles also have a hyperlink to the corresponding glycan structures in the KEGG GLYCAN database. Although the biosynthesis pathway of the N-glycan precursor described here is well conserved among a wide range of eukaryotes, some organisms are known to lack a part of the pathway (see below). One of the features of the KEGG PATHWAY is to be able to show such differences in genes and pathways among organisms. When an organism is selected from the pull-down menu at the top,

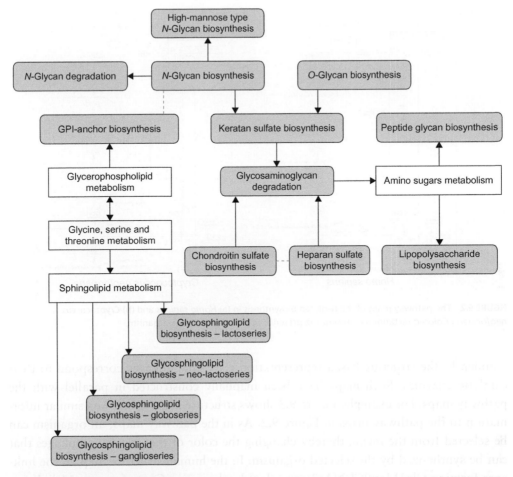

**FIGURE 9.1**   The overall relationship of 15 pathway maps for glycan biosynthesis and metabolism (shaded) and some related metabolic pathway maps.

the boxes of the corresponding enzymes are colored and their hyperlinks are changed to reflect the organism-specific pathway. Figure 9.2b shows the pathway displayed by selecting *Cryptococcus neoformans*. Unlike the human pathway, the last three enzymes that transfer glucoses are not colored. No sequences similar to these proteins (ALG6, 8, and 10) have been detected in the *C. neoformans* genome, which was sequenced in 2005. In addition, a previous study identified the structure of Man9GlcNAc2 in vivo and in vitro in this organism [3], indicating that enzymatic reactions proceed along the pathway with colored enzymes and stop at the second structure from the bottom. In this way, the KEGG PATHWAY can provide specific pathways depending on genes that the selected organism possesses in its genome.

## Glycan Structure Maps

The pathways described above represent the process of glycan biosynthesis as consecutive reactions. On the other hand, another representation is possible, in which the glycosyltransferases synthesize glycans by adding monosaccharides in a stepwise

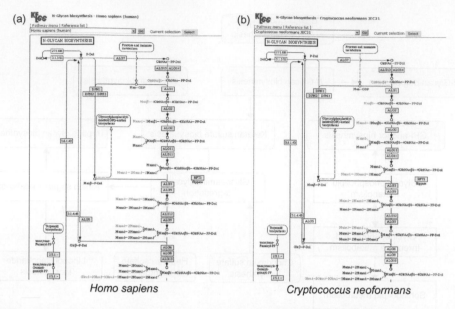

**FIGURE 9.2**   The pathway maps of the *N*-glycan biosynthesis in (a) *Homo sapiens* and (b) *Cryptococcus neoformans*. Colored rectangles represent the presence of the enzyme in the organism.

manner. In the structure-based representation glycosidic linkages correspond to their catalytic enzymes. Such maps have been manually constructed in parallel with the pathway maps. For example, Figure 9.3 shows structure maps containing similar information to the pathway maps in Figure 9.2. As in the pathway maps, an organism can be selected from the menu, thereby changing the color of the glycosidic linkages that can be synthesized by the selected organism. In the human structure map all the linkages forming Glc3Man9GlcNAc2 are colored, whereas in *C. neoformans* 11 linkages forming Man9GlcNAc2 are colored. A feature of the structure map is that structural differences are intuitively understandable.

We also developed the Composite Structure Map (CSM), which is a type of structure map [4] that represents all possible variations of glycan structures in a tree format. The tree was computed by taking all the glycan structures in the KEGG GLYCAN structure database and superimposing them into one unified structure. Each monosaccharide is represented by a symbol according to the standards set down by the Consortium for Functional Glycomics. Using this tool it is possible to color the edges corresponding to genes and colors stored in a local file. Thus, experimental results, such as microarray gene expression data, can be displayed using this tool.

## Genes Related to Glycan Biosynthesis

Rapid improvements in sequencing techniques have allowed us to determine a variety of genome sequences. Sequencing of about 200 genomes was completed in 2007 alone, and a large number of genome projects are in progress. Such genomic information enables the pathway map and the structure map to represent systemic differences

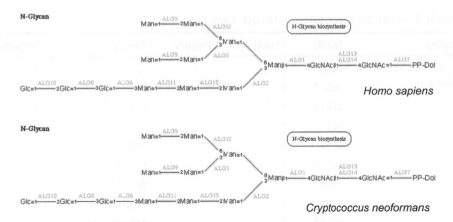

**FIGURE 9.3** The structure maps of the *N*-glycan precursor in *Homo sapiens* and *Cryptococcus neoformans*. Colored edges represent the presence of the enzyme in the organism.

in organisms. We describe below how newly determined genomes are integrated into the database and how the data connections are constructed.

## Definition of Ortholog Groups

KEGG stores all nucleotide and amino acid sequences in complete genomes generated from publicly available resources, mostly NCBI RefSeq [5]. Biological features of the genes and proteins obtained from other resources are annotated by manual and computational methods, such as KAAS (KEGG Automatic Annotation Server) [6]. Proper annotations can yield useful information even regarding non-model organisms. One of the most important annotations for linking genomes to biological systems is the KEGG ORTHOLOGY (KO) system, in which orthologous genes are defined as an ortholog group.

Many KO groups of proteins related to glycans are defined, such as glycosyltransferases, glycosidases, and glycan-binding proteins. In particular, ortholog groups of glycosyltransferases were comprehensively defined, including 20 ortholog groups of sialyltransferases and 11 ortholog groups of fucosyltransferases. Table 9.2 shows the hierarchical classification of the KO groups, which were recently updated following a comprehensive analysis of glycosyltransferase families. This classification divides glycosyltransferases into six functional categories, each of which contains several families. For example, the category of *N*-glycan precursor biosynthesis contains 13 ALG families and the DPM1 family, and the category of GPI-anchor biosynthesis contains 5 PIG (phosphatidylinositol glycan) families. Each family consists of one or more KO groups that share sequence similarity with each other. The classification system is available in the KEGG BRITE database (http://www.genome.jp/kegg/brite.html).

## Orthologous Information for the Pathway and the Structure Maps

Genes in newly sequenced genomes are assigned K numbers that indicate the specific ortholog group. Once K numbers are assigned to genes, the genes automatically

Table 9.2  A list of glycosyltransferase families

| Category | Family | Orthologs | Category | Family | Orthologs |
|---|---|---|---|---|---|
| N-Glycan Precursor biosynthesis | ALG7 | 1 | Glycan extension | α4 | 4 |
| | ALG13 | 1 | | β3 | 17 |
| | ALG14 | 1 | | GLCAT | 4 |
| | ALG1 | 1 | | β4 | 12 |
| | ALG2 | 1 | | β4GALNT | 2 |
| | ALG11 | 1 | | MGAT1 | 3 |
| | DPM1 | 1 | | MGAT3 | 1 |
| | ALG3 | 1 | | MGAT4 | 1 |
| | ALG12 | 1 | | MGAT5 | 2 |
| | ALG9 | 1 | | ABO | 4 |
| | ALG5 | 1 | | GCNT | 4 |
| | ALG6 | 1 | | LARGE | 2 |
| | ALG8 | 1 | | EXTL | 5 |
| | ALG10 | 1 | | KTR | 3 |
| GPI-anchor biosynthesis | PIGA | 1 | | MNN1 | 5 |
| | PIGM | 1 | | MNN9 | 3 |
| | PIGV | 1 | | ATXYLT | 1 |
| | PIGB | 1 | Terminal extension | ST | 20 |
| | PIGZ | 1 | | FUT1 | 1 |
| O-Glycan and glycolipid biosynthesis | UGCG | 1 | | FUT3 | 8 |
| | OGT | 1 | | FUT8 | 1 |
| | XYLT | 1 | Unclassified | Polysaccharides | 5 |
| | POMT | 1 | | Bacterial lipopolysaccharides | 6 |
| | POFUT | 1 | | Other glycosyltransferases | 8 |
| | GALNT | 1 | | Sulfotransferases | 34 |
| Others | ATXT | 3 | | | |
| | UGT | 7 | | | |
| | MGD | 1 | | | |

become linked to pathway maps and structure maps because the maps correspond to existing ortholog groups. Thus, organism-specific pathways can be displayed, reflecting their genomic information. This system can provide functional information based on the genomes of organisms on which few experiments have been performed.

It is well known that there is a great diversity of glycan structures among organisms. The ortholog information also provides an important clue for understanding the differences in glycans among organisms. ALG7, which catalyzes the transfer of the first GlcNAc residue to dolichol phosphate, is of very old origin and is conserved in almost all eukaryotes and some prokaryotes. On the other hand, as described above, ALG6, ALG8, and ALG10, which catalyze the transfer of the glucose residue, are not encoded in some parasite genomes, such as *Entamoeba histolytica*, *Plasmodium falciparum*, and *C. neoformans* [3]. The absence of the enzymes seems to lead to differences in structures and functions. In fact, a recent study demonstrated that the smaller *N*-glycan precursors lack the quality control functions of glycoprotein folding and the endoplasmic reticulum (ER)-associated degradation of proteins [7]. Thus, this system is helpful to reveal structural differences and to predict functional differences among organisms.

# Glycan Structures

## Glycan Structure Database

In contrast to the high-throughput and high-quality sequence data for nucleic acids and proteins, glycan structures with full linkage information are low-throughput data. We cannot comprehensively elucidate glycan structures in a cell, such as the genome and the transcriptome, because of the difficulties in determining the types of glycosidic linkages. Nevertheless, a number of glycan structures have been determined and stored in multiple databases. The GLYCAN database includes about 11 000 glycan structures, mostly derived from CarbBank [8], KEGG PATHWAY, and the literature. Each entry has a unique structure, which is identified by an accession number starting with the letter "G." Each entry is annotated with various structural properties, such as the composition, mass, class, and links to the original CarbBank entries. The entries in the GLYCAN database can be searched using keywords (e.g., accession number, commonly used names, or composition) or using a structure search. In addition, an application programming interface (API) is available to allow programmers to access all data in KEGG, including glycan information, from external programs.

## Tools for Structure Search and Drawing Glycans

The structure search tool is called KCaM, which utilizes a dynamic programming technique and a theoretically proven efficient algorithm for finding the maximum common subtree between two trees [9]. KCaM provides two types of structure searches, an exact matching algorithm and an approximate matching algorithm. The former aligns both monosaccharides and glycosidic linkages without any gaps, whereas the latter aligns monosaccharides allowing gaps in the alignment. The exact matching algorithm thus uses stricter criteria for identifying similar structures. Both types of search provide local and global options. The local approximate matching algorithm does not penalize gaps in unaligned regions, whereas the global version does. Thus, only conserved regions can be found using the local approximate matching algorithm. The local exact matching algorithm simply finds the first largest matching subtree, although the global algorithm attempts to find as many matching subtrees as possible.

As a guideline, local exact matching should be applied to search for specific structures related to the query. In contrast, local approximate matching can be applied to search for general structures related to the query.

Another useful tool is KegDraw, which is a freely available Java application for creating drawings of compound and glycan structures. KegDraw consists of two drawing modes: compound mode for drawing chemical compounds in a similar way as ChemDraw, and glycan mode for drawing glycan structures with monosaccharide units. In glycan mode, glycan structures can be drawn in a variety of ways. The simplest method is selecting monosaccharides and glycosidic linkages from popup menus one by one (Figure 9.4). Uncommon types of monosaccharides and linkages can be entered in the text box. It is also possible to use a structure stored in KEGG GLYCAN as a template; the structure can be imported using its accession number. Predefined template structures and a bracket symbol are also available. KegDraw handles files for input in the KCF (KEGG Chemical Function) format [10] and for output in the KCF format, PNG Image, and the LINUCS (LInear Notation for Unique description of Carbohydrate Sequences) format [11]. Glycan structures drawn in KegDraw can be used as queries to search through KEGG GLYCAN and CarbBank.

# Bioinformatics Approach to Glycomics Research

## From Genomic Information to Chemical Information

We described KEGG GLYCAN resources from the viewpoint of three types of information, namely genomic, chemical, and systemic functional information. Currently, genomic information is the most abundant and the most comprehensive. In particular, many glycosyltransferase substrate specificities have been experimentally characterized. With such genomic information as the point of departure, we attempted first to

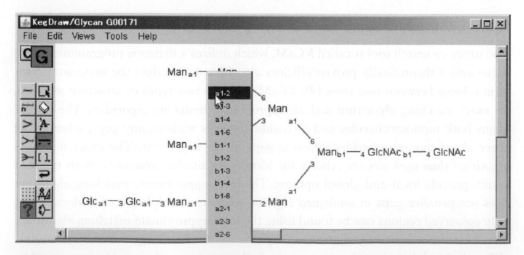

**FIGURE 9.4** An example of inputting a glycan structure using KegDraw. The structure can be drawn by selecting monosaccharides and glycosidic linkages from popup menus.

determine the repertoire of glycosyltransferases in various organisms, and then to elucidate differences in the synthesizable glycans among organisms.

More concretely, we investigated 36 eukaryotic genomes to obtain all the glycosyltransferases and defined 53 families based on sequence similarity. Using the repertoire of the families, we examined differences among the major glycan structures, including the *N*-glycan precursor and the GPI-anchor core structure among the organisms. As a result, glycosyltransferases seem to be quite different among families in the degree of conservation and in the number of paralogs. For example, glycosyltransferases involved in the *N*-glycan precursor or the GPI-anchor biosynthesis are well conserved in a wide range of organisms except for a few parasites and algae, which are likely to synthesize smaller structures due to missing enzymes. At the same time, conserved glycosyltransferases appear to have few paralogs. On the other hand, glycosyltransferases for middle or terminal extensions, such as sialyltransferases, are present only in a fraction of the organisms and have many paralogs. The different trends in glycosyltransferase evolution should correspond to differences in glycan structures and functions.

## Predicting Glycan Structures from Gene Expression Data

Several powerful experimental instruments for glycan purification and analysis have been developed and successively improved, such as high-performance liquid chromatography and mass spectrometry. However, even with these advances, it is still difficult to experimentally determine detailed glycan structures because of their complicated forms. Bioinformatics methods may be used to integrate different types of data to obtain structural information.

Due to this background, we have developed a method to predict glycan structures from gene expression data, utilizing KEGG resources, such as the reaction library of glycosyltransferases and the glycan structure database [12,13]. The method is based on the expression levels of genes encoding glycosyltransferases and the co-occurrence score of the glycosyltransferase reactions. The co-occurrence score is designed to represent how frequently two glycosyltransferase reactions occur simultaneously. To calculate the score, we first extracted all the *N*-glycan structures from KEGG GLYCAN. Second, each glycan was broken down into reaction components, where a reaction component consists of two adjacent monosaccharides and their linkage. Finally, we obtained co-occurrence scores between all the combinations of reaction components by computing the correlation coefficient of reaction components that appeared in a glycan structure.

The predictor calculates candidate scores from the combination of the expression levels and the co-occurrence score, thereby presenting predicted glycans according to their scores. Currently, target structures are *N*-glycans stored in the database as well as intermediate structures, estimated from existing structures. This prediction program is available in GECS (Gene Expression to Chemical Structure), which is a collection of prediction methods linking genomic or transcriptomic contents of genes to chemical structures of biosynthetic substances (http://www.genome.ad.jp/tools/gecs/).

# Algorithms for Feature Extraction of Glycan Structures

## Capturing Complex Patterns in Glycans Using Probabilistic Models

A large number of algorithms for analysis of nucleotide and protein sequences have been developed, whereas algorithms and tools to analyze glycan structures are still limited. However, there are many areas of research on tree structures using models in theoretical computer science or probabilistic models such as hidden Markov models (HMM) [14] and stochastic context-free grammars [15]. Several approaches to glycan structures based on such algorithms have recently been proposed. We introduce two probabilistic models, known as PSTMM (probabilistic sibling-dependent tree Markov model) [16] and OTMM (ordered tree Markov model) [17].

HMM is a widely used Markov model for protein sequence analysis, as well as for processing strings such as speech recognition and natural language processing. PSTMM is an extension of the HMM for glycans, namely labeled ordered trees. In this model, a vertex (a monosaccharide) depends not only on the parent, as in the HMM, but also on the immediately elder sibling. PSTMM thus incorporates a hidden dependency pattern that is able to capture relationships that may exist across siblings or even further away. In addition, PSTMM can approach multiple tree alignment from the perspective of using a probabilistic model for aligning tree structures, much as HMMs apply to multiple sequence alignment.

Although computational experiments showed that PSTMM was able to capture complex patterns of both synthetically generated trees and real glycans derived from KEGG GLYCAN, there remained two problems to be solved in the model. First, applying PSTMM to large-sized data is intractable due to time and space complexity. Second, for small data sets, such as those found in glycobiology, PSTMM would overfit the data. OTMM overcomes these problems by introducing limited dependencies into the model. In PSTMM, every child depends on its parent, whereas the parent–child dependencies in OTMM are limited to the dependencies of the eldest siblings on their parents. The effectiveness of OTMM was evaluated using a variety of data sets and was found to significantly reduce the computation time. The performance analysis also indicated that OTMM avoided overfitting the training data, while preserving the predictive power of PSTMM. Figure 9.5 shows the state transition diagram of an OTMM that was trained from N-glycans. A solid line corresponds to a parent–child state transition while a dashed line represents a sibling state transition. This diagram indicates that the model captures N-glycan extension and terminal patterns, including GlcNAc, Gal, and sialic acids, as well as the core structure, which starts with GlcNAc and Man.

## Capturing Complex Patterns in Glycans using Data Mining Techniques

PSTMM and OTMM work well for aligning multiple glycan trees and find patterns located at almost the same position in different glycans. However, conserved patterns are often found at different branches or leaves in glycans. Therefore, a method, called

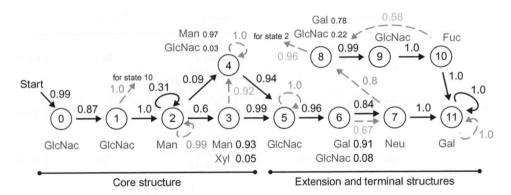

**FIGURE 9.5** The state transition diagram of an ordered tree Markov model (OTMM) that was trained from *N*-glycans. A solid line and a dashed line represent a parent–child state transition and a sibling state transition, respectively. Numbers in circles and above lines indicate states and probabilities of state transitions, respectively. A typical *N*-glycan starts with state 0 for the first GlcNAc, followed by state 1 for the second GlcNAc.

"alpha-closed frequent subtree miming" has been developed for mining motifs and significant patterns which appear in different positions in multiple glycans [18].

This method is based on frequent subtree mining, which is an extension of frequent item set mining, a traditional framework in the field of data mining. In particular, the alpha-closed method was extended from two powerful methods, namely mining closed frequent subtrees and mining maximal frequent subtrees [19]. The method of closed frequent subtrees can remove redundant subtrees when frequencies of appearance of subtrees that are in the inclusion relation are equal to each other. On the other hand, the method of maximal frequent subtrees removes all the redundant subtrees that are in the inclusion relation excluding the largest frequent subtrees. The alpha-closed method can flexibly adjust the removal level of redundant subtrees by varying the parameter alpha. In other words, alpha-closed is a generalized method integrating the closed method and the maximal method. If alpha = 0, the alpha-closed method is equivalent to mining closed frequent subtrees, while if alpha = 1, it is equivalent to mining maximal frequent subtrees. The only criterion to rank frequent subtrees obtained in the method is their frequencies. However, frequencies might not be suitable, since they are easily biased by the frequencies of monosaccharides. Thus, obtained subtrees are ranked according to *P*-values from statistical hypothesis testing.

As an example of applying the alpha-closed method to glycan data, we present the top five subtrees in the analysis using all glycan structures in KEGG GLYCAN (Figure 9.6). They were obtained with alpha = 0.4 and were ranked according to Fisher's exact test. It is clear that all these top ranked subtrees have individual biological significance. The first-ranked and the fifth-ranked subtrees are Lewis x and Lewis a, respectively. The other three subtrees are core structures of *O*-glycan and glycosphingolipid. Core structures of *N*-glycan and known other motifs also appear in the following ranks. These results indicate that the method allows us to capture biologically significant motifs. The method can also be applied to a larger

| Rank | Substructures | Name | Frequency | P-value |
|------|---------------|------|-----------|---------|
| 1 | 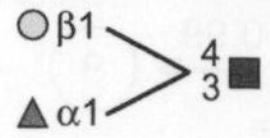 ○β1 ▲α1 >4/3■ | Lewis x | 381 | 1.6e-46 |
| 2 | 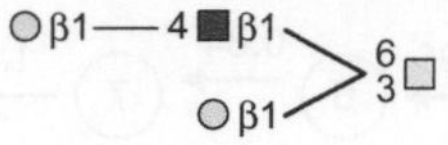 ○β1—4■β1 ○β1 >6/3□ | O-glycan core | 164 | 1.1e-40 |
| 3 |  ○β1—3■β1—3○β1—4● | Glycosphingolipid core lactoseries | 109 | 5.0e-26 |
| 4 |  ○β1—3■β1—3○ | Glycosphingolipid core lactoseries | 233 | 5.6e-26 |
| 5 | 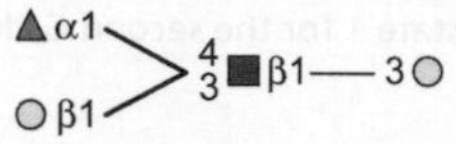 ▲α1 ○β1 >4/3■β1—3○ | Lewis a | 83 | 8.2e-26 |

**FIGURE 9.6** Top five frequent subtrees obtained using the alpha-closed method. Monosaccharides are represented by the symbols defined by the Nomenclature Committee of the Consortium for Functional Glycomics.

data set. If we come to obtain a comprehensive glycan repertoire for a cell, the method will be useful to find motifs that are specifically expressed in particular cells such as cancer cells.

## Concluding Remarks

Bioinformatics methods for glycans have been developed mostly as extensions of those for proteins and nucleic acids; namely, extensions from linear chain molecules to branched chain molecules. There are, however, additional features that need to be considered for glycan bioinformatics. We emphasized the importance of integrated analysis of pathways, genes, and glycan structures. This is especially true in order to make full use of the genome sequence data being generated for an increasing number of organisms. All protein sequences can be obtained directly from the genome sequence once a proper gene model is determined. As we indicated here, all glycan structures (backbone structures excluding modifications) may also be obtained indirectly from the genome sequence through our knowledge on biosynthetic pathways and enzymes.

Glycan structures are likely to be less fixed than protein sequences determined by template-based biosynthesis, which would make the understanding of structure–function relationships more difficult. We introduced new algorithms, PSTMM and OTMM, for extraction of glycan structural motifs. These probabilistic models are suitable for analyzing populations of structures, but their success depends on the experimental data for use as a training data set. Thus, bioinformatics approaches have to be developed in conjunction with well-designed experiments for data acquisition.

There are already successful bioinformatics methods for analyzing experimental data, such as interpretation of MS data [20]. We expect that bioinformatics contributes to improving the efficiency of experiments and that experiments will be designed with consideration of subsequent data accumulation and utilization.

# References

1. Kanehisa M, Araki M, Goto S, Hattori M, Hirakawa M, Itoh M, Katayama T, Kawashima S, Okuda S, Tokimatsu T, Yamanishi Y. KEGG for linking genomes to life and the environment. *Nucleic Acids Res* 2008;**36**:D480–4.

2. Hashimoto K, Goto S, Kawano S, Aoki-Kinoshita KF, Ueda N, Hamajima M, Kawasaki T, Kanehisa M. KEGG as a glycome informatics resource. *Glycobiology* 2006;**16**(5):63R–70R.

3. Samuelson J, Banerjee S, Magnelli P, Cui J, Kelleher DJ, Gilmore R, Robbins PW. The diversity of dolichol-linked precursors to Asn-linked glycans likely results from secondary loss of sets of glycosyltransferases. *Proc Natl Acad Sci USA* 2005;**102**(5):1548–53.

4. Hashimoto K, Kawano S, Goto S, Aoki-Kinoshita KF, Kawashima M, Kanehisa M. A global representation of the carbohydrate structures: a tool for the analysis of glycan. *Genome Inform* 2005;**16**(1):214–22.

5. Pruitt KD, Tatusova T, Maglott DR. NCBI reference sequences (RefSeq): a curated non-redundant sequence database of genomes, transcripts and proteins. *Nucleic Acids Res* 2007;**35**:D61–5.

6. Moriya Y, Itoh M, Okuda S, Yoshizawa AC, Kanehisa M. KAAS: an automatic genome annotation and pathway reconstruction server. *Nucleic Acids Res* 2007;**35**:W182–5.

7. Banerjee S, Vishwanath P, Cui J, Kelleher DJ, Gilmore R, Robbins PW, Samuelson J. The evolution of N-glycan-dependent endoplasmic reticulum quality control factors for glycoprotein folding and degradation. *Proc Natl Acad Sci USA* 2007;**104**(28):11676–81.

8. Doubet S, Bock K, Smith D, Darvill A, Albersheim P. The complex carbohydrate structure database. *Trends Biochem. Sci.* 1989;**14**(12):475–7.

9. Aoki-Kinoshita KF, Ueda N, Mamitsuka H, Kanehisa M. ProfilePSTMM: capturing tree-structure motifs in carbohydrate sugar chains. *Bioinformatics* 2006;**22**(14):e25–34.

10. Hattori M, Okuno Y, Goto S, Kanehisa M. Development of a chemical structure comparison method for integrated analysis of chemical and genomic information in the metabolic pathways. *J Am Chem Soc* 2003;**125**(39):11853–65.

11. Bohne-Lang A, Lang E, Forster T, von der Lieth CW. LINUCS: linear notation for unique description of carbohydrate sequences. *Carbohydr Res* 2001;**336**(1):1–11.

12. Kawano S, Hashimoto K, Miyama T, Goto S, Kanehisa M. Prediction of glycan structures from gene expression data based on glycosyltransferase reactions. *Bioinformatics* 2005;**21**(21):3976–82.

13. Suga A, Yamanishi Y, Hashimoto K, Goto S, Kanehisa M. An improved scoring scheme for predicting glycan structures from gene expression data. *Genome Inform* 2007;**18**(1):237–46.

14. Rabiner LJ, Juang BH. An introduction to hidden Markov models. *IEEE ASSP Mag* 1986;**3**:4–16.

15. Lari K, Young SJ. The estimation of stochastic context-free grammars using the inside-outside algorithm. *Computer Speech and Language* 1990;**4**:35–6.

16. Aoki KF, Yamaguchi A, Ueda N, Akutsu T, Mamitsuka H, Goto S, Kanehisa M. KCaM (KEGG Carbohydrate Matcher): a software tool for analyzing the structures of carbohydrate sugar chains. *Nucleic Acids Res* 2004;**32**:W267–72.

17. Hashimoto K, Aoki-Kinoshita KF, Ueda N, Kanehisa M, Mamitsuka H. A new efficient probabilistic model for mining labeled ordered trees applied to glycobiology. *ACM TKDD* 2008;**2**(1). Article No. 6.

18. Hashimoto K, Takigawa I, Shiga M, Kanehisa M, Mamitsuka H. Mining significant tree patterns in carbohydrate sugar chains. *Bioinformatics* 2008;**24**(16):i167–73.

19. Chi Y, Xia Y, Yang Y, Muntz RR. Mining Closed and Maximal Frequent Subtrees from Databases of Labeled Rooted Trees. *IEEE TKDE* 2005;**17**(2):190–202.

20. Goldberg D, Sutton-Smith M, Paulson J, Dell A. Automatic annotation of matrix-assisted laser desorption/ionization N-glycan spectra. *Proteomics* 2005;**5**(4):865–75.

# 10

# Carbohydrate Structure Databases

R. René Ranzinger[a], Stephan Herget[a], Thomas Lütteke[a,b] and Martin Frank[a]

[a]*DEUTSCHES KREBSFORSCHUNGSZENTRUM (GERMAN CANCER RESEARCH CENTRE), HEIDELBERG, GERMANY*
[b]*JUSTUS-LIEBIG-UNIVERSITY GIEßEN, INSTITUTE OF BIOCHEMISTRY AND ENDOCRINOLOGY, GIEßEN, GERMANY*

## Introduction

Significant improvements in experimental analytical methods over recent years—particularly in carbohydrate (glycan) analysis by mass spectrometry [1–4]—have led to a tremendous increase in the amount of carbohydrate structure data generated. The availability of databases and tools to store, retrieve, and analyze these data in an efficient way is of fundamental importance to progress in glycomics [1,5,6]. Consequently, over the past few years, there has also been a substantial increase in both the development, and use, of informatics tools and databases in glycomics [7]. However, bioinformatics for glycomics ("glycoinformatics") is still in its infancy when compared with bioinformatics for genomics or proteomics [8–12]. Importantly, no comprehensive and curated database of carbohydrate structures currently exists.

The present situation in glycoinformatics is characterized by the existence of multiple disconnected and incompatible islands of experimental data, data resources, and specific applications, managed by various consortia, institutions, or local groups [5,13]. Nevertheless a variety of valuable resources are available to glycoscientists: a few major databases and many smaller, more specialized databases or other resources (e.g., static HTML pages, software tools, etc.) exist [14]. Some of the major, well-established databases are described in this chapter, and two important new database initiatives are introduced. Two practical use cases of the databases and tools are also described. Additional discussions about glycomic databases are found in Chapters 8 and 9.

At present there is hardly any cross-linking between the established carbohydrate databases [15]. This is mainly due to the fact that the various databases use different sequence formats to encode carbohydrate structures [16]. (Note: In this

**Handbook of Glycomics**
Copyright © 2009 by Elsevier Inc. All rights of reproduction in any form reserved.

chapter we use the term "carbohydrate sequence format" for computer-readable text formats that encode carbohydrate structures. From the sequence formats a variety of graphical representations of a carbohydrate structure can be generated for display in user interfaces.) Although many of the formats used to encode carbohydrate structure are based on monosaccharide residue names that are similar to the IUPAC recommendations [17], there are still some differences between the residue names of the various sequence formats and between the various ways of encoding the glycosidic linkages (Figure 10.1). Because of this, translation tools for the different formats are necessary to establish cross-links between the databases.

A first step to overcome this problem was the agreement in 2006 to use Glyde [18] as a common exchange format for information on carbohydrate structure [5]. In the context of the EUROCarbDB design study (described below) a new carbohydrate

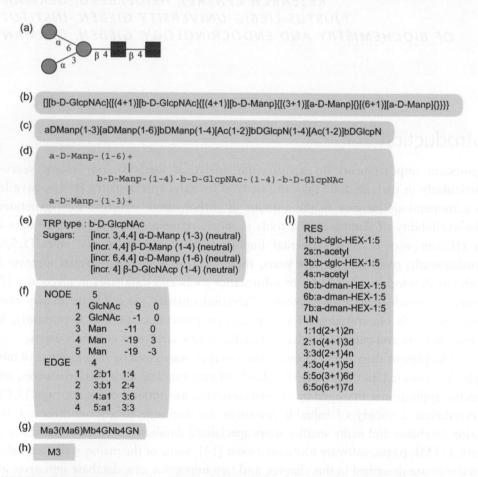

**FIGURE 10.1** Sequence formats used by different carbohydrate databases. Compilation of the different sequence formats used by the carbohydrate structure databases for the example of the *N*-glycan core. (a) Pictorial representation, as used by the CFG (http://www.functionalglycomics.org/static/consortium/Nomenclature.shtml), with reducing end at the right. (b) LINUCS encoding used in GLYCOSCIENCES.de. (c) BCSDB encoding. (d) ASCII 2D graph as employed in CCSD (CarbBank). (e) Notation used by GlycoBase of the Université des Sciences et Technologies de Lille. (f) KEGG Chemical Function format (KCF) used by KEGG. (g) Linear Code® used in the CFG database. (h) Oxford notation used in GlycoBase from NIBRT. (i) GlycoCT encoding used in EUROCarbDB and GlycomeDB. (See color plate 10.)

sequence format, named GlycoCT [16], for use in databases has been developed as well as a reference database for monosaccharide notation (www.monosaccharidedb. org). Recently, as a result of a close collaboration between Glyde and EUROCarbDB developers, the Glyde-II format was created as the current standard exchange format for carbohydrate structures.

From the user's point of view, the lack of cross-links between carbohydrate databases means that they have to visit different database web portals in order to get all the available information on a specific carbohydrate structure. In addition, they might have to learn the different local query options, some of which require knowledge of the encoding of the residues in the respective database. To address this inconvenient situation, the meta-database GlycomeDB [19] has been developed as a single query point for several databases.

# Current Established Carbohydrate Structure Databases

At the time of writing, there are nine major database projects dedicated to the storage of carbohydrate structures: seven of these follow an open access policy, while two more are commercial and thus follow a more restricted access model (see Table 10.1 for an overview).

The open access databases have different capabilities to perform queries, as shown in Table 10.2. Most of the initiatives offer structural searches to some extent, while the additional query functions are highly diverse, indicating a specialization of the individual databases. In the following, the open access initiatives shown in Table 10.1 are briefly described.

### Complex Carbohydrate Structure Database (CCSD, CarbBank)

The Complex Carbohydrate Structure Database (CCSD; often referred to as CarbBank in reference to its query software) is the mother of all the modern carbohydrate structure databases. It was developed and maintained for more than 10 years by the Complex Carbohydrate Research Centre of the University of Georgia (USA) [20,21]. The CCSD was the largest endeavor during the 1990s to collect the structures of carbohydrates, mainly through retrospective manual extraction from literature. The main aim of the CCSD was to catalogue all publications in which complex carbohydrate structures were reported. Unfortunately, funding for the CCSD stopped during the second half of the 1990s, after which the database was not further developed and no longer updated, although an online version of the database is still available. Nevertheless, with about 50 000 records relating to approximately 15 000 different carbohydrate sequences, the CCSD is still one of the largest repositories of carbohydrate-related data, from which subsets have been incorporated in almost all recent open access databases. CCSD uses the extended IUPAC format (2D ASCII graph) for encoding of carbohydrate structures (Figure 10.1d).

Table 10.1 Overview of the nine major carbohydrate structure databases

| Name | URL | Policy | Records | Size | Scope | Exp. data |
|---|---|---|---|---|---|---|
| Bacterial Carbohydrate Structure Database (BCSDB) | http://www.glyco.ac.ru/bcsdb/start.shtml | Open access | 8883 | 5756 | Mainly bacteria | NMR |
| CCSD (CarbBank) | http://www.boc.chem.uu.nl/sugabase/CarbBank.html | Open access | 49897 | 14594 | All species | |
| CFG Glycan Database | http://www.functionalglycomics.org/glycomics/molecule/jsp/carbohydrate/carbMoleculeHome.jsp | Open access | 8595 | 6278 | Mainly mammalia | MS Glycan array |
| GLYCOSCIENCES.de | http://www.glycosciences.de/sweetdb | Open access | 23179 | 15594 | All species | 3D data NMR |
| GlycoBase (Lille) | http://glycobase.univ-lille1.fr/base/ | Open access | 154 | 154 | Amphibia | NMR |
| GlycoBase (Dublin) | http://glycobase.nibrt.ie/cgi-bin/public/glycobase.cgi | Open access | 377 | 352 | Mainly mammalia | HPLC |
| KEGG Glycan | http://www.genome.jp/kegg | Open access | 11161 | 9948 | All species | Pathways |
| GlycosuiteDB | https://glycosuite.proteomesystems.com/glycosuite/glycodb | Commercial | >9000 | >000 | Mainly mammalia | MS |
| Glycomics Database | http://www.glycominds.com | Commercial | Unknown | >000 | Unknown | |

The size of a database is denoted in two ways: with the number of records in the particular initiative (Records) and with the number of distinct carbohydrate sequences without aglycons (Size). The latter number is derived from GlycomeDB in the case of the open access databases, or else from personal communication or other sources. The column Scope details the main taxonomical sources of the initiatives. The last column shows the type of experimental data that are stored in the database. NMR, nuclear magnetic resonance (spectroscopy); MS, mass spectrometry; HPLC, high-performance liquid chromatography.

**Table 10.2 The query functionalities of the different open access databases in glycomics**

| Name | Structural query options | Annotation and extended query functionality |
|---|---|---|
| Bacterial Carbohydrate Structure Database (BCSDB) | Substructure<br>Exact structure<br>Enhancements for repeating units | Bibliography<br>Taxonomy<br>NMR data |
| Complex Carbohydrate Structure Database (CCSD, CarbBank) | Substructure<br>Exact structure | Bibliography<br>Taxonomy<br>Experimental methods |
| CFG Glycan Database | Substructure<br>Molecular weight<br>Composition | Taxonomy<br>Additional biological annotations |
| GLYCOSCIENCES.de | Substructure<br>Exact structure<br>Composition<br>Classification<br>Formula | Bibliography<br>Taxonomy<br>NMR data<br>PDB data<br>MS data |
| GlycoBase (Lille) | Substructure<br>Exact structure | Taxonomy<br>NMR data |
| GlycoBase (Dublin) | Structure groups | Source of the sequence |
| KEGG | Substructure<br>Similarity search<br>Composition<br>Classification | Links to other KEGG databases |

## Bacterial Carbohydrate Structure Database (BCSDB)

The aim of the Bacterial Carbohydrate Structure Database (BCSDB) is to provide a database of structural, bibliographic, and related information on bacterial carbohydrates [22]. While the majority of its current data content is derived from retrospective literature analysis and direct submission, part of the data is compiled from a translation of CCSD data. The BCSDB uses a monosaccharide encoding, which is in some aspects similar to the extended IUPAC format. Structures are encoded in a linear fashion using brackets (Figure 10.1c). A special focus is on repeating units, which are common in bacterial polysaccharides.

## CFG Glycan Database

As part of the Consortium for Functional Glycomics (CFG), which is aimed at elucidating the roles of carbohydrate–protein interactions in cell surface biology, a carbohydrate structure database was generated [23]. This repository was built starting from structures in the CCSD and curated structures obtained from a commercial database (developed by Glycominds Ltd., Lod, Israel). To this database, carbohydrate structures synthesized by the consortium members have been added. In addition to their Glycan Database, the CFG initiative also provides databases on glycosyltransferases and glycan-binding proteins (GBP). The monosaccharide encoding used in the CFG database, in which monosaccharide building blocks are given an abbreviated form, with the most common forms being represented by a single letter code, and the carbohydrate sequence format of the database (Linear Code®, Figure 10.1(g) was developed by Glycominds Ltd. [24].

## GLYCOSCIENCES.de

The GLYCOSCIENCES.de database [25]—the former SweetDB [26]—was established at the end of the 1990s at the German Cancer Research Center (DKFZ) and includes most of the structures from the CCSD and SUGABASE [27] (a nuclear magnetic resonance (NMR) spectra database for carbohydrates). In addition, the database was updated with manually selected structures and new NMR spectra from the literature. The CCSD nomenclature is used as the encoding scheme for the monosaccharides. For encoding of carbohydrate structures, the GLYCOSCIENCES.de portal uses a sequence format called Linear Notation for Unique Description of Carbohydrate Sequences (LINUCS) [28] (Figure 10.1b). The GLYCOSCIENCES.de portal also provides direct access to carbohydrate structures contained in the Protein Data Bank (PDB) [29–31].

## KEGG Glycan

The Kyoto Encyclopedia of Genes and Genomes (KEGG) [32], an integrated knowledge database for biological sequences, contains a Glycan database built on carbohydrate structures from the CCSD and in-house curation efforts [33]. The Glycan database offers a range of tools from similarity search and expression profile analysis to links to manually drawn metabolic pathways. The database employs the KCF format as the sequence encoding format [34] (Figure 10.1f). A detailed description of the KEGG Glycan database is given in Chapter 9.

## GlycoBase (Lille) and GlycoBase (Dublin)

Considerably smaller than the databases described above, these two databases have specialized in collecting experimental data for carbohydrates. While GlycoBase (Lille) focuses on NMR data, GlycoBase (Dublin) [35] provides access to HPLC relevant data. Both databases use proprietary sequence formats (Figure 10.1e,h).

## Other Carbohydrate-Related Databases and Resources

In addition to the databases described above, there are also a number of other, specialized, databases [14] (see also www.eurocarbdb.org/links for an overview) which incorporate information about carbohydrate structure and/or biological interactions. Examples of these are GlycoEpitope (http://www.glyco.is.ritsumei.ac.jp/epitope/) [36] which stores data on antibodies to carbohydrate epitopes, SugarBindDB

(sugarbinddb.mitre.org/) a database of known carbohydrate ligands for pathogens, and the databases provided by the Japan Consortium for Glycobiology and Glycotechnology (www.jcggdb.jp): GlycoGene DB [37], Lectin Frontier DB and GlycoProtein DB.

Among the classical protein databases, the Universal Protein Resource (UniProt) (www.uniprot.org) [38,39] provides information on glycosylation of proteins; however, only limited information on the carbohydrates that are attached to the glycosylation sites. Information on glycosylation of a protein is displayed in the sequence annotation sections and is encoded as a "feature table (FT)" entry ("170" denotes the index of the glycosylation site; more details can be found at www.uniprot.org/manual/carbohyd) [40]:

> FT CARBOHYD 170 170 N-linked (GlcNAc . . .)(Potential).

The query options in UniProt are tailored for proteins. No query option for carbohydrate structures is available.

With regard to carbohydrate three-dimensional (3D) structure, the two major databases where experimentally determined 3D structures of carbohydrates are stored are the Cambridge Structural Database (CSD, http://www.ccdc.cam.ac.uk/products/csd/) and the Protein Data Bank (PDB, http://www.rcsb.org/pdb/). Many crystal structures of oligosaccharides are also accessible through the Glyc03D web interface (http://www.cermav.cnrs.fr/glyc03d/). The PDB (www.rcsb.org) [41] or wwPDB (www.wwpdb.org) [31,42] currently contains more than 50 000 3D structures of biomolecules, of which about 3700 contain carbohydrates. Using the PDB homepage to query for carbohydrates is limited to name searches. Due to the lack of a consistently used nomenclature for carbohydrates in PDB files it is difficult to find the entries of interest. However, the GLYCOSCIENCES.de web portal [30,43] and the Glycoconjugate Data Bank: Structures (www.glycostructures.jp) [44] offer convenient ways to search for carbohydrate structures in the PDB.

Most of the carbohydrates in the PDB are either connected covalently to a (glycol-)protein, or the carbohydrate forms a complex with a lectin, enzyme, or antibody. Isolated carbohydrates are only rarely found in the PDB. When looking at the (PDB) carbohydrate structures one has to keep in mind that sometimes only fragments of the original carbohydrates present when growing crystals for X-ray crystallography may be resolved. Additionally the 3D structures of the carbohydrates in the PDB do not always meet high quality standards [43,45]; therefore one has to look at the structures with care. Despite these limitations the PDB is an important source of information on carbohydrate 3D structure [30].

## New Database Initiatives

### EUROCarbDB—Development of an Infrastructure for Storage of Carbohydrate Structure and Analytical Data

The lack of a comprehensive database for storage and retrieval of carbohydrate structures is regarded as the biggest deficiency in glycomics and glycobiology research at this time [5]. The existence of such a database is critical for the progress of the field, and the long-term maintenance of such a database has to be guaranteed. The traditional approach would be to establish a centrally maintained database, where data entry and quality checks are performed by database curators who are experts in the field. However, such a central database infrastructure depends critically on continuous funding and there is a high risk that the database would no longer be updated when funding for database curation stops, as was the case with the CCSD. The EUROCarbDB project (www.eurocarbdb.org) is a design study that aims to create the foundations for a new infrastructure of distributed databases and bioinformatics tools where scientists themselves can upload carbohydrate structure-related data. At the time of writing the EUROCarbDB contains only a few test data sets, uploaded by the developers, therefore in this section the concepts and aims of a future database will be described.

The EUROCarbDB project, initiated by Willi von der Lieth, who is regarded as a pioneer of modern glycoinformatics [5], and funded by the European Union between 2005 and 2009, involves a consortium of eight research groups covering the areas of informatics development, and carbohydrate analysis by MS, HPLC, and NMR. Fundamental ethics of the project are that all data are freely accessible and all provided tools are open source. A prototype of a database application has been developed that can store carbohydrate structures plus additional data such as biological context (organism, tissue, disease, etc.), and literature references. Primary experimental data (MS, HPLC, and NMR) that can serve as evidence or reference data for the carbohydrate structure in question can be uploaded as well.

Depending on the analytical method used it may not be possible to determine a carbohydrate structure completely and some "unknowns" (e.g., linkage type, location of terminal sialic acids) or "fuzziness" (e.g., statistically distributed sulfate substituents in glycosaminoglycans) remain. It was decided to include such underdetermined structures in the EUROCarbDB in order to support the current and future glycomics projects that are driven by the tremendous progress in glycan analysis by MS. A new encoding format (GlycoCT, Figure 10.1i) [16] has been developed that meets all requirements for encoding these carbohydrate structures in EUROCarbDB. Since EUROCarbDB supports decentralized data input, it is of particular importance that procedures are put in place to reduce data entry errors. GlycanBuilder [46], an intuitive software tool for building and displaying carbohydrate structures, has been developed to facilitate the graphical input of carbohydrate sequences and to make this step less error-prone.

The EUROCarbDB project aims to support the analytical process of carbohydrate structure determination from the spectrometer to database storage. Therefore, the project includes the development of algorithms and programs that enable rapid and reliable semi-automatic interpretation of MS, HPLC, and NMR data. Currently available tools are: Glyco-Peakfinder [47]—a web service allowing de novo determination of glycan compositions from their mass spectra signals; GlycoWorkbench [48]—a tool for the computer-assisted annotation of mass spectra; AutoGU [35]—a tool to assist the interpretation and assignment of HPLC-glycan profiles; ProSpectND—a versatile NMR spectra processing and visualization package; and CASPER [49,50]—an increment rule based tool to estimate [1]H and [13]C NMR shifts of carbohydrates that can be used to speed up peak assignment using the ccpNMR analysis package (www.ccpn.ac.uk). By using the freely provided tools of EUROCarbDB, data analysis can be performed more efficiently. It is also easier to upload the data into the database, because the tools ensure that the data are in the proper format. No significant extra work should be required for the upload process. Once the database contains enough high-quality reference data, the tools may be able to make efficient use of these data to speed up the process of spectral peak assignment. The goal is that many scientists will upload their analytical data to the EUROCarbDB database as supplementary material during the process of scientific publication of their results.

## GlycomeDB—Integration of Carbohydrate Structure Databases

As pointed out already, one of the greatest problems in glycobiology, at this time, is the lack of a unified and complete database for carbohydrate structures providing a single entry point for structural queries. This complicates the situation for scientists in a number of fields, including structure elucidation, comparative analysis, or development of bioinformatics applications, who are interested in finding previously recorded carbohydrate structures. The existing structure data are scattered among several databases, which are not connected to each other, and each of these databases uses a proprietary encoding scheme for monosaccharides and carbohydrate topology. Currently there are seven freely available and accessible carbohydrate structure databases (Table 10.1), each of which stores a different but partially overlapping set of structures and annotated information, such as species information or experimental data. Figure 10.1 shows a compilation of the different sequence formats utilized by the different carbohydrate structure databases. Because of the incompatibility of the sequence formats, there were, until recently, no implementations that allowed communication and comparison between the databases. The only exception was a cross-database search between the BCSDB and GLYCOSCIENCES.de [15].

In 2005, a new initiative was begun to overcome the isolation of the carbohydrate structure databases and to create a comprehensive index of all available structures with references back to the original databases. To achieve this goal, all structures of the freely available databases were translated to the GlycoCT sequence format [16], and stored in a new database, the GlycomeDB [19]. A schematic representation of the concept of GlycomeDB, which currently contains 34445 unique carbohydrate structures in the GlycoCT encoding, is shown in Figure 10.2. In addition to the structures, the available species information was also integrated into the database. A total of 14857 structures were found to have a taxonomic annotation, and 1845 different taxons were referenced. Also included in GlycomeDB are the access keys (database IDs) of the origin of the data. The integration process is performed incrementally on a weekly basis, updating the GlycomeDB with the newest structures available in the associated databases. The GlycomeDB database is freely available for download from the project homepage (www.glycome-db.org). To foster the usage of this database, a web portal was created providing several structural and annotation searches (Table 10.3). The web portal can be seen as a carbohydrate search engine for data in all open access databases, since it is possible to access the original data set by the stored database ID.

GlycomeDB supports a number of different input options for a structure query, allowing the user to enter a structure or substructure using several different input tools, such as the GlycanBuilder [48] from EUROCarbDB, the DrawRings of the RINGS portal (rings.t.soka.ac.jp), or textual input. Each structure returned as a result can be viewed in detailed mode, which shows all species annotations assigned to the structure, all original database access keys (database ID), all aglycon residues which are attached to the structure in the integrated databases, and a list of known

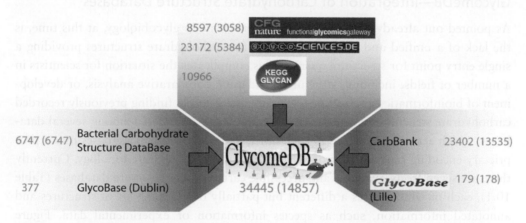

**FIGURE 10.2** The GlycomeDB concept. GlycomeDB integrates and stores the structure and species information from seven open access carbohydrate structure databases. For GlycomeDB and for each database the number of unique sequences in the database specific format (as of November 2008) is given. The numbers in parentheses show the number of sequences with at least one species annotation in the database. (See color plate 11.)

## Table 10.3  List of the search options implemented in the GlycomeDB web portal

| Search | Description |
|---|---|
| Search by database ID | Search for all entries of a carbohydrate using the ID of one of the integrated databases or a GlycomeDB structure ID. In addition, it is possible to make a search for all carbohydrates occurring in any one of the integrated databases |
| Exact structure search | Search for a structure based on the exact definition of this structure |
| Substructure search | Search for all structures containing a defined substructure |
| Similarity search | Search for all structures that are similar to a given structure |
| Maximum Common Substructure (MCS) search | A search based on a defined structure: all GlycomeDB structures containing the defined structure as well as any substructures thereof are retrieved |
| Search by species | Search for all structures with a given species annotation. This search also allows retrieval of structures based on higher taxonomic classes such as genus or kingdom |

motifs which can be found in the structure. The original access keys of the integrated databases are implemented as hyperlinks, which allow the user to access the web pages of the original database. In addition, it is possible to generate images of the structure in different output formats and file formats. It is also possible to download the sequence encoding of the structure in LINUCS, Glyde-II, and in GlycoCT.

Very often a user of a database is interested to search for structures not only by one search criteria, such as substructure or species, but by a combination of them. In contrast to most other carbohydrate structure databases, GlycomeDB allows for the refinement or expansion of a search result set by another search. This feature is called the Complex Query System and allows the generation of subsets, or a union set, between the two search results. It is also possible to generate the complementary set of a result set. A screen shot of the input interface for the Complex Query System, which is part of each result page of GlycomeDB, is shown in Figure 10.3. An example of using the Complex Query System is given in Case Study 1 below. With this feature it is possible to formulate any complex request to GlycomeDB based on structural properties or species annotations. With the Complex Query System the GlycomeDB has the most powerful carbohydrate search engine of the freely available databases.

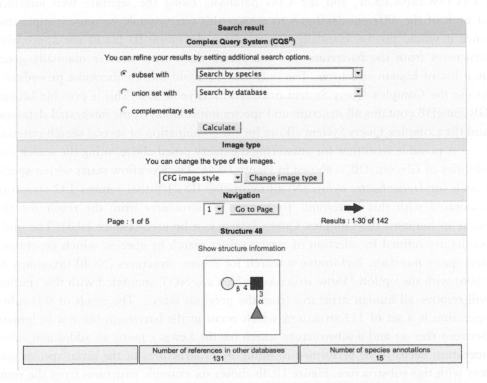

FIGURE 10.3  Search result page of the GlycomeDB. Screenshot of the input options for the Complex Query System which are shown in the header of each result search page. The three options for manipulating the result set are shown: the generation of a subset, or a union set, between the result set and a new search, and the calculation of the complementary set. (See color plate 12.)

# Case Studies

The following two case studies demonstrate how the existing database resources can be used to answer specific scientific questions.

## Case Study 1—A Search for Species-Specific Carbohydrates

The carbohydrate structures displayed by bacteria can differ substantially, in both monosaccharide content and in structure, from those displayed on mammalian cells [51,52]. However, a number of significant carbohydrate epitopes, including the Lewis and blood group antigens, are found in both bacteria and humans [53–55]. In the bacterium *Helicobacter pylori*, for example, which colonizes human gastric mucosa causing active inflammation, the O-chains of the bacterial lipopolysaccharide are commonly composed of internal Lewis x motifs with terminal Lewis x or Lewis y motifs [55]. In a study of the Lewis x-containing glycoforms expressed by *H. pylori*, a scientific query of a database may be: "Find all carbohydrate structures from *Helicobacter pylori* which do not occur in human and which contain a Lewis x motif." To answer this query it would be necessary to access the structural contents of several databases, since bacterial structures can be found in the BCSDB, the CCSD, and GLYCOSCIENCES.de and human structures in the CCSD, GLYCOSCIENCES.de, and the CFG database. Using the separate web interfaces of each of the databases to answer this query would be quite complicated, since it would involve manually writing down the entire ID list of the appropriate structures from the bacterium and comparing them with another manually generated list of human structures. The only way to avoid this cumbersome procedure is to use the Complex Query System provided by GlycomeDB. This is possible because GlycomeDB contains all structure and species information of the integrated databases and the Complex Query System allows for the combination of several search criteria.

A possible workflow for answering the query posed above, using the search possibilities of GlycomeDB, is shown in Figure 10.4a. The workflow starts with a species search for *Helicobacter pylori* (NCBI taxonomy ID 210) that returns 142 structures annotated with this bacterium. To remove all structures from the result list that occur in humans, the Complex Query System can be used (Figure 10.3). The initial results are refined by selection of a new subset search by species, which generates a new query interface. Performing a search for human structures (NCBI taxonomy ID 9606) with the option "Show structures which are NOT annotated with this species" will remove all human structures from the previous search. The result of this subset operation is a set of 115 structures which occur in the bacterium but not in human. Between this set and a substructure search for the Lewis x motif an additional subset operation needs to be performed resulting in 27 structures of the given species context with this substructure. Figure 10.4b shows six example structures from the result set of the GlycomeDB complex query search. By using the Complex Query System the formerly complex task, which would have involved the use of several databases and manual work, can be done using the web interface of a single database. Since

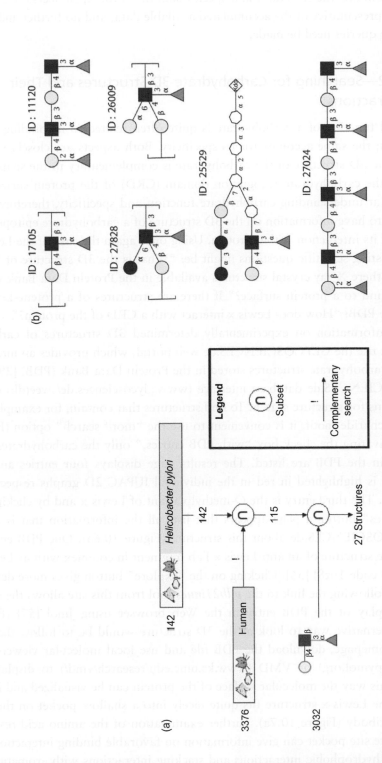

**FIGURE 10.4** Example using the GlycomeDB Complex Query System (Case Study 1). Task: Retrieve glycan structures containing the Lewis x trisaccharide epitope that are found in *Helicobacter pylori* but not in *Homo sapiens*. (a) Workflow for the complex query in GlycomeDB: the results of the "species search" for *Helicobacter pylori* (142 structures) are refined using the Complex Query System by, first, a subset search for all non-human structures, and second, by a substructure search for the Lewis x epitope, resulting in 27 structures matching the represented search criteria. The numbers before each of the search symbols show the total numbers of structures in GlycomeDB matching the represented search criteria. The numbers in the middle represent the intermediate result after a subset operation. (b) Six of the 27 retrieved structures are shown with their GlycomeDB structure ID. Three of these are indicated as polysaccharides (enclosed with square brackets). In all structures the Lewis x trisaccharide can be seen. (See color plate 13.)

GlycomeDB contains the structural and species data of all the open access databases the result is representative of the accumulated available data, and no further, independent, database queries need be made.

## Case Study 2—Searching for Carbohydrate 3D Structures and Their Protein Interactions

The biological function of a carbohydrate is quite often related to its binding affinity to proteins; the same accounts for its specificity. Both aspects are closely related to how well the 3D structure of the carbohydrate is complementary to the shape and properties of the carbohydrate recognition domain (CRD) of the protein surface. In studies aimed at understanding carbohydrate function and specificity, therefore, it is of great value to have information on the 3D structure of a carbohydrate epitope, and particularly of its interaction with a protein. Using once again the case of the Lewis x epitope, interesting scientific questions might be: "What is the 3D structure of Lewis x?" and "Are there X-ray crystal structures available in the Protein Data Bank where Lewis x is bound to a protein surface?" If there are structures of a protein–Lewis x complex in the PDB: "How does Lewis x interact with a CRD of the protein?"

To get information on experimentally determined 3D structures of carbohydrates one can use the GLYCOSCIENCES.de web portal, which provides an interface to search for carbohydrate structures stored in the Protein Data Bank (PDB) [29–31]. The GLYCOSCIENCES.de database interface (www.glycosciences.de/sweetdb) offers different options for structure search. To find structures that contain, for example, the Lewis x trisaccharide motif, it is convenient to use the "motif search" option (Figure 10.5a). By activating the check box "with PDB entries," only the carbohydrates that are available in the PDB are listed. The results page displays four entries and the Lewis x motif is highlighted in red in the individual IUPAC 2D graphs respectively (Figure 10.5b). The third entry is the O-methyl variant of Lewis x and by clicking on the "pdb-entries" button a page appears that lists all the information that is available in GLYCOSCIENCES.de about this structure (Figure 10.6a). One PDB entry is available: "The structure of an anti-Lewis x Fab fragment in complex with its Lewis x antigen" (PDB code 1uz8) [56]. Clicking on the "explore" button gives more detailed information. Following the link to the *pdb2linucs* tool from this site allows the direct interactive display of the PDB entry in the Web browser using Jmol [57] (Figure 10.6b). An alternative way to look at the 3D structure would be to follow the link to the PDB homepage, download the PDB file and use local molecular viewers like PyMol (www.pymol.org) or VMD (www.ks.uiuc.edu/research/vmd/) to display the structure. In this way the molecular surface of the protein can be visualized and it can be seen that the Lewis x structure fits quite nicely into a shallow pocket on the surface of the antibody (Figure 10.7a). Further examination of the amino acid residues lining the active site pocket can give information on favorable binding interactions. It is known that hydrophobic interactions and stacking interactions with aromatic side chains of tryptophan (Trp), tyrosine (Tyr), phenylalanine (Phe), and histidine (His)

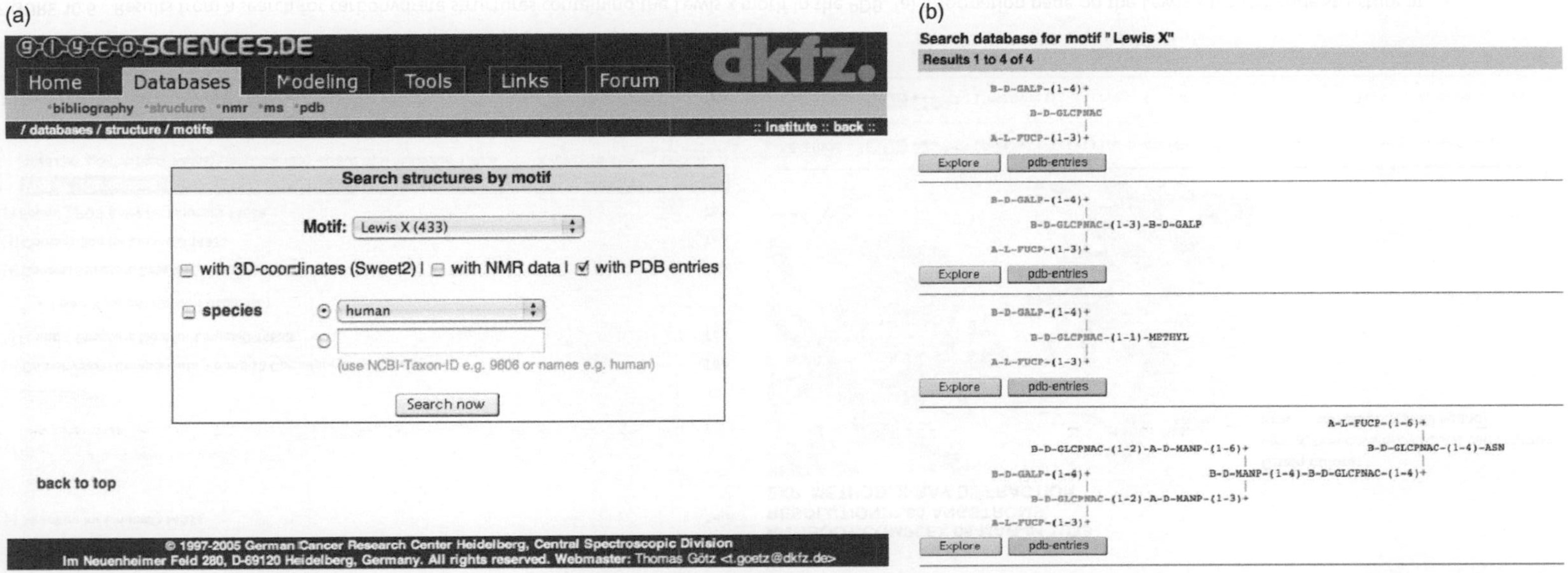

**FIGURE 10.5** Searching for three-dimensional structures of carbohydrates in the Protein Data Bank (PDB) using GLYCOSCIENCES.de tools. (a) Motif search interface. (b) Results page offering access to the PDB entries. (See color plate 14.)

**FIGURE 10.6**  Results from a search for carbohydrate structures containing the Lewis x motif in the PDB. (a) Information page on the Lewis x trisaccharide structure at GLYCOSCIENCES.de. (b) *pdb2linux* display of the PDB entry 1uz8. (See color plate 15.)

(a)

(b)

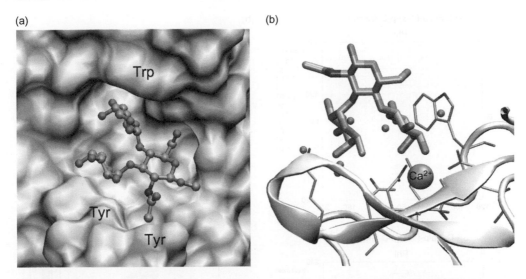

**FIGURE 10.7** (a) Lewis x in the binding site of anti-Lewis x Fab fragment (PDB code 1uz8). (b) Lewis x in the binding site of scavenger receptor C-type lectin (PDB code 20×9). VMD [65] was used to generate the figures from the PDB files. (See color plate 16.)

contribute significantly to the binding affinity of carbohydrates [30,58–61]. By visual inspection of the Lewis x–antibody complex one can see that stacking of the Gal*p* residue with a tryptophan and hydrophobic interaction of the methyl group of Glc*p*NAc with two tyrosines occurs.

Metal ion-dependent binding is also a quite common feature of carbohydrate–protein complexes. Following the first entry in Figure 10.5b leads to the PDB structure of Lewis x bound to scavenger receptor C-type lectin (SRCL). Here the binding is mediated by a $Ca^{2+}$ ion (PDB code 20×9) [62] (Figure 10.7b).

To check whether the Lewis x conformation in the PDB entry 1uz8 represents a low-energy structure, one can use the *carp* tool (www.glycosciences.de/tools/carp) that compares the values of the glycosidic linkage torsions $\phi$ and $\psi$ with other PDB data or with calculated disaccharide conformational maps derived from GlycoMapsDB [63]. The conformations of the glycosidic linkages of 1uz8 are located in the low-energy area of the corresponding disaccharide maps (Figure 10.8a) representing the linkages. However, they are not located exactly in the global energy minimum calculated for the disaccharides. The reason for this is that the Gal and Fuc residues are attached to the GlcNAc at directly adjacent positions, which causes some unfavorable interaction between those residues, resulting in a reduced accessible conformational space. This difference can be seen by comparing the disaccharide maps with conformational maps which have been calculated for the Lewis x trisaccharide that can be accessed through GlycoMapsDB directly (www.glycosciences.de/modeling/glycomapsdb/) (Figure 10.8b).

## Conclusion and Outlook

The worldwide efforts to generate and store carbohydrate-related information in databases have increased significantly over recent years. An important impetus was the

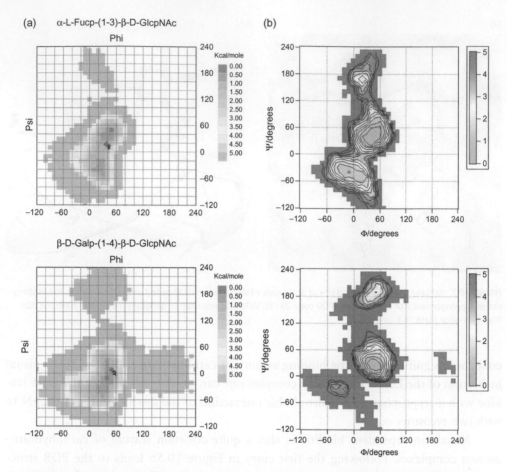

**FIGURE 10.8** Analysis of the glycosidic torsion angles of carbohydrates using the *carp* tool. (a) Results of the *carp* analysis for PDB entry 1uz8. Glycosidic linkage torsion values ($\phi$ and $\psi$) of carbohydrates in the crystal structure are indicated by crosses in the plot. Calculated conformational maps for the disaccharides representing the linkages are displayed in the background. (b) Conformational map for the Lewis[x] trisaccharide derived from GlycoMapsDB. (See color plate 17.)

establishment of the Consortium for Functional Glycomics, the first large-scale glycomics project that clearly emphasized the need for informatics to manage and automatically annotate the vast amount of experimental data generated by glycomics research. Currently, a variety of databases are freely available over the Internet, maintained by groups in the USA, Japan, and Europe. It has also been announced that GlycoSuiteDB, developed in Australia, will soon join the open access database community.

The field of glycoinformatics faces two significant challenges at this time. One is the lack of connectivity between the current data resources with their proprietary encoding of carbohydrate structure: this ideally requires linking these resources through a common interface. This issue has now been addressed to a large extent by GlycomeDB, which provides a powerful central search interface for carbohydrate structures in all major open access carbohydrate structure databases. The other challenge is to generate a unified, comprehensive, thoroughly curated, and sustainable database for carbohydrate structures identified in biological samples.

At the moment it is not possible for scientists to find out quickly whether a particular carbohydrate structure has already been characterized in a biological context (organism, tissue, etc.) because no comprehensive database of all known carbohydrate structures exists. After the CCSD project stopped in the mid-1990s, there was no further concerted effort to input all published carbohydrate structures into a single database. Data collection from literature was only performed within specific scopes, for example in the BCSDB [15], KEGG [33], and GlycoSuiteDB [64]. The consequence is that there are many carbohydrate structures published over the last 10 years that are not available in any database. An important task in the coming years will be to close this gap, to extract the missing data from the scientific literature and input it into an appropriate database.

Once there is a database which accurately represents current knowledge, ensuring the continuance of carbohydrate structure input will be critical to its success. Data entry and quality checks performed by database curators who are experts in the field would be the ideal approach to keep a database up-to-date, however this requires significant continuous funding. Automated updating of carbohydrate databases from electronic publications using text-mining algorithms is difficult because branched carbohydrate structures are often only illustrated in the figures of scientific articles and therefore cannot be located by algorithms that search the manuscript texts. An option for solving the maintenance problem is to offer scientists who generate the data the possibility to enter the data into a publicly available database themselves. Such an approach is the aim of the EUROCarbDB project. Similar procedures are routinely used in genomics and proteomics research, allowing scientists to directly enter structural data and biological annotations during the publication process. Rapid progress in glycobiology will depend on the acceptance of such procedures by glycoscientists [5].

Drawing from the experience of the first major carbohydrate sequence/literature database, the CCSD [20,21], the maintenance of a central carbohydrate structure database cannot depend solely on grant funding. The better solution would be for the database to become an integral part of the infrastructures provided by organizations like the NCBI or the EBI. In such infrastructures the currently missing cross-links between carbohydrate and protein databases could also be more easily established.

The existence of a centralized, comprehensive, carbohydrate structure database will offer many advantages to glycomics research. First of all it would allow researchers to find out easily whether a particular carbohydrate has been found before in a given biological context. A unique identifier (ID) could be assigned to a carbohydrate structure that could be used in scientific publications as a reference. A carbohydrate ID would also substantially increase the ability to annotate and cross-reference data with other bioinformatics resources. Additionally a comprehensive data collection, used in conjunction with analytical tools for MS and NMR annotation, would increase the speed (and accuracy) of spectral annotation. Finally the existence of a comprehensive carbohydrate structure database would make data mining projects [51,52] more reliable and bioinformatics approaches to guide research more feasible.

# References

1. Haslam SM, Julien S, Burchell JM, et al. Characterizing the glycome of the mammalian immune system. *Immunol Cell Biol* 2008;**86**:564–73.

2. Haslam SM, North SJ, Dell A. Mass spectrometric analysis of N- and O-glycosylation of tissues and cells. *Curr Opin Struct Biol* 2006;**16**:584–91.

3. Geyer H, Geyer R. Strategies for analysis of glycoprotein glycosylation. *Biochim Biophys Acta* 2006;**1764**:1853–69.

4. Harvey DJ, Dwek RA, Rudd PM. Determining the structure of glycan moieties by mass spectrometry. Curr Protoc Protein Sci 2006;Chapter 12:Unit 12 17.

5. Packer NH, von der Lieth C-W, Aoki-Kinoshita KF, Lebrilla CB, Paulson JC, Raman R, Rudd P, Sasisekharan R, Taniguchi N, York WS. Frontiers in glycomics: Bioinformatics and biomarkers in disease. An NIH White Paper prepared from discussions by the focus groups at a workshop on the NIH campus, Bethesda MD (September 11–13, 2006). *Proteomics* 2008;**8**:8–20.

6. Royle L, Campbell MP, Radcliffe CM, White DM, Harvey DJ, Abrahams JL, Kim YG, Henry GW, Shadick NA, Weinblatt ME, Lee DM, Rudd PM, Dwek RA. HPLC-based analysis of serum N-glycans on a 96-well plate platform with dedicated database software. *Anal Biochem* 2008;**376**:1–12.

7. von der Lieth C-W, Lutteke T, Frank M. The role of informatics in glycobiology research with special emphasis on automatic interpretation of MS spectra. *Biochim Biophys Acta* 2006;**1760**:568–77.

8. von der Lieth C-W, Bohne-Lang A, Lohmann KK, Frank M. Bioinformatics for glycomics: status, methods, requirements and perspectives. *Brief Bioinform* 2004;**5**:164–78.

9. Kersey P, Apweiler R. Linking publication, gene and protein data. *Nat Cell Biol* 2006;**8**:1183–9.

10. Velankar S, McNeil P, Mittard-Runte V, et al. E-MSD: an integrated data resource for bioinformatics. *Nucleic Acids Res* 2005;**33**:D262–5.

11. Whitfield EJ, Pruess M, Apweiler R. Bioinformatics database infrastructure for biotechnology research. *J Biotechnol* 2006;**124**:629–39.

12. Mulder NJ, Kersey P, Pruess M, Apweiler R. In silico characterization of proteins: UniProt, InterPro and Integr8. *Mol Biotechnol* 2008;**38**:165–77.

13. von der Lieth C-W. Databases and Informatics for Glycobiology and Glycomics. In: Kamerling JP, editor. *Comprehensive Glycoscience—from Chemistry to Systems Biology*. New York: Elsevier; 2007, p. 329–46.

14. Lutteke T. Web resources for the glycoscientist. *Chembiochem* 2008;**9**:2155–60.

15. Toukach P, Joshi HJ, Ranzinger R, Knirel Y, von der Lieth CW. Sharing of worldwide distributed carbohydrate-related digital resources: online connection of the Bacterial Carbohydrate Structure DataBase and GLYCOSCIENCES.de. *Nucleic Acids Res* 2007;**35**:D280–6.

16. Herget S, Ranzinger R, Maass K, von der Lieth C-W. GlycoCT-a unifying sequence format for carbohydrates. *Carbohydr Res* 2008;**343**:2162–71.

17. McNaught AD. Nomenclature of carbohydrates (recommendations 1996). *Adv Carbohydr Chem Biochem* 1997;**52**:43–177.

18. Sahoo SS, Thomas C, Sheth A, Henson C, York WS. GLYDE—an expressive XML standard for the representation of glycan structure. *Carbohydr Res* 2005;**340**:2802–7.

19. Ranzinger R, Herget S, Wetter T, von der Lieth C-W. GlycomeDB—integration of open-access carbohydrate structure databases. *BMC Bioinformatics* 2008;**9**:384.

20. Doubet S, Bock K, Smith D, Darvill A, Albersheim P. The Complex Carbohydrate Structure Database. *Trends Biochem Sci* 1989;**14**:475–7.

21. Doubet S, Albersheim P. CarbBank. *Glycobiology* 1992;**2**:505.

22. Toukach FV, Knirel YA. New database of bacterial carbohydrate structures. *Glycoconjugate J* 2005;**22**:216–17.

23. Raman R, Venkataraman M, Ramakrishnan S, Lang W, Raguram S, Sasisekharan R. Advancing glycomics: Implementation strategies at the consortium for functional glycomics. *Glycobiology* 2006;**16**:82R–90R.

24. Banin E, Neuberger Y, Altshuler Y, et al. A novel Linear Code((R)) nomenclature for complex carbohydrates. *TIGG* 2002;**14**:127–37.

25. Lutteke T, Bohne-Lang A, Loss A, Goetz T, Frank M, von der Lieth CW. GLYCOSCIENCES.de: an Internet portal to support glycomics and glycobiology research. *Glycobiology* 2006;**16**:71R–81R.

26. Loss A, Bunsmann P, Bohne A, Loss A, Schwarzer E, Lang E, von der Lieth CW. SWEET-DB: an attempt to create annotated data collections for carbohydrates. *Nucleic Acids Res* 2002;**30**:405–8.

27. van Kuik JA, Vliegenthart JF. Databases of complex carbohydrates. *Trends Biotechnol* 1992;**10**:182–5.

28. Bohne-Lang A, Lang E, Forster T, von der Lieth C-W. LINUCS: linear notation for unique description of carbohydrate sequences. *Carbohydr Res* 2001;**336**:1–11.

29. Lutteke T, Frank M, von der Lieth C-W. Carbohydrate Structure Suite (CSS): analysis of carbohydrate 3D structures derived from the PBD. *Nucleic Acids Res* 2005;**33**:D242–6.

30. Lutteke T, von der Lieth C-W. The protein data bank (PDB) as a versatile resource for glycobiology and glycomics. *Biocatalysis and Biotransformation* 2006;**24**:147–55.

31. Berman H, Henrick K, Nakamura H, Markley JL. The worldwide Protein Data Bank (wwPDB): ensuring a single, uniform archive of PDB data. *Nucleic Acids Res* 2007;**35**:D301–3.

32. Ogata H, Goto S, Sato K, Fujibuchi W, Bono H, Kanehisa M. KEGG: Kyoto Encyclopedia of Genes and Genomes. *Nucleic Acids Res* 1999;**27**:29–34.

33. Hashimoto K, Goto S, Kawano S, Aoki-Kinoshita KF, Ueda N, Hamajima M, Kawasaki T, Kanehisa M. KEGG as a glycome informatics resource. *Glycobiology* 2006;**16**:63R–70R.

34. Aoki KF, Yamaguchi A, Ueda N, Akutsu T, Mamitsuka H, Goto S, Kanehisa M. KCaM (KEGG Carbohydrate Matcher): a software tool for analyzing the structures of carbohydrate sugar chains. *Nucleic Acids Res* 2004;**32**:W267–72.

35. Campbell MP, Royle L, Radcliffe CM, Dwek RA, Rudd PM. GlycoBase and autoGU: tools for HPLC-based glycan analysis. *Bioinformatics* 2008;**24**:1214–16.

36. Kawasaki T, Nakao H, Takahashi E, Tominaga T. GlycoEpitope: the integrated database of carbohydrate antigens and antibodies. *TIGG* 2006;**18**:267–72.

37. Narimatsu H. Construction of a human glycogene library and comprehensive functional analysis. *Glycoconjugate J* 2004;**21**:17–24.

38. Apweiler R, Bairoch A, Wu CH, Barker WC, Boeckmann B, Ferro S, Gasteiger E, Huang H, Lopez R, Magrane M, Martin MJ, Natale DA, O'Donovan C, Redaschi N, Yeh LS. UniProt: the Universal Protein knowledgebase. *Nucleic Acids Res* 2004;**32**:D115–19.

39. The Universal Protein Resource (UniProt) 2009. Nucleic Acids Res 2008.

40. Jung E, Veuthey AL, Gasteiger E, Bairoch A. Annotation of glycoproteins in the SWISS-PROT database. *Proteomics* 2001;**1**:262–8.

41. Berman HM, Westbrook J, Feng Z, Gilliland G, Bhat TN, Weissig H, Shindyalov IN, Bourne PE. The Protein Data Bank. *Nucleic Acids Res* 2000;**28**:235–42.

42. Henrick K, Feng Z, Bluhm WF, Dimitropoulos D, Doreleijers JF, Dutta S, Flippen-Anderson JL, Ionides J, Kamada C, Krissinel E, Lawson CL, Markley JL, Nakamura H, Newman R, Shimizu Y, Swaminathan J, Velankar S, Ory J, Ulrich EL, Vranken W, Westbrook J, Yamashita R, Yang H, Young J, Yousufuddin M, Berman HM. Remediation of the protein data bank archive. *Nucleic Acids Res* 2008;**36**:D426–33.

43. Lutteke T, Frank M, von der Lieth C-W. Data mining the protein data bank: automatic detection and assignment of carbohydrate structures. *Carbohydr Res* 2004;**339**:1015–20.

44. Nakahara T, Hashimoto R, Nakagawa H, Monde K, Miura N, Nishimura S. Glycoconjugate Data Bank: Structures—an annotated glycan structure database and N-glycan primary structure verification service. *Nucleic Acids Res* 2008;**36**:D368–71.

45. Crispin M, Stuart DI, Jones EY. Building meaningful models of glycoproteins. *Nat Struct Mol Biol* 2007;**14**:354. discussion 354–355.

46. Ceroni A, Dell A, Haslam SM. The GlycanBuilder: a fast, intuitive and flexible software tool for building and displaying glycan structures. *Source Code Biol Med* 2007;**2**:3.

47. Maass K, Ranzinger R, Geyer H, von der Lieth CW, Geyer R. "Glyco-peakfinder"—de novo composition analysis of glycoconjugates. *Proteomics* 2007;**7**:4435–44.

48. Ceroni A, Maass K, Geyer H, Geyer R, Dell A, Haslam SM. GlycoWorkbench: A Tool for the Computer-Assisted Annotation of Mass Spectra of Glycans. *J Proteome Res* 2008;**7**:1650–9.

49. Jansson PE, Kenne L, Widmalm G. Casper—a Computer-Program Used for Structural-Analysis of Carbohydrates. *J Chem Inf Comput Sci* 1991;**31**:508–16.

50. Loss A, Stenutz R, Schwarzer E, von der Lieth C-W. GlyNest and CASPER: two independent approaches to estimate 1H and 13C NMR shifts of glycans available through a common web-interface. *Nucleic Acids Res* 2006;**34**:W733–7.

51. Herget S, Toukach PV, Ranzinger R, Hull WE, Knirel YA, von der Lieth CW. Statistical analysis of the Bacterial Carbohydrate Structure Data Base (BCSDB): Characteristics and diversity of bacterial carbohydrates in comparison with mammalian glycans. *BMC Struct Biol* 2008;**8**:35.

52. Werz DB, Ranzinger R, Herget S, Adibekian A, von der Lieth CW, Seeberger PH. Exploring the structural diversity of mammalian carbohydrates ("glycospace") by statistical databank analysis. *ACS Chem Biol* 2007;**2**:685–91.

53. Moran AP, Prendergast MM, Appelmelk BJ. Molecular mimicry of host structures by bacterial lipopolysaccharides and its contribution to disease. *FEMS Immunol Med Microbiol* 1996;**16**:105–15.

54. Yuki N. Carbohydrate mimicry: a new paradigm of autoimmune diseases. *Curr Opin Immunol* 2005;**17**:577–82.

55. Moran AP. Relevance of fucosylation and Lewis antigen expression in the bacterial gastroduodenal pathogen Helicobacter pylori. *Carbohydr Res* 2008;**343**:1952–65.

56. van Roon AM, Pannu NS, de Vrind JP, van der Marel GA, van Boom JH, Hokke CH, Deelder AM, Abrahams JP. Structure of an anti-Lewis x Fab fragment in complex with its Lewis x antigen. *Structure* 2004;**12**:1227–36.

57. Jmol: an open-source Java viewer for chemical structures in 3D. http://www.jmol.org/.

58. Toone EJ. Structure and Energetics of Protein Carbohydrate Complexes. *Curr Opin Struct Biol* 1994;**4**:719–28.

59. Poveda A, Jimenez-Barbero J. NMR studies of carbohydrate-protein interactions in solution. *Chem Soc Rev* 1998;**27**:133–43.

60. Siebert H-C, von der Lieth C-W, Gilleron M, et al. Carbohydrate–Protein Interaction. In: Gabius H-J, Gabius S, editors. *Glycosciences: Status and Perspectives*: Chapman & Hall; 1997, p. 291–310.

61. del Carmen Fernandez-Alonso M, Canada FJ, Jimenez-Barbero J, Cuevas G. Molecular recognition of saccharides by proteins. Insights on the origin of the carbohydrate-aromatic interactions. *J Am Chem Soc* 2005;**127**:7379–86.

62. Feinberg H, Taylor ME, Weis WI. Scavenger receptor C-type lectin binds to the leukocyte cell surface glycan Lewis(x) by a novel mechanism. *J Biol Chem* 2007;**282**:17250–8.

63. Frank M, Lutteke T, von der Lieth C-W. GlycoMapsDB: a database of the accessible conformational space of glycosidic linkages. *Nucleic Acids Res* 2007;**35**:287–90.

64. Cooper CA, Harrison MJ, Wilkins MR, Packer NH. GlycoSuiteDB: a new curated relational database of glycoprotein glycan structures and their biological sources. *Nucleic Acids Res* 2001;**29**:332–5.

65. Humphrey W, Dalke A, Schulten K. VMD: Visual molecular dynamics. *J Mol Graphics* 1996;**14**:33–8.

SECTION
V

# Glycomes

# 11

# Glycomics of the Immune System

Pierre Redelinghuys and Paul R. Crocker
*COLLEGE OF LIFE SCIENCES, UNIVERSITY OF DUNDEE,
DUNDEE, SCOTLAND, UK*

This chapter will provide an overview regarding our current understanding of the immune cell glycan repertoire (glycome), how this repertoire changes during the activation and differentiation of immune cells, and how this influences their interactions with various endogenous lectins. The significance of these interactions within the immune response will be discussed.

## Glycosylation in the Immune System

The immune system is made up of many different cell types derived from hematopoietic stem cells that undergo complex differentiation pathways. Some cells are short-lived with high turnover, like neutrophils and myeloid dendritic cells (DCs), others are long-lived like macrophages and memory lymphocytes. The main task of the immune system is to sense and eliminate potential pathogens and the innate immune system deals with the vast majority of these. Failure to do this leads to activation of the adaptive immune system consisting of clonally expanding populations of T and B lymphocytes whose exquisite specificity runs the risk of autoimmunity. The expansion and contraction of the innate and adaptive immune system is tightly regulated via cell–cell interactions, cytokines, and chemokines (Figure 11.1).

As with all cells in the body, the cells of the immune system are coated in a thick layer of sugars, the glycocalyx, and the effector molecules they secrete are variably glycosylated. It is becoming increasingly clear that immune cell differentiation, activation, and death are accompanied by dramatic changes in glycosylation patterns due to the complex and dynamic expression of various glycosidases and glycosyltransferases. This leads to the generation of glycoproteins and glycolipids with considerable heterogeneity in the nature and extent of their glycan modifications. This form of posttranslational modification occurs principally in the Golgi apparatus and involves a group of enzymes,

**Handbook of Glycomics**
Copyright © 2009 by Elsevier Inc. All rights of reproduction in any form reserved.

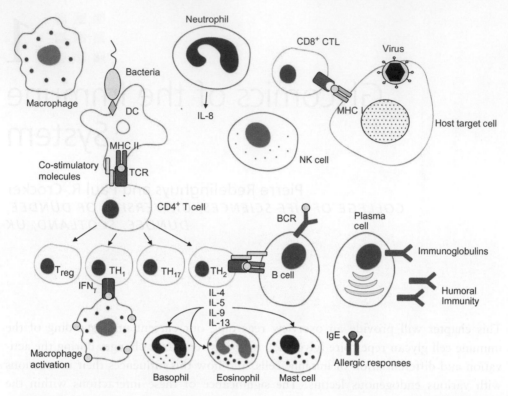

**FIGURE 11.1** Hemopoietic stem cell differentiation gives rise to an extensive and diverse repertoire of cells mediating innate and adaptive immune responses.

the glycosyltransferases. These enzymes are type II transmembrane proteins specifically transferring activated sugar nucleotide donors like CMP-sialic acid, UDP-*N*-acetylgalactosamine, UDP-*N*-acetylglucosamine, UDP-galactose, and GDP-fucose to glycoconjugate acceptors [1].

Of particular importance in the immune system is the addition of sialic acid residues to the non-reducing termini of oligosaccharides mediated by a family of ~20 sialyltransferase genes. Sialic acids are nine-carbon carboxylated sugars occurring in specific linkages (α2,3, α2,6 or α2,8) at the non-reducing terminal ends of *N*- and *O*-linked oligosaccharides, glycosaminoglycans, and glycolipids. Mammals mainly express *N*-acetyl neuraminic acid (Neu5Ac), *N*-glycolyl neuraminic acid (Neu5Gc), and 5,(7)9-*N*,*O*-diacetylneuraminic acid (Neu5,(7)9Ac2). Sialylation can mask ligands for lectins such as galectins, it can create ligands for lectins such as siglecs, and can also non-specifically prevent cell–cell interactions due to its electronegative charge.

## Mammalian Lectins of the Immune System

Over 100 lectins have been characterized in mammalian genomes, many of which are expressed in the immune system. These lectins can be classified according to whether they mediate primarily glycan interactions with host tissues, or whether they are more important as pathogen recognition molecules. Since this chapter deals with immune cell glycomics rather than pathogen glycomics, more emphasis will be placed on immune system lectins recognizing host glycans.

Table 11.1 summarizes some of the major families of lectins discussed in this chapter. They are organized into families based on structural features and include I-type lectins (predominantly siglecs), C-type lectins, and S-type lectins like galectins. It should also be pointed out that many antibodies and some T cell receptors have been described which bind glycans, but these are not normally classified as lectins. In addition to those listed in Table 11.1, there are many other lectin-like molecules expressed in the immune system such as hyaluronan-binding proteins, chitin-binding lectins and probably many others remaining to be discovered. We refer you to the "Genomics Resource for Animal Lectins" for a more complete overview of mammalian lectins in general and families of immune-associated lectins not covered specifically in this chapter (http://www.imperial.ac.uk/research/animallectins).

## Current Approaches in the Study of Immune Cell Glycomes

The cellular glycomes of the immune system have been investigated using multiple approaches, most of which are discussed in detail in other chapters of this book. The Consortium for Functional Glycomics has adopted these strategies to understand the nature and functions of immune system glycomics and the interested reader is encouraged to refer to its website for further information (http://www.functionalglycomics. org/static/index.shtml).

Matrix-assisted laser desorption/ionization-MS (MALDI-MS) and electrospray (ES)-MS approaches have been developed, allowing the rapid, high-sensitivity sequencing of complex mixtures of *N*- and *O*-glycans isolated from various leukocyte populations including T and B lymphocytes, neutrophils, natural killer (NK) cells, macrophages, and dendritic cells. These approaches have enabled the determination of the terminal sequences and branching patterns of oligosaccharides and, in particular, the changes in the glycan repertoire during cell maturation, differentiation, and activation.

The analysis of the expression of the glycosyltransferases underlying the observed changes in immune cell surface glycan repertoires has involved the use of quantitative real time polymerase chain reaction (PCR) as well as whole genome and custom gene microarrays containing probes for various glycosyltransferases.

The glycan-binding specificities of various endogenous lectins of the immune system have been determined using glycan arrays. Here, carbohydrate-based interactions have been analyzed in a high-throughput chip-based format with natural, chemically or enzymatically synthesized oligosaccharides covalently or non-covalently coupled to multi-well plates.

Of particular importance is the identification of the ligands for endogenous lectins. Here, approaches have included the use of recombinant lectins (e.g., chimeric

## Table 11.1 Some families of lectins involved in innate and adaptive immunity

| Lectin | Expression | Glycan specificity and function | Representative structures |
|---|---|---|---|
| I-type lectins: Siglecs | Restricted mostly to leukocytes of the innate and adaptive immune systems | Sialic acids, e.g. Neu5Acα2,3Galβ1,4GlcNAc; Neu5Acα2,6Galβ1,4GlcNAc; Neu5Acα2,8Neu5Ac | Sialoadhesin<br>CD22 |
| Non-CD33-related siglecs, e.g. sialoadhesin, CD22, MAG | | Diverse functions, including cell–cell and cell–matrix interactions, pathogen-binding (sialoadhesin), myelin maintenance (MAG) and inhibitory signaling (CD22) | |
| CD33-related siglecs, e.g. Siglecs-7, -8, -14 | | Functioning mostly as inhibitory, activating and endocytic receptors. Recognition of sialylated pathogens | Siglec-7 |
| C-type lectins: Selectins, e.g. selectins E, L, and P | Activated endothelial cells, myeloid cells, lymphocytes | Sialyl Lewis x (sLe^x) (Neu5Acα 2,3Galβ1,4[Fucα1,3]GlcNAc); sialyl 6-sulfo Lewis x (Neu5Ac α2,3Galβ1,4[Fucα1,3]GlcNAcβ [6-SO3-]) Tethering and rolling of leukocytes in homing and inflammation | |
| Macrophage mannose receptors | DCs, Langerhans cells, monocytes, macrophages | Mannose, fucose, N-acetylglucosamine, sLe^x and 3-sulfo-Le^x. Binds bacteria, viruses, fungi, and various endogenous ligands | |
| Type II receptors, e.g. DC-SIGN, MGL | Dendritic cells, macrophages, B cell subsets | Mannosylated and fucosylated glycans such as Lewis x (Galβ1,4[Fucα1,3]GlcNAcβ). Pattern recognition receptor for pathogens like HIV-1 | DC-SIGN |

*(Continued)*

**Table 11.1 Continued**

| Lectin | Expression | Glycan specificity and function | Representative structures |
|---|---|---|---|
| S-type lectins: Galectins | | β-Galactoside-containing glycans with repeating LacNAc units (Galβ1,4GlcNAc) as disaccharides at the terminii of complex-type N-glycans or as repeating units within in a poly-N-acetyllactosamine chain on N- or O-glycans. Varying functions including proapoptotic, antiapoptotic, antiadhesive, anti-inflammatory | |
| Prototype galectins, e.g. Galectins-1, -2, -5, -7, -10, -11, -13, -14, -15 | Lymph nodes, spleen, thymus, macrophages, B cells, T cells, DCs, tumors, GIT, erythrocytes, eosinophils, basophils, lens, placenta | Galectin-1: Binds matrix glycoproteins and cell surface receptors like CD2, CD3, GM1, CD7, CD43, CD45, and matrix glycoproteins like laminin and fibronectin. Proapoptotic towards T cells and B cells, pro- and antiadhesive properties, anti-inflammatory | |
| Chimera-type galectins, e.g. Galectin-3 | Tumor cells, macrophages, epithelial cells, fibroblasts, activated T cells | Galectin-3: Binds matrix glycoproteins like laminin and fibronectin. Antiapoptotic towards T cells and breast cancer. Pro-inflammatory | |
| Bivalent tandem repeat-type galectins, e.g. Galectins-4, -6, -8, -9, and -12 | Liver, prostate, kidney, cardiac muscle, lung and brain, thymus, T cells, kidney, Hodgkin's lymphoma, adipocytes | Galectin-9: Proapoptotic towards thymocytes | |

Fc proteins) coupled to agarose and the identification of ligands from labeled (tritiated, iodinated, or biotinylated) or unlabeled whole cell preparations or cellular fractions like membranes. Furthermore, various candidate ligand approaches have been used which typically involve the immunoprecipitation of potential ligands followed by lectin blots or vice versa.

# Cellular Glycomes of the Innate Immune System

## Neutrophils

Neutrophils are the most abundant leukocytes in human peripheral blood and they play a key role in innate immunity against bacterial pathogens. Neutrophils phagocytose extracellular bacteria and release antimicrobial substances during early infection, allowing time for the adaptive immune response to complete pathogen clearance. Activated neutrophils also produce and release chemokines like interleukin 8 (IL-8), attracting additional neutrophils and other immune cells such as monocytes, macrophages, dendritic cells (DCs), and T cells through the release of other chemokines and antimicrobial peptides. Neutrophil recruitment to infected or damaged sites proceeds via several steps: (1) circulating neutrophils are captured and induced to roll on the stimulated endothelium via selectin–glycan interactions; (2) neutrophil $\beta_2$ integrins are activated which immobilize cells on the endothelium. Selectins also play an important role here (Table 11.1); (3) neutrophils migrate through the endothelium by degrading the extracellular matrix via secreted matrix metalloproteinases (MMPs). Migration to infected or inflamed tissue leads to phagocytosis and engulfment of pathogens such as Gram-positive and Gram-negative bacteria. Lipopolysaccharides from Gram-negative bacteria induce neutrophil activation and this is accompanied by an early downregulation of L-selectin and an upregulation of the integrin CD11b/CD18 (Mac-1/CR-3) on the cell surface which is important in mediating neutrophil adhesion to the vascular endothelium [2].

Neutrophils have a unique glycosylation pattern and unlike other leukocytes, also express non-sialylated Lewis x (Le$^x$) glycans on the cell surface. This is likely due to the differential expression of fucosyltransferases IV and VII which incorporate a fucose through an $\alpha$3 linkage on the GlcNAc moiety of a non-sialylated lactosamine to form a Le$^x$ glycan [3,4]. Neutrophil Le$^x$ (Gal$\beta$(1,4)[Fuc$\alpha$(1,3)]GlcNAc) and sialyl Lewis x (sLe$^x$) (Neu5Ac$\alpha$(2,3)Gal$\beta$(1,4)[Fuc$\alpha$(1,3)]GlcNAc) structures are predominantly found on tetra-antennary $N$-linked glycans consisting of poly-$N$-acetyllactosamine chains with variable $\alpha$1,3-linked fucose substitutions. The sialyltransferases ST3GalIV and ST3GalVI are largely responsible for sLe$^x$ biosynthesis in neutrophils [5].

E-selectin is transiently expressed on vascular endothelial cells in response to inflammatory agents [6]. The major E-selectin ligand on mouse neutrophils has been identified as ESL-1, which harbors sLe$^x$ on its $N$-glycans. E-selectin also binds sLe$^x$ expressed on the integrins CD11b/CD18 as well as on the neutrophil L-selectin itself. CD44, a hyaluronan-binding cell surface glycoprotein, mediates E-selectin-dependent rolling of neutrophils. P-selectin also plays a crucial role in the migration of neutrophils to sites of inflammation through its binding to sLe$^x$ and sulfated tyrosines presented on neutrophil P-selectin glycoprotein ligand-1 (PSGL-1). Additional selectin ligands on neutrophils may also include sialyl Lewis a (Neu5Ac$\alpha$(2,3)Gal$\beta$(1,3)(Fuc$\alpha$(1,4))GlcNAc-R and sulfated Lewis a. Human neutrophils express Siglecs-5 and -9 which are masked at the cell surface by *cis*-interactions with neutrophil-associated glycans. Each of these siglecs has a distinct preference for sialylated glycans, in particular Siglec-9 has a preference for sLe$^x$ structures. *Cis*-interactions between Siglec-9 and sLe$^x$ may compete with *trans*-interactions between these glycans and P- and E-selectin expressed on inflamed endothelial cells.

CD18 is expressed as a heterodimer with CD11b forming the $\beta_2$ integrin Mac-1 on neutrophils. In neutrophils it is involved in pathogen recognition via complement fixation and redirects pathogens to phagosomes for destruction. Mac-1 also functions in the regulation of cell survival and apoptosis where its signaling is induced by its binding to ligands such as ICAM-1. Binding to ICAM-1 also plays an important role in mediating neutrophil adhesion to endothelial cells and their subsequent migration [7]. Galectin-8 has been identified as a modulator of neutrophil function where it induces the adhesion of peripheral blood neutrophils through binding to Mac-1. Through its interaction with Mac-1, galectin-8 is able to induce superoxide production in peripheral blood neutrophils. Galectin-3-mediated ligand clustering induces neutrophil phagocytosis, the production of reactive oxygen species, the secretion of IL-8 and protease release, while the binding of galectin-1 to an oligosaccharide on the $\alpha$-subunit of Mac-1 activates Mac-1 by promoting a direct interaction between Mac-1 and the substratum to which the neutrophil adheres.

Carcinoembryonic antigen-related adhesion molecule 1 (CD66a/CEACAM1) comprises four Ig-like domains, a transmembrane and cytoplasmic domain. It is highly glycosylated and in granulocytes it harbors high mannose and fucose oligosaccharides including non-sialylated Le$^x$ groups as well as highly complex *N*-linked glycans. Consequently, it is the major granulocyte membrane glycoprotein which binds Le$^x$ antibodies and the presence of these Le$^x$ groups are rather specific [8]. During homeostasis, neutrophils and DCs are located in different compartments (bloodstream vs. peripheral tissues, respectively). However, during infection, both cell types accumulate at sites of inflammation where they physically interact (Figure 11.2) [9]. When neutrophils become activated, Mac-1 and CEACAM1 are upregulated where they bind DCs via the C-type lectin DC-SIGN. Here, DC-SIGN expression is highest in immature DCs such that interactions with neutrophils will induce their maturation. The binding of DC-SIGN to CEACAM1 is dependent on high levels of Le$^x$ glycans on CEACAM1. Here, binding mediates intercellular adhesion via a Ca$^{2+}$-independent homophilic interaction which regulates proliferation and delays the induction of neutrophil apoptosis. The lifespan of neutrophils is also prolonged via DC-SIGN–Mac-1 interactions which depend upon the expression of Le$^x$, Le$^y$, and high mannose structures on Mac-1. The interaction between neutrophils and DC-SIGN-positive DCs is important in generating a robust Th1 response through the release of IL-12 by DCs. Thus, neutrophils may regulate adaptive immune responses indirectly through interactions with DCs. The DC–neutrophil DC-SIGN–CEACAM1-mediated interaction may also enable DCs to transfer captured bacteria to neutrophils which would lead to their efficient internalization and destruction. Phagocytosis of these pathogens by neutrophils may supply pathogen-derived peptides, enabling DCs to activate pathogen-specific

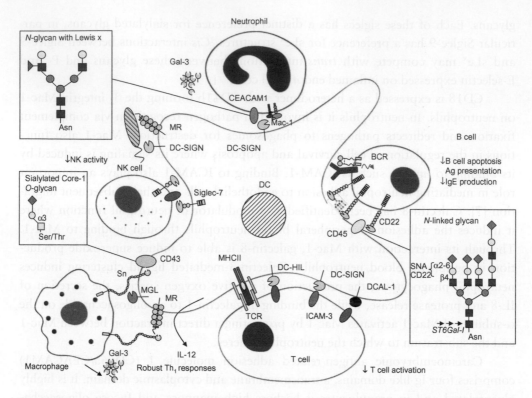

**FIGURE 11.2** Dendritic cells occupy a central position in immunity where they bridge the gap between innate and adaptive responses. This role is mediated by multiple interactions with numerous cell types, with many being glycan and lectin dependent.

T cells (Figure 11.2). CEACAM1 may mediate the uptake of damaged granulocytes or cell debris during inflammatory responses to bacterial infection and as such the interaction between CEACAM1 and DC-SIGN is important in regulating self and non-self recognition, modulating the immune response. CEACAM1 may also be involved in neutrophil activation where it associates with the protein tyrosine kinases lyn and hck. Galectin-3, released by macrophages under inflammatory conditions, also binds CEACAM1 via its lactosamine glycan structures and enhances the oxidative burst in activated neutrophils [10].

Transmigrating neutrophils secrete MMP-9 in order to degrade matrix proteins [7]. Galectin-8 has been shown to accelerate the processing of proMMP-9 mediated by MMP-3 via an interaction with the oligosaccharide chains of these two MMPs.

## Eosinophils

The accumulation of eosinophils in tissues and the release of their mediators is a prominent feature of allergic and inflammatory diseases like asthma, atopic dermatitis, and allergic rhinitis. While galectin-9 is a potent and selective eosinophil chemoattractant [7], P-selectin serves as a mediator of selective eosinophil migration in allergic disease [11] since eosinophils show enhanced binding to P-selectin. The P-selectin ligand on eosinophils is PSGL-1 and eosinophils express significantly more PSGL-1 than do neutrophils. Murine Siglec-F and human Siglec-8, though not orthologous, are expressed selectively on eosinophils and exhibit similar ligand preferences for 6′-sulfo-sLe$^x$ and may thus have similar functional roles. In this regard, Siglec-F is expressed on mature circulating eosinophils as well as on activated CD4$^+$ T cells (following allergen exposure) and is a useful marker for studying eosinophil turnover and function. During allergic airways inflammation, Siglec-F expression and that of its ligands is increased. Ligand binding results in the induction of eosinophil apoptosis suggesting that Siglec-F (and possibly Siglec-8) negatively regulate allergic responses [12].

## Natural Killer Cells

Following activation by cytokines like IL-2, NK cells undergo alterations in cell surface glycosylation with distinct patterns as compared to other cells of lymphoid origin. In humans, a sialyl SSEA-1 (sLe$^x$-i) antigen has been identified as a cell surface marker of NK cells [13]. NK cells also selectively bind the α2,3Neu5Ac-specific plant lectin *Maackia amurensis* leukoagglutinin (MALII). Human NK cells express CD56, a form of NCAM carrying α2,8-linked polysialic acid chains, which may be important in modulating cellular interactions, and CD57, an acidic glycoprotein with a core chain comprising polylactosamine. A subset of NK cells expresses sulfated L-selectin ligands which may play a role in the homing and trafficking of these cells [14] while activated effector murine NK cells express a cell surface glycolipid, ganglio-*n*-tetraosylceramide, asialo-GM1. In humans, NK cells express Siglec-7 which has a ligand binding preference for α2,8-linked disialic acid. Siglec-7 is masked at the NK cell surface and this could be due to *cis*-interactions with glycoproteins carrying this modification. In support of this, antibodies reactive with α2,8-linked disialic acid were shown to react strongly with NK cells, NK cells consistently expressed mRNA for sialyltransferase ST8SiaVI, which transfers one sialic acid onto sialylated O-glycan acceptors and MALDI-TOF analysis of total NK cell glycans revealed the presence of an O-linked glycan containing α2, 8-linked disialic acid [15].

## Macrophages

Macrophages are a very heterogeneous group of bone marrow-derived cells present in nearly all tissues of the body. They are recruited from the blood as monocytes and rapidly differentiate into tissue macrophages both under steady-state conditions as well as during acute and chronic inflammatory responses. Macrophages are long-lived and are able to modulate gene expression according to the local environment and the specific functions they play in different tissues such as the central nervous system (CNS), liver, and lymphoid tissues. At portals of entry such as gut and skin, macrophages play a key role in pathogen surveillance and, together with DCs, are involved in regulation of adaptive immune responses, becoming rapidly activated in response to a wide diversity of pathogen-associated and host tissue-derived molecules (Figure 11.1).

Macrophages typically express a rich abundance of soluble and membrane-bound lectins, including C-type lectins, galectins, and siglecs which are important in both pathogen and host cell interactions. Macrophage glycosylation is also important in their interactions with lectins. For example, galectin-3 is a novel monocyte and macrophage chemoattractant [7]. Furthermore, galectins form $\beta(1,6)$ N-acetylglucosaminyltransferase V (GnTV)-dependent glycoprotein lattices on macrophages, a requirement for maintaining sufficient cell surface cytokine receptor density and to drive motility and phagocytosis [16]. Galectin-1 binding to specific saccharide ligands also differentially regulates Fc$\gamma$ receptor I-dependent phagocytosis and inhibits major histocompatibility complex (MHC) class II-dependent antigen presentation. The siglec sialoadhesin is a large, extended molecule possessing 17 Ig domains (Table 11.1). It is expressed constitutively by tissue-resident macrophages and its expression is upregulated in inflammatory macrophages where it plays a role in their proinflammatory activity. Sialoadhesin binds glycoconjugates containing the terminal oligosaccharides Neu5Ac$\alpha$(2,3)Gal$\beta$(1,3)GalNAc or Neu5Ac$\alpha$(2,3)Gal$\beta$(1,3) which are expressed by another extended sialylated molecule, the sialomucin CD43 [17] (Figure 11.2). Sialoadhesin can also function as a target for other lectins via its N-linked glycans. This was first seen using recombinant forms of the cysteine-rich domain of the mannose receptor which selectively binds GalNAc-4-SO$_4$-modified N-linked glycans. Similar interactions of sialoadhesin's N-linked glycans have been observed with the macrophage galactose-type-N-acetylgalactosamine-specific lectin 1 (MGL-1), expressed on dendritic cells (Figure 11.2) [18].

## Dendritic Cells

Dendritic cells (DCs) are key regulators of the immune response, where they constantly monitor the extracellular space for foreign proteins, presenting them on MHC II molecules and less efficiently by cross-presentation on MHC I molecules (Figure 11.1). Different DC lineages can develop from separate precursors or may represent different activation states of a single subtype. There are different functional phenotypes based on cytokine and metabolite production and the expression of cell surface markers and thus DCs are able to adapt to various environmental stimuli. DC subtypes include plasmacytoid DCs (pDCs), Langerhans cells, and interstitial DCs, each requiring their own set of cytokines and growth factors. Human Langerhans cells and interstitial DCs are derived from a common precursor while pDCs develop along a separate pathway [19].

The differentiation of monocytes into immature DCs as well as their subsequent pathogen/cytokine-induced maturation in the peripheral tissues is accompanied by the transcriptional regulation of various glycosyltransferases. Monocytes and immature DCs show the presence of abundant mono-, di-, and tri-sialylated tri- and tetra-antennary N-glycans. Some of these are decorated with Lewis-type fucose or elongated with poly-N-acetyllactosamine. The dominant O-glycan structure is the abundant sialyl-T Ag (Neu5Ac$\alpha$(2,3)Gal$\beta$(1,3)GalNAc) found on sialomucins like PSGL-1 and CD43 which may serve as ligands for the macrophage-associated siglec sialoadhesin ($\alpha$2,3-linked sialic acid-specific) (Figure 11.2) [20,21]. DCs and macrophages coexist in the lymph nodes, tonsils, and spleen and a sialoadhesin-mediated interaction between these cells may enhance their intercellular contact and induce or modulate DC signaling (Figure 11.2). Monocyte differentiation into immature DCs induces an upregulation of core-2-Galβ(1,3)GalNAcβ(1,6)GlcNAc-transferases (C2GnTs), C2GnT1 and C2GnT2, such that they express glycoproteins harboring core 2 O-glycans. Core 2 O-glycans on PSGL-1 serve as substrates for the generation of sLe^x which in turn serves as a ligand for E- and P-selectins. These selectins are expressed on the vascular endothelium and facilitate the migration of monocytes and immature DCs from the blood and their extravasation into the peripheral tissues [22] (Figure 11.3).

The subsequent maturation of DCs is accompanied by a decrease in the expression and activity of C2GnT1 and C2GnT2 such that mature DCs carry predominantly core 1 O-glycans. Since sLe^x is carried on the core 2 O-glycans of PSGL-1, while retaining their expression of PSGL-1, they lose their expression of sLe^x, thus preventing any further interactions with selectins-E and -P. Thus, the regulation of the core 1/core 2 ratio drives the expression of sLe^x in these cells. DC maturation also results in an increase in ST3GalII and ST6GalNAcII mRNA and while the density of α2,3-linked sialic acid remains largely unchanged, that of α2,6-linked sialic acid decreases in mature DCs. Maturing DCs also show an increase in the expression of poly-N-acetyllactosamine-elongated glycans due to the upregulation of β4GalT-4, -5, and β3GnT-2

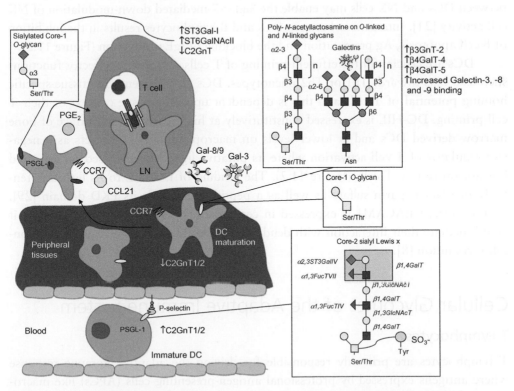

**FIGURE 11.3** The migration, activation, and trafficking of dendritic cells is accompanied by distinct changes in the cell surface glycan repertoire which influences interactions with selectins in the vasculature as well as with chemokines and galectins among others. The principal glycan structures as well as their corresponding glycosyltransferases are shown.

[21] (Figure 11.3). This may account for the increased binding of galectins-3 and -8. Galectin-3 binding has multiple effects on DCs including modulation of cell adhesion, chemoattraction, cell activation, proliferation, and apoptosis [23]. Galectin-8 binding is associated with extracellular matrix adhesion and integrin-related cytoskeletal reorganization [24,25]. Furthermore, galectin-1 and -9 binding to DCs induces their maturation and results in a strong proinflammatory capacity [26,27]. In the activated/mature state, the density of MHC molecules on DCs is upregulated together with costimulatory molecules such as CD80, CD83, and CD86 and increased IL-2 production.

Mature DCs leave the site of inflammation and migrate towards the draining lymph nodes, a process dependent on chemokines like CCL21 and their cognate receptors on DCs such as CCR7 [28]. Prostaglandin $E_2$ (PGE$_2$) is essential for the CCR7-mediated migration of mature DCs and it induces a rapid upregulation of ST3GalI and ST6GalNAcII as well as a downregulation of C2GnT1. This resulting sialylation of core 1 O-glycans is required to permit efficient migration of mature DCs possibly via the binding of CCL21 also to PSGL-1. In order to minimize Ag uptake, C-type lectin receptor expression on the maturing DC cell surface is downregulated (Figure 11.3).

Sialoadhesin, CD22, and Siglec-7 show strong binding to immature DCs and this is further upregulated during their maturation. Siglec-7, also expressed on DCs, may interact in *cis* with DC ligands thereby regulating DC activation. *Trans*-interactions between DCs and NK cells may enable the Siglec-7-mediated down-modulation of NK cell activity [21]. Interactions between DCs and B lymphocytes results in the inhibition of B cell apoptosis, Ag presentation and the blocking of IgE production (Figure 11.2).

DCs are crucial for the efficient priming of T cells, influencing effector functions such as a bias towards Th1 or Th2 phenotypes. DCs also influence the tissue-specific homing potential of T cells and this is dependent upon the tissue context of their T cell priming. DC-HIL is expressed constitutively at high levels on the surface of bone marrow-derived DCs and at lower levels on macrophages. DC-HIL acts as a negative regulator of T cell activation where its putative ligand is expressed on activated but not on resting T cells (Figure 11.2). The binding of DC-HIL to its ligand potentially involves heparin sulfate as well as a peptide that binds its PKD domain [29]. Furthermore, CEACAM1 is expressed in certain subpopulations of T cells where it could mediate their interaction with dendritic cells and in so doing regulate self/non-self recognition [8].

# Cellular Glycomes of the Adaptive Immune System

## T Lymphocytes

T lymphocytes are primarily responsible for driving the adaptive immune response where antigens expressed by professional antigen-presenting cells (APCs) like macrophages and dendritic cells (DCs) trigger their activation, expansion, differentiation, and effector function through engagement of an antigen-specific T cell receptor (TCR) (Figure 11.1). There are two main subsets of T lymphocytes, namely CD4$^+$ T cells and

CD8$^+$ T cells. CD4$^+$ T cells recognize antigenic peptides bound to the class II MHC on APCs and they develop into several functionally distinct effectors known as T helper (Th) cells. Th1 cells are necessary for the clearance of intracellular pathogens and are associated with delayed-type hypersensitivity reactions, Th1$_7$ cells are associated with tissue inflammation and the exacerbation of autoimmunity [30], while Th2 cells are critical in the response to extracellular pathogens and promote allergic disease. Here, they secrete cytokines such as IL-4, -5, -9, and -13 which drive the development of mast cells, basophils, and eosinophils, cells which release mediators such as histamine in response to immunoglobulin E (IgE). T regulatory cells (T$_{regs}$) are a distinct subset of T cells characterized by expression of the FOXP3 transcription factor that play a crucial role in regulating effector T cell functions. Host cells expressing viral, tumor, and transplantation antigens bound to class I MHC are targeted by the second T cell subset, CD8$^+$ T lymphocytes (Figure 11.1).

The differentiation, maturation, trafficking, and activation of T lymphocyte subsets is accompanied by programmed microheterogeneous changes in cell surface glycosylation, specific for individual glycan chains and the glycoproteins on which they are found [31].

## *T Lymphocyte Development in the Thymus*

T cell differentiation and maturation is a tightly regulated process during which bone marrow lymphoid precursors undergo a transit from the thymic cortex to the thymic medulla leading to the elimination of those cells with a B lymphocyte potential and the removal of self-reactive thymocytes (recognizing self-antigens). Non-self, antigen-reactive CD4$^+$ and CD8$^+$ T cells are finally selected and then proceed to play their respective roles in mediating adaptive immune responses. During this process thymocytes may be classified phenotypically as being double negative (DN), single positive (SP), or double positive (DP) based upon their expression of specific cell surface markers. The thymic cortex contains immature double positive (DP) (CD4$^+$CD8$^+$) and double negative (DN) (CD4$^-$CD8$^-$) thymocytes (Figure 11.4a) [32]. DN thymocytes, which are the most immature subset, are further subdivided into four stages (DN1–DN4). The final stage, having undergone rearrangement of the T cell receptor β (TCRβ) differentiates into DP cells which then undergo thymic selection leading to the emergence of mature SP CD4$^+$ and CD8$^+$ T cells.

Immature cortical DN thymocytes show increased expression of primarily three types of glycosyltransferase, namely core-1-β1,3-galactosyltransferase (C1β1,3GalT or T-synthase), core-2-Galβ1,3GalNAcβ1,6GlcNAc-transferase (C2GnT) as well as β1,3-N-acetylglucosaminyl (GlcNAc) transferases also known as Fringe (Fng) proteins. C1β1,3GalT expression generates cell surface Galβ1,3GalNAc core 1 O-glycans which are detectable with the plant lectin peanut agglutinin (PNA), rendering the cells PNA$^+$ (Figure 11.4b). PNA also binds the core 2 O-glycans (Galβ1,3GalNAcβ1,6GlcNAc) generated by a marked increase in C2GnT expression [33]. The Fng protein Lunatic Fringe (L-Fng), which adds N-acetylglucosamine (GlcNAc) residues to O-fucose (Figure 11.4b) is highly expressed in immature DN thymocytes, following which its

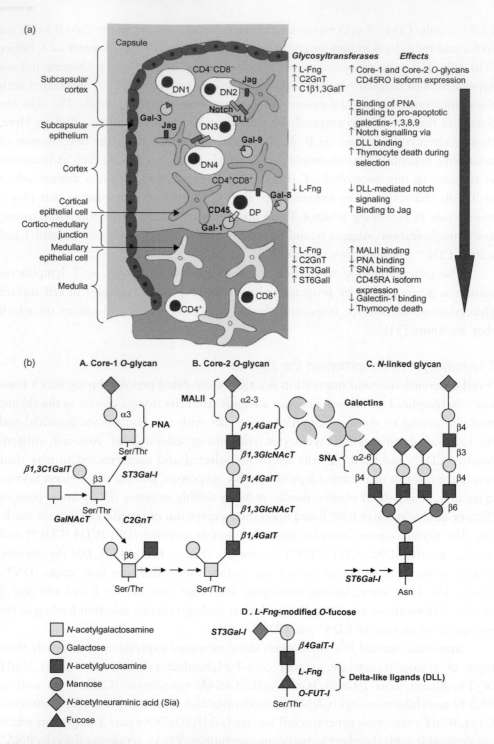

**FIGURE 11.4** T lymphocyte development in the thymus (a). Bone marrow precursors migrate as double negative (DN) and double positive (DP) cells from the thymic cortex to the medulla (shown by red arrow). Thymic selection eventually gives rise to CD4+ and CD8+ T cells. This transit is associated with changes in the expression of various glycosyltransferases, changes in the glycan repertoire of thymocytes (b) and influences interactions with endogenous lectins such as galectins and Notch receptors. These interactions ensure the T cell potential of thymocytes and serve to eliminate self-reactive cells.

expression decreases dramatically in DP thymocytes and increases again upon maturation into SP CD4$^+$ and CD8$^+$ T cells (Figure 11.4a). These glycan modifications occur on several cell surface glycoproteins including CD2–CD4, CD7 and CD8, CD43 and CD45 as well as Notch receptors.

Like L-Fng, Notch-1 expression is high in DN thymocytes and low in DP thymocytes where its signaling regulates cell fate choices, survival, and proliferation during intrathymic T cell development. L-Fng modifies the extracellular domain of Notch receptors, altering their specificity for ligands including Delta-like (DLL) 1, 3, and 4 and Jagged (Jag) 1 and 2 which are expressed in stromal cells of the thymic epithelium as well as thymocytes where they regulate lateral Notch-1 signaling via thymocyte–thymocyte interactions. In the DN stage of development, L-Fng and Notch expression is high and L-Fng modifications of Notch enhances interactions with DLL, activating Notch signaling (Figure 11.4a,b). At this stage, Notch-1 interactions with Jag, expressed throughout the cortex, are inhibited through an elongation of the L-Fng-generated disaccharide by β4 galactosyltransferase I (β4GalT-I) (Figure 11.4a,b). In DP thymocytes, L-Fng and Notch expression is reduced, allowing DP cells to interact with Jag, modulating TCR signaling during positive selection. By switching off Notch-1 signaling in DP cells, the low L-Fng expression enhances the generation of mature T cells (Figure 11.4a) [34].

The increased expression of core 1 and core 2 O-glycans in immature cortical thymocytes is accompanied by the isoform-specific expression of cell surface glycoproteins like CD43 and CD45. While CD43 occurs as a 130 kDa isoform at this stage [35], the receptor tyrosine phosphatase CD45 is expressed in its under-glycosylated, low molecular weight RO isoform (CD45RO) [36]. The increased expression of core 2 O-glycans on CD45RO renders it susceptible to the transient carbohydrate-dependent binding of the pro-apoptotic galectin-1 [35], which is expressed as an active dimer by thymic epithelial cells [37]. Dimeric galectin-1 binding induces the multivalent cross-linking of cell surface receptors, receptor clustering, inactivation, and apoptosis of both DP thymocytes and more immature DN thymocytes [38]. The process is marked by the reorganization of CD7, CD43, and CD45 into membrane microdomains with the CD3/CD45 complex segregating from CD7/CD43 to membrane blebs on dying cells. Galectin-1-mediated CD45 inactivation reduces its protein tyrosine phosphatase activity [33], separates this activity from the kinase activities of CD7/CD43 and activates the death signal via tyrosine phosphorylation. Other galectins involved in the induction of thymocyte apoptosis include galectins-3, -8, and -9 [33,39] and the interactions of immature cortical thymocytes with these pro-apoptotic galectins may play a role in thymic selection leading to the elimination of self-reactive thymocytes.

Maturation and differentiation of thymocytes into SP CD4$^+$ and CD8$^+$ T cells in the medulla is accompanied by a general increase in cell surface sialylation [36] driven by several sialyltransferases including the α2,3 sialyltransferase ST3GalI [32] which sialylates core 1 O-glycans and α2,6 sialyltransferase ST6GalI which sialylates N-linked glycans (Figure 11.4b). However, here there is a reduced expression of C2GnT, decreasing the prevalence of core 2 O-glycans [36] (Figure 11.4a).

ST3GalI generates Neu5Acα(2,3)Galβ(1,3)GalNAc [40] on core 1 O-glycans which is detectable with the plant lectin MALII (specific for sialic acid α2,3 linked to galactose). Masking of the Galβ(1,3)GalNAc PNA glycan recognition site reduces binding of PNA as compared with cortical thymocytes (Figure 11.4a,b). Moreover, sialylation of the same site abrogates the binding of pro-apoptotic thymic galectins, rendering mature medullary thymocytes resistant to galectin-mediated death. While both medullary and cortical thymocytes express similar levels of ST6GalI mRNA, medullary thymocytes express the higher molecular weight RA isoform of CD45 (CD45RA), which is a preferred acceptor substrate for ST6GalI [31,41]. This results in unique α2,6 sialylation of Galβ(1,4)GlcNAc in the N-linked glycans of CD45RA to generate Neu5Acα(2,6)Galβ(1,4)GlcNAc which may be detected by the increased binding of the α(2,6)Neu5Ac-specific plant lectin *Sambucus nigra* agglutinin (SNA). Like the α2,3 sialylation of core 1 O-glycans, α2,6 sialylation reduces the binding of pro-apoptotic galectin-1 by masking its underlying LacNAc ligand, contributing to a resistance to galectin-induced cell death (Figure 11.4a,b). Furthermore, increased sialylation of mature medullary thymocyte glycoproteins raises their negative charge and prevents the formation of clusters that would normally promote galectin-induced apoptosis. Thus, the resistance of mature medullary thymocytes to galectin-induced apoptosis contributes to the process of thymic selection, leading to mature antigen-reactive T cells entering the peripheral circulation. Changes in cell surface sialylation may impact upon other cell surface glycoproteins such as the heavily O-glycosylated CD8, which is the co-receptor for MHC class I-restricted CD8$^+$ T cells (Figure 11.1) [36]. Upregulation of ST3GalI in maturing medullary thymocytes diminishes the binding of CD8 to MHC I as compared to immature thymocytes [42], influencing the efficiency of thymic selection by controlling self-reactivity [43]. Furthermore, increased O-linked sialylation may affect the conformation or clustering of CD8αβ dimers, limiting their signaling while their access to MHC I may be influenced by the sialylation of other more abundant T cell glycoproteins like CD43 and CD45.

Following their maturation, naïve T cells recirculate continuously through-out the body, migrating through the specialized endothelium of secondary lymphoid organs such as lymph nodes, Peyer's patches and the spleen. Upon encountering cognate antigens presented by APCs in the presence of appropriate co-stimulatory molecules, they become primed and activated [44]. Following priming, T cells migrate through postcapillary venules and infiltrate target peripheral tissues until they reach antigenic sites [45]. The role of glycosylation in this process of homing and trafficking will be covered later in the chapter.

## T Lymphocyte Activation: the Expansion and Contraction Phases of the Adaptive Immune Response

In mature naïve T cells, the smaller 115 kDa isoform of CD43 carries disialylated core 1 O-tetrasaccharides Neu5Acα(2,3)Galβ(1,3)(Neu5Acα(2,6))GalNAc-Ser/Thr serving as a pan T cell marker. These may be detected by increased binding of the plant lectin SNA and they may form ligands for the B lymphocyte-specific siglec

CD22 [46] (Figure 11.5a). While naïve T cells do not express core 2 *O*-glycans, in activated T cells there is branching into larger core 2 hexasaccharide structures: Neu5Acα(2,3)Galβ(1,3)(Neu5Acα2,3Galβ(1,4)GlcNAcβ(1,6))GalNAc-Ser/Thr giving rise to the activation-associated 130 kDa isoform of CD43 [36]. This conversion is due to the increased expression of C2GnT and a decrease in the activity of ST6GalI (Figure 11.5b). These enzymes compete for the same Galβ(1,3)GalNAc-Ser/Thr acceptor substrate to form the alternative branch points in these structures. T cells thus acquire more complex hexasaccharides in the early phase following activation [47]. Furthermore, both CD43 isoforms bind the α2,3 sialic acid-specific siglec sialoadhesin and the extensively negatively charged sialylated *O*-glycans of CD43 may contribute to its role in negatively regulating T cell adhesion and activation. Changes in the complexity of *O*-glycans on CD43 also greatly influence its interactions with ligands on B cells.

ST6GalNAcIV expression is rapidly induced following murine CD8$^+$ T lymphocyte activation [48] which may protect these cells from galectin-1-induced apoptosis early during the immune response (Figure 11.5b) [33]. On the other hand, ST6GalI expression is reduced as detected by a decreased binding of SNA. Murine CD4$^+$ cells differ from their CD8$^+$ counterparts in that they show increased 9-*O* acetylation of α2,6-linked sialic acids. Furthermore, their polarization into Th1 and Th2 phenotypes is associated with the differentially regulated expression of ST6GalI, leading to SNA$^{low}$ and SNA$^{high}$ phenotypes respectively. T$_{reg}$ cells generally show a greater level of sialylation [28,30] where it may be a means of regulating tolerance and immune cell responsiveness. Moreover, these cells overexpress galectins-1 and -10 which preferentially bind and induce the death of SNA$^{low}$ Th1 cells, skewing the immune response to a Th2 cytokine profile, a key aspect of their immunosuppressive activity (Figure 11.5b) [30].

Later in activation, there is a global desialylation of core 1 *O*-glycans (Neu5Acα(2,3)Galβ(1,3)GalNAc-Ser/Thr) to generate asialo core 1 *O*-glycans (Galβ(1,3)GalNAc-Ser/Thr) which are recognized by the plant lectin PNA on CD8, CD43, and CD45 [49]. These changes correspond to a reduced expression of ST6GalI and an increased expression of the galactosyltransferase α1,3GalT (Figure 11.5c). In the case of CD45, there may be de novo synthesis of CD45 harboring asialo core 1 *O*-glycans rather than desialylation of the existing glycoprotein. CD45 desialylation may be isoform dependent. For example, in activated CD4$^+$ T cells, while the core 1 *O*-glycans of the CD45RB isoform are sialylated, reduced sialylation of other CD45 isoforms on activated T cells may promote antigen recognition, TCR signaling, and cytokine production [31].

β(1,6)*N*-Acetylglucosaminyltransferase V (GlcNAcTV; GnTV) is a key enzyme in the *N*-glycosylation of T cells. Unlike the differential expression of C2GnT observed during thymocyte maturation and T lymphocyte activation, the expression of GnTV and its resulting oligosaccharide structure remains constant in thymocytes and naïve peripheral T cells. However, following T cell activation, GnTV activity is upregulated, initiating GlcNAcβ(1,6) branching of *N*-glycans on the TCR to generate more complex polylactosamine structures which are recognized by galectins-1 and -3 [50]

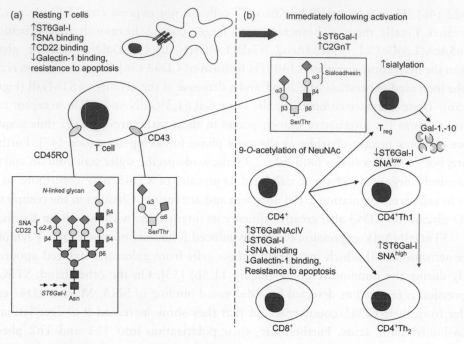

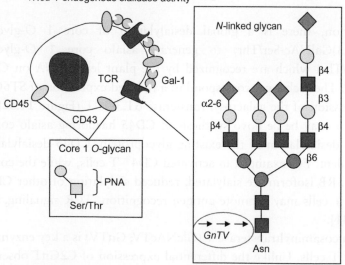

**FIGURE 11.5** Changes in the expression of glycosyltransferases and the glycan repertoire accompanies the activation of T lymphocytes following the presentation of antigens by antigen-presenting cells (APCs). Resting T lymphocytes (a), T lymphocytes and their subsets soon after activation (b) and T lymphocytes later after activation (c). C2GnT and the α2,6 sialyltransferases are largely responsible for these changes and differences exist between CD8⁺ and CD4⁺ T cells as well as between the various subsets of Th cells. Later, activation ST6Gal-I expression decreases while GnTV expression increases, creating galectin ligands on the T cell receptor (TCR).

(Figure 11.5c). Here, galectin binding creates a cell surface lattice which limits the lateral mobility of the TCR, restraining its cross-linking and clustering. Furthermore, the larger size of GnTV-modified N-glycans may also restrict the topography and spacing of TCR clusters in the plane of the membrane. This raises the threshold for T cell activation (strength of signal required to induce T cell responsiveness to antigen) by restricting TCR recruitment to sites of antigen presentation [33]. As discussed previously for their role in the induction of apoptosis in thymocytes, the formation of these galectin–glycoprotein lattices plays an important role in cell signaling and receptor turnover, determining the choice among cell survival, proliferation, and differentiation and regulating the decisions between tolerance and immune responsiveness.

Endogenous sialidase activity increases in immune cells during activation or differentiation. There are four genetically distinct forms of mammalian sialidase, each with a predominant cellular localization (plasma membrane-associated, lysosomal or cytosolic) and substrate specificity. *Neu-1* is a lysosomal sialidase occurring as a multi-enzyme complex that includes β-galactosidase and PPCA, a protein that protects and activates *Neu-1*. Following T cell activation, *Neu-1* is upregulated and subsequently desialylates core 1 O-glycans on glycoproteins and on gangliosides such as GM1 and GM3, not only in lysosomes but also in extralysosomal locations at the periphery [51] (Figure 11.5c).

Following a successful immune response, T lymphocytes undergo apoptosis, eliminating 90% of previously activated antigen-specific T cells and generating a new population of memory cells. In the absence of recent TCR stimulation, when immune signals are waning and when IL-2 levels are limiting, peripheral CD8$^+$ T cells undergo caspase-dependent apoptosis, a process which is regulated by the re-emergence of galectin-susceptible asialo-core 1 O-glycans (Galβ(1,3)GalNAc-Ser/Thr) which may be observed as an increase in binding to PNA. This change occurs as a result of a reduction in ST3GalI function via a posttranscriptional regulatory mechanism and due to increased *Neu-1* sialidase activity following activation [52,53]. By binding to these asialo-core 1 O-glycans, both galectins-1 and -9 contribute to the elimination of activated antigen-specific T cells in the resolution phase of the immune response.

The elongation of lactosamine (Galβ(1,4)GlcNAcβ(1,3)) sequences, the increased activity of C2GnT and the resulting increase in core 2 O-glycans on cell surface glycoproteins such as CD7, CD43, and CD45 together with reduced α2,6 sialylation of CD45, increases binding to galectin-1 which also leads to the induction of cell death [36]. Differential glycosylation of these cell surface glycoproteins can selectively regulate the survival of Th1, Th2, and Th17 effector cells. While Th1 and Th17 cells express a glycan profile required for galectin-1 signaling, cell death, and termination of the inflammatory response, differential α2,6 sialylation of N- and O-glycans on Th2 cell glycoproteins protects these cells from galectin-1 [30]. Thus, galectin-1 negatively regulates Th1 and Th17-mediated immune responses in an autocrine manner by leading to their preferential elimination.

The minor subset of T cells that are destined to become memory cells promote their survival by reducing the expression of core 2 O-glycans, thus avoiding interactions with apoptosis-inducing core 2 *O*-specific glycan-binding proteins like galectins [33]. Once they become memory T cells the activity of C2GnT increases and consequently there is increased expression of cell surface core 2 *O*-glycans. Furthermore, differentiation of activated effector CD8$^+$ T cells into viable memory cells occurs with a significant resialylation of core 1 *O*-glycans [52].

## T Lymphocyte Trafficking

CD4$^+$ and CD8$^+$ effector and memory T cells migrate to a greater extent to non-lymphoid tissues and inflammatory foci but exhibit distinct migratory traits. Homing requires the activation of appropriate integrins via the chemokine-mediated stimulation of chemokine receptors as well as interactions between selectins and their ligands [54,55]. Mature recirculating naïve T lymphocytes home to the secondary lymphoid organs like the peripheral lymph nodes and enter via specialized endothelial cells lining the high endothelial venules (HEVs). The L-selectin ligand on these cells is sialyl 6-sulfo Lewis x (Neu5Ac$\alpha$2,3Gal$\beta$1,4[Fuc$\alpha$1,3]GlcNAc$\beta$[6-SO$_3^-$]) and this glycan plays multiple roles in the homing of naïve T cells to the lymph nodes as well as helper memory T cells to the skin and to the gut. Naïve T cells are CD45RA$^+$ and express CCR7, the receptor for the secondary lymphoid chemokine (SLC/CCL21). Activated effector T cells do not express CCR7 and are L-selectin$^{-/low}$ and as such do not recirculate efficiently through lymph nodes, while memory T cells express these lymph node homing receptors and are thus capable of recirculation through these tissues. A reduction in sialyl 6-sulfo Le$^x$ occurs in part due to a posttranslational modification of the sialic acid moiety via conversion to de-*N*-acetyl sialic acid and further to lactamized cyclic sialic acid via sialic acid cyclase. Cyclic sialyl 6-sulfo Le$^x$ no longer binds any of the selectins. This serves as a rapid negative feedback regulation for selectin-mediated cell adhesion. Stimulation of chemokine receptors by chemokines following selectin-mediated binding of lymphocytes to the endothelium also activates sialic acid cyclase [14].

Increased expression of C2GnT, sialyltransferases, and fucosyltransferases such as FucTVII leads to the generation of the sialyl Le$^x$ oligosaccharide (Neu5Ac$\alpha$(2,3)Gal$\beta$(1,4)[Fuc$\alpha$(1,3)]GlcNAc) (sLe$^x$) which serves as a high-affinity ligand for selectins P and E, upregulated on endothelial cells exposed to inflammatory mediators in various tissues. Polarized CD4$^+$ Th1 and Th2 cells show different selectin-mediated trafficking patterns. While Th1 cells display the sLe$^x$ epitope, Th2 cells do not and this results in the differential recruitment of Th1 cells to sites of inflammation. Furthermore, in CD4$^+$ T cells FucTVII is upregulated by IL-12 and repressed by IL-4 while in CD8$^+$ cells, IL-2 enhances expression of a P-selectin ligand via FucTVII and C2GnT while this is inhibited by IL-4 [36].

## B Lymphocytes

B lymphocytes exist as different subsets, the majority belonging to the recirculating pool of follicular "B2" B cells, but in addition there are the T cell independent

"B1" and marginal zone B cell subsets. The best-characterized B cell lectin is CD22/Siglec-2 which functions as an inhibitory receptor regulating multiple B lymphocyte functions such as survival and the threshold of cellular activation. The preferred ligands for CD22 are Neu5Acα(2,6)Galβ(1,4)GlcNAc (human) or Neu5Gcα(2,6)Galβ-(1,4)GlcNAc (mouse) [56]. These are generated by ST6GalI, the expression of which in B cells is crucial for mediating CD22 functions since B cells lacking this enzyme are anergic. Additional expression of 6-O-sulfotransferases generates α2,6-sialylated 6-sulfated LacNAc glycans which also confer significant CD22 binding [57]. The binding affinity of this siglec appears to depend on the number and spatial arrangement of α2,6-sialylated moieties [58]. In the cell membrane, CD22 interacts homotypically with other CD22 molecules in a glycan-dependent manner, whereas its interaction with the B cell receptor appears to be non-carbohydrate dependent and increased in the absence of ST6GalI [59,60]. CD22 harbors immunoreceptor tyrosine-based inhibitory motifs (ITIMs) in its cytosolic domain. Ligation of the B cell receptor increases ITIM phosphorylation on CD22 by the src family kinase lyn leading to recruitment of the tyrosine phosphatase SHP-1 (Src homology 2 domain-containing protein tyrosine phosphatase-1) and the downregulation of BCR signaling [18]. In a non-lymphoid microenvironment, the level of α2,6 sialylated carbohydrates is low, favoring *cis*-associations of CD22 and negative regulation. When a B cell enters an activated lymphoid microenvironment that is rich is α2,6 sialic acid, CD22 may be drawn away from IgM through *trans*-interactions resulting in augmented BCR-mediated signaling (Figure 11.2). Following activation, B cells maintain their sialylated *N*-linked glycans due to increased expression of ST6GalI and decreased expression of α1,3GalT.

## Conclusions

The rapidly evolving field of glycomics and its associated technologies, combined with analyses of genetically mutated mice, has enabled a greater understanding of the critical role that glycosylation plays in the regulation of the innate and adaptive immune responses and has revealed the enormous complexity of the immune cell glycan repertoire. The identification and characterization of the ligands of endogenous immune-associated lectins and the significance of these interactions is considered an important goal, particularly given the potential for therapeutic interventions in the treatment of infectious, neoplastic, and inflammatory diseases.

## References

1. Sperandio M. Selectins and glycosyltransferases in leukocyte rolling *in vivo. FEBS J* 2006;**273**(19): 4377–89.

2. Soler-Rodriguez AM, Zhang H, Lichenstein HS, Qureshi N, Niesel DW, Crowe SE, Peterson JW, Klimpel GR. Neutrophil activation by bacterial lipoprotein versus lipopolysaccharide: differential requirements for serum and CD14. *J Immunol* 2000;**164**(5):2674–83.

3. van Gisbergen KP, Sanchez-Hernandez M, Geijtenbeek TB, van Kooyk Y. Neutrophils mediate immune modulation of dendritic cells through glycosylation-dependent interactions between Mac-1 and DC-SIGN. *J Exp Med* 2005;**201**(8):1281–92.

4. Bogoevska V, Horst A, Klampe B, Lucka L, Wagener C, Nollau P. CEACAM1, an adhesion molecule of human granulocytes, is fucosylated by fucosyltransferase IX and interacts with DC-SIGN of dendritic cells via Lewis x residues. *Glycobiology* 2006;**16**(3):197–209.

5. Marino JH, Hoffman M, Meyer M, Miller KS. Sialyltransferase mRNA abundances in B cells are strictly controlled, correlated with cognate lectin binding, and differentially responsive to immune signaling in vitro. *Glycobiology* 2004;**14**(12):1265–74.

6. Muthing J, Spanbroek R, Peter-Katalinic J, Hanisch FG, Hanski C, Hasegawa A, Unland F, Lehmann J, Tschesche H, Egge H. Isolation and structural characterization of fucosylated gangliosides with linear poly-N-acetyllactosaminyl chains from human granulocytes. *Glycobiology* 1996;**6**(2):147–56.

7. Nishi N, Shoji H, Seki M, Itoh A, Miyanaka H, Yuube K, Hirashima M, Nakamura T. Galectin-8 modulates neutrophil function via interaction with integrin alphaM. *Glycobiology* 2003;**13**(11):755–63.

8. Lucka L, Fernando M, Grunow D, Kannicht C, Horst AK, Nollau P, Wagener C. Identification of Lewis x structures of the cell adhesion molecule CEACAM1 from human granulocytes. *Glycobiology* 2005;**15**(1):87–100.

9. Ludwig IS, Geijtenbeek TB, van Kooyk Y. Two way communication between neutrophils and dendritic cells. *Curr Opin Pharmacol* 2006;**6**(4):408–13.

10. Feuk-Lagerstedt E, Jordan ET, Leffler H, Dahlgren C, Karlsson A. Identification of CD66a and CD66b as the major galectin-3 receptor candidates in human neutrophils. *J Immunol* 1999;**163**(10):5592–8.

11. Symon FA, Lawrence MB, Williamson ML, Walsh GM, Watson SR, Wardlaw AJ. Functional and structural characterization of the eosinophil P-selectin ligand. *J Immunol* 1996;**157**(4):1711–19.

12. McMillan SJ, Crocker PR. CD33-related sialic-acid-binding immunoglobulin-like lectins in health and disease. *Carbohydr Res* 2008;**343**(12):2050–6.

13. McCoy Jr. JP, Chambers WH. Carbohydrates in the functions of natural killer cells. *Glycobiology* 1991;**1**(4):321–8.

14. Kannagi R. Regulatory roles of carbohydrate ligands for selectins in the homing of lymphocytes. *Curr Opin Struct Biol* 2002;**12**(5):599–608.

15. Avril T, North SJ, Haslam SM, Willison HJ, Crocker PR. Probing the cis interactions of the inhibitory receptor Siglec-7 with alpha2,8-disialylated ligands on natural killer cells and other leukocytes using glycan-specific antibodies and by analysis of alpha2,8-sialyltransferase gene expression. *J Leukoc Biol* 2006;**80**(4):787–96.

16. Rabinovich GA, Toscano MA, Jackson SS, Vasta GR. Functions of cell surface galectin-glycoprotein lattices. *Curr Opin Struct Biol* 2007;**17**(5):513–20.

17. van den Berg TK, Nath D, Ziltener HJ, Vestweber D, Fukuda M, van Die I, Crocker PR. Cutting edge: CD43 functions as a T cell counterreceptor for the macrophage adhesion receptor sialoadhesin (Siglec-1). *J Immunol* 2001;**166**(6):3637–40.

18. Crocker PR, Paulson JC, Varki A. Siglecs and their roles in the immune system. *Nat Rev Immunol* 2007;**7**(4):255–66.

19. van Vliet SJ, van Liempt E, Geijtenbeek TB, van Kooyk Y. Differential regulation of C-type lectin expression on tolerogenic dendritic cell subsets. *Immunobiology* 2006;**211**(6–8):577–85.

20. Crocker PR, Kelm S, Dubois C, Martin B, McWilliam AS, Shotton DM, Paulson JC, Gordon S. Purification and properties of sialoadhesin, a sialic acid-binding receptor of murine tissue macrophages. *Embo J* 1991;**10**(7):1661–9.

21. Bax M, Garcia-Vallejo JJ, Jang-Lee J, North SJ, Gilmartin TJ, Hernandez G, Crocker PR, Leffler H, Head SR, Haslam SM, Dell A, van Kooyk Y. Dendritic cell maturation results in pronounced changes in glycan expression affecting recognition by siglecs and galectins. *J Immunol* 2007;**179**(12):8216–24.

22. Julien S, Grimshaw MJ, Sutton-Smith M, Coleman J, Morris HR, Dell A, Taylor-Papadimitriou J, Burchell JM. Sialyl-Lewis(x) on P-selectin glycoprotein ligand-1 is regulated during differentiation and maturation of dendritic cells: a mechanism involving the glycosyltransferases C2GnT1 and ST3Gal I. *J Immunol* 2007;**179**(9):5701–10.

23. Dumic J, Dabelic S, Flogel M. Galectin-3: an open-ended story. *Biochim Biophys Acta* 2006;**1760**(4): 616–35.

24. Levy Y, Auslender S, Eisenstein M, Vidavski RR, Ronen D, Bershadsky AD, Zick Y. It depends on the hinge: a structure-functional analysis of galectin-8, a tandem-repeat type lectin. *Glycobiology* 2006;**16**(6):463–76.

25. Zick Y, Eisenstein M, Goren RA, Hadari YR, Levy Y, Ronen D. Role of galectin-8 as a modulator of cell adhesion and cell growth. *Glycoconj J* 2004;**19**(7–9):517–26.

26. Fulcher JA, Hashimi ST, Levroney EL, Pang M, Gurney KB, Baum LG, Lee B. Galectin-1-matured human monocyte-derived dendritic cells have enhanced migration through extracellular matrix. *J Immunol* 2006;**177**(1):216–26.

27. Levroney EL, Aguilar HC, Fulcher JA, Kohatsu L, Pace KE, Pang M, Gurney KB, Baum LG, Lee B. Novel innate immune functions for galectin-1: galectin-1 inhibits cell fusion by Nipah virus envelope glycoproteins and augments dendritic cell secretion of proinflammatory cytokines. *J Immunol* 2005;**175**(1):413–20.

28. Jenner J, Kerst G, Handgretinger R, Muller I. Increased alpha2,6-sialylation of surface proteins on tolerogenic, immature dendritic cells and regulatory T cells. *Exp Hematol* 2006;**34**(9):1212–18.

29. Chung JS, Sato K, Dougherty II, Cruz Jr. PD, Ariizumi K. DC-HIL is a negative regulator of T lymphocyte activation. *Blood* 2007;**109**(10):4320–7.

30. Toscano MA, Bianco GA, Ilarregui JM, Croci DO, Correale J, Hernandez JD, Zwirner NW, Poirier F, Riley EM, Baum LG, Rabinovich GA. Differential glycosylation of TH1, TH2 and TH-17 effector cells selectively regulates susceptibility to cell death. *Nat Immunol* 2007;**8**(8):825–34.

31. Hernandez JD, Klein J, Van Dyken SJ, Marth JD, Baum LG. T-cell activation results in microheterogeneous changes in glycosylation of CD45. *Int Immunol* 2007;**19**(7):847–56.

32. Gillespie W, Paulson JC, Kelm S, Pang M, Baum LG. Regulation of alpha 2,3 sialyltransferase expression correlates with conversion of peanut agglutinin (PNA)+ to PNA- phenotype in developing thymocytes. *J Biol Chem* 1993;**268**(6):3801–4.

33. Hernandez JD, Baum LG. Ah, sweet mystery of death! Galectins and control of cell fate. *Glycobiology* 2002;**12**(10):127R–136R.

34. Visan I, Yuan JS, Tan JB, Cretegny K, Guidos CJ. Regulation of intrathymic T-cell development by Lunatic Fringe- Notch1 interactions. *Immunol Rev* 2006;**209**:76–94.

35. Nguyen JT, Evans DP, Galvan M, Pace KE, Leitenberg D, Bui TN, Baum LG. CD45 modulates galectin-1-induced T cell death: regulation by expression of core 2 O-glycans. *J Immunol* 2001;**167**(10):5697–707.

36. Daniels MA, Hogquist KA, Jameson SC. Sweet 'n' sour: the impact of differential glycosylation on T cell responses. *Nat Immunol* 2002;**3**(10):903–10.

37. Brewer CF, Miceli MC, Baum LG. Clusters, bundles, arrays and lattices: novel mechanisms for lectin-saccharide-mediated cellular interactions. *Curr Opin Struct Biol* 2002;**12**(5):616–23.

38. Amano M, Galvan M, He J, Baum LG. The ST6Gal I sialyltransferase selectively modifies N-glycans on CD45 to negatively regulate galectin-1-induced CD45 clustering, phosphatase modulation, and T cell death. *J Biol Chem* 2003;**278**(9):7469–75.

39. Tribulatti MV, Mucci J, Cattaneo V, Aguero F, Gilmartin T, Head SR, Campetella O. Galectin-8 induces apoptosis in the CD4(high)CD8(high) thymocyte subpopulation. *Glycobiology* 2007;**17**(12):1404–12.

40. Balcan E, Tuglu I, Sahin M, Toparlak P. Cell surface glycosylation diversity of embryonic thymic tissues. *Acta Histochem* 2008;**110**(1):14–25.

41. Baum LG, Derbin K, Perillo NL, Wu T, Pang M, Uittenbogaart C. Characterization of terminal sialic acid linkages on human thymocytes. Correlation between lectin-binding phenotype and sialyltransferase expression. *J Biol Chem* 1996;**271**(18):10793–9.

42. Starr TK, Daniels MA, Lucido MM, Jameson SC, Hogquist KA. Thymocyte sensitivity and supramolecular activation cluster formation are developmentally regulated: a partial role for sialylation. *J Immunol* 2003;**171**(9):4512–20.

43. Kao C, Sandau MM, Daniels MA, Jameson SC. The sialyltransferase ST3Gal-I is not required for regulation of CD8-class I MHC binding during T cell development. *J Immunol* 2006;**176**(12):7421–30.

44. Mora JR, von Andrian UH. T-cell homing specificity and plasticity: new concepts and future challenges. *Trends Immunol* 2006;**27**(5):235–43.

45. Marelli-Berg FM, Cannella L, Dazzi F, Mirenda V. The highway code of T cell trafficking. *J Pathol* 2008;**214**(2):179–89.

46. Stamenkovic I, Sgroi D, Aruffo A, Sy MS, Anderson T. The B lymphocyte adhesion molecule CD22 interacts with leukocyte common antigen CD45RO on T cells and alpha 2–6 sialyltransferase, CD75, on B cells. *Cell* 1991;**66**(6):1133–44.

47. Piller F, Piller V, Fox RI, Fukuda M. Human T-lymphocyte activation is associated with changes in O-glycan biosynthesis. *J Biol Chem* 1988;**263**(29):15146–50.

48. Kaufmann M, Blaser C, Takashima S, Schwartz-Albiez R, Tsuji S, Pircher H. Identification of an alpha2, 6-sialyltransferase induced early after lymphocyte activation. *Int Immunol* 1999;**11**(5):731–8.

49. Comelli EM, Sutton-Smith M, Yan Q, Amado M, Panico M, Gilmartin T, Whisenant T, Lanigan CM, Head SR, Goldberg D, Morris HR, Dell A, Paulson JC. Activation of murine CD4$^+$ and CD8$^+$ T lymphocytes leads to dramatic remodeling of N-linked glycans. *J Immunol* 2006;**177**(4):2431–40.

50. Demetriou M, Granovsky M, Quaggin S, Dennis JW. Negative regulation of T-cell activation and autoimmunity by Mgat5 N-glycosylation. *Nature* 2001;**409**(6821):733–9.

51. Nan X, Carubelli I, Stamatos NM. Sialidase expression in activated human T lymphocytes influences production of IFN-gamma. *J Leukoc Biol* 2007;**81**(1):284–96.

52. Priatel JJ, Chui D, Hiraoka N, Simmons CJ, Richardson KB, Page DM, Fukuda M, Varki NM, Marth JD. The ST3Gal-I sialyltransferase controls CD8$^+$ T lymphocyte homeostasis by modulating O-glycan biosynthesis. *Immunity* 2000;**12**(3):273–83.

53. Van Dyken SJ, Green RS, Marth JD. Structural and mechanistic features of protein O glycosylation linked to CD8$^+$ T cell apoptosis. *Mol Cell Biol* 2007;**27**(3):1096–111.

54. Matsumoto M, Atarashi K, Umemoto E, Furukawa Y, Shigeta A, Miyasaka M, Hirata T. CD43 functions as a ligand for E-Selectin on activated T cells. *J Immunol* 2005;**175**(12):8042–50.

55. Knibbs RN, Craig RA, Natsuka S, Chang A, Cameron M, Lowe JB, Stoolman LM. The fucosyltransferase FucT-VII regulates E-selectin ligand synthesis in human T cells. *J Cell Biol* 1996;**133**(4):911–20.

56. Nitschke L. The role of CD22 and other inhibitory co-receptors in B-cell activation. *Curr Opin Immunol* 2005;**17**(3):290–7.

57. Kimura N, Ohmori K, Miyazaki K, Izawa M, Matsuzaki Y, Yasuda Y, Takematsu H, Kozutsumi Y, Moriyama A, Kannagi R. Human B-lymphocytes express alpha2–6-sialylated 6-sulfo-N-acetyllactosamine serving as a preferred ligand for CD22/Siglec-2. *J Biol Chem* 2007;**282**(44):32200–7.

58. Kniep B, Schakel K, Nimtz M, Schwartz-Albiez R, Schmitz M, Northoff H, Vilella R, Gramatzki M, Rieber EP. Differential expression of alpha2–6 sialylated polylactosamine structures by human B and T cells. *Glycobiology* 1999;**9**(4):399–406.

59. Collins BE, Blixt O, Bovin NV, Danzer CP, Chui D, Marth JD, Nitschke L, Paulson JC. Constitutively unmasked CD22 on B cells of ST6Gal I knockout mice: novel sialoside probe for murine CD22. *Glycobiology* 2002;**12**(9):563–71.

60. Hennet T, Chui D, Paulson JC, Marth JD. Immune regulation by the ST6Gal sialyltransferase. *Proc Natl Acad Sci USA* 1998;**95**(8):4504–9.

# 12

# Mouse and Human Glycomes

Simon J. North, Sara Chalabi, Mark Sutton-Smith, Anne Dell, and Stuart M. Haslam

*DIVISION OF MOLECULAR BIOSCIENCES, FACULTY OF NATURAL SCIENCES, IMPERIAL COLLEGE LONDON, UK*

## Introduction

Glycomics, as defined in previous chapters, is the scientific effort to explore, characterize and study all of the carbohydrate structures produced by a defined system [1–3]. This can be performed at the level of the whole organism (such as *Caenorhabditis elegans* [4] or *Saccharomyces cerevisiae* [5], or an individual cell or tissue type.

In this chapter, we discuss the efforts made towards the whole-system analysis of the N- and O-glycomes of mouse and human tissues. Knowing the glycome of these tissues is a vital step towards investigating the roles that glycans play in cell communication.

### The Consortium for Functional Glycomics

Increasingly, efforts towards such large undertakings as defining mammalian glycomes are being coordinated through major international consortia. The most mature and influential of these is the Consortium for Functional Glycomics (CFG) (http://www.functionalglycomics.org), established in 2001 with the overarching goal to elucidate the roles of carbohydrate–protein interactions in cell communication at the cell surface. The CFG is organized into seven scientific "cores," which provide resources, tools, reagents, and databases to support biologically relevant research being carried out by participating investigators from the international glycobiology community.

The primary objective of the Analytical Glycotechnology Core (Core-C) is to acquire and disseminate glycan structural data which will empower the scientific community in its efforts to determine the structures of glycan-binding protein (GBP) ligands. To this end Core-C has implemented and optimized methodologies originally developed at Imperial College London for the glycomic characterization of tissues

**Handbook of Glycomics**
Copyright © 2009 by Elsevier Inc. All rights of reproduction in any form reserved.

from glycosyltransferase knockout mice [6,7]. To date, Core-C has performed glycan profiling of 108 tissues from wild-type and selected knockout mouse strains and 63 human tissues from five anonymous tissue donors and has deposited the acquired glycomic data in open access CFG databases (http://www.functionalglycomics.org/glycomics/publicdata/glycoprofiling.jsp).

A growing number of purified cell populations originating mainly from immortalized cell lines and human/mouse immune cell populations have also been subjected to glycomic analyses [8,9] (see also publicly available data on the CFG website), though this goes beyond the scope of this chapter.

## Glycomic Strategy

Mass spectrometry (MS) has remained at the forefront of glycobiological analysis for more than two decades, with strategies having been developed and refined for the glycomic analysis of total glycan populations [10–14]. Figure 12.1 illustrates the outline of a typical glycomics experiment such as the ones performed to produce the data included within this chapter. It comprises sample preparation, mass fingerprinting, and a number of supplementary experimental or in silico activities that serve to inform further on the structural data, so feeding back into the original analysis. Similar approaches are utilized by other laboratories, though the specifics (i.e., the choice of derivatization and/or MS technology) may vary.

## Interpretation of Glycomic Data

The matrix-assisted laser desorption/ionization (MALDI) analyses of the permethylated N- and O-linked glycan pools from specific defined systems provide highly sensitive mass profiles. The assignments made within these profiles should be viewed primarily as unequivocal monosaccharide unit compositions for each peak. Knowledge of the specific biosynthetic pathways utilized by the system being analyzed is then used to provide the most informed suggestion for the structural identity of the glycoprotein-derived N-linked and O-linked glycans. For example, since N- and O-glycans are constructed from well-defined pathways, the observation of a specific structural motif leads to the expectation of other related intermediate structures along the same pathway. This can also be indicative of the relative activities of the various glycosyltransferases present in the sample. The symbolic nomenclature used to annotate the resultant spectra is that used by the CFG (Figure 12.2).

Even with very well-informed biosynthetic knowledge, the possibility of alternative sequences for these putative assignments cannot initially be ruled out. Wherever possible, therefore, electrospray (ES)-MS/MS or MALDI-time-of-flight (TOF)-TOF-MS/MS, in conjunction with additional data derived from experiments such as linkage analysis and enzyme digests, are applied to differentiate alternative possibilities. In most cases, Core-C obtains structural cues from the fragmentation analysis of signals within the spectrum. This can lead, for example, to the establishment of the propensity for a cell or tissue type to produce antennal Galβ1–4GlcNAc (LacNAc) extensions

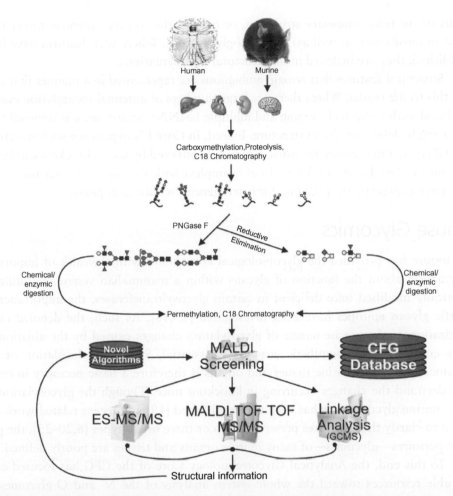

**FIGURE 12.1** A simplified schematic representation of the glycomic strategy employed by Core-C for the Consortium for Functional Glycomics (CFG). Glycoproteins are extracted from human or murine tissues and purified before the release of the specific subpopulations of glycans. These glycan populations can then either be directly analyzed or subjected to chemical or enzymatic digestions in order to elucidate further structural information. The glycan pools are derivatized via permethylation in order to optimize the sensitivity of detection of the constituents and direct subsequent MS/MS fragmentations. A range of analytical techniques can be employed, but typically an initial screening by matrix-assisted laser desorption ionization (MALDI)-MS is performed as a first step. These results are used to select informed targets for further analysis by MS/MS techniques and supplementary experiments such as linkage analysis. Structural information gathered is then used to refine the original screening information. All data are deposited within the various databases curated by the CFG [15,16]. (See color plate 18.)

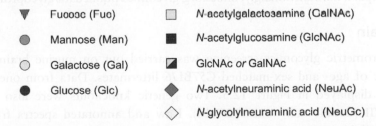

**FIGURE 12.2** Symbol nomenclature as outlined by the Consortium for Functional Glycomics Nomenclature Committee (May 2004). Full documentation available from: http://glycomics.scripps.edu/CFGnomenclature.pdf. (See color plate 19.)

versus tri- or tetra-antennary structures, or fucosylation on the antennae rather than (or as in most cases, as well as) on the *N*-glycan core. Where such features have been established, they are included in the structural representations.

Structural features that remain ambiguous are represented in a manner that conveys this to the reader. Where there are large numbers of antennal fucosylation events, combined with sialic acid capping and multiple LacNAc extensions, it is impossible to give a single definitive glycan structure. Indeed, in Core-C's experience such structures inevitably exist in multiple isoforms, which are conveyed by use of brackets within the assignments (see Figure 12.3 for visual examples) with the intention that the reader can better appreciate the potential for heterogeneity within such peaks.

## Mouse Glycomics

The mouse has provided the glycobiological community with a wealth of important information about the function of glycans within a mammalian system. By utilizing genetically modified mice deficient in certain glycosyltransferases, the importance of specific glycan epitopes have been investigated [17,18]. As such, the detailed characterization of the precise nature of glycosylation changes caused by the ablation of genes encoding for biosynthetic enzymes is essential. Structural elucidation of the glycomes of normal murine tissues and organs is therefore a basic necessity in order to understand the changes occurring in knockout mice. Though the glycosylation of some murine glycoproteins has been well established [19] and recent related work has served to clarify the structures present on two or three organ types [6,20–23], the glycan repertoires—glycomes—of many murine organs and tissues are poorly defined.

To this end, the Analytical Glycotechnology Core of the CFG has focused considerable resources toward the whole-system analysis of the *N*- and *O*-glycomes of murine tissues with the intention of laying a foundation of structural knowledge as a resource for the scientific community. In this section, we discuss the findings resulting from the analysis of ten murine organ and tissue types (Figure 12.4) from pairs of age- and sex-matched C57BL/6 wild-type littermates. These data were also supplemented by the glycomic screening of various knockout mice. Though these experiments are referred to in the text, a detailed discussion of the findings is beyond the scope of the chapter. However, all of the raw data files and annotated spectra are freely available online at http://www.functionalglycomics.org/glycomics/publicdata/glycoprofiling.jsp.

### Mouse Brain

Mass spectrometric glycomics screening was carried out on murine brain samples from a pair of age- and sex-matched C57BL/6 littermates. Data from one of these analyses is displayed in Figure 12.5. Two genetic knockouts were also analyzed (FucTIV+VII (double null) and ST3Gal-I). Raw and annotated spectra from these data are available online, together with the wild-type analyses presented here, from http://www.functionalglycomics.org/glycomics/publicdata/glycoprofiling.jsp. A discussion of these results in relation to the gene-expression profiles is also available [24].

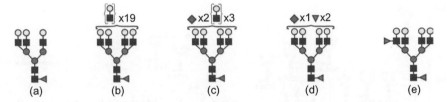

(a)          (b)          (c)          (d)          (e)

**FIGURE 12.3**   Representation of structural ambiguity. (a) Typical representation of a triantennary structure. For convenience, only a single branching pattern is shown. Also, this composition could correspond to a biantennary structure with an extended antenna. Further experiments would be required to distinguish these structural features. (b) Example of the use of brackets to convey both structural identity and variation. The cartoon indicates the presence of multiple (19) polylactosamine extensions but Core-C is not indicating the exact length of an extended antenna nor on which antenna or antennae they are found. (c) Example of a structure with both sialic acid capping and LacNAc extensions. The sialic acids are able to cap unextended antennae as well as the longer polyLacNAc type, so in the absence of more definitive structural evidence these moieties are represented as shown. (d) Example of a structure with both antennal fucosylation and sialylation. It is biosynthetically possible that the fucosylation and sialylation are on the same or different antennae. In the absence of any further structural evidence for a specific sample these residues are represented as shown. (e) Example of a structure with antennal fucosylation. Linkages are not indicated on the cartoons nor should they be directly inferred from them. For example it is not inferred that the fucosylated antenna is exclusively on the 3-linked or 6-linked mannose of the core.

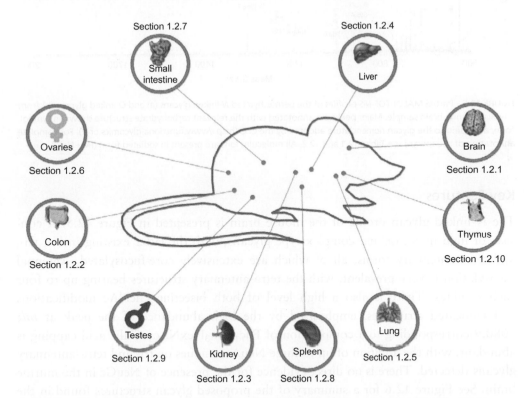

**FIGURE 12.4**   The mouse, illustrating the organs and tissues whose glycomes are covered within this chapter.

Earlier work was successful in analyzing the rodent brain [20,21,25], though it was carried out using fast atom bombardment (FAB)-MS [26], as opposed to the more modern MALDI instrumentation. As a consequence of the reduced sensitivity the high molecular weight components found here were not observed.

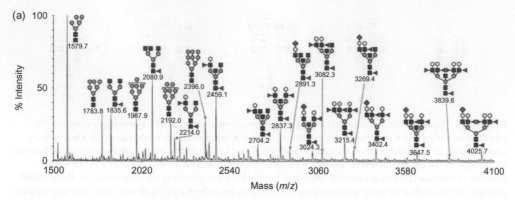

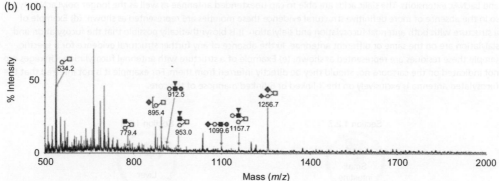

**FIGURE 12.5**   Partial MALDI-TOF-MS profiles of the permethylated *N*-linked glycans (a) and *O*-linked glycans (b) from a typical murine brain sample. Major peaks are annotated with the relevant carbohydrate structure shown in symbol form, according to the glycan nomenclature adopted by the CFG (http://www.functionalglycomics.org/). For complete annotation of the spectra see Tables 12.1 and 12.2. All molecular ions are present in sodiated form ([M+Na]$^+$).

## Key Features

The *N*-linked glycan profile of the mouse brain is presented in Figure 12.5. It consists of high mannose and complex type glycans, with the latter existing in bi-, tri-, and tetra-antennary forms, all of which are extensively core-fucosylated. Antennal fucosylation is very prevalent, with the tetra-antennary structures bearing up to four such residues. There is also a high level of both bisecting GlcNAc modifications and truncated structures, emphasized by the high abundance of the peak at *m/z* 2080.9 corresponding to a composition of $Fuc_1Hex_3HexNAc_5$. Sialic acid capping is abundant, with the addition of up to three NeuAc residues on tri- and tetra-antennary glycans detected. There is no direct evidence for the presence of NeuGc in the murine brain. See Figure 12.6 for a summary of the proposed glycan structures found in the *N*-glycan profile of murine brain.

The spectrum of mouse brain-derived *O*-linked glycans (Figure 12.5b) consists of both core 1 and core 2 type *O*-glycans, capped with sialic acid, as well as more unusually two signals consistent with *O*-mannosyl structures. These structures were first observed in a rat brain proteoglycan in 1979 [27] and later shown to be present on cranin (dystroglycan), a more well-defined mucin-like glycoprotein, purified from

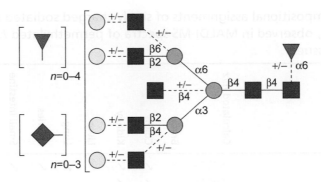

**FIGURE 12.6**   Proposed complex *N*-glycan structures from mouse brain.

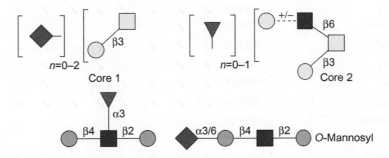

**FIGURE 12.7**   Proposed *O*-glycan structures from mouse brain.

brain tissue [28,29]. These studies showed that the fucosylated structure consists of a Lewis x (Le$^x$) epitope (Galβ1–4(Fucα1–3)GlcNAc-R) 3-linked to the mannose residue [30]. The high abundance of these particular structures relative to other related mucin-type glycans within the brain is intriguing. Figure 12.7 shows a summary of the proposed core 1 and core 2 glycans present in the *O*-glycan profile of murine brain.

For a complete summary of the glycans observed in both the *N*-linked and *O*-linked spectra, see Tables 12.1 and 12.2.

## Mouse Colon

Mass spectrometric glycomics screening was carried out on murine colon samples from a pair of age- and sex-matched C57/BL/6 littermates. Data from one of these analyses are shown in Figure 12.8. Two genetic knockouts were also analyzed (FucTIV+VII (double null) and ST3Gal-I). Raw and annotated spectra from these data are available online, together with the wild-type analyses presented here, from http://www.functionalglycomics.org/glycomics/publicdata/glycoprofiling.jsp. Earlier work was successful in analyzing the *O*-glycans of the murine colon [20,21,25], though it was carried out using FAB-MS [26], as opposed to the more modern MALDI instrumentation. Hence the less abundant and/or high molecular weight components found here were not observed.

**Table 12.1** Compositional assignments of singly charged sodiated molecular ions, [M+Na]$^+$, observed in MALDI-MS spectra of permethylated *N*-glycans from mouse tissues

| | Calculated *m/z* | Brain | Colon | Kidney | Liver | Lung | Ovaries | Small Intestine | Spleen | Testes | Thymus |
|---|---|---|---|---|---|---|---|---|---|---|---|
| Hex$_5$ HexNAc$_2$ | 1579.78 | ✓ | ✓ | ✓ | ✓ | ✓ | ✓ | ✓ | ✓ | ✓ | ✓ |
| Fuc$_1$ Hex$_3$ HexNAc$_3$ | 1590.80 | | ✓ | ✓ | | | | | ✓ | ✓ | |
| Hex$_4$ HexNAc$_3$ | 1620.81 | | | | | | | | | | |
| Hex$_3$ HexNAc$_4$ | 1661.84 | ✓ | ✓ | ✓ | ✓ | | | | | ✓ | |
| Hex$_6$ HexNAc$_2$ | 1783.88 | ✓ | ✓ | ✓ | ✓ | ✓ | ✓ | ✓ | ✓ | ✓ | ✓ |
| Fuc$_1$ Hex$_4$ HexNAc$_3$ | 1794.90 | ✓ | ✓ | | | | | | | | |
| Fuc$_1$ Hex$_3$ HexNAc$_4$ | 1835.93 | ✓ | ✓ | ✓ | ✓ | ✓ | ✓ | ✓ | ✓ | ✓ | |
| Hex$_4$ HexNAc$_4$ | 1865.94 | ✓ | ✓ | | | ✓ | ✓ | | | | |
| Hex$_3$ HexNAc$_5$ | 1906.96 | ✓ | ✓ | ✓ | ✓ | | | ✓ | | ✓ | |
| Hex$_7$ HexNAc$_2$ | 1987.98 | ✓ | ✓ | ✓ | ✓ | ✓ | | ✓ | ✓ | | |
| Fuc$_1$ Hex$_5$ HexNAc$_3$ | 1999.00 | ✓ | | | | | | | | | |
| Hex$_6$ HexNAc$_3$ | 2029.01 | | | | | ✓ | ✓ | | | | |
| Fuc$_1$ Hex$_4$ HexNAc$_4$ | 2040.03 | ✓ | ✓ | ✓ | ✓ | ✓ | ✓ | ✓ | ✓ | ✓ | |
| Hex$_5$ HexNAc$_4$ | 2070.04 | ✓ | ✓ | | | ✓ | ✓ | | ✓ | | ✓ |
| Fuc$_1$ Hex$_3$ HexNAc$_5$ | 2081.05 | ✓ | ✓ | ✓ | ✓ | ✓ | | | ✓ | ✓ | |
| Hex$_4$ HexNAc$_5$ | 2111.06 | | ✓ | | ✓ | | | ✓ | | | |
| Hex$_8$ HexNAc$_2$ | 2192.08 | ✓ | ✓ | | ✓ | | ✓ | ✓ | ✓ | ✓ | ✓ |
| Fuc$_1$ Hex$_6$ HexNAc$_3$ | 2203.10 | | | | ✓ | | | | | | |
| Fuc$_2$ Hex$_4$ HexNAc$_4$ | 2214.11 | ✓ | ✓ | ✓ | | | | | | | |
| NeuGc$_1$ Hex$_5$ HexNAc$_3$ | 2216.09 | | | | ✓ | ✓ | | | | | |
| Fuc$_1$ Hex$_5$ HexNAc$_4$ | 2244.13 | ✓ | ✓ | ✓ | ✓ | ✓ | ✓ | ✓ | ✓ | ✓ | ✓ |
| NeuGc$_1$ Hex$_4$ HexNAc$_4$ | 2257.12 | | | ✓ | | | | | | | |
| Hex$_6$ HexNAc$_4$ | 2274.14 | | ✓ | | | | | ✓ | | | |
| Fuc$_1$ Hex$_4$ HexNAc$_5$ | 2285.15 | ✓ | ✓ | ✓ | ✓ | | ✓ | | ✓ | ✓ | |
| Hex$_5$ HexNAc$_5$ | 2315.16 | | ✓ | | | | | ✓ | | | |
| Hex$_9$ HexNAc$_2$ | 2396.18 | ✓ | ✓ | ✓ | ✓ | ✓ | ✓ | ✓ | ✓ | ✓ | ✓ |
| Fuc$_2$ Hex$_5$ HexNAc$_4$ | 2418.21 | ✓ | ✓ | | ✓ | | ✓ | | | | |
| NeuGc$_1$ Hex$_6$ HexNAc$_3$ | 2420.19 | | | | ✓ | ✓ | | | | | |
| NeuAc$_1$ Hex$_5$ HexNAc$_4$ | 2431.21 | | | | ✓ | ✓ | | | | | ✓ |
| NeuGc$_1$ Fuc$_1$ Hex$_4$ HexNAc$_4$ | 2431.21 | | | | ✓ | ✓ | ✓ | ✓ | | | |
| Fuc$_1$ Hex$_6$ HexNAc$_4$ | 2448.22 | | | | ✓ | ✓ | | | | | |
| Fuc$_2$ Hex$_4$ HexNAc$_5$ | 2459.24 | ✓ | ✓ | ✓ | | | ✓ | | | | |
| NeuGc$_1$ Hex$_5$ HexNAc$_4$ | 2461.22 | | | | ✓ | ✓ | | | | | |
| Hex$_7$ HexNAc$_4$ | 2478.23 | | | | ✓ | | | | | | |

*(Continued)*

## Table 12.1  Continued

| | Calculated *m/z* | Brain | Colon | Kidney | Liver | Lung | Ovaries | Small Intestine | Spleen | Testes | Thymus |
|---|---|---|---|---|---|---|---|---|---|---|---|
| Fuc$_1$ Hex$_5$ HexNAc$_5$ | 2489.25 | | ✓ | ✓ | | ✓ | ✓ | | ✓ | ✓ | ✓ |
| Fuc$_1$ Hex$_4$ HexNAc$_6$ | 2530.28 | | ✓ | | | | | | | | |
| Fuc$_3$ Hex$_5$ HexNAc$_4$ | 2592.30 | ✓ | | ✓ | | | | | | | |
| Hex$_{10}$ HexNAc$_2$ | 2600.28 | | | | | ✓ | | | ✓ | ✓ | |
| NeuAc$_1$ Fuc$_1$ Hex$_5$ HexNAc$_4$ | 2605.30 | | | | ✓ | ✓ | ✓ | ✓ | | ✓ | ✓ |
| Fuc$_2$ Hex$_6$ HexNAc$_4$ | 2622.31 | ✓ | | | | ✓ | | | | | |
| Fuc$_3$ Hex$_4$ HexNAc$_5$ | 2633.33 | | | ✓ | | | | | | | |
| NeuAc$_1$ Hex$_6$ HexNAc$_4$ | 2635.31 | | | | ✓ | ✓ | ✓ | ✓ | ✓ | ✓ | ✓ |
| NeuGc$_1$ Fuc$_1$ Hex$_5$ HexNAc$_4$ | 2635.31 | | | | ✓ | ✓ | ✓ | ✓ | ✓ | ✓ | ✓ |
| NeuAc$_1$ Fuc$_1$ Hex$_4$ HexNAc$_5$ | 2646.33 | ✓ | | | | | | | | | |
| Fuc$_1$ Hex$_7$ HexNAc$_4$ | 2652.32 | | ✓ | | ✓ | ✓ | | ✓ | | | |
| Fuc$_2$ Hex$_5$ HexNAc$_5$ | 2663.34 | ✓ | | ✓ | | ✓ | | | | | |
| NeuAc$_1$ Hex$_6$ HexNAc$_4$ | 2665.32 | | | | ✓ | | | ✓ | | | |
| Fuc$_1$ Hex$_6$ HexNAc$_5$ | 2693.35 | | ✓ | | | ✓ | | | ✓ | ✓ | |
| Fuc$_2$ Hex$_4$ HexNAc$_6$ | 2704.37 | ✓ | ✓ | | | | | | | | |
| NeuAc$_1$ Hex$_4$ HexNAc$_6$ | 2717.36 | | | | | | | ✓ | | | |
| Fuc$_1$ Hex$_5$ HexNAc$_6$ | 2734.38 | | ✓ | | | | | | | | |
| Fuc$_1$ Hex$_4$ HexNAc$_7$ | 2775.40 | | ✓ | | | | | | | | |
| NeuAc$_1$ Fuc$_2$ Hex$_5$ HexNAc$_4$ | 2779.39 | ✓ | | | | | | | | | |
| NeuAc$_1$ Fuc$_1$ Hex$_6$ HexNAc$_4$ | 2809.40 | | | | ✓ | ✓ | | ✓ | ✓ | ✓ | ✓ |
| NeuAc$_1$ NeuGc$_1$ Hex$_5$ HexNAc$_4$ | 2822.39 | | | | | ✓ | | | | | |
| Fuc$_3$ Hex$_5$ HexNAc$_5$ | 2837.43 | ✓ | ✓ | ✓ | | | | | | | |
| NeuGc$_1$ Fuc$_1$ Hex$_6$ HexNAc$_4$ | 2839.41 | | | | ✓ | ✓ | | | | | ✓ |
| NeuAc$_1$ Fuc$_1$ Hex$_5$ HexNAc$_5$ | 2850.43 | ✓ | | | | | | | ✓ | | |
| NeuGc$_2$ Hex$_5$ HexNAc$_4$ | 2852.40 | | | | ✓ | ✓ | ✓ | | ✓ | ✓ | ✓ |
| Fuc$_2$ Hex$_6$ HexNAc$_5$ | 2867.44 | | | | | | | | | ✓ | |
| NeuAc$_1$ Fuc$_1$ Hex$_4$ HexNAc$_6$ | 2891.45 | ✓ | | | | | | ✓ | | | |
| Fuc$_1$ Hex$_7$ HexNAc$_5$ | 2897.45 | | | | | | | | ✓ | | |
| Fuc$_2$ Hex$_5$ HexNAc$_6$ | 2908.47 | ✓ | | | | | | | | | |
| Fuc$_1$ Hex$_6$ HexNAc$_6$ | 2938.48 | | ✓ | | | | | ✓ | | | |
| NeuAc$_2$ Fuc$_1$ Hex$_5$ HexNAc$_4$ | 2966.47 | ✓ | | | ✓ | ✓ | ✓ | | ✓ | ✓ | ✓ |
| Fuc$_1$ Hex$_5$ HexNAc$_7$ | 2979.50 | | ✓ | | | | | | | | |
| NeuAc$_1$ NeuGc$_1$ Fuc$_1$ Hex$_5$ HexNAc$_4$ | 2996.48 | | | | ✓ | | ✓ | | | ✓ | ✓ |
| NeuAc$_1$ Fuc$_2$ Hex$_5$ HexNAc$_5$ | 3024.51 | ✓ | | | | | | | | | |
| NeuGc$_2$ Fuc$_1$ Hex$_5$ HexNAc$_4$ | 3026.49 | | | | ✓ | ✓ | ✓ | ✓ | ✓ | ✓ | ✓ |

*(Continued)*

## Table 12.1 Continued

| | Calculated m/z | Brain | Colon | Kidney | Liver | Lung | Ovaries | Small Intestine | Spleen | Testes | Thymus |
|---|---|---|---|---|---|---|---|---|---|---|---|
| Fuc$_3$ Hex$_6$ HexNAc$_5$ | 3041.53 | | | ✓ | | ✓ | | | | | |
| NeuAc$_1$ Fuc$_1$ Hex$_6$ HexNAc$_5$ | 3054.52 | | | | | ✓ | | | | | ✓ |
| NeuGc$_1$ Fuc$_2$ Hex$_5$ HexNAc$_5$ | 3054.52 | | | | | ✓ | | | | | |
| Fuc$_3$ Hex$_5$ HexNAc$_6$ | 3082.56 | ✓ | ✓ | | | | | | | | |
| NeuGc$_1$ Fuc$_1$ Hex$_6$ HexNAc$_5$ | 3084.54 | | | | | ✓ | | | | | ✓ |
| NeuGc$_1$ Fuc$_1$ Hex$_6$ HexNAc$_5$ | 3084.54 | | | | | ✓ | | | | | |
| Fuc$_2$ Hex$_6$ HexNAc$_6$ | 3112.57 | | | | ✓ | | | | | | |
| Fuc$_2$ Hex$_5$ HexNAc$_7$ | 3153.59 | | | ✓ | | | | | | | |
| Fuc$_1$ Hex$_6$ HexNAc$_7$ | 3183.60 | | | | | | | ✓ | | | |
| Fuc$_4$ Hex$_6$ HexNAc$_5$ | 3215.62 | ✓ | | ✓ | | | | | | | |
| NeuAc$_2$ Hex$_6$ HexNAc$_5$ | 3241.61 | | | | | | | | | ✓ | |
| NeuAc$_1$ Fuc$_2$ Hex$_5$ HexNAc$_6$ | 3269.64 | ✓ | | | | | | | | | |
| Fuc$_3$ Hex$_6$ HexNAc$_6$ | 3286.66 | | | ✓ | | | | | | | |
| NeuGc$_2$ Hex$_6$ HexNAc$_5$ | 3301.63 | | | | ✓ | | | | | | |
| Fuc$_1$ Hex$_9$ HexNAc$_5$ | 3305.65 | | | | | ✓ | | | | | |
| NeuAc$_1$ Fuc$_1$ Hex$_5$ HexNAc$_7$ | 3340.68 | | | | | | | ✓ | | | |
| NeuAc$_1$ Hex$_6$ HexNAc$_7$ | 3370.69 | | | | | | | ✓ | | | |
| NeuGc$_1$ Fuc$_1$ Hex$_5$ HexNAc$_7$ | 3370.69 | | | | | | | ✓ | | | |
| NeuAc$_1$ Fuc$_3$ Hex$_6$ HexNAc$_5$ | 3402.70 | ✓ | | | | | | | | | |
| NeuAc$_2$ Fuc$_1$ Hex$_6$ HexNAc$_5$ | 3415.70 | | | | | ✓ | | | | ✓ | |
| NeuAc$_1$ NeuGc$_1$ Fuc$_1$ Hex$_6$ HexNAc$_5$ | 3445.71 | | | | | ✓ | | | | | |
| Fuc$_4$ Hex$_6$ HexNAc$_6$ | 3460.75 | ✓ | | ✓ | | | | | | | |
| NeuGc$_1$ Fuc$_2$ Hex$_7$ HexNAc$_5$ | 3462.72 | | | | | ✓ | | | | | |
| NeuAc$_1$ Fuc$_1$ Hex$_8$ HexNAc$_5$ | 3462.72 | | | | | ✓ | | | | | |
| NeuGc2 Fuc1 Hex6 HexNAc5 | 3475.72 | | | | ✓ | ✓ | | | | | |
| NeuAc1 NeuGc1 Hex7 HexNAc5 | 3475.72 | | | | ✓ | ✓ | | | | | |
| Fuc$_3$Hex$_7$HexNAc$_6$ | 3490.75 | | | ✓ | | | | | | | |
| NeuGc$_1$ Fuc$_1$ Hex$_8$ HexNAc$_5$ | 3492.73 | | | | | ✓ | | | | | |
| Fuc$_3$ Hex$_6$ HexNAc$_7$ | 3531.78 | | | ✓ | | | | | | | |
| NeuAc$_2$ Fuc$_2$ Hex$_6$ HexNAc$_5$ | 3589.79 | ✓ | | | | | | | | | |
| NeuAc$_1$ Fuc$_3$ Hex$_6$ HexNAc$_6$ | 3647.83 | ✓ | | | | | | | | | |
| NeuAc$_1$ NeuGc$_1$ Fuc$_1$ Hex$_7$ HexNAc$_5$ | 3649.81 | | | | | ✓ | | | | | |
| NeuAc$_2$ Hex$_8$ HexNAc$_5$ | 3649.81 | | | | | ✓ | | | | | |
| Fuc$_4$ Hex$_7$ HexNAc$_6$ | 3664.84 | | | ✓ | | | | | | | |
| NeuGc$_2$ Fuc$_1$ Hex$_7$ HexNAc$_5$ | 3679.82 | | | | ✓ | ✓ | | | | | ✓ |

*(Continued)*

## Table 12.1 Continued

| | Calculated *m/z* | Brain | Colon | Kidney | Liver | Lung | Ovaries | Small Intestine | Spleen | Testes | Thymus |
|---|---|---|---|---|---|---|---|---|---|---|---|
| NeuGc$_3$ Hex$_6$ HexNAc$_5$ | 3692.81 | | | | ✓ | ✓ | | | | ✓ | ✓ |
| NeuAc$_2$ Fuc$_1$ Hex$_5$ HexNAc$_7$ | 3701.85 | | | | | | | ✓ | | | |
| Fuc$_4$ Hex$_6$ HexNAc$_7$ | 3705.87 | | | ✓ | | | | | | | |
| Fuc$_3$ Hex$_7$ HexNAc$_7$ | 3735.88 | | | ✓ | | | | | | | |
| NeuAc$_3$ Fuc$_1$ Hex$_6$ HexNAc$_5$ | 3776.87 | | | | | | | | | ✓ | |
| NeuGc$_2$ Hex$_6$ HexNAc$_7$ | 3791.88 | | | | | | | ✓ | | | |
| NeuAc$_2$ NeuGc$_1$ Fuc$_1$ Hex$_6$ HexNAc$_5$ | 3806.88 | | | | | | | ✓ | | ✓ | |
| Fuc$_5$ Hex$_7$ HexNAc$_6$ | 3838.93 | ✓ | | | | | | | | | |
| NeuAc$_2$ Fuc$_1$ Hex$_7$ HexNAc$_6$ | 3864.92 | | | | | | | | ✓ | | |
| NeuGc$_3$ Fuc$_1$ Hex$_6$ HexNAc$_5$ | 3866.90 | | | | ✓ | ✓ | | | | | ✓ |
| NeuAc$_1$ Fuc$_3$ Hex$_6$ HexNAc$_7$ | 3892.96 | ✓ | | | | | | | | | |
| Fuc$_4$ Hex$_7$ HexNAc$_7$ | 3909.97 | | | ✓ | | | | | | | |
| Fuc$_3$ Hex$_8$ HexNAc$_7$ | 3939.98 | | | ✓ | | | | | | | |
| NeuGc$_2$ Hex$_7$ HexNAc$_7$ | 3995.98 | | | | | | | ✓ | | | |
| NeuAc$_1$ Fuc$_4$ Hex$_7$ HexNAc$_6$ | 4026.02 | ✓ | | | | | | | | | |
| NeuGc$_3$ Fuc$_2$ Hex$_6$ HexNAc$_5$ | 4040.99 | | | | | ✓ | | | | | |
| NeuAc$_2$ NeuGc$_1$ Hex$_7$ HexNAc$_6$ | 4082.02 | | | | | | | ✓ | | ✓ | |
| Fuc$_5$ Hex$_7$ HexNAc$_7$ | 4084.06 | | | ✓ | | | | | | | |
| NeuAc$_1$ NeuGc$_2$ Hex$_6$ HexNAc$_7$ | 4153.06 | | | | | | | ✓ | | | |
| NeuAc$_2$ Fuc$_3$ Hex$_7$ HexNAc$_6$ | 4213.10 | ✓ | | | | | | | | | |
| NeuGc$_3$ Fuc$_1$ Hex$_7$ HexNAc$_6$ | 4316.13 | | | | | ✓ | | | | | |
| NeuAc$_1$ NeuGc$_2$ Hex$_8$ HexNAc$_6$ | 4316.13 | | | | | ✓ | | | | | |
| NeuGc$_2$ Fuc$_1$ Hex$_9$ HexNAc$_6$ | 4333.15 | | | | | ✓ | | | | | |
| NeuAc$_3$ Fuc$_2$ Hex$_7$ HexNAc$_6$ | 4400.19 | ✓ | | | | | | | | | |

## Key Features

Figure 12.8a displays a selected mass profile of the N-glycans released from mouse colon. It consists of high mannose and bi-, tri-, and tetra-antennary complex glycans. The most abundant signals in the spectrum correspond to truncated biantennary structures, with this theme of part-processed glycans continuing through to the larger oligosaccharides. Where capping does occur, it is by Gal residues, rather than the sialic acids, of which there is little evidence. Bisecting GlcNAc is a common feature amongst the N-linked glycans and antennal fucosylation is also detected, with up to two such residues modifying the higher mass structures. See Figure 12.9 for a summary of the proposed glycan structures found in the N-glycan profile of murine colon.

Table 12.2 Compositional assignments of singly charged sodiated molecular ions, [M+Na]$^+$, observed in MALDI-MS spectra of permethylated O-glycans from mouse tissues. Tissues indicated with an asterisk (*) have their data drawn from previous work [21]

| Composition | Calculated m/z | Brain | Colon | Kidney | Liver | Lung* | Ovaries* | Small Intestine* | Spleen | Testes* | Thymus |
|---|---|---|---|---|---|---|---|---|---|---|---|
| Hex$_1$HexNAc$_1$-itol | 534.29 | ✓ | ✓ | | | ✓ | | ✓ | ✓ | ✓ | ✓ |
| Hex$_1$HexNAc$_2$-itol | 779.41 | ✓ | ✓ | ✓ | | ✓ | | ✓ | ✓ | ✓ | ✓ |
| NeuAc$_1$Hex$_1$HexNAc$_1$-itol | 895.46 | ✓ | ✓ | ✓ | | ✓ | ✓ | ✓ | ✓ | ✓ | ✓ |
| Fuc$_1$ Hex$_2$ HexNAc$_1$-itol | 912.48 | ✓ | | | | | | | | | |
| NeuGc$_1$ Hex$_1$ HexNAc$_1$-itol | 925.47 | | ✓ | ✓ | | ✓ | ✓ | ✓ | | | |
| Fuc$_1$ Hex$_1$ HexNAc$_2$-itol | 953.50 | ✓ | ✓ | | | | | ✓ | | | |
| Hex$_2$HexNAc$_2$-itol | 983.51 | | ✓ | ✓ | | ✓ | | ✓ | ✓ | ✓ | |
| Hex$_1$HexNAc$_3$-itol | 1024.54 | | ✓ | | | | | | ✓ | ✓ | |
| NeuAc$_1$ Fuc$_1$ Hex$_1$ HexNAc$_1$-itol | 1069.55 | | | | | | | ✓ | | | |
| NeuAc$_1$Hex$_2$HexNAc$_1$-itol | 1099.56 | ✓ | | | | | | | | | |
| NeuAc$_1$ Hex$_1$ HexNAc$_2$-itol | 1140.59 | | | | | | | ✓ | | | ✓ |
| Fuc$_1$Hex$_2$HexNAc$_2$-itol | 1157.60 | ✓ | ✓ | ✓ | | ✓ | | | ✓ | ✓ | ✓ |
| NeuGc$_1$ Hex$_1$ HexNAc$_2$-itol | 1170.60 | | | | | | | | ✓ | | |
| Hex$_3$HexNAc$_2$-itol | 1187.61 | ✓ | | | | | | | ✓ | | ✓ |
| Hex$_2$ HexNAc$_3$-itol | 1228.64 | | | | | | | | | | ✓ |
| NeuAc$_2$Hex$_1$HexNAc$_1$-itol | 1256.63 | ✓ | ✓ | ✓ | | ✓ | ✓ | ✓ | ✓ | ✓ | ✓ |
| NeuGc$_1$ NeuAc$_1$ Hex$_1$ HexNAc$_1$-itol | 1286.65 | | ✓ | ✓ | | ✓ | ✓ | | ✓ | ✓ | |
| NeuGc$_2$ Hex$_1$ HexNAc$_1$-itol | 1316.66 | | ✓ | ✓ | | ✓ | ✓ | | ✓ | ✓ | |
| NeuAc$_1$Hex$_2$HexNAc$_2$-itol | 1344.69 | | | | | | | | ✓ | | ✓ |
| NeuGc$_1$ Hex$_2$ HexNAc$_2$-itol | 1374.70 | | | ✓ | | | | | ✓ | | |
| NeuAc$_1$Hex$_1$HexNAc$_3$-itol | 1385.72 | | | | | | | | | | ✓ |
| Fuc$_1$ Hex$_2$ HexNAc$_3$-itol | 1402.73 | ✓ | ✓ | | | | | | | | |
| Hex$_3$HexNAc$_3$-itol | 1432.74 | ✓ | | | | | | | ✓ | | ✓ |
| NeuAc$_2$Hex$_3$HexNAc$_1$-itol | 1589.82 | | | | | | | ✓ | | | ✓ |
| NeuAc$_2$Hex$_3$HexNAc$_2$-itol | 1950.99 | | | | | | | ✓ | | | |
| NeuAc$_1$Hex$_3$HexNAc$_4$-itol | 2039.04 | | | | | | | ✓ | | | |
| NeuAc$_2$Hex$_2$HexNAc$_4$-itol | 2196.12 | | | | | | | ✓ | | | |

The O-glycan profile of mouse colon is shown in Figure 12.8b. Core 1 and core 2 type structures are observed. The core 1 glycans are modified by both NeuGc and NeuAc, with all combinations up to a sum total of two sialic acids detected. Core 2 glycans are detected with a range of possible modifications. Fucose-containing structures

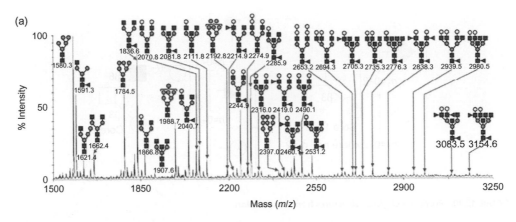

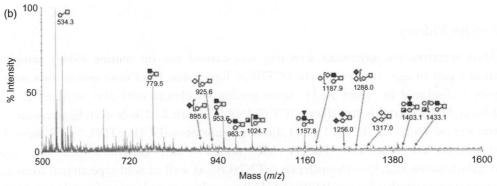

**FIGURE 12.8**   Partial MALDI-TOF-MS profiles of the permethylated *N*-linked glycans (a) and *O*-linked glycans (b) from a typical murine colon sample. Major peaks are annotated with the relevant carbohydrate structure shown in symbol form, according to the glycan nomenclature adopted by the CFG (http://www.functionalglycomics.org/). For complete annotation of the spectra see Tables 12.1 and 12.2. All molecular ions are present in sodiated form ([M+Na]$^+$).

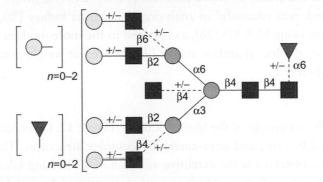

**FIGURE 12.9**   Proposed complex *N*-glycan structures from mouse colon.

are also detected, at *m/z* 1157.8 and 1403.1, which are suspected to be arranged in a Le$^x$ conformation. Figure 12.10 shows a summary of the proposed core 1 and core 2 glycans present in the *O*-glycan profile of murine colon.

For a complete summary of the glycans observed in both the *N*-linked and *O*-linked spectra, see Tables 12.1 and 12.2.

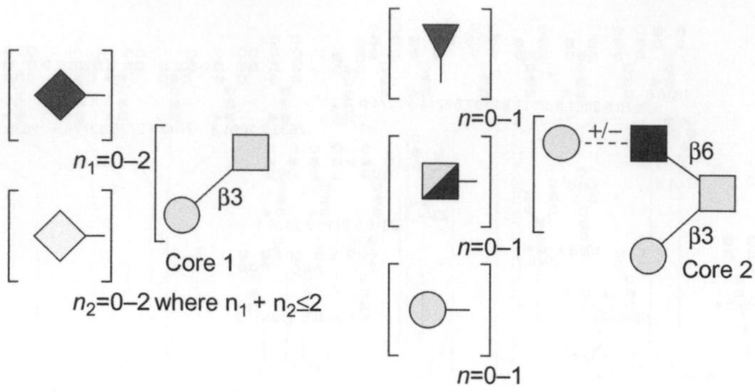

**FIGURE 12.10**   Proposed *O*-glycan structures from mouse colon.

## Mouse Kidney

Mass spectrometric glycomics screening was carried out on murine kidney samples from a pair of age- and sex-matched C57BL/6 littermates. Data from one of these analyses is displayed in Figure 12.11. Seven genetic knockouts were also analyzed (β-1,4 *N*-acetylgalactosaminyltransferase (CT GalNAcT), core 2 β1,6-*N*-acetylglucosaminyltransferase (core 2 GlcNAcT), α(1,3)fucosyltransferase-IV (FucTIV), α(1,3)fucosyltransferase-VII (FucTVII), FucTIV+VII (double null), galectin-3 (Gal 3), and α-galactosidase α(2,3)sialyltransferase (ST3Gal-I)), as well as wild-type animal from an alternative genetic background (129×1/SvJ). Raw and annotated spectra from these data are available online, together with the wild-type analyses presented here, from http://www. functionalglycomics.org/glycomics/publicdata/glycoprofiling.jsp. A discussion of these results in relation to the gene-expression profiles is also available [24], as well as some recent analyses of the kidney by liquid chromatography (LC)-MS [31].

Earlier work was successful in analyzing the murine kidney [20,21,25], though it was carried out using FAB-MS [26], as opposed to the more modern MALDI instrumentation. Hence the less abundant and/or high molecular weight components found here were not observed.

### Key Features

The *N*-linked glycan profile of the murine kidney (Figure 12.11a) contains high mannose glycans and bi-, tri-, and tetra-antennary complex structures. The most notable structural feature observed is the extensive addition of a bisecting GlcNAc residue β4 to the core mannose, a feature which is further confirmed by GC-MS linkage data showing high levels of 3,4,6-linked mannose (data not shown). The antennae are heavily fucosylated, with up to four fucose residues attached in the larger structures. Galactose capping of the structures was not detected and sialylation only occurs on low abundance glycans. The most intense signal observed belongs to a biantennary, core fucosylated, asialo structure modified with a bisecting GlcNAc and two antennal fucoses at *m/z* 2837.4. These findings confirm and expand upon what has been observed previously [20,21], and have since been confirmed in unrelated work by

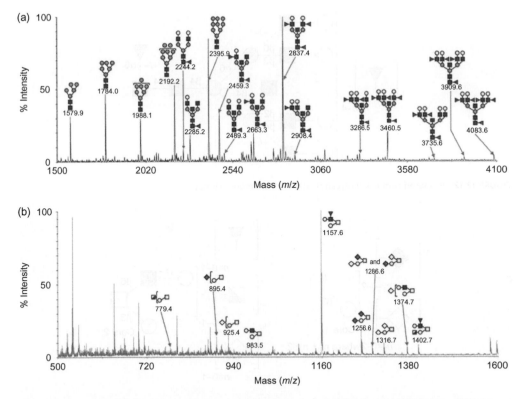

**FIGURE 12.11**   Partial MALDI-TOF-MS profiles of the permethylated *N*-linked glycans (a) and *O*-linked glycans (b) from a typical murine kidney sample. Major peaks are annotated with the relevant carbohydrate structure shown in symbol form, according to the glycan nomenclature adopted by the CFG (http://www.functionalglycomics.org/). For complete annotation of the spectra see Tables 12.1 and 12.2. All molecular ions are present in sodiated form ([M+Na]$^+$).

other groups [31]. See Figure 12.12 for a summary of the proposed glycan structures found in the *N*-glycan profile of mouse kidney.

The *O*-glycan profile of the mouse kidney (Figure 12.11b) contains both core 1 and core 2 type *O*-linked glycans, with sialylation being observed on both the core 1 and core 2 archetypes. By far the most abundant structure is a putative core 2 glycan, extended by a single galactose and decorated by a fucose residue, assumed to be in a Le$^x$ arrangement. Figure 12.13 shows a summary of the proposed core 1 and core 2 glycans present in the *O*-glycan profile of murine kidney.

For a complete summary of the glycans observed in both the *N*-linked and *O*-linked spectra, see Tables 12.1 and 12.2.

## Mouse Liver

Mass spectrometric *N*-glycomics screening was carried out on murine liver samples from a pair of age- and sex-matched C57BL/6 littermates. Data from one of these analyses are displayed in Figure 12.14. Raw and annotated spectra from these data are available online, from http://www.functionalglycomics.org/glycomics/publicdata/

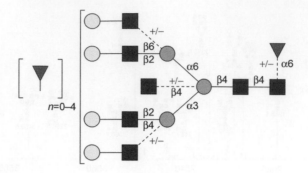

**FIGURE 12.12**  Proposed complex *N*-glycan structures from mouse kidney.

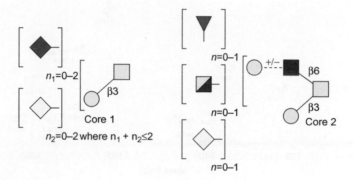

**FIGURE 12.13**  Proposed *O*-glycan structures from mouse kidney.

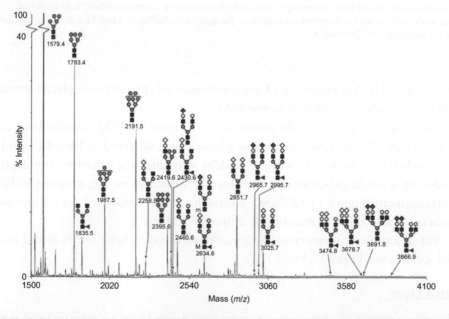

**FIGURE 12.14**  Partial MALDI-TOF-MS profiles of the permethylated *N*-linked glycans from a typical murine liver sample. Major peaks are annotated with the relevant carbohydrate structure shown in symbol form, according to the glycan nomenclature adopted by the CFG (http://www.functionalglycomics.org/). Where MS/MS analysis has informed upon the relative abundance of glycoforms, major (M) and minor (m) components are indicated. For complete annotation of the spectrum see Table 12.1. All molecular ions are present in sodiated form ([M+Na]$^+$).

glycoprofiling.jsp. A discussion of these results in relation to the gene expression profiles is also available [24].

Earlier work was successful in analyzing the N-glycans of the murine liver [20,21,25], though it was carried out using FAB-MS [26], as opposed to the more modern MALDI instrumentation.

### Key Features

The mouse liver N-glycan profile (Figure 12.14a) contains high mannose and bi-, tri-, and tetra-antennary glycans. The complex glycans predominantly contain fucose on their cores, with no evidence (compositional, GC-MS or MS/MS) being observed for the presence of antennal fucose. Extensive sialylation by NeuGc is observed, with the structures being capped with up to three such sialic acid residues. NeuAc capping is also detected, along with galactose terminations, but both features are at a much lower abundance relative to NeuGc containing structures. No evidence of bisecting GlcNAc is observed, confirmed by the absence of 3,4,6-linked mannose within GC-MS linkage experiments (data not shown). The most prevalent complex structure in the spectrum is a non-core-fucosylated, biantennary glycan bearing two NeuGc terminal groups (*m/z* 2851.7). See Figure 12.15 for a summary of the proposed glycan structures found in the N-glycan profile of mouse liver.

For a complete summary of the glycans observed in the N-linked spectrum, see Tables 12.1 and 12.2.

## Mouse Lung

Mass spectrometric glycomics screening was carried out on murine lung samples from a pair of age- and sex-matched C57BL/6 littermates. Data from one of these analyses are displayed in Figure 12.16. Raw and annotated spectra from these data are available online, from http://www.functionalglycomics.org/glycomics/publicdata/glycoprofiling. jsp. A discussion of these results in relation to the gene-expression profiles is also available [24].

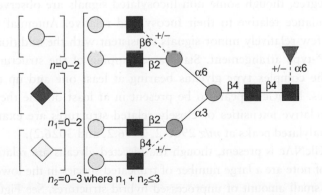

**FIGURE 12.15**  Proposed complex *N*-glycan structures from mouse liver.

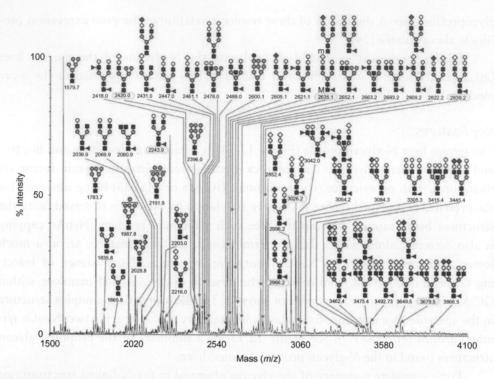

**FIGURE 12.16**   Partial MALDI-TOF-MS profiles of the permethylated *N*-linked glycans from a typical murine lung sample. Major peaks are annotated with the relevant carbohydrate structure shown in symbol form, according to the glycan nomenclature adopted by the CFG (http://www.functionalglycomics.org/). Structures confirmed by further analyses have their masses shaded. Where MS/MS analysis has informed upon the relative abundance of glycoforms, major (M) and minor (m) components are indicated. For complete annotation of the spectrum see Table 12.1. All molecular ions are present in sodiated form ([M+Na]⁺).

Earlier work was successful in analyzing the *O*-glycans of the murine lung [20,21,25].

## Key Features

The *N*-glycan profile of mouse lung (Figure 12.16) consists of high mannose and bi-, tri-, and tetra-antennary complex structures. The latter are core fucosylated to a reasonably high degree, though some non-fucosylated signals are observed, typically at 75–100% abundance relative to their fucosylated relative. Antennal fucosylation is limited, with a few relatively minor signals consistent with the addition of up to two fucoses in a Le^x-type arrangement. Sialic acid capping of the structures is plentiful, with most of the complex type glycans bearing at least one and up to three of the terminal residues. NeuGc appears to be present in at least double the abundance of NeuAc, when relative intensities of closely related structures are examined (see the three variably sialylated peaks at *m/z* 2966.2, 2996.2, and 3026.2).

Bisecting GlcNAc is present, though the bisected glycans are relatively low level. Other features of note are a large number of truncated species in the lower mass region, together with a small amount of unprocessed hybrid structures. See Figure 12.17 for a summary of the proposed glycan structures found in the *N*-glycan profile of murine lung.

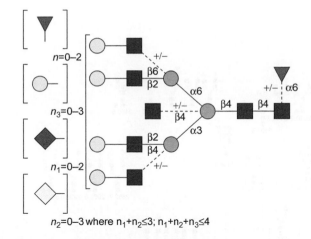

$n=0-2$

$n_3=0-3$

$n_1=0-2$

$n_2=0-3$ where $n_1+n_2\leq3$; $n_1+n_2+n_3\leq4$

**FIGURE 12.17**  Proposed complex *N*-glycan structures from mouse lung.

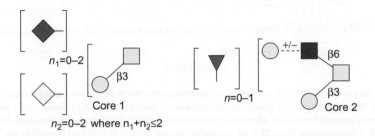

$n_1=0-2$

$n_2=0-2$ where $n_1+n_2\leq2$

Core 1

$n=0-1$

Core 2

**FIGURE 12.18**  Proposed *O*-glycan structures from mouse lung (summarized from previous work [20,21]).

Core 1 and core 2 type *O*-glycans are both present in the *O*-linked profile of mouse lung [20,21]. The core 1 subtype is readily sialylated with up to two NeuAc or NeuGc residues. The core 2 glycans are modified with fucose (thought to be in Le$^x$ form). Figure 12.18 shows a summary of the proposed core 1 and core 2 glycans present in the *O*-glycan profile of murine lung.

For a complete summary of the glycans observed in both the *N*-linked and *O*-linked spectra, see Tables 12.1 and 12.2.

## Mouse Ovaries

Mass spectrometric glycomics screening was carried out on murine ovaries from a pair of age- and sex-matched C57BL/6 littermates. Data from one of these analyses are displayed in Figure 12.19. Raw and annotated spectra from these data are available online, from http://www.functionalglycomics.org/glycomics/publicdata/glycoprofiling.jsp. Earlier work was successful in analyzing the *O*-glycans of murine ovaries [20,21,25].

### *Key Features*

The *N*-glycan profile from murine ovaries is displayed in Figure 12.19. Owing to the diminutive physical size of the organ derived from a single animal, there are relatively

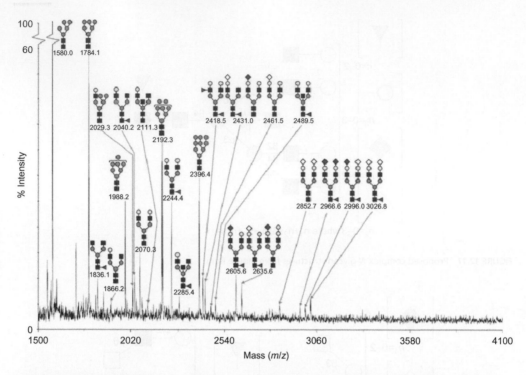

**FIGURE 12.19** Partial MALDI-TOF-MS profiles of the permethylated *N*-linked glycans from a typical sample of murine ovaries. Major peaks are annotated with the relevant carbohydrate structure shown in symbol form, according to the glycan nomenclature adopted by the CFG (http://www.functionalglycomics.org/). For complete annotation of the spectrum see Table 12.1. All molecular ions are present in sodiated form ([M+Na]⁺).

few structures discerned. However, despite these limitations, the quality of the spectrum is sufficient to identify both high mannose and biantennary complex type glycans. These structures are capped with both NeuAc and NeuGc sialic acids, in an approximately even fashion. They also exhibit an appreciable level of bisecting GlcNAc and some limited antennal fucosylation. See Figure 12.20 for a summary of the proposed glycan structures found in the *N*-glycan profile of murine ovaries.

Only core 1 type structures were observed within the *O*-linked profile of mouse ovaries [21]. This subtype is readily sialylated with up to two NeuAc or NeuGc residues. Figure 12.21 shows a summary of the proposed core 1 glycans present in the *O*-glycan profile of murine ovaries.

For a complete summary of the glycans observed in both the *N*-linked and *O*-linked spectra, see Tables 12.1 and 12.2.

## Mouse Small Intestine

Mass spectrometric glycomics screening was carried out on murine small intestine samples from a pair of age- and sex-matched C57BL/6 littermates. Data from one of these analyses are displayed in Figure 12.22. Three genetic knockouts were also analyzed (FucTIV+VII (double null), Gal 3, and ST3Gal-I). Raw and annotated spectra from these data are available online, together with the wild-type analyses presented

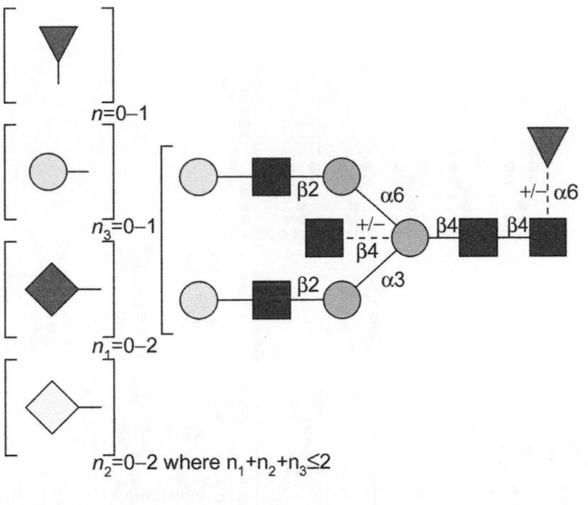

**FIGURE 12.20**   Proposed complex *N*-glycan structures from mouse ovaries.

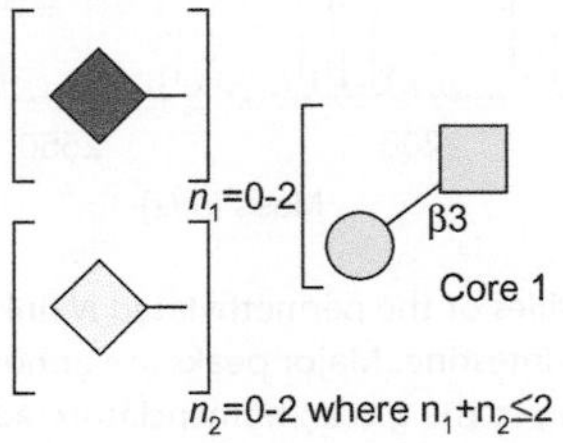

**FIGURE 12.21**   Proposed *O*-glycan structures from mouse ovaries (summarized from previous work [21]).

here, from http://www.functionalglycomics.org/glycomics/publicdata/glycoprofiling. jsp. Earlier work was successful in analyzing the *O*-glycans of the murine small intestine [20,21,25], though it was carried out using FAB-MS [26], as opposed to the more modern MALDI instrumentation.

## *Key Features*

The *N*-linked glycan profile generated from a typical mouse small intestine sample is shown in Figure 12.22. High mannose and hybrid structures are present, along with bi-, tri-, and tetra-antennary complex type. These structures are seen to core fucosylate to a high degree. An unusually high level of bisecting GlcNAc is observed, with bisected peaks accounting for the majority of higher mass structures. This extensive action of GlcNAcTIII is consistent with the relatively high levels of 3,4,6-linked mannose within GC-MS linkage analyses (data not shown); indeed, the most abundant complex structure is that belonging to a biantennary, core fucosylated, bisected glycan (*m/z* 2489.1). Sialylation of the antennae is present (*n* = 0–4), though at relatively low levels when compared with other tissues. NeuGc is slightly more common than NeuAc. See Figure 12.23 for a summary of the proposed glycan structures found in the *N*-glycan profile of murine small intestine.

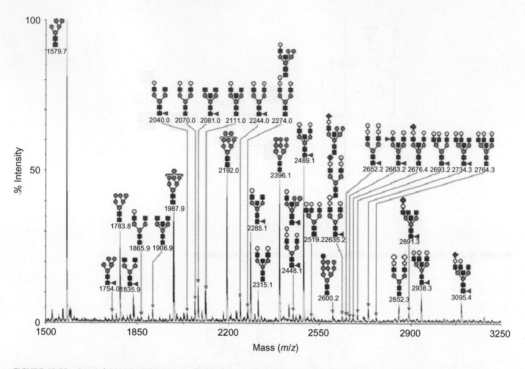

**FIGURE 12.22** Partial MALDI-TOF-MS profiles of the permethylated *N*-linked glycans (a) and *O*-linked glycans (b) from a typical sample of murine small intestine. Major peaks are annotated with the relevant carbohydrate structure shown in symbol form, according to the glycan nomenclature adopted by the CFG (http://www.functionalglycomics.org/). For complete annotation of the spectra see Tables 12.1 and 12.2. All molecular ions are present in sodiated form ([M+Na]⁺).

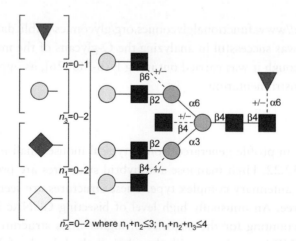

**FIGURE 12.23** Proposed complex *N*-glycan structures from mouse small intestine.

Core 1 and core 2 type *O*-glycans are both present in the *O*-linked profile of mouse small intestine [21]. The core 1 subtype is readily sialylated with up to two NeuAc residues, while the core 2 glycans are modified with a wider range of decorations, including fucose (thought to be in Le$^x$ form), NeuAc (up to two residues) and

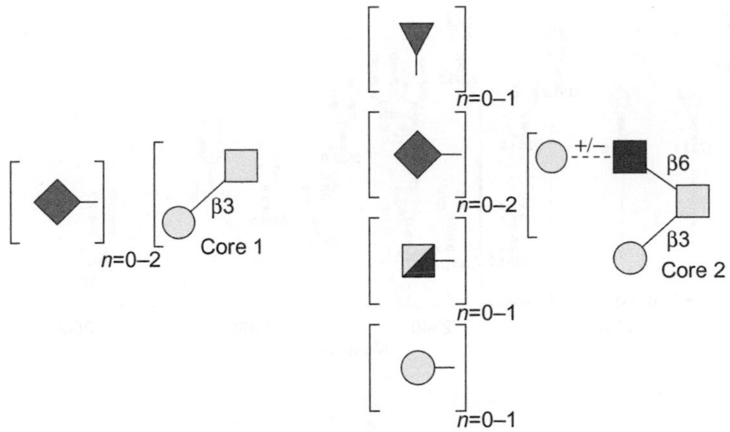

**FIGURE 12.24** Proposed *O*-glycan structures from mouse small intestine (summarized from previous work [21]).

galactose. Figure 12.24 shows a summary of the proposed core 1 and core 2 glycans present in the *O*-glycan profile of murine small bowel.

For a complete summary of the glycans observed in both the *N*-linked and *O*-linked spectra, see Tables 12.1 and 12.2.

## Mouse Spleen

Mass spectrometric glycomics screening was carried out on murine spleen samples from a pair of age- and sex-matched C57BL/6 littermates. Data from one of these *N*-glycome analyses are displayed in Figure 12.25a. Six genetic knockouts were also analyzed (CT GalNAcT, FucTIV, FucTVII, FucTIV+VII (double null), Gal 3, and ST3Gal-I), as well as a wild-type animal from an alternative genetic background (129×1/SvJ), the *O*-glycan analysis from which is displayed in Figure 12.25b. Raw and annotated spectra from these data are available online, together with the wild-type analyses presented here, from http://www.functionalglycomics.org/glycomics/publicdata/glycoprofiling.jsp. A discussion of these results in relation to the gene expression profiles is also available [24], as well as some recent analyses of spleen-derived immune cell populations [32].

Earlier work was successful in analyzing the *N*-glycans from the murine spleen [20,21,25], though it was carried out using FAB-MS [26], as opposed to the more modern MALDI instrumentation.

*Key Features*

The *N*-glycan profile from a typical murine spleen is shown in Figure 12.25a. The presence of both high mannose and bi- and tri-antennary complex glycans is discernable. The latter is modified by a reasonable level of bisecting GlcNAc and readily capped by either sialic acid (NeuAc or NeuGc) or galactose, with NeuGc appearing to be most common of the three followed by Gal and NeuAc. See Figure 12.26 for a

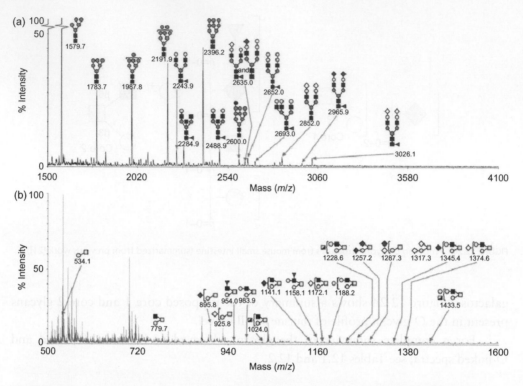

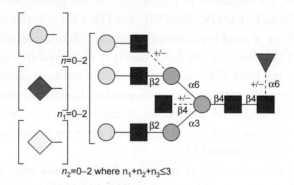

**FIGURE 12.25** Partial MALDI-TOF-MS profiles of the permethylated N-linked glycans (a) and O-linked glycans (b) from a typical sample of murine spleen. Major peaks are annotated with the relevant carbohydrate structure shown in symbol form, according to the glycan nomenclature adopted by the CFG (http://www.functionalglycomics.org/). For complete annotation of the spectra see Tables 12.1 and 12.2. All molecular ions are present in sodiated form ([M +Na]⁺).

**FIGURE 12.26** Proposed complex N-glycan structures from mouse spleen.

summary of the proposed glycan structures found in the N-glycan profile of murine spleen.

The O-glycan spectrum of murine spleen derived from a 129×1/SvJ wild-type animal is shown in Figure 12.25b. The core 1 subtype is readily sialylated with up to two NeuAc or NeuGc residues, while the core 2 glycans are modified with a wider range of capping groups, including fucose (thought to be in Lewis x form), sialic acid

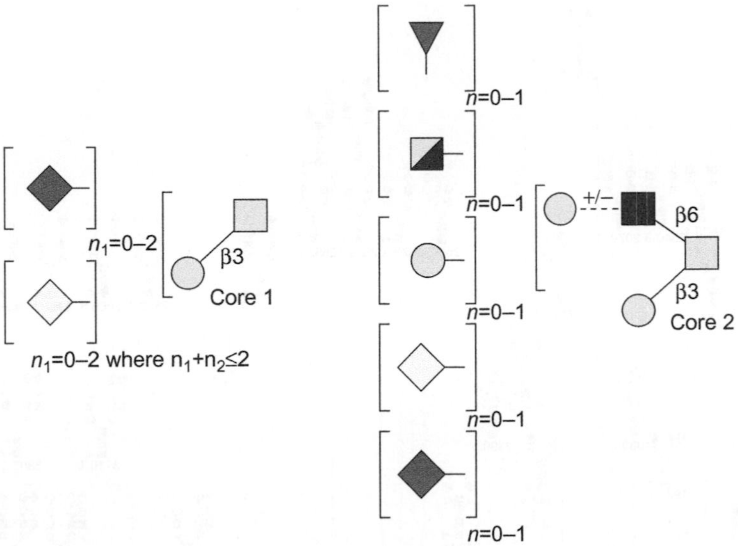

**FIGURE 12.27**   Proposed *O*-glycan structures from mouse spleen.

(either NeuAc or NeuGc) and galactose. Figure 12.27 shows a summary of the proposed core 1 and core 2 glycans present in the *O*-glycan profile of murine spleen.

For a complete summary of the glycans observed in both the *N*-linked and *O*-linked spectra, see Tables 12.1 and 12.2.

## Mouse Testes

Mass spectrometric glycomics screening was carried out on murine testes samples from a pair of age- and sex-matched C57BL/6 littermates. Data from one of these analyses are displayed in Figure 12.28. Three genetic knockouts were also analyzed (FucTIV+VII (double null), Gal 3, and ST3Gal-I). Raw and annotated spectra from these data are available online, together with the wild-type analyses presented here, from http://www.functionalglycomics.org/glycomics/publicdata/glycoprofiling. jsp. Earlier work was successful in analyzing the *O*-glycans of the murine testes [20,21,25], though it was carried out using FAB-MS [26], as opposed to the more modern MALDI instrumentation.

### Key Features

The *N*-linked glycan profile generated from a typical mouse testes sample is shown in Figure 12.28. High mannose and hybrid structures are present, along with bi-, tri-, and tetra-antennary complex type. These structures are seen to core fucosylate to a high degree. A reasonable level of bisecting GlcNAc is observed, with the most abundant complex structure belonging to a truncated biantennary, core-fucosylated, bisected glycan lacking $\beta$4-linked galactose (*m/z* 2081.3). Sialylation of the antennae is extensive ($n = 0$–3), with both NeuAc and NeuGc observed. Some galactose terminations are detected, though these are relatively minor in terms of relative abundance

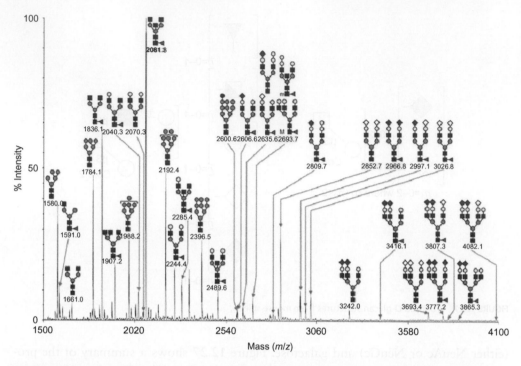

**FIGURE 12.28**   Partial MALDI-TOF-MS profiles of the permethylated N-linked glycans (a) and O-linked glycans (b) from a typical sample of murine testes. Major peaks are annotated with the relevant carbohydrate structure shown in symbol form, according to the glycan nomenclature adopted by the CFG (http://www. functionalglycomics.org/). Structures confirmed by further analyses have their masses shaded. Where MS/MS analysis has informed upon the relative abundance of glycoforms, major (M) and minor (m) components are indicated. For complete annotation of the spectra see Tables 12.1 and 12.2. All molecular ions are present in sodiated form ([M +Na]⁺).

compared to the sialic acid capped antennae. See Figure 12.29 for a summary of the proposed glycan structures found in the N-glycan profile of murine testes.

Core 1 and core 2 type O-glycans are both present in the O-linked profile of mouse testes [21]. The core 1 subtype is readily sialylated with up to two NeuAc or NeuGc residues, while the core 2 glycans are observed to be modified with fucose (thought to be in Lewis x form). Figure 12.30 shows a summary of the proposed core 1 and core 2 glycans present in the O-glycan profile of murine testes.

For a complete summary of the glycans observed in both the N-linked and O-linked spectra, see Tables 12.1 and 12.2.

## Mouse Thymus

Mass spectrometric glycomics screening was carried out on murine thymus samples from a pair of age- and sex-matched C57BL/6 littermates. Data from one of these analyses are displayed in Figure 12.31. Seven genetic knockouts were also analyzed (CT GalNAcT, core 2 GlcNAcT, FucTIV, FucTVII, FucTIV+VII (double null), Gal 3, and ST3Gal-I), as well as a wild-type animal from an alternative genetic background (129×1/SvJ). Raw and annotated spectra from these data are available online, together

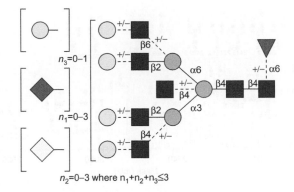

**FIGURE 12.29**   Proposed complex *N*-glycan structures from mouse testes.

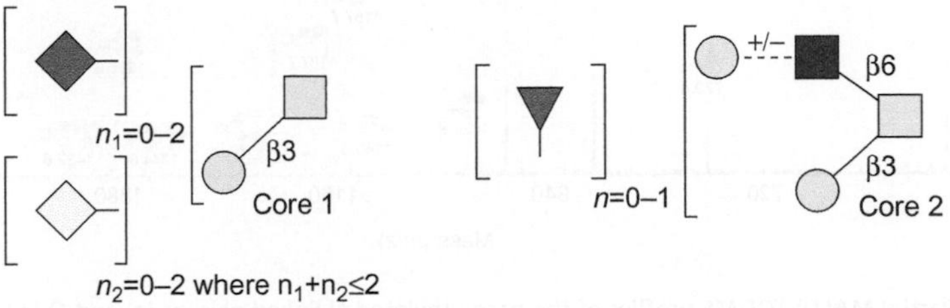

**FIGURE 12.30**   Proposed *O*-glycan structures from mouse testes (summarized from previous work [21]).

with the wild-type analyses presented here, from http://www.functionalglycomics. org/glycomics/publicdata/glycoprofiling.jsp. A discussion of these results in relation to the gene expression profiles is also available [24].

## Key Features

Figure 12.31a shows the *N*-linked glycosylation profile for a typical mouse thymus sample. Both high mannose and complex glycans are observed, with the latter being present in bi- and tri-antennary forms. There are no observable antennal fucosylation or bisecting GlcNAc residues present in the analysis. The most dominant feature is the high level of sialylation by both NeuAc and NeuGc, and up to three such moieties are present on the larger glycans. The relative levels of the two residues, determined by observation of two closely related structures (such as the two disialyl, bianten-nary, core-fucosylated *N*-glycans at *m/z* 2966.5 and 3026.5) indicates that NeuGc is noticeably more common than NeuAc. A small amount of α-galactose capping is also observed. See Figure 12.32 for a summary of the proposed glycan structures found in the *N*-glycan profile of murine thymus.

A typical *O*-glycan profile from murine thymus is shown in Figure 12.31b, demon-strating the presence of both core 1 and core 2 type *O*-linked glycans within the sample. The core 1 structures are capped with up to two sialic acid residues of the NeuAc type, while the core 2 glycans are modified with a wider range of decorations, including fucose (thought to be in Le^x form), NeuAc, galactose, and GalNAc. Figure 12.33 shows

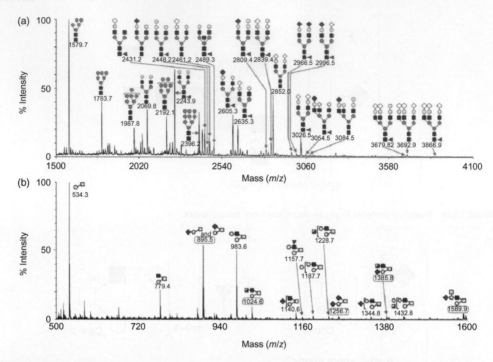

**FIGURE 12.31**  Partial MALDI-TOF-MS profiles of the permethylated *N*-linked glycans (a) and *O*-linked glycans (b) from a typical sample of murine thymus. Major peaks are annotated with the relevant carbohydrate structure shown in symbol form, according to the glycan nomenclature adopted by the CFG (http://www.functionalglycomics.org/). Structures confirmed by further analyses have their masses shaded. For complete annotation of the spectra see Tables 12.1 and 12.2. All molecular ions are present in sodiated form ([M +Na]⁺).

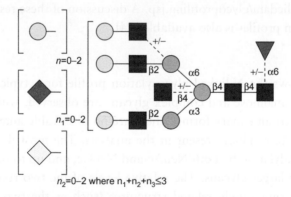

**FIGURE 12.32**  Proposed complex *N*-glycan structures from mouse thymus.

a summary of the proposed core 1 and core 2 glycans present in the *O*-glycan profile of murine thymus.

For a complete summary of the glycans observed in both the *N*-linked and *O*-linked spectra, see Tables 12.1 and 12.2.

## Summary of the Mouse Glycome

See Tables 12.1 and 12.2.

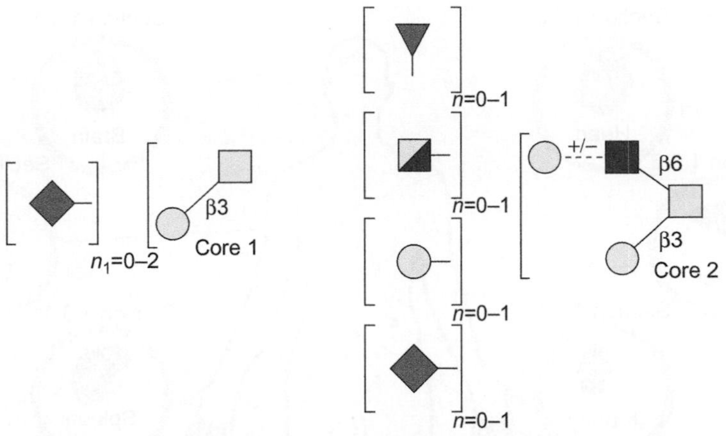

**FIGURE 12.33**   Proposed *O*-glycan structures from mouse thymus.

# Human Glycomics

There is growing recognition of the role glycosylation plays in a variety of human diseases including cancers, congenital muscular dystrophies, autoimmune diseases, and neurological disorders [33–37]. In a number of these cases, the evidence remains patchy or circumstantial. It is clear that detailed knowledge of the structure and nature of the glycans and receptors involved in such interactions could provide improved diagnostic and therapeutic approaches. To this end, the Analytical Glycotechnology Core-C of the CFG has focused considerable resources toward the whole-system analysis of the *N*- and *O*-glycomes of human tissues with the intention of laying a foundation of structural knowledge as a resource for the scientific community. In this section, we discuss the findings resulting from the analysis of 11 tissue types (Figure 12.34) from a total of 10 anonymous human donors. Raw data files and annotated spectra are available online at http://www.functionalglycomics.org/glycomics/publicdata/glycoprofiling.jsp.

## Human Brain

The CFG has not given high priority to screening human brain tissues because earlier work has provided compelling evidence for very close similarity between rodent and human brain *N*- and *O*-glycomes. For illustrative purposes data from one of these studies is described below. These data originate from a study into childhood ataxia with central nervous system hypomyelination (CACH) [38], and were obtained using FAB-MS, as opposed to the more modern MALDI instrumentation. Hence the less abundant and/or high molecular weight components are not observed. Despite this, the data capture the characteristic features of human brain glycosylation.

### *Key Features*

The *N*-linked glycan profile of the human brain is presented in Figure 12.35a. It consists of high mannose and complex type glycans, with the latter being of the biantennary type.

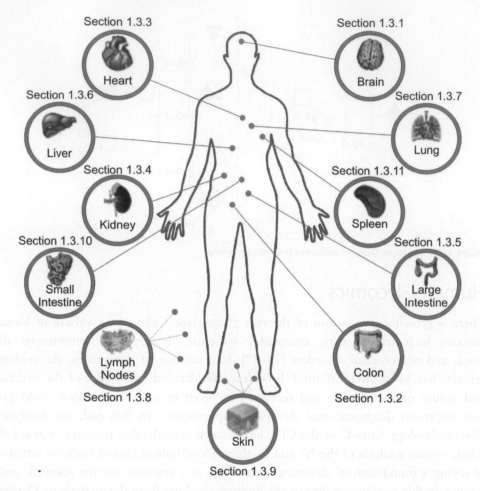

**FIGURE 12.34** The human body, illustrating the organs and tissues whose glycomes are covered within this chapter.

The technology used to produce the spectra limits the mass range, making the observation of larger glycoforms unlikely. Previous structural studies [21,25] have shown that the human and rodent brains exhibit similar glycosylation profiles. It is therefore reasonable to surmise that, as in the murine brain tissues examined earlier, the human brain will contain larger tri- and tetra-antennary glycans bearing the same structural features that would have been observed had the technology been available at the time of study.

The N-glycans exhibit large amounts of bisecting GlcNAc and antennal fucose. There is also a notably high proportion of highly truncated structures present in the profile. See Figure 12.36 for a summary of the proposed glycan structures found in the N-glycan profile of human brain.

The spectrum of human brain-derived O-linked glycans (Figure 12.35b) consists of both core 1 and core 2 type O-glycans, capped with sialic acid, as well as more unusually a signal consistent with an O-mannosyl structure. As described earlier in the mouse, the high abundance of this particular structure relative to other related mucin-type glycans within the brain is especially intriguing. Figure 12.37 shows a summary of the proposed core 1 and core 2 glycans present in the O-glycan profile of human brain.

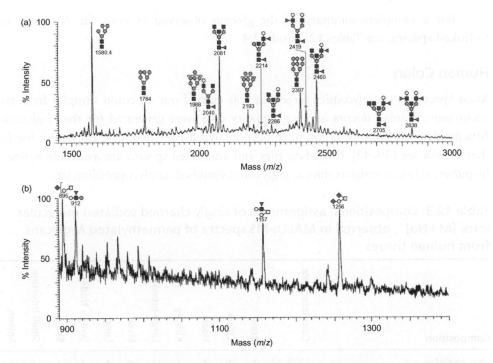

**FIGURE 12.35** Partial fast atom bombardment (FAB)-MS profiles of the permethylated *N*-linked glycans (a) and *O*-linked glycans (b) from a typical human brain sample. Major peaks are annotated with the relevant carbohydrate structure shown in symbol form, according to the glycan nomenclature adopted by the CFG (http://www.functionalglycomics.org/). For complete annotation of the spectra see Tables 12.3 and 12.4. All molecular ions are present in sodiated form ([M+Na]+).

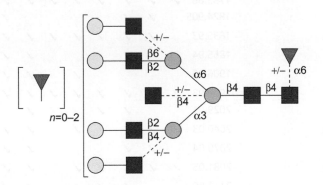

**FIGURE 12.36** Proposed complex *N*-glycan structures from human brain.

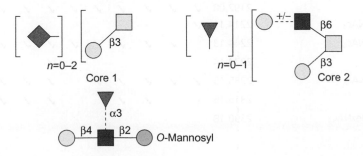

**FIGURE 12.37** Proposed *O*-glycan structures from human brain.

For a complete summary of the glycans observed in both the *N*-linked and *O*-linked spectra, see Tables 12.3 and 12.4.

## Human Colon

Mass spectrometric glycomics screening was carried out on colon samples from two anonymous human donors and high-quality data were generated for the *N*-glycome. Mucins derived from the colon have been analyzed elsewhere in great detail; for further details see [39–43]. Raw data files and annotated spectra are available online at http://www.functionalglycomics.org/glycomics/publicdata/glycoprofiling.jsp.

**Table 12.3** Compositional assignments of singly charged sodiated molecular ions, [M+Na]+, observed in MALDI-MS spectra of permethylated *N*-glycans from human tissues

| Composition | Calculated m/z | Brain | Colon | Heart | Kidney | Large Intestine | Liver | Lung | Lymph Nodes | Skin | Small Intestine | Spleen |
|---|---|---|---|---|---|---|---|---|---|---|---|---|
| Hex$_5$ HexNAc$_2$ | 1579.78 | ✓ | ✓ | ✓ | ✓ | ✓ | ✓ | ✓ | ✓ | ✓ | ✓ | ✓ |
| Fuc$_1$ Hex$_3$ HexNAc$_3$ | 1590.80 | | | | | | | ✓ | | | | |
| Hex$_4$ HexNAc$_3$ | 1620.81 | | | | | | | ✓ | | | | |
| Hex$_3$ HexNAc$_4$ | 1661.84 | | | | | | | | | | | |
| Hex$_6$ HexNAc$_2$ | 1783.88 | ✓ | ✓ | ✓ | ✓ | ✓ | ✓ | ✓ | ✓ | ✓ | ✓ | ✓ |
| Hex$_5$ HexNAc$_3$ | 1824.905 | | ✓ | ✓ | | | | | ✓ | ✓ | | |
| Fuc$_1$ Hex$_3$ HexNAc$_4$ | 1835.93 | | ✓ | ✓ | ✓ | ✓ | ✓ | ✓ | ✓ | ✓ | ✓ | ✓ |
| Hex$_4$ HexNAc$_4$ | 1865.94 | | ✓ | ✓ | | | | ✓ | ✓ | ✓ | | |
| Hex$_3$ HexNAc$_5$ | 1906.96 | | | | | ✓ | | ✓ | ✓ | ✓ | | ✓ | ✓ |
| Hex$_7$ HexNAc$_2$ | 1987.98 | ✓ | ✓ | ✓ | ✓ | ✓ | ✓ | ✓ | ✓ | ✓ | ✓ | ✓ |
| Hex$_6$ HexNAc$_3$ | 2029.01 | | | ✓ | ✓ | | | ✓ | | | | ✓ |
| Fuc$_1$ Hex$_4$ HexNAc$_4$ | 2040.03 | ✓ | ✓ | ✓ | ✓ | ✓ | | ✓ | ✓ | ✓ | ✓ | ✓ |
| Hex$_5$ HexNAc$_4$ | 2070.04 | ✓ | ✓ | | | | | ✓ | ✓ | ✓ | | ✓ |
| Fuc$_1$ Hex$_3$ HexNAc$_5$ | 2081.05 | ✓ | ✓ | ✓ | ✓ | ✓ | | ✓ | ✓ | ✓ | ✓ | ✓ |
| Hex$_4$ HexNAc$_5$ | 2111.06 | | | ✓ | ✓ | ✓ | | ✓ | ✓ | ✓ | ✓ | ✓ |
| NeuAc$_1$ Fuc$_1$ Hex$_4$ HexNAc$_3$ | 2156.07 | | | ✓ | ✓ | | | | | ✓ | | ✓ |
| Hex$_8$ HexNAc$_2$ | 2192.08 | ✓ | ✓ | ✓ | ✓ | ✓ | ✓ | ✓ | ✓ | ✓ | ✓ | ✓ |
| Fuc$_2$ Hex$_4$ HexNAc$_4$ | 2214.11 | ✓ | | | | | | | | | | |
| Fuc$_1$ Hex$_5$ HexNAc$_4$ | 2244.13 | | ✓ | ✓ | ✓ | ✓ | ✓ | ✓ | ✓ | ✓ | ✓ | ✓ |
| Hex$_6$ HexNAc$_4$ | 2274.14 | | | | | | | ✓ | | | | |
| Fuc$_1$ Hex$_4$ HexNAc$_5$ | 2285.15 | ✓ | ✓ | ✓ | ✓ | ✓ | | ✓ | ✓ | ✓ | ✓ | ✓ |
| Hex$_5$ HexNAc$_5$ | 2315.16 | | | ✓ | ✓ | | | ✓ | ✓ | ✓ | | ✓ |
| NeuAc$_1$ Hex$_6$ HexNAc$_3$ | 2390.18 | | | | | | | | | | | ✓ |

(Continued)

## Table 12.3 Continued

| Composition | Calculated m/z | Brain | Colon | Heart | Kidney | Large Intestine | Liver | Lung | Lymph Nodes | Skin | Small Intestine | Spleen |
|---|---|---|---|---|---|---|---|---|---|---|---|---|
| Hex$_9$ HexNAc$_2$ | 2396.18 | ✓ | ✓ | | ✓ | ✓ | ✓ | ✓ | | ✓ | ✓ | ✓ |
| NeuAc$_1$ Fuc$_1$ Hex$_4$ HexNAc$_4$ | 2401.20 | | | | | | | ✓ | ✓ | | | ✓ |
| Fuc$_2$ Hex$_5$ HexNAc$_4$ | 2418.21 | ✓ | | | ✓ | ✓ | | ✓ | | | | ✓ |
| NeuAc$_1$ Hex$_5$ HexNAc$_4$ | 2431.21 | | ✓ | | ✓ | ✓ | ✓ | ✓ | ✓ | ✓ | ✓ | ✓ |
| Fuc$_1$ Hex$_6$ HexNAc$_4$ | 2448.219 | ✓ | ✓ | | | | | ✓ | | | | ✓ |
| Fuc$_2$ Hex$_4$ HexNAc$_5$ | 2459.24 | ✓ | | | | | | | | | | |
| NeuAc$_1$ Hex$_4$ HexNAc$_5$ | 2472.24 | | | | | | | ✓ | ✓ | | | ✓ |
| Hex$_7$ HexNAc$_4$ | 2478.23 | | ✓ | | | | | | | | | |
| Fuc$_1$ Hex$_5$ HexNAc$_5$ | 2489.25 | | ✓ | | ✓ | ✓ | ✓ | ✓ | | ✓ | ✓ | ✓ |
| Hex$_6$ HexNAc$_5$ | 2519.26 | | ✓ | | ✓ | | | ✓ | | ✓ | | |
| NeuAc$_1$ Fuc$_1$ Hex$_5$ HexNAc$_4$ | 2605.30 | ✓ | ✓ | | ✓ | ✓ | ✓ | ✓ | ✓ | ✓ | ✓ | ✓ |
| Fuc$_2$ Hex$_6$ HexNAc$_4$ | 2622.31 | | | | | ✓ | | | | | ✓ | |
| NeuGc$_1$ Fuc$_1$ Hex$_5$ HexNAc$_4$ | 2635.31 | | | | | ✓ | | | | | ✓ | |
| NeuAc$_1$ Fuc$_1$ Hex$_4$ HexNAc$_5$ | 2646.33 | | | | | | | | | | | ✓ |
| Fuc$_1$ Hex$_7$ HexNAc$_4$ | 2652.319 | ✓ | ✓ | | | | | | | ✓ | | ✓ |
| Fuc$_2$ Hex$_5$ HexNAc$_5$ | 2663.34 | | | | | ✓ | ✓ | ✓ | | | ✓ | |
| NeuAc$_1$ Hex$_5$ HexNAc$_5$ | 2676.34 | | | | | | | ✓ | ✓ | | | ✓ |
| Fuc$_1$ Hex$_6$ HexNAc$_5$ | 2693.35 | | ✓ | | ✓ | ✓ | | ✓ | ✓ | ✓ | ✓ | |
| Fuc$_2$ Hex$_4$ HexNAc$_6$ | 2704.37 | ✓ | | | | | | | | | | |
| Fuc$_1$ Hex$_5$ HexNAc$_6$ | 2734.38 | | | | | ✓ | | | | ✓ | | ✓ |
| NeuAc$_1$ Fuc$_2$ Hex$_5$ HexNAc$_4$ | 2779.39 | | ✓ | | | ✓ | | | ✓ | | | |
| NeuAc$_2$ Hex$_5$ HexNAc$_4$ | 2792.38 | ✓ | ✓ | | ✓ | ✓ | ✓ | ✓ | ✓ | ✓ | ✓ | ✓ |
| Fuc$_3$ Hex$_5$ HexNAc$_5$ | 2837.43 | ✓ | | | | ✓ | | | | | ✓ | |
| NeuAc$_1$ Fuc$_1$ Hex$_5$ HexNAc$_5$ | 2850.43 | ✓ | ✓ | | ✓ | | | ✓ | ✓ | ✓ | | ✓ |
| Fuc$_2$ Hex$_6$ HexNAc$_5$ | 2867.44 | | | | | ✓ | | ✓ | | | | |
| NeuAc$_1$ Hex$_6$ HexNAc$_5$ | 2880.44 | | | | ✓ | ✓ | | ✓ | ✓ | ✓ | | |
| Fuc$_1$ Hex$_6$ HexNAc$_6$ | 2938.48 | | | | | ✓ | | ✓ | ✓ | ✓ | | |
| NeuAc$_2$ Fuc$_1$ Hex$_5$ HexNAc$_4$ | 2966.47 | ✓ | ✓ | | ✓ | ✓ | ✓ | ✓ | ✓ | ✓ | ✓ | ✓ |
| Hex$_6$ HexNAc$_7$ | 3009.51 | | | | | | | | | | | ✓ |
| Fuc$_4$ Hex$_5$ HexNAc$_5$ | 3011.52 | | | | | | | | | | ✓ | |
| NeuAc$_1$ Fuc$_1$ Hex$_7$ HexNAc$_4$ | 3013.491 | ✓ | | | | | | | | | | ✓ |
| NeuAc$_1$ Fuc$_2$ Hex$_5$ HexNAc$_5$ | 3024.51 | | | | | | | | | ✓ | | ✓ |
| Fuc$_3$ Hex$_6$ HexNAc$_5$ | 3041.53 | | | | ✓ | ✓ | | | | | ✓ | ✓ |
| NeuAc$_1$ Fuc$_1$ Hex$_6$ HexNAc$_5$ | 3054.52 | ✓ | ✓ | | ✓ | ✓ | ✓ | ✓ | ✓ | ✓ | ✓ | ✓ |
| Fuc$_1$ Hex$_8$ HexNAc$_5$ | 3101.55 | | | | | | | | | | | ✓ |

*(Continued)*

## Table 12.3 Continued

| Composition | Calculated m/z | Brain | Colon | Heart | Kidney | Large Intestine | Liver | Lung | Lymph Nodes | Skin | Small Intestine | Spleen |
|---|---|---|---|---|---|---|---|---|---|---|---|---|
| $Fuc_2 Hex_6 HexNAc_6$ | 3112.57 | | | | ✓ | | | | | | | |
| $Fuc_1 Hex_7 HexNAc_6$ | 3142.58 | | ✓ | ✓ | ✓ | | | ✓ | ✓ | ✓ | | ✓ |
| $Fuc_5 Hex_5 HexNAc_5$ | 3185.61 | | | | ✓ | | | | | | ✓ | |
| $NeuAc_2 Fuc_1 Hex_5 HexNAc_5$ | 3211.60 | ✓ | ✓ | ✓ | | | | ✓ | ✓ | | | ✓ |
| $NeuAc_1 Fuc_2 Hex_6 HexNAc_5$ | 3228.61 | | | | | | | ✓ | | | | |
| $NeuAc_2 Hex_6 HexNAc_5$ | 3241.61 | | | | ✓ | ✓ | ✓ | ✓ | ✓ | ✓ | | ✓ |
| $Fuc_4 Hex_5 HexNAc_6$ | 3256.65 | | | | ✓ | | | | | | | |
| $Fuc_3 Hex_6 HexNAc_6$ | 3286.66 | | | | ✓ | ✓ | | | | | | |
| $NeuAc_1 Fuc_1 Hex_6 HexNAc_6$ | 3299.65 | ✓ | ✓ | ✓ | | | | ✓ | ✓ | | | ✓ |
| $Fuc_2 Hex_7 HexNAc_6$ | 3316.67 | | | | ✓ | | | | | | | |
| $NeuAc_1 Hex_7 HexNAc_6$ | 3329.66 | | | ✓ | ✓ | | | ✓ | ✓ | | | ✓ |
| $Fuc_2 Hex_6 HexNAc_7$ | 3357.69 | | | | | | | | | | | ✓ |
| $Fuc_1 Hex_7 HexNAc_7$ | 3387.70 | | | | | | | | | ✓ | | |
| $NeuAc_1 Fuc_3 Hex_6 HexNAc_5$ | 3402.70 | | | | ✓ | | | ✓ | | | | |
| $NeuAc_2 Fuc_1 Hex_6 HexNAc_5$ | 3415.70 | ✓ | ✓ | ✓ | ✓ | ✓ | ✓ | ✓ | ✓ | ✓ | ✓ | ✓ |
| $Fuc_4 Hex_6 HexNAc_6$ | 3460.75 | | | | ✓ | ✓ | | | | | ✓ | |
| $NeuAc_1 Fuc_2 Hex_6 HexNAc_6$ | 3473.74 | | | ✓ | ✓ | | | | | | | |
| $Fuc_3Hex_7HexNAc_6$ | 3490.75 | | | | ✓ | | | | | | | |
| $NeuAc_1 Fuc_1 Hex_7 HexNAc_6$ | 3503.75 | ✓ | ✓ | ✓ | ✓ | ✓ | ✓ | ✓ | ✓ | ✓ | ✓ | ✓ |
| $NeuAc_1 Fuc_1 Hex_6 HexNAc_7$ | 3544.78 | | | | | | | | | | | ✓ |
| $Fuc_2 Hex_7 HexNAc_7$ | 3561.79 | | | | ✓ | | | | | | | |
| $NeuAc_2 Fuc_2 Hex_6 HexNAc_5$ | 3589.79 | | ✓ | | | | | ✓ | | | | ✓ |
| $Fuc_1 Hex_8 HexNAc_7$ | 3591.80 | | | | | | | | | | | |
| $NeuAc_3 Hex_6 HexNAc_5$ | 3602.78 | ✓ | ✓ | | | ✓ | ✓ | ✓ | ✓ | ✓ | | ✓ |
| $Fuc_5 Hex_6 HexNAc_6$ | 3634.83 | | | | | ✓ | | | | | ✓ | |
| $NeuAc_1 Fuc_3 Hex_6 HexNAc_6$ | 3647.83 | | | ✓ | | | | | | | | |
| $NeuAc_2 Fuc_1 Hex_6 HexNAc_6$ | 3660.82 | | | | | | | ✓ | | ✓ | | ✓ |
| $Fuc_4 Hex_7 HexNAc_6$ | 3664.84 | | | ✓ | | | | | | | | |
| $NeuAc_1 Fuc_2 Hex_7 HexNAc_6$ | 3677.84 | | | ✓ | ✓ | | | ✓ | | | | ✓ |
| $NeuAc_2 Hex_7 HexNAc_6$ | 3690.84 | | | ✓ | ✓ | | | | | ✓ | | ✓ |
| $NeuAc_1 Fuc_2 Hex_6 HexNAc_7$ | 3718.87 | | | | | | | | | | | ✓ |
| $Fuc_3 Hex_7 HexNAc_7$ | 3735.88 | | | ✓ | | | | | | | | |
| $NeuAc_1 Fuc_1 Hex_7 HexNAc_7$ | 3748.88 | | | ✓ | ✓ | | | | | ✓ | | ✓ |
| $NeuAc_3 Fuc_1 Hex_6 HexNAc_5$ | 3776.87 | ✓ | ✓ | ✓ | ✓ | | | ✓ | ✓ | ✓ | ✓ | ✓ |

*(Continued)*

## Table 12.3 Continued

| Composition | Calculated *m/z* | Brain | Colon | Heart | Kidney | Large Intestine | Liver | Lung | Lymph Nodes | Skin | Small Intestine | Spleen |
|---|---|---|---|---|---|---|---|---|---|---|---|---|
| NeuAc$_1$ Hex$_8$ HexNAc$_7$ | 3778.89 | | | | | | ✓ | | | | | |
| Fuc$_6$ Hex$_6$ HexNAc$_6$ | 3808.92 | | | | | ✓ | | | | | ✓ | |
| NeuAc$_1$ Fuc$_4$ Hex$_6$ HexNAc$_6$ | 3821.92 | | | | ✓ | | | | | | | |
| NeuAc$_2$ Fuc$_2$ Hex$_6$ HexNAc$_6$ | 3834.91 | | | | | | | | | | | ✓ |
| Fuc$_5$ Hex$_7$ HexNAc$_6$ | 3838.93 | | | | | ✓ | | | | | | |
| NeuAc$_1$ Fuc$_3$ Hex$_7$ HexNAc$_6$ | 3851.93 | | | | | ✓ | | | | | | |
| NeuAc$_2$ Fuc$_1$ Hex$_7$ HexNAc$_6$ | 3864.92 | ✓ | ✓ | ✓ | | | ✓ | | | ✓ | ✓ | ✓ |
| Hex$_9$ HexNAc$_8$ | 3866.94 | | | | | | ✓ | | ✓ | | | |
| NeuAc$_1$ Fuc$_3$ Hex$_6$ HexNAc$_7$ | 3892.96 | | | | | | | | | | | ✓ |
| Fuc$_4$ Hex$_7$ HexNAc$_7$ | 3909.97 | | | | | ✓ | | | | | ✓ | |
| NeuAc$_1$ Fuc$_2$ Hex$_7$ HexNAc$_7$ | 3922.97 | | | | | ✓ | | | | | | |
| NeuAc$_3$ Fuc$_2$ Hex$_6$ HexNAc$_5$ | 3950.96 | | | | | ✓ | | | ✓ | | | ✓ |
| NeuAc$_1$ Fuc$_1$ Hex$_8$ HexNAc$_7$ | 3952.98 | ✓ | | | | | ✓ | ✓ | | ✓ | | |
| Fuc$_7$ Hex$_6$ HexNAc$_6$ | 3983.01 | | | | | ✓ | | | | | ✓ | |
| NeuAc$_2$ Fuc$_3$ Hex$_6$ HexNAc$_6$ | 4009.00 | | | | | ✓ | | | | | | |
| NeuAc$_1$ Fuc$_4$ Hex$_7$ HexNAc$_6$ | 4026.02 | | | | | ✓ | | | | | | ✓ |
| NeuAc$_2$Fuc$_2$Hex$_7$HexNAc$_6$ | 4039.01 | | | ✓ | ✓ | | | | | ✓ | | ✓ |
| Fuc$_1$ Hex$_9$ HexNAc$_8$ | 4041.03 | | | | | | | ✓ | | | | |
| NeuAc$_3$ Hex$_7$ HexNAc$_6$ | 4052.01 | | | | | | | | | ✓ | | ✓ |
| Fuc$_6$ Hex$_6$ HexNAc$_7$ | 4054.05 | | | | | ✓ | | | | | ✓ | |
| Fuc$_5$ Hex$_7$ HexNAc$_7$ | 4084.06 | | | | | ✓ | ✓ | | | | ✓ | |
| NeuAc$_2$ Fuc$_3$ Hex$_7$ HexNAc$_7$ | 4097.06 | | | | | ✓ | | | | | | |
| NeuAc$_2$ Fuc$_1$ Hex$_7$ HexNAc$_7$ | 4110.05 | | | | | | | | | ✓ | | ✓ |
| NeuAc$_3$ Fuc$_3$ Hex$_6$ HexNAc$_5$ | 4125.05 | ✓ | | | | | | | | | | ✓ |
| NeuAc$_2$ Hex$_8$ HexNAc$_7$ | 4140.06 | | | ✓ | | | | | | | | |
| NeuAc$_2$ Fuc$_4$ Hex$_6$ HexNAc$_6$ | 4183.09 | | | | | ✓ | | | | | | |
| NeuAc$_3$ Fuc$_1$ Hex$_7$ HexNAc$_6$ | 4226.10 | | ✓ | | | ✓ | ✓ | ✓ | ✓ | ✓ | ✓ | ✓ |
| Fuc$_6$ Hex$_7$ HexNAc$_7$ | 4258.15 | | | | | ✓ | | | | | ✓ | |
| NeuAc$_2$ Fuc$_2$ Hex$_7$ HexNAc$_7$ | 4284.14 | | | | | | | | | | | ✓ |
| NeuAc$_2$ Fuc$_2$ Hex$_8$ HexNAc$_7$ | 4314.15 | ✓ | ✓ | | | | ✓ | ✓ | ✓ | ✓ | | ✓ |
| NeuAc$_3$ Fuc$_2$ Hex$_7$ HexNAc$_6$ | 4400.19 | | | | | | | | | ✓ | | ✓ |
| NeuAc$_1$ Fuc$_1$ Hex$_9$ HexNAc$_8$ | 4402.20 | | | | | | ✓ | ✓ | | | | |
| NeuAc$_4$ Hex$_7$ HexNAc$_6$ | 4413.18 | | | | | | | | | ✓ | | ✓ |
| Fuc$_7$ Hex$_7$ HexNAc$_7$ | 4432.24 | | | | | ✓ | | | | | ✓ | |

*(Continued)*

## Table 12.3 Continued

| Composition | Calculated $m/z$ | Brain | Colon | Heart | Kidney | Large Intestine | Liver | Lung | Lymph Nodes | Skin | Small Intestine | Spleen |
|---|---|---|---|---|---|---|---|---|---|---|---|---|
| $NeuAc_2 Fuc_2 Hex_8 HexNAc_7$ | 4488.24 | | | | | | | | | ✓ | | |
| $NeuAc_3 Fuc_3 Hex_7 HexNAc_6$ | 4574.28 | | | | | | | | | | | |
| $NeuAc_4 Fuc_1 Hex_7 HexNAc_6$ | 4587.27 | | | ✓ | | ✓ | ✓ | ✓ | ✓ | | ✓ | ✓ |
| $Fuc_8 Hex_7 HexNAc_7$ | 4606.33 | | | | | ✓ | | | | ✓ | | |
| $NeuAc_3 Fuc_1 Hex_8 HexNAc_7$ | 4675.32 | | | ✓ | | | | ✓ | | | | |
| $NeuAc_2 Fuc_3 Hex_7 HexNAc_8$ | 4703.36 | | | | | | | | | | | |
| $Fuc_6 Hex_8 HexNAc_8$ | 4707.38 | | | | | ✓ | | | | | | |
| $NeuAc_2 Fuc_1 Hex_9 HexNAc_8$ | 4763.38 | ✓ | | | | | ✓ | ✓ | | ✓ | | ✓ |
| $Fuc_9 Hex_7 HexNAc_7$ | 4780.42 | | | | | ✓ | | | | | ✓ | |
| $NeuAc_3 Fuc_2 Hex_8 HexNAc_7$ | 4849.41 | | | | | | | | | ✓ | | ✓ |
| $NeuAc_1 Fuc_1 Hex_{10} HexNAc_9$ | 4851.43 | | | | | | | ✓ | | | | |
| $Fuc_7 Hex_8 HexNAc_8$ | 4881.46 | | | | | ✓ | | | | | ✓ | |
| $NeuAc_3 Fuc_5 Hex_7 HexNAc_6$ | 4922.46 | | | | | | | | | | | ✓ |
| $NeuAc_4 Fuc_3 Hex_7 HexNAc_6$ | 4935.45 | | | | | | | | | | | ✓ |
| $NeuAc_2 Fuc_2 Hex_9 HexNAc_8$ | 4937.47 | | | | | | | | | ✓ | | |
| $NeuAc_4 Fuc_1 Hex_8 HexNAc_7$ | 5036.50 | | | | | | | | | | | ✓ |
| $NeuAc_3 Fuc_1 Hex_9 HexNAc_8$ | 5124.55 | | | | | | | | | | | ✓ |
| $NeuAc_2 Fuc_1 Hex_{10} HexNAc_9$ | 5212.60 | | | | | | | | | | | ✓ |
| $NeuAc_3 Fuc_2 Hex_9 HexNAc_8$ | 5298.64 | | | | | | | | | | | ✓ |
| $NeuAc_1 Fuc_4 Hex_{10} HexNAc_9$ | 5373.70 | | | | | | | | | | | ✓ |
| $NeuAc_4 Fuc_3 Hex_8 HexNAc_7$ | 5384.68 | | | | | | | | | | | ✓ |
| $NeuAc_4 Fuc_1 Hex_9 HexNAc_8$ | 5485.72 | | | | | | | | | | | ✓ |
| $NeuAc_3 Fuc_1 Hex_{10} HexNAc_9$ | 5573.78 | | | | | | | | | | | ✓ |
| $NeuAc_4 Fuc_2 Hex_9 HexNAc_8$ | 5659.81 | | | | | | | | | | | ✓ |
| $NeuAc_3 Fuc_2 Hex_{10} HexNAc_9$ | 5747.87 | | | | | | | | | | | ✓ |
| $NeuAc_4 Fuc_3 Hex_9 HexNAc_8$ | 5833.90 | | | | | | | | | | | ✓ |
| $NeuAc_4 Fuc_1 Hex_{10} HexNAc_9$ | 5934.95 | | | | | | | | | | | ✓ |
| $NeuAc_3 Fuc_1 Hex_{11} HexNAc_{10}$ | 6023.00 | | | | | | | | | | | ✓ |
| $NeuAc_4 Fuc_2 Hex_{10} HexNAc_9$ | 6109.04 | | | | | | | | | | | ✓ |
| $NeuAc_3 Fuc_2 Hex_{11} HexNAc_{10}$ | 6197.09 | | | | | | | | | | | ✓ |
| $NeuAc_4 Fuc_1 Hex_{11} HexNAc_{10}$ | 6384.18 | | | | | | | | | | | ✓ |
| $NeuAc_3 Fuc_1 Hex_{12} HexNAc_{11}$ | 6472.23 | | | | | | | | | | | ✓ |
| $NeuAc_4 Fuc_2 Hex_{11} HexNAc_{10}$ | 6558.27 | | | | | | | | | | | ✓ |
| $NeuAc_3 Fuc_2 Hex_{12} HexNAc_{11}$ | 6646.32 | | | | | | | | | | | ✓ |

Table 12.4 Compositional assignments of singly charged sodiated molecular ions, [M+Na]$^+$, observed in MALDI-MS spectra of permethylated *O*-glycans from human tissues

| Composition | Calculated m/z | Brain | Colon | Heart | Kidney | Large Intestine | Liver | Lung | Lymph Nodes | Skin | Spleen |
|---|---|---|---|---|---|---|---|---|---|---|---|
| Hex$_1$HexNAc$_1$-itol | 534.29 | | | ✓ | ✓ | ✓ | | ✓ | ✓ | ✓ | ✓ |
| Fuc$_1$ Hex$_1$ HexNAc$_1$-itol | 708.38 | | | | | ✓ | | ✓ | | | |
| Hex$_2$ HexNAc$_1$-itol | 738.39 | | | | | ✓ | | | | | |
| Hex$_1$HexNAc$_2$-itol | 779.41 | | | ✓ | ✓ | ✓ | ✓ | ✓ | | ✓ | ✓ |
| NeuAc$_1$Hex$_1$HexNAc$_1$-itol | 895.46 | ✓ | | ✓ | ✓ | ✓ | ✓ | ✓ | ✓ | ✓ | ✓ |
| Fuc$_1$ Hex$_2$ HexNAc$_1$-itol | 912.48 | ✓ | | | | | | | | | |
| Fuc$_1$ Hex$_1$ HexNAc$_2$-itol | 953.50 | | | | | | | | | | |
| Hex$_2$HexNAc$_2$-itol | 983.51 | | | ✓ | ✓ | ✓ | | ✓ | | ✓ | ✓ |
| NeuAc$_1$ Fuc$_1$ Hex$_1$ HexNAc$_1$-itol | 1069.55 | | | | | ✓ | | | | | |
| Fuc$_1$Hex$_2$HexNAc$_2$-itol | 1157.60 | ✓ | | | | ✓ | | | ✓ | | ✓ |
| Fuc$_1$ Hex$_1$ HexNAc$_3$-itol | 1198.63 | | | | | ✓ | | | | | |
| NeuAc$_2$Hex$_1$HexNAc$_1$-itol | 1256.63 | | | ✓ | ✓ | ✓ | ✓ | ✓ | | | |
| Fuc$_2$ Hex$_2$ HexNAc$_2$-itol | 1331.69 | | | | | ✓ | | | ✓ | | |
| NeuAc$_1$Hex$_2$HexNAc$_2$-itol | 1344.69 | | | ✓ | | | | ✓ | ✓ | ✓ | ✓ |
| Fuc$_1$ Hex$_2$ HexNAc$_3$-itol | 1402.73 | | | | | ✓ | | | | ✓ | |
| Fuc$_3$ Hex$_2$ HexNAc$_2$-itol | 1505.78 | | | | | ✓ | | | | | |
| NeuAc$_1$Fuc$_1$Hex$_2$HexNAc$_2$-itol | 1518.78 | | | | | | | ✓ | | | |
| Fuc$_2$ Hex$_3$ HexNAc$_2$-itol | 1535.79 | | | | | ✓ | | | | | |
| Fuc$_2$ Hex$_2$ HexNAc$_3$-itol | 1576.82 | | | | | ✓ | | | | | |
| NeuAc$_3$Hex$_1$HexNAc$_1$-itol | 1617.81 | | | ✓ | ✓ | | ✓ | ✓ | | | ✓ |
| Hex$_3$HexNAc$_4$-itol | 1677.87 | | | | | | | ✓ | | | |
| NeuAc$_2$Hex$_2$HexNAc$_2$-itol | 1705.86 | | | ✓ | ✓ | ✓ | ✓ | ✓ | | | |
| Fuc$_2$ Hex$_3$ HexNAc$_3$-itol | 1780.92 | | | | | ✓ | | | | | |
| NeuAc$_2$Fuc$_1$Hex$_2$HexNAc$_2$-itol | 1879.95 | | | | | | | ✓ | | | |
| Fuc$_3$ Hex$_3$ HexNAc$_3$-itol | 1955.01 | | | | | ✓ | | | | | |
| Fuc$_2$ Hex$_3$ HexNAc$_4$-itol | 2026.05 | | | | | ✓ | | | | | |
| NeuAc$_2$Fuc$_1$Hex$_3$HexNAc$_2$-itol | 2084.05 | | | | | | | ✓ | | | |
| Fuc$_4$ Hex$_3$ HexNAc$_3$-itol | 2129.10 | | | | | ✓ | | | | | |
| Fuc$_3$ Hex$_3$ HexNAc$_4$-itol | 2200.14 | | | | | ✓ | | | | | |
| NeuAc$_2$Fuc$_1$Hex$_4$HexNAc$_2$-itol | 2288.15 | | | | | | | ✓ | | | |

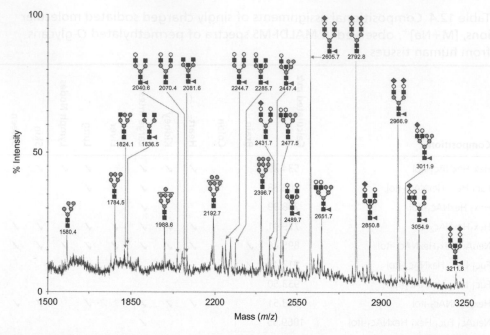

**FIGURE 12.38** Partial MALDI-TOF-MS profile of the permethylated *N*-linked glycans from a typical human colon sample. Major peaks are annotated with the relevant carbohydrate structure shown in symbol form, according to the glycan nomenclature adopted by the CFG (http://www.functionalglycomics.org/). For complete annotation of the spectrum see Tables 12.3 and 12.4. All molecular ions are present in sodiated form ([M+Na]$^+$).

## Key Features

The human colon N-linked glycan profile (Figure 12.38) contains high mannose, hybrid and bi-, tri-, and tetra-antennary complex type glycans. The hybrid structures are observed at a low abundance, though they are more prevalent than in many other human samples. Low levels of bisecting structures are present, confirmed by low levels of 3,4,6-linked mannose within GC-MS linkage studies of the colon samples (data not shown). There is evidence of some antennal fucosylation, with up to two of the deoxy-hexose structures decorating the antennae of the complex N-glycans. It is possible that they are present in sialyl Le$^x$ (sLe$^x$) form, though MS/MS confirmation of this was not obtained so this should be borne in mind. The most abundant glycan signals in the spectrum belong to the biantennary complex glycans, mono- and disialylated and with/without core fucose (*m/z* 2605.7, 2792.8, and 2966.9). See Figure 12.39 for a summary of the proposed glycan structures found in the N-glycan profile of human liver.

For a complete summary of the glycans observed in the N-linked spectra, see Tables 12.3 and 12.4.

## Human Heart

Mass spectrometric glycomics screening was carried out on heart samples from eight anonymous human donors for *N*- and *O*-linked glycan populations. Raw data files and annotated spectra are available online at http://www.functionalglycomics.org/glycomics/publicdata/glycoprofiling.jsp.

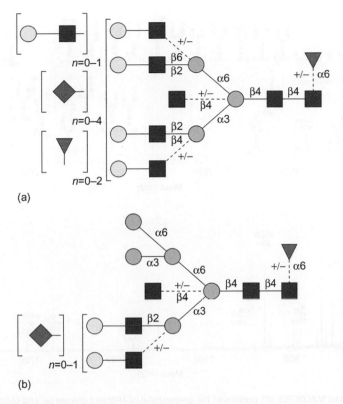

**FIGURE 12.39**  Proposed complex (a) and hybrid (b) *N*-glycan structures from human colon.

## Key Features

The human heart *N*-linked glycan profile (Figure 12.40a) consists of high mannose, bi-, tri-, and tetra-antennary complex and limited hybrid glycans. An atypically large number of truncated structures are also observed in the low mass region. Core fucosylation is present, though it is not as prevalent as in other tissues, with a number of relatively abundant non-fucosylated complex *N*-glycans being found within the tissue. Antennal fucosylation is limited, with some very low abundance peaks accounting for singly fucosylated antennae. Similarly, there are compositions present that are consistent with bisecting structures, but at very low abundance, an assertion which is confirmed by low levels of 3,4,6-linked mannose within GC-MS linkage studies of the heart (data not shown). Sialylation, however, is widespread, with most structures being capped by at least one, and up to four NeuAc residues. Biantennary glycans account for the most intense signals in the spectra, differentially core fucosylated and capped with one or two sialic acids (*m/z* 2605.9, 2792.9, and 2967.0). See Figure 12.41 for a summary of the proposed glycan structures found in the *N*-glycan profile of human heart.

Figure 12.40b shows a typical *O*-glycan spectrum of the human heart. Core 1 and core 2 type glycans are observed, with sialylation on each. The most abundant glycans are indicative of this, with very intense signals corresponding to mono- and disialyl core 1 structures (*m/z* 895.7 and 1257.0, respectively). Figure 12.42 shows a summary of the proposed core 1 and core 2 glycans present in the *O*-glycan profile of human heart.

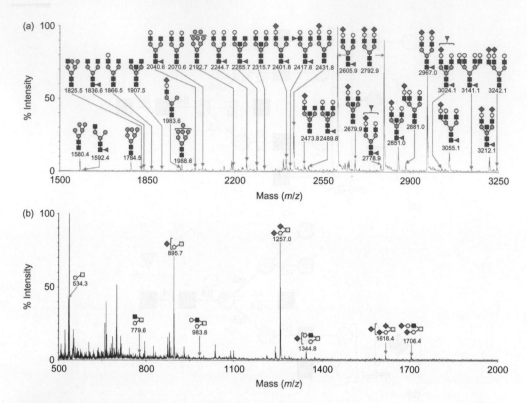

**FIGURE 12.40** Partial MALDI-TOF-MS profiles of the permethylated *N*-linked glycans (a) and *O*-linked glycans (b) from a typical human heart sample. Major peaks are annotated with the relevant carbohydrate structure shown in symbol form, according to the glycan nomenclature adopted by the CFG (http://www.functionalglycomics.org/). For complete annotation of the spectra see Tables 12.3 and 12.4. All molecular ions are present in sodiated form ([M+Na]+).

For a complete summary of the glycans observed in both the *N*-linked and *O*-linked spectra, see Tables 12.3 and 12.4.

## Human Kidney

Mass spectrometric glycomics screening was carried out on kidney biopsies from four anonymous human donors for *N*- and *O*-linked glycan populations. Raw data files and annotated spectra are available online at http://www.functionalglycomics.org/glycomics/publicdata/glycoprofiling.jsp.

### Key Features

The *N*-linked glycan profile of the human kidney (Figure 12.43a) contains high mannose glycans and bi-, tri-, and tetra-antennary complex structures. The most notable structural feature observed is the extensive addition of a bisecting GlcNAc residue β4 to the core mannose, a feature which is further confirmed by GC-MS data showing high levels of 3,4,6-linked mannose (data not shown). Concerning the antennae, there is evidence for limited LacNAc extensions in the higher mass glycans. They are also heavily fucosylated, with up to four fucose residues attached in the larger structures. Finally,

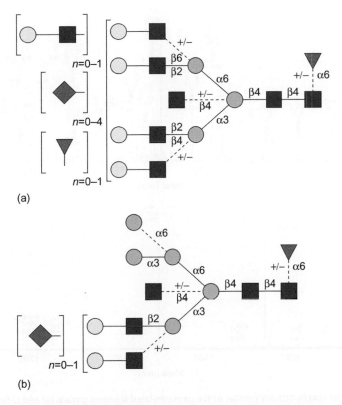

(a)

(b)

**FIGURE 12.41** Proposed complex (a) and hybrid (b) *N*-glycan structures from human heart.

Core 1                    Core 2

**FIGURE 12.42** Proposed *O*-glycan structures from human heart.

sialylation is prevalent, with up to four NeuAc residues capping the antennae. The most intense signals observed belong to a biantennary, non-core fucosylated, disialylated structure at *m/z* 2791.4 and a core-fucosylated, biantennary glycan with a bisecting GlcNAc at *m/z* 2488.5. Though there are one or two observed structures whose compositions would suggest the presence of sialyl Le$^x$, these are present in very low abundance, relative to the structures where the sialylation occurs on non-fucosylated arms. This is in keeping with the general observation that human cells and tissues prefer to sialylate non-fucosylated antennae where possible [9]. See Figure 12.44 for a summary of the proposed glycan structures found in the *N*-glycan profile of human kidney.

The *O*-glycan profile of the human kidney (Figure 12.43b) contains both core 1 and core 2 type *O*-linked glycans, with sialylation being observed on both core archetypes. By far the most abundant structures are the monosialylated (*m/z* 895.7)

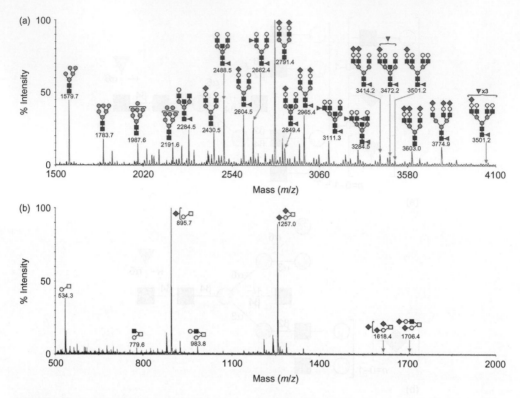

**FIGURE 12.43** Partial MALDI-TOF-MS profiles of the permethylated *N*-linked glycans (a) and *O*-linked glycans (b) from a typical human kidney sample. Major peaks are annotated with the relevant carbohydrate structure shown in symbol form, according to the glycan nomenclature adopted by the CFG (http://www.functionalglycomics.org/). For complete annotation of the spectra see Tables 12.3 and 12.4. All molecular ions are present in sodiated form ([M+Na]⁺).

and disialylated (*m/z* 1257.0) core 1 species. Figure 12.45 shows a summary of the proposed core 1 and core 2 glycans present in the *O*-glycan profile of human kidney.

For a complete summary of the glycans observed in both the *N*-linked and *O*-linked spectra, see Tables 12.3 and 12.4.

## Human Large Bowel

Mass spectrometric glycomics screening was carried out on samples of large bowel tissue from five anonymous human donors for *N*- and *O*-linked glycan populations. Raw data files and annotated spectra are available online at http://www.functionalglycomics.org/glycomics/publicdata/glycoprofiling.jsp.

### *Key Features*

The samples of human large intestine possess *N*-glycan profiles comprising high mannose and complex type oligosaccharides (Figure 12.46a). Bi-, tri-, and tetra-antennary structures are present, with a high degree of both core fucosylation and bisecting GlcNAc moieties, the former confirmed through MS/MS studies and the latter being verified by the identification of a relatively high level of 3,4,6-linked mannose within

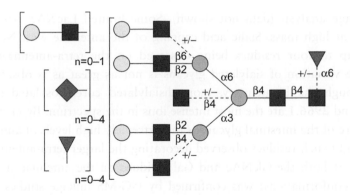

**FIGURE 12.44** Proposed complex *N*-glycan structures from human kidney.

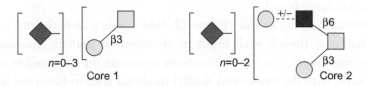

**FIGURE 12.45** Proposed *O*-glycan structures from human kidney.

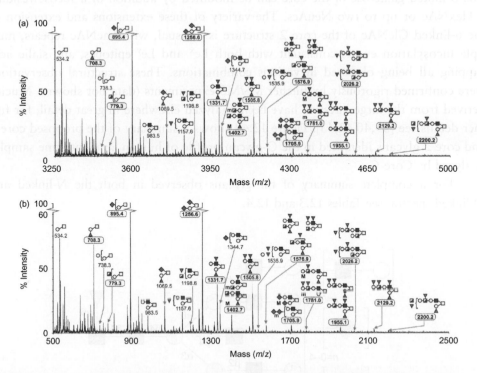

**FIGURE 12.46** Partial MALDI-TOF-MS profiles of the permethylated *N*-linked glycans (a) and *O*-linked glycans (b) from a typical human large bowel sample. Major peaks are annotated with the relevant carbohydrate structure shown in symbol form, according to the glycan nomenclature adopted by the CFG (http://www.functionalglycomics. org/). Structures confirmed by further analyses have their masses shaded. Where MS/MS analysis has informed upon the relative abundance of glycoforms, major (M) and minor (m) components are indicated. For complete annotation of the spectra see Tables 12.3 and 12.4. All molecular ions are present in sodiated form ([M+Na]$^+$).

GC-MS linkage analysis (data not shown). Some limited LacNAc extension of the arms is seen at high mass. Sialic acid capping of the antennae with NeuAc is common, with up to four residues being observed on the tetra-antennary structures. However, the variation of sialylated glycans is not as great as is observed in other tissues, although the biantennary, mono/disialylated corefucosylated structures at $m/z$ 2605.0 and 2966.1 are the most intense ions in the spectrum. By far the most distinctive feature of the intestinal glycans is the extremely high levels of antennal fucose, with up to eight such residues observed decorating the larger tetra-antennary glycans. Fucosylation of both the GlcNAc and Gal residues on the antennae (most likely in Le$^x$ and Le$^y$ conformations) was confirmed by GC-MS linkage studies and MS/MS analyses of smaller related structures at $m/z$ 2837.1 and 3185.2 (data not shown). See Figure 12.47 for a summary of the proposed glycan structures found in the N-glycan profile of human large bowel.

The O-linked glycan profile (Figure 12.46b) consists of both core 1 and core 2 type carbohydrates, though with much more variation than is commonly seen in other organs and tissues. The core 1 structures dominate the spectrum in terms of relative abundance, with the mono- and disialyl modified glycans being the largest peaks present. The core 2 structures, though present in smaller quantity, are very diverse. The 3-linked galactose of the core can be modified by addition of a fucose residue, a HexNAc or up to two NeuAcs. The variety of these extensions and extension of the 6-linked GlcNAc of the core 2 structure is unusual, with LacNAc repeats, multiple fucosylation events consistent with both Le$^x$ and Le$^y$ epitopes, and sialic acid capping all being observed in various combinations. These structural observations were confirmed rigorously by further MS/MS experiments (data not shown). Mucins derived from the large intestine have been analyzed elsewhere in great detail; for further details see [39,40,42,43]. Figure 12.48 shows a summary of the proposed core 1 and core 2 glycans identified in the O-glycan profile of human large intestine samples analyzed by Core-C.

For a complete summary of the glycans observed in both the N-linked and O-linked spectra, see Tables 12.3 and 12.4.

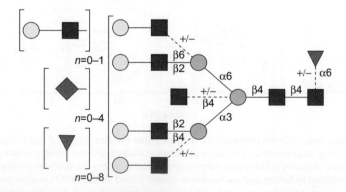

**FIGURE 12.47** Proposed complex N-glycan structures from human large bowel.

## Human Liver

Mass spectrometric glycomics screening was carried out on liver biopsies from four anonymous human donors for *N*- and *O*-linked glycan populations. Raw data files and annotated spectra are available online at http://www.functionalglycomics.org/glycomics/publicdata/glycoprofiling.jsp.

### *Key Features*

The liver samples exhibited very similar structural profiles, with the exception of one donor, whose *N*-glycan profile exhibited structures previously linked with hepatocellular carcinoma and cirrhosis [44,45]. This contrasting example will be discussed in greater detail in the section on Glycome Comparisons below.

The human liver *N*-glycan profile (Figure 12.49a) contains high mannose and bi-, tri-, and tetra-antennary glycans. The complex glycans predominantly contain fucose on their cores, with no evidence (GC-MS or MS/MS) being observed for the presence of antennal fucose or bisecting GlcNAc. Extensive sialylation by NeuAc is observed, with the structures being capped with up to four sialic acid residues. Structures consistent with polyLacNAc extensions were also observed. The most prevalent complex structures in the spectrum are variably core fucosylated biantennary glycans bearing one or two NeuAc terminal groups (*m/z* 2792.5, 2605.4, and 2966.6). See Figure 12.50 for a summary of the proposed glycan structures found in the *N*-glycan profile of human liver.

The *O*-glycan profile of the human liver (Figure 12.49b) contains both core 1 and core 2 type *O*-linked glycans, extensively sialylated with up to three NeuAc residues being observed. There are also peaks consistent with galactose ($n = 1$–3) and fucose extension to the core 2 type glycans, an assertion which is confirmed by MS/MS studies. The most abundant glycan species is a disialylated core 1 structure at *m/z* 1256.5. Figure 12.51 shows a summary of the proposed core 1 and core 2 glycans present in the *O*-glycan profile of human liver.

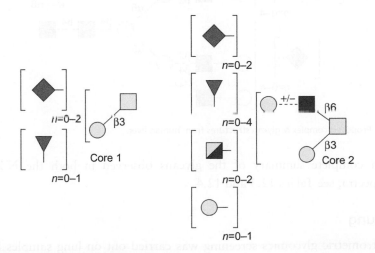

**FIGURE 12.48**   Proposed *O*-glycan structures from human large bowel.

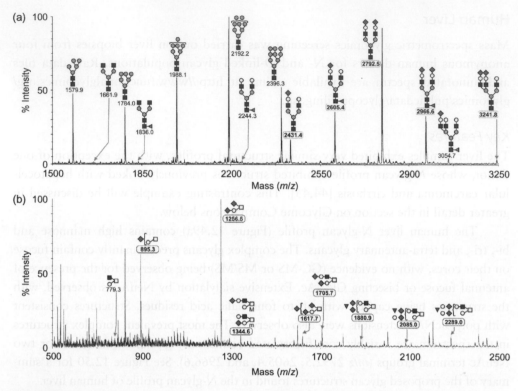

**FIGURE 12.49** Partial MALDI-TOF-MS profiles of the permethylated *N*-linked glycans (a) and *O*-linked glycans (b) from a typical human liver sample. Major peaks are annotated with the relevant carbohydrate structure shown in symbol form, according to the glycan nomenclature adopted by the CFG (http://www.functionalglycomics.org/). Structures confirmed by further analyses have their masses shaded. For complete annotation of the spectra see Tables 12.3 and 12.4. All molecular ions are present in sodiated form ([M+Na]$^+$).

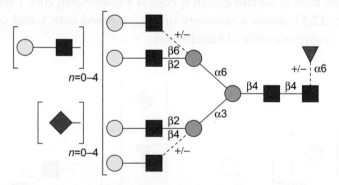

**FIGURE 12.50** Proposed complex *N*-glycan structures from human liver.

For a complete summary of the glycans observed in both the *N*-linked and *O*-linked spectra, see Tables 12.3 and 12.4.

## Human Lung

Mass spectrometric glycomics screening was carried out on lung samples from eight anonymous human donors for *N*- and *O*-linked glycan populations. Raw data files

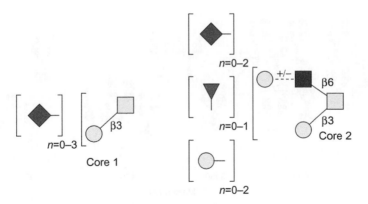

**FIGURE 12.51**   Proposed *O*-glycan structures from human liver.

and annotated spectra are available online at http://www.functionalglycomics.org/glycomics/publicdata/glycoprofiling.jsp.

## Key Features

The *N*-glycan profile of human lung (Figure 12.52a) consists of high mannose and bi-, tri-, and tetra-antennary complex structures. The latter are core fucosylated to a fairly high level, though some non-fucosylated signals are observed. Antennal fucosylation is limited, with a few relatively minor signals consistent with the addition of up to two fucoses in a Le$^x$ type arrangement. Sialic acid capping of the structures is common, with up to four NeuAc residues being observed on the larger tetra-antennary molecules. Some polylactosamine extensions ($n = 0$–$2$) are also observed in the high mass region. Bisecting GlcNAc is present, confirmed by low levels of 3,4,6-linked mannose within GC-MS linkage experiments (data not shown), though the bisected glycans are relatively low level when compared to related species (*m/z* 3416.1 vs. 3661.1 for example). Other features of note are a large number of truncated species in the lower mass region, together with a small amount of unprocessed hybrid structures. Differentially core fucosylated biantennary glycans account for the most intense signals in the spectra, capped with one or two sialic acids (*m/z* 2605.8, 2792.9, and 2967.0). See Figure 12.53 for a summary of the proposed glycan structures found in the *N*-glycan profile of human lung.

Core 1 and core 2 type *O*-glycans are both present in the *O*-linked profile of human lung (Figure 12.52b). Each core subtype is readily sialylated with up to two (core 2) or three (core 1) NeuAc residues. The core 2 glycans are also modified with fucose (thought to be in Le$^x$ form) and optionally extended by a single LacNAc repeat. The core 1 glycans appear as the most intense signals in the spectrum, with the naked core 1 structure and its mono- and disialyl versions (*m/z* 534.5, 895.4, and 1256.7) dominating the less intense core 2 signals. *O*-Glycan structures from airway mucins have been analyzed elsewhere in great detail, demonstrating that the mucins contain a variety of core 1, core 2, core 3, and core 4 glycans, modified by sulfation, sialylation

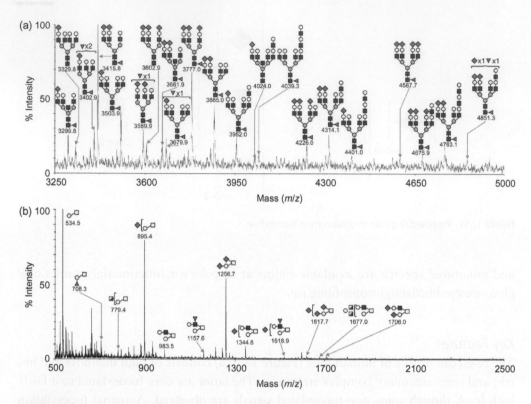

**FIGURE 12.52** Partial MALDI-TOF-MS profiles of the permethylated *N*-linked glycans (a) and *O*-linked glycans (b) from a typical human lung sample. Major peaks are annotated with the relevant carbohydrate structure shown in symbol form, according to the glycan nomenclature adopted by the CFG (http://www.functionalglycomics.org/). For complete annotation of the spectra see Tables 12.3 and 12.4. All molecular ions are present in sodiated form ([M+Na]+).

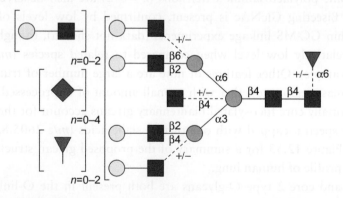

**FIGURE 12.53** Proposed complex *N*-glycan structures from human lung.

and fucosylation events [46–48]. Figure 12.54 shows a summary of the proposed core 1 and core 2 glycans identified in the *O*-glycan profile of human lung samples analyzed by Core-C.

For a complete summary of the glycans observed in both the *N*-linked and *O*-linked spectra, see Tables 12.3 and 12.4.

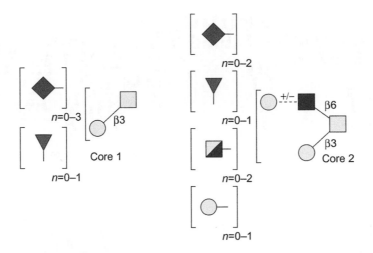

**FIGURE 12.54**    Proposed *O*-glycan structures from human lung.

## Human Abdominal Lymph Nodes

Mass spectrometric glycomics screening was carried out on abdominal lymph node samples from four anonymous human donors for *N*- and *O*-linked glycan populations. Raw data files and annotated spectra are available online at http://www.functionalglycomics. org/glycomics/publicdata/glycoprofiling.jsp.

### Key Features

The *N*-linked glycan profile derived from human abdominal lymph nodes (Figure 12.55a) consists of high mannose, bi-, tri-, and tetra-antennary complex and some minor hybrid glycans. Core fucosylation is present on virtually all of the observed glycans, though there are a number of relatively abundant non-fucosylated complex *N*-glycans found within the tissue. Antennal fucosylation is very limited, with some very low abundance peaks accounting for singly fucosylated arms. Similarly, there is some bisecting GlcNAc observed, but these peaks are only present at very low abundance, confirmed by the low relative intensity of the 3,4,6-linked mannose within GC-MS linkage studies (data not shown). In contrast to these minor structural characteristics, sialylation is very extensive, with the antennae of the complex *N*-glycans being capped with up to four NeuAc residues. Biantennary glycans account for the most intense signals in the spectra, differentially core fucosylated and capped with one or two sialic acids (*m/z* 2606.4, 2794.5, and 2967.6). See Figure 12.56 for a summary of the proposed glycan structures found in the *N*-glycan profile of human abdominal lymph nodes.

Figure 12.55b shows a typical *O*-glycan spectrum of human abdominal lymph nodes. Core 1 and core 2 type glycans are observed, with ample sialylation on each. Signals corresponding to mono- and disialyl core 1 structures (*m/z* 895.7 and 1257.0) are both present at relative abundances higher than that of their non-substituted relative (*m/z* 534.2). Figure 12.57 shows a summary of the proposed core 1 and core 2 glycans present in the *O*-glycan profile of human abdominal lymph nodes.

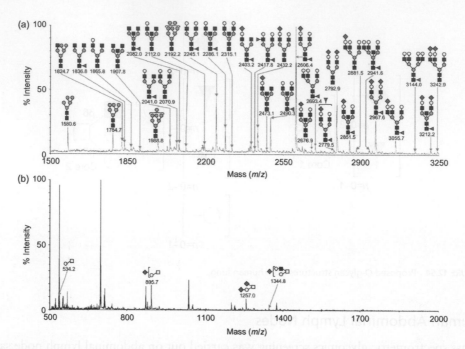

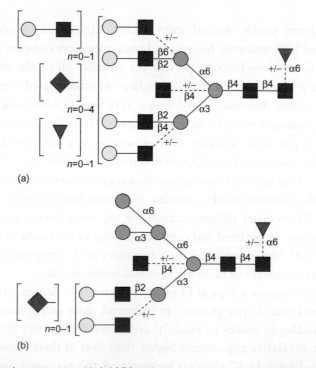

**FIGURE 12.55**    Partial MALDI-TOF-MS profiles of the permethylated *N*-linked glycans (a) and *O*-linked glycans (b) from a typical human abdominal lymph node sample. Major peaks are annotated with the relevant carbohydrate structure shown in symbol form, according to the glycan nomenclature adopted by the CFG (http://www.functionalglycomics.org/). For complete annotation of the spectra see Tables 12.3 and 12.4. All molecular ions are present in sodiated form ([M+Na]⁺).

**FIGURE 12.56**    Proposed complex (a) and hybrid (b) *N*-glycan structures from human abdominal lymph nodes (for hybrid signals, see Table 12.3).

**FIGURE 12.57**  Proposed *O*-glycan structures from human abdominal lymph nodes.

For a complete summary of the glycans observed in both the *N*-linked and *O*-linked spectra, see Tables 12.3 and 12.4.

## Human Skin

Mass spectrometric glycomics screening was carried out on skin samples from four anonymous human donors for *N*- and *O*-linked glycan populations. Raw data files and annotated spectra are available online at http://www.functionalglycomics.org/glycomics/publicdata/glycoprofiling.jsp.

### Key Features

The *N*-linked glycan profile derived from a typical human skin sample (Figure 12.58a) consists of high mannose and bi-, tri-, and tetra-antennary complex glycans. Core fucosylation is present on virtually all of the observed glycans, though there are a number of relatively abundant non-fucosylated complex *N*-glycans found within the tissue. Though there are some signals consistent with larger glycans bearing single antennal fucose residues, overall the level of this structural feature is very low. There is a relatively high amount of bisecting GlcNAc; indeed, the most abundant complex peak in the spectrum is an asialo, core fucosylated biantennary structure bisected by a the attachment of a GlcNAc residue 4-linked to the core mannose. This is further confirmed by GC-MS linkage data showing an abundance of 3,4,6-linked mannose (data not shown). Sialylation of the complex glycans is extensive, with virtually all of these glycan structures carrying at least one and up to four NeuAc residues. See Figure 12.59 for a summary of the proposed glycan structures found in the *N*-glycan profile of human skin.

Figure 12.58b shows a typical *O*-glycan spectrum of human skin. Core 1 and core 2 type glycans are observed, with signals corresponding to the capping of the core 1 structure with one or two NeuAc residues present at *m/z* 895.4 and 1256.6. Figure 12.60 shows a summary of the proposed core 1 and core 2 glycans present in the *O*-glycan profile of human skin.

For a complete summary of the glycans observed in both the *N*-linked and *O*-linked spectra, see Tables 12.3 and 12.4.

## Human Small Bowel

Mass spectrometric glycomics screening was carried out on small bowel samples from five anonymous human donors and high quality data were generated for the *N*-glycome. Mucins from the small intestine have been analyzed elsewhere in great detail; for further

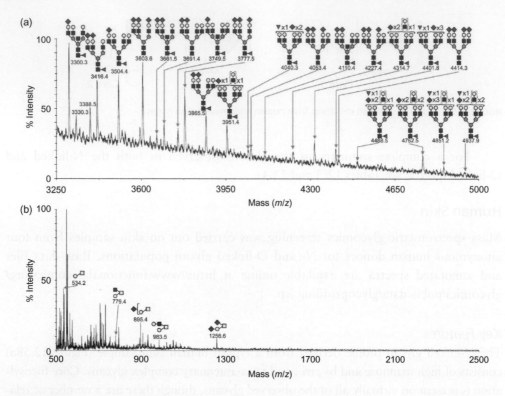

**FIGURE 12.58** Partial MALDI-TOF-MS profiles of the permethylated *N*-linked glycans (a) and *O*-linked glycans (b) from a typical human skin sample. Major peaks are annotated with the relevant carbohydrate structure shown in symbol form, according to the glycan nomenclature adopted by the CFG (http://www.functionalglycomics.org/). For complete annotation of the spectra see Tables 12.3 and 12.4. All molecular ions are present in sodiated form ([M+Na]⁺).

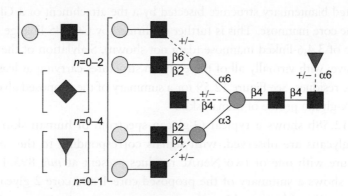

**FIGURE 12.59** Proposed complex *N*-glycan structures from human skin.

**FIGURE 12.60** Proposed *O*-glycan structures from human skin.

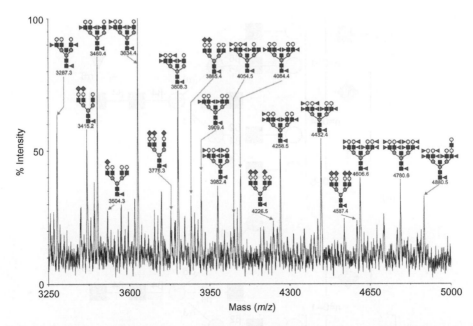

**FIGURE 12.61**   Partial MALDI-TOF-MS profiles of the permethylated *N*-linked glycans from a typical human small bowel sample. Major peaks are annotated with the relevant carbohydrate structure shown in symbol form, according to the glycan nomenclature adopted by the CFG (http://www.functionalglycomics.org/). For complete annotation of the spectrum see Table 12.3. All molecular ions are present in sodiated form ([M+Na]⁺).

details see [39,40,42,43]. Raw data files and annotated spectra are available online at http://www.functionalglycomics.org/glycomics/publicdata/glycoprofiling.jsp.

*Key Features*

The *N*-linked glycan profile generated from a typical human small bowel sample is shown in Figure 12.61a. High mannose and hybrid structures are present, along with bi-, tri-, and tetra-antennary complex type. These structures are almost exclusively core fucosylated. An unusually high level of bisecting GlcNAc is observed, with bisected peaks accounting for the majority of higher mass structures. This extensive action of GlcNAcTIII is consistent with the relatively high levels of 3,4,6-linked mannose within GC-MS linkage analyses (data not shown). Sialylation of the antennae is present ($n = 1–4$), though at relatively low levels when compared with other tissues. The most intense signal in the spectrum belongs to an asialo, biantennary core fucosylated complex glycan (*m/z* 2243.9).

As with the large bowel, the small bowel's most distinctive characteristic is the very high levels of antennal fucose. Most of the high mass glycans have been modified by multiple fucosylation events, with up to eight residues attached to a single glycan. Biosynthetic limitations, together with evidence from GC-MS analyses and MS/MS studies of smaller related species at *m/z* 2837.1 and 3185.2 (data not shown) indicates that the fucose residues are present on both the Gal and GlcNAc residues of the antennae in Le<sup>x</sup> and Le<sup>y</sup> conformations. See Figure 12.62 for a summary of the proposed glycan structures found in the *N*-glycan profile of human small bowel.

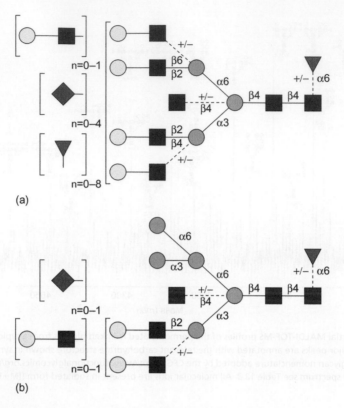

(a)

(b)

**FIGURE 12.62** Proposed complex (a) and hybrid (b) N-glycan structures from human small bowel (for hybrid signals, see Table 12.3).

For a complete summary of the glycans observed in the N-linked spectrum, see Table 12.3.

## Human Spleen

Mass spectrometric glycomics screening was carried out on spleen samples from five anonymous human donors for N- and O-linked glycan populations. Raw data files and annotated spectra are available online at http://www.functionalglycomics.org/glycomics/publicdata/glycoprofiling.jsp.

### Key Features

The samples of human spleen possess N-glycan profiles comprising high mannose, hybrid, and complex type oligosaccharides (Figure 12.63a). Bi-, tri-, and tetra-antennary structural variations of the latter are present, with a high degree of core fucosylation. Some bisecting GlcNAc containing glycans are observed, but at relatively low abundance, verified by the identification of a relatively low level of 3,4,6-linked mannose within GC-MS linkage analysis (data not shown). Sialic acid capping is very abundant ($n = 0-4$), with many of the most abundant structures observed bearing

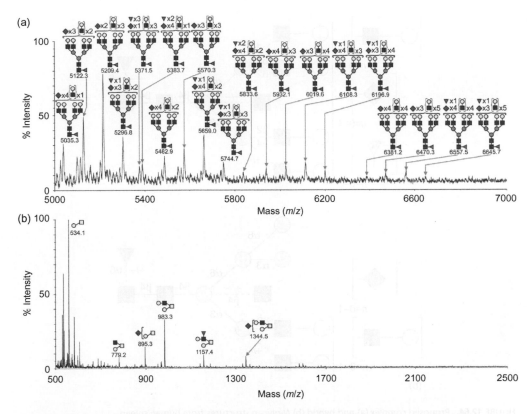

**FIGURE 12.63** Partial MALDI-TOF-MS profiles of the permethylated *N*-linked glycans (a) and *O*-linked glycans (b) from a typical human spleen sample. Major peaks are annotated with the relevant carbohydrate structure shown in symbol form, according to the glycan nomenclature adopted by the CFG (http://www.functionalglycomics.org/). For complete annotation of the spectra see Tables 12.3 and 12.4. All molecular ions are present in sodiated form ([M+Na]$^+$).

two or more NeuAc residues. Some antennal fucosylation is also present (up to three residues), with many peaks carrying at least one additional fucose on their arms. The most notable feature of the profile is the large number of polylactosamine extensions present across the whole spectrum, with some of the biggest glycans being extended by up to five LacNAc repeats (*m/z* 6645.7). The most intense signals in the spectrum belong to bi- and tri-antennary fully sialylated structures (*m/z* 2792.0 and 3775.9).

The *O*-glycan profile for a typical human spleen sample is shown in Figure 12.63b). Both core 1 and core 2 type structures are observed, each of which can be modified by a single NeuAc. In contrast to many human organs, the spleen appears to produce a relatively high level of core 2 *O*-glycans with signals corresponding to a Gal extended core 2 (*m/z* 983.3) and a Le$^x$ type core 2 glycan (*m/z* 1157.4) being particularly intense. However, despite their relatively high abundance, the largest peak in the profile is the unsubstituted core 1 glycan. Figure 12.65 shows a summary of the proposed core 1 and core 2 glycans present in the *O*-glycan profile of human spleen.

For a complete summary of the glycans observed in both the *N*-linked and *O*-linked spectra, see Tables 12.3 and 12.4.

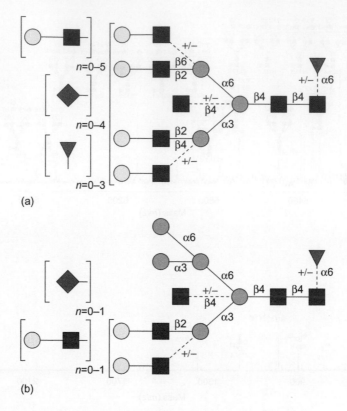

**FIGURE 12.64**   Proposed complex (a) and hybrid (b) *N*-glycan structures from human spleen.

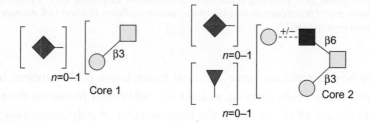

**FIGURE 12.65**   Proposed *O*-glycan structures from human spleen.

## Summary of the Human Glycome

See Tables 12.3 and 12.4.

# Glycome Comparisons

We present here a short section giving some examples of the most obvious similarities and differences observed amongst the glycomic profiles of murine and human tissues. More rigorous data mining is outside the scope of this chapter but suffice to say that the CFG's glycomic databases are a powerful resource for future efforts in systems glycobiology.

## Human vs. Mouse Variation

Though much of the *N*- and *O*-glycosylation biosynthetic pathway is conserved between mammals, there are a few well defined differences. It is known for example that the Gal-α-Gal epitope, a common terminal group in most mammals, is absent from Old World monkeys (monkeys of Asia and Africa), apes, and humans [49]. Furthermore, NeuGc is barely detectable in healthy human tissues [50]. Interestingly neither Gal-α-Gal nor NeuGc are expressed at significant levels in the mouse brain. Indeed it has been known for some time that the mouse and the human express similar glycan epitopes in the brain [21,25]. Beyond this specific example, the question remains as to how similar the overall glycomes of the two species are. From Core-C's data, it can be concluded that within the organs studied there are clear examples of both similarity and difference. The kidney epitomizes this dichotomy. Figure 12.66 shows a comparison between the *N*-linked glycans of a mouse and human kidney. The murine kidney has a very distinctive bisecting GlcNAc and antennal fucosylation of its *N*-glycans, with very little sialylation. In comparison, the human kidney exhibits similar bisected and fucosylated structures (see Table 12.3), but the most abundant glycans by far are capped with NeuAc.

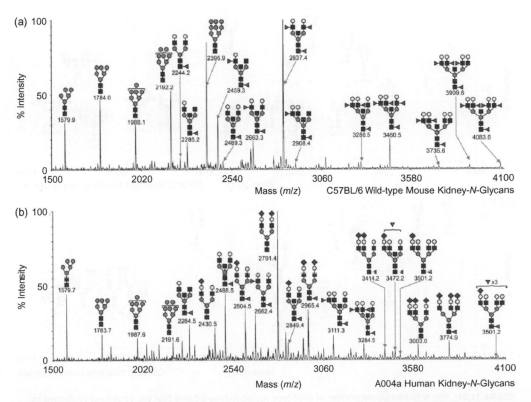

**FIGURE 12.66**   Comparison of *N*-linked glycan profiles derived from wild-type mouse (a) and human (b) kidney tissue.

## Differences Among Tissue Glycomes

It has been known for more than 20 years that glycoproteins expressed in different cell types are often differently glycosylated [51]. The tools of modern glycomics are enabling the expression of particular glycan epitopes to be studied in a systematic way with the aim of unraveling their biological significance. There have been a number of recent glycomic studies exploring cell-type specific glycosylation, with evidence that the main cause is the differential expression of glycosyltransferases involved in the synthesis of terminal structures within these cell types or tissues [21,24,52,53]. As an example, Figure 12.67 shows a comparison between two tissues derived from a single C57BL/6 wild-type mouse, the kidney (a) and liver (b). The difference between the profiles of the two tissues is very pronounced. While the liver sialylates the antennae of its N-linked glycans, the kidney bisects with GlcNAc and fucosylates the antennae. These results, together with microarray data produced from the same animals, form part of a coordinated study carried out by the CFG [24]. The glycogene expression profiles for the kidney show a very high level of the fucosyltransferases responsible for antennal fucose addition, correlating very well with the observed structural data.

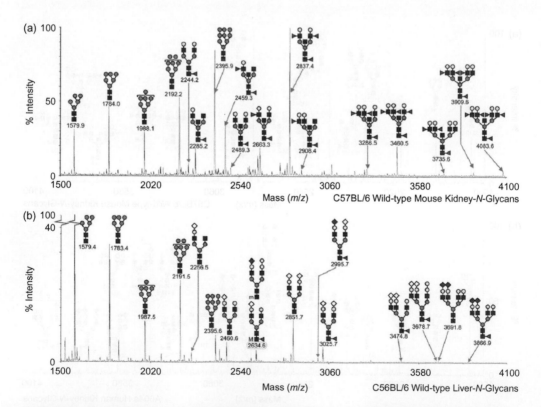

**FIGURE 12.67**   The N-linked glycan profiles of wild-type mouse kidney (a) and liver (b). Structures confirmed by further analyses have their masses shaded. Where MS/MS analysis has informed upon the relative abundance of glycoforms, major (M) and minor (m) components are indicated.

In contrast the liver shows a lower level of fucosyltransferase expression and a concomitant higher level of expression of the sialyltransferases responsible for α2,3-linked sialic acid.

## Variation in Single Tissues of Single Species

As has been outlined in the introductory sections of this chapter, as well as other chapters in this volume, part of the importance of glycomic analyses lies in the potential for diagnostic and therapeutic applications. Modification to the glycan complement affects protein folding and stability and interferes with the various carbohydrate-binding interactions. As a result, a great many physiological and pathological events can be affected by aberrant glycosylation, including host–pathogen interactions, cell growth, cell differentiation, cell trafficking, transmembrane signaling, and tumor invasion. It is clear, therefore, that abnormal glycosylation patterns can serve as biomarkers for differential states such as those observed in disease progression. Figure 12.68 gives an example of one such case from Core-C's human glycomic studies. The lower panel (Figure 12.68b) shows a profile that is considered normal and healthy and it is shared by the majority of the anonymous samples that were investigated. It consists

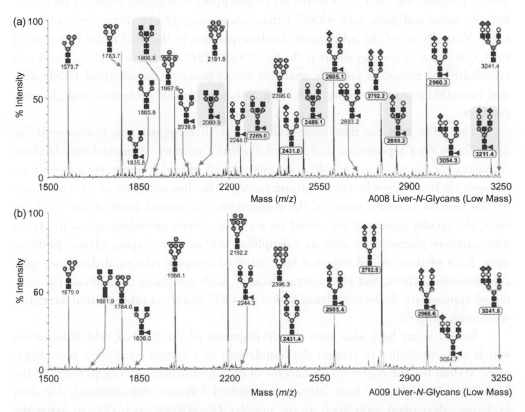

**FIGURE 12.68** Comparison of *N*-linked glycan profiles derived from two human liver samples, (a) potentially diseased and (b) healthy. Structures confirmed by further analyses have their masses shaded. Unusual structures observed are highlighted.

mainly of complex glycans, bi-, tri-, and tetra-antennary with a large amount of NeuAc sialylation. No evidence is seen for bisecting GlcNAc.

The upper panel (Figure 12.68a) shows the N-glycan profile derived from a potentially diseased liver. It is remarkably similar to the healthy organ, save for two key differences—the presence of a large amount of bisecting GlcNAc, and an increase in the level of core fucosylation, structural features that have been implicated in hepatocellular carcinoma and cirrhosis [44,45].

## Data Repositories

The rapid expansion of the related "omics" fields—genomics, proteomics, glycomics—has clearly demonstrated how the unrestricted dissemination of scientific data can accelerate research and open new avenues of investigation. Of course, central to this open access approach is the need for well curated and detailed databases.

This requirement in the field of glycobiology was first recognized in the 1980s and led to the creation of the Complex Carbohydrate Structure Database (CCSD), often referred to by the name of its interface software, *CarbBank* [54,55]. The funding for this project was unfortunately lost and updates finally ceased during the late 1990s. Despite this, the CCSD is still the largest publicly available resource for carbohydrate structural data, with 49 897 entries comprising 23 118 distinct glycan structures. Virtually all of the subsequent database projects in the field make some use of the CCSD data. A good example is the *GLYCOSCIENCES.de* portal [56], which was established to provide a publicly available access point to the archived CCSD data and to enable cross-referencing with the growing number of proteomic and glycomic resources.

The CFG was the first large-scale project to recognize the requirement for open access data repositories and has established a number of interrelated database resources for mammalian carbohydrate analysis, all of which are freely available on the web via http://www.functionalglycomics.org/. The dissemination of data produced by the various scientific cores of the consortium is a central tenet of the CFG. As such, the results generated are stored on a central server, providing access to glycan array, mouse phenotyping and, as exemplified within this chapter, glycan profiling data. Each of these is tied together by a series of complex relational databases (glycan structures, GBPs, and glycosyltransferases), with interfaces to facilitate the linking of appropriate data—both from within the CFG and from other related databases worldwide [15,16].

Recent years have also seen the development of the EUROCarbDB initiative, which was established to cement the foundations of a system to allow a more organized system for the collection and access of glycan information. The partners of the EUROCarbDB initiative have already established formats and standards for data exchange, developed tools such as the popular GlycoWorkbench [57] to assist the interpretation of experimental data, and designed a database architecture that could be used to store and retrieve glycan data [58].

Other important and valuable carbohydrate related resources include the Kyoto Encyclopedia of Genes and Genomes (KEGG) [59,60] and the GlycoEpitope database [61]. An excellent review of web-based bioinformatics resources (which includes coverage of analytical tools such as Cartoonist [62], as well as databases) was produced in 2004 by Claus-Wilhelm von der Lieth and remains a very valuable reference material for the area [2].

## Current Limitations and the Future

As with any aspect of science that is dependent upon technology, the clearest limitation—and most often the greatest source for advancement—is the technology itself. The level to which we can study the glycomes of organisms is presently determined by the power of the techniques that we employ. A straightforward example is contained within this chapter, with the data acquired from the MALDI-TOF instrumentation providing greater sensitivity and mass ranges than that which was produced by the similar but older FAB mass spectrometers. This trend continues, with the application of state-of-the-art MALDI-TOF-TOF technology to glycomic analysis further extending our reach in terms of sensitivity and mass range [9]. It is likely that the advancement of such analytical techniques and instrumentation will continue to further the field, increasing the depth of study possible, perhaps to one day allow us to perform total glycome analysis on a single cell or tiny amounts of intact glycoconjugates, identifying sites of glycosylation, site-specific heterogeneity and precise configurational information, all at the touch of a button. Of course, the volume of data that will be generated by such endeavors will also require that databases, automation, and data analysis tools keep pace with the development of the experimental side.

The significance of glycomic analyses, such as those presented within this chapter, is vastly enhanced when used in conjunction with other complementary analyses. One such example of this more holistic approach is referred to in the section on Data Repositories, with the use of microarray data to enhance the mass spectrometric profiling data [8,24]. Monitoring the expression of glycogenes in this manner, together with the profiling of the resultant glycome, is a step on the road towards being able to predict the glycome from the pattern of biosynthetic enzymes expressed. These elements make up the basis for burgeoning complementary systems biology approaches, which take the glycomic screening and microarray data and marry it to data integration and modeling approaches.

Such is the magnitude of the undertaking involved in these coordinated studies, the need for appropriate analytical standards is of increasing importance. To this end, the Human Disease Glycomics/Proteomics Initiative (HGPI) (http://www.hgpi.jp), sponsored by the Human Proteome Organisation (HUPO) has undertaken to establish experimental standards and common data formats for the diagnostic screening of glycans [63]. Similarly, the need for large-scale international consortia to coordinate and focus scientific endeavors in the field of glycomics is obvious, though it is critical that they continue to embrace the open access resource and data-sharing policies. With such a complex field, availability of the data is essential to progress.

## Acknowledgments

The glycan analyses were performed by the Analytical Glycotechnology Core of the Consortium for Functional Glycomics (GM62116). Anne Dell was a Biotechnology and Biological Sciences Research Council (BBSRC) Professorial Fellow.

# References

1. Feizi T, Mulloy B. Carbohydrates and glycoconjugates. Glycomics: the new era of carbohydrate biology. *Curr Opin Struct Biol* 2003;**13**(5):602–4.

2. von der Lieth CW, Bohne-Lang A, Lohmann KK, Frank M. Bioinformatics for glycomics: status, methods, requirements and perspectives. *Brief Bioinform* 2004;**5**(2):164–78.

3. Haslam SM, North SJ, Dell A. Mass spectrometric analysis of N- and O-glycosylation of tissues and cells. *Curr Opin Struct Biol* 2006;**16**(5):584–91.

4. Haslam S, Gems D, Morris HR, Dell A. The glycomes of Caenorhabditis elegans and other model organisms. *Biochemical Society Symposium* 2001;**69**:117–34. University of York, Portland Press.

5. Gemmill TR, Trimble RB. Overview of N- and O-linked oligosaccharide structures found in various yeast species. *Biochim Biophys Acta* 1999;**1426**(2):227–37.

6. Akama TO, Nakagawa H, Wong NK, Sutton-Smith M, Dell A, Morris HR, et al. Essential and mutually compensatory roles of α-mannosidase II and α-mannosidase IIx in N-glycan processing in vivo in mice. *Proc Natl Acad Sci USA* 2006;**103**(24):8983–8.

7. Wang Y, Tan J, Sutton-Smith M, Ditto D, Panico M, Campbell RM, et al. Modeling human congenital disorder of glycosylation type IIa in the mouse: conservation of asparagine-linked glycan-dependent functions in mammalian physiology and insights into disease pathogenesis. *Glycobiology* 2001;**11**(12):1051–70.

8. Bax M, Garcia-Vallejo JJ, Jang-Lee J, North SJ, Gilmartin TJ, Hernandez G, et al. Dendritic cell maturation results in pronounced changes in glycan expression affecting recognition by siglecs and galectins. *J Immunol* 2007;**179**(12):8216–24.

9. Babu P, North SJ, Jang-Lee J, Chalabi S, Mackerness K, Stowell SR, et al. Structural characterisation of neutrophil glycans by ultra sensitive mass spectrometric glycomics methodology. *Glycoconj J* 2008[Epub ahead of print].

10. Jang-Lee J, North SJ, Sutton-Smith M, Goldberg D, Panico M, Morris H, et al. Glycomic profiling of cells and tissues by mass spectrometry: fingerprinting and sequencing methodologies. *Methods Enzymol* 2006;**415**:59–86.

11. Sutton-Smith M, Dell A. Analysis of carbohydrates/glycoproteins by MS. In: Cellis JE, editor. *Cell Biology: a laboratory handbook*: Academic Press; 2006, p. 415–25.

12. Harvey DJ. Proteomic analysis of glycosylation: structural determination of N- and O-linked glycans by mass spectrometry. *Expert Rev Proteomics* 2005;**2**(1):87–101.

13. Morelle W, Michalski JC. Glycomics and mass spectrometry. *Curr Pharm Des* 2005;**11**(20):2615–45.

14. Dell A, Chalabi S, Hitchen PG, Jang-Lee J, Ledger V, North SJ, et al. Mass Spectrometry of Glycoprotein Glycans: Glycomics and Glycoproteomics. In: Kamerling JP, editor. *Comprehensive Glycoscience*: Elsevier; 2007, p. 69–100.

15. Raman R, Raguram S, Venkataraman G, Paulson JC, Sasisekharan R. Glycomics: an integrated systems approach to structure-function relationships of glycans. *Nat Methods* 2005;**2**(11):817–24.

16. Raman R, Venkataraman M, Ramakrishnan S, Lang W, Raguram S, Sasisekharan R. Advancing glycomics: implementation strategies at the consortium for functional glycomics. *Glycobiology* 2006;**16**(5):82R–90R.

17. Stanley P. Biological consequences of overexpressing or eliminating N-acetylglucosaminyltransferase-TIII in the mouse. *Biochim Biophys Acta* 2002;**1573**(3):363–8.

18. Martin LT, Marth JD, Varki A, Varki NM. Genetically altered mice with different sialyltransferase deficiencies show tissue-specific alterations in sialylation and sialic acid 9-O-acetylation. *J Biol Chem* 2002;**277**(36):32930–8.

19. Yamashita K, Hitoi A, Tateishi N, Higashi T, Sakamoto Y, Kobata A. The structures of the carbohydrate moieties of mouse kidney gamma-glutamyltranspeptidase: occurrence of X-antigenic determinants and bisecting N-acetylglucosamine residues. *Arch Biochem Biophys* 1985;**240**(2):573–82.

20. Sutton-Smith M, Morris HR, Dell AA. Rapid mass spectrometric strategy suitable for the investigation of glycan alterations in knockout mice. *Tetrahedron-Asymmetry* 2000;**11**(2):363–9.

21. Sutton-Smith M, Morris HR, Grewal PK, Hewitt JE, Bittner RE, Goldin E, et al. MS screening strategies: investigating the glycomes of knockout and myodystrophic mice and leukodystrophic human brains. *Biochemical Society Symposium* 2002;**69**:105–15.

22. Parry S, Ledger V, Tissot B, Haslam SM, Scott J, Morris HR, et al. Integrated mass spectrometric strategy for characterizing the glycans from glycosphingolipids and glycoproteins: direct identification of sialyl Le$^x$ in mice. *Glycobiology* 2007;**17**(6):646–54.

23. Parry S, Hadaschik D, Blancher C, Kumaran MK, Bochkina N, Morris HR, et al. Glycomics investigation into insulin action. *Biochim Biophys Acta* 2006;**1760**(4):652–68.

24. Comelli EM, Head SR, Gilmartin T, Whisenant T, Haslam SM, North SJ, et al. A focused microarray approach to functional glycomics: transcriptional regulation of the glycome. *Glycobiology* 2006;**16**(2):117–31.

25. Albach C, Klein RA, Schmitz B. Do rodent and human brains have different N-glycosylation patterns? *Biol Chem* 2001;**382**(2):187–94.

26. Dell A. F.A.B.-mass spectrometry of carbohydrates. *Adv Carbohydr Chem Biochem* 1987;**45**:19–72.

27. Finne J, Krusius T, Margolis RK, Margolis RU. Novel mannitol-containing oligosaccharides obtained by mild alkaline borohydride treatment of a chondroitin sulfate proteoglycan from brain. *J Biol Chem* 1979;**254**(20):10295–300.

28. Yuen CT, Chai W, Loveless RW, Lawson AM, Margolis RU, Feizi T. Brain contains HNK-1 immunoreactive O-glycans of the sulfoglucuronyl lactosamine series that terminate in 2-linked or 2,6-linked hexose (mannose). *J Biol Chem* 1997;**272**(14):8924–31.

29. Chiba A, Matsumura K, Yamada H, Inazu T, Shimizu T, Kusunoki S, et al. Structures of sialylated O-linked oligosaccharides of bovine peripheral nerve alpha-dystroglycan. The role of a novel O-mannosyl-type oligosaccharide in the binding of alpha-dystroglycan with laminin. *J Biol Chem* 1997;**272**(4):2156–62.

30. Smalheiser NR, Haslam SM, Sutton-Smith M, Morris HR, Dell A. Structural analysis of sequences O-linked to mannose reveals a novel Lewis x structure in cranin (dystroglycan) purified from sheep brain. *J Biol Chem* 1998;**273**(37):23698–703.

31. Hashii N, Kawasaki N, Itoh S, Nakajima Y, Kawanishi T, Yamaguchi T. Alteration of N-glycosylation in the kidney in a mouse model of systemic lupus erythematosus: relative quantification of N-glycans using an isotope-tagging method. *Immunology* 2008.

32. Haslam SM, Julien S, Burchell JM, Monk CR, Ceroni A, Garden OA, et al. Characterizing the glycome of the mammalian immune system. *Immunol Cell Biol* 2008;**86**(7):564–73.

33. Varki A. Sialic acids in human health and disease. *Trends Mol Med* August 2008;**14**(8):351–60.

34. Zhao YY, Takahashi M, Gu JG, Miyoshi E, Matsumoto A, Kitazume S, et al. Functional roles of N-glycans in cell signaling and cell adhesion in cancer. *Cancer Sci* July 2008;**99**(7):1304–10.

35. Vogt G, Vogt B, Chuzhanova N, Julenius K, Cooper DN, Casanova JL. Gain-of-glycosylation mutations. *Curr Opin Genet Dev* 2007;**17**(3):245–51.

36. Endo T, Toda T. Glycosylation in congenital muscular dystrophies. *Biol Pharm Bull.* December 2003;**26**(12):1641–7.

37. Freeze HH. Genetic defects in the human glycome. *Nat Rev Genet* July 2006;**7**(7):537–51.

38. Schiffmann R, Boespflug-Tanguy O. An update on the leukodsytrophies. *Curr Opin Neurol* 2001;**14**(6):789–94.

39. Robbe C, Capon C, Coddeville B, Michalski JC. Structural diversity and specific distribution of O-glycans in normal human mucins along the intestinal tract. *Biochem J* 2004;**384**(Pt 2):307–16.

40. Robbe C, Capon C, Maes E, Rousset M, Zweibaum A, Zanetta JP, et al. Evidence of regio-specific glycosylation in human intestinal mucins: presence of an acidic gradient along the intestinal tract. *J Biol Chem* 2003;**278**(47):46337–48.

41. Capon C, Maes E, Michalski JC, Leffler H, Kim YS. Sd(a)-antigen-like structures carried on core 3 are prominent features of glycans from the mucin of normal human descending colon. *Biochem J* 2001;**358**(Pt 3):657–64.

42. van Klinken BJ, Dekker J, van Gool SA, van Marle J, Buller HA, Einerhand AW. MUC5B is the prominent mucin in human gallbladder and is also expressed in a subset of colonic goblet cells. *Am J Physiol* 1998;**274**(5 Pt 1):G871–8.

43. Robbe-Masselot C, Maes E, Rousset M, Michalski JC, Capon C. Glycosylation of human fetal mucins: a similar repertoire of O-glycans along the intestinal tract. *Glycoconj J* 2008;**26**(4):397–413.

44. Block TM, Comunale MA, Lowman M, Steel LF, Romano PR, Fimmel C, et al. Use of targeted glycoproteomics to identify serum glycoproteins that correlate with liver cancer in woodchucks and humans. *Proc Natl Acad Sci USA* 2005;**102**(3):779–84.

45. Liu XE, Desmyter L, Gao CF, Laroy W, Dewaele S, Vanhooren V, et al. N-glycomic changes in hepatocellular carcinoma patients with liver cirrhosis induced by hepatitis B virus. *Hepatology* 2007;**46**(5):1426–35.

46. Morelle W, Sutton-Smith M, Morris HR, Davril M, Roussel P, Dell A. FAB-MS characterization of sialyl Lewis x determinants on polylactosamine chains of human airway mucins secreted by patients suffering from cystic fibrosis or chronic bronchitis. *Glycoconj J* 2001;**18**(9):699–708.

47. Xia B, Royall JA, Damera G, Sachdev GP, Cummings RD. Altered O-glycosylation and sulfation of airway mucins associated with cystic fibrosis. *Glycobiology* 2005;**15**(8):747–75.

48. Thomsson KA, Carlstedt I, Karlsson NG, Karlsson H, Hansson GC. Different O-glycosylation of respiratory mucin glycopeptides from a patient with cystic fibrosis. *Glycoconj J* 1998;**15**(8):823–33.

49. Macher BA, Galili U. The Galalpha1,3Galbeta1,4GlcNAc-R (alpha-Gal) epitope: a carbohydrate of unique evolution and clinical relevance. *Biochim Biophys Acta* 2008;**1780**(2):75–88.

50. Varki A. N-glycolylneuraminic acid deficiency in humans. *Biochimie* 2001;**83**(7):615–22.

51. Kobata A. The third chains of living organisms–a trail of glycobiology that started from the third floor of building 4 in NIH. *Arch Biochem Biophys* 2004;**426**(2):107–21.

52. Wang D, Liu S, Trummer BJ, Deng C, Wang A. Carbohydrate microarrays for the recognition of cross-reactive molecular markers of microbes and host cells. *Nat Biotechnol* 2002;**20**(3):275–81.

53. Fukui S, Feizi T, Galustian C, Lawson AM, Chai WG. Oligosaccharide microarrays for high-throughput detection and specificity assignments of carbohydrate-protein interactions. *Nature Biotechnology* 2002;**20**(10):1011–17.

54. Doubet S, Albersheim P. CarbBank. *Glycobiology* 1991;**2**(6):505.

55. Doubet S, Bock K, Smith D, Darvill A, Albersheim P. The Complex Carbohydrate Structure Database. *Trends Biochem Sci* 1989;**14**(12):475–7.

56. Lutteke T, Bohne-Lang A, Loss A, Goetz T, Frank M, von der Lieth C-W. GLYCOSCIENCES.de: an Internet portal to support glycomics and glycobiology research. *Glycobiology* 2006;**16**(5):71R–81R.

57. Ceroni A, Maass K, Geyer H, Geyer R, Dell A, Haslam SM. GlycoWorkbench: a tool for the computer-assisted annotation of mass spectra of glycans. *J Proteome Res* 2008;**7**(4):1650–9.

58. Design Studies Related to the Development of Distributed, Web-based European Carbohydrate Databases (EUROCarbDB). [cited]; Available from: http://www.eurocarbdb.org/.

59. Hashimoto K, Goto S, Kawano S, Aoki-Kinoshita KF, Ueda N, Hamajima M, et al. KEGG as a glycome informatics resource. *Glycobiology* 2006;**16**(5):63R–70R.

60. Kanehisa M, Goto S, Hattori M, Aoki-Kinoshita KF, Itoh M, Kawashima S, et al. From genomics to chemical genomics: new developments in KEGG. *Nucl Acids Res* 2006;**34**(Suppl. 1):D354–7.

61. Kawasaki T, Nakao H, Takahashi E, Tominaga T. GlycoEpitope: the integrated database of carbohydrate antigens and antibodies. *Trends in Glycoscience and Glycotechnology* 2006;**18**(102):267–72.

62. Goldberg D, Sutton-Smith M, Paulson J, Dell A. Automatic annotation of matrix-assisted laser desorption/ionization N-glycan spectra. *Proteomics* 2005;**5**(4):865–75.

63. Taniguchi N, Nakamura K, Narimatsu H, von der Lieth CW, Paulson J. Human Disease Glycomics/Proteome Initiative Workshop and the 4th HUPO Annual Congress. *Proteomics* 2006;**6**(1):12–13.

# 13

# Genetic and Structural Analysis of the Glycoprotein and Glycolipid Glycans of *Drosophila melanogaster*

Mary Sharrow[a], Kazuhiro Aoki[a], Sarah Baas[a,b],
Mindy Porterfield[a,c], and Michael Tiemeyer[a,b]

[a]*COMPLEX CARBOHYDRATE RESEARCH CENTER,
THE UNIVERSITY OF GEORGIA, ATHENS, GA, USA*
[b]*DEPARTMENT OF BIOCHEMISTRY AND MOLECULAR BIOLOGY,
THE UNIVERSITY OF GEORGIA, ATHENS, GA, USA*
[c]*DEPARTMENT OF CHEMISTRY,
THE UNIVERSITY OF GEORGIA, ATHENS, GA, USA*

## Introduction, Overview, and Scope

Complexity is a relative term. Complex problems. Complex organisms. Complex glycosylation. It is the antithesis of simple. Simple problems. Simple organisms. Simple glycosylation. However, close examination of the simple frequently reveals unexpected complexity. Such is the unfolding paradigm in functional and structural analyses of the glycans of invertebrates. This contribution to the *Handbook* will focus on the glycans of *Drosophila melanogaster*, an organism selected by a growing number of investigators as a system of choice for breaking new ground in glycobiology. Repeatedly over the past two decades, discoveries in *Drosophila* glycobiology have found structural or functional parallels in vertebrates, including mammals. In some cases, these common themes have evolved out of genetic characterization of interesting phenotypes and in other instances targeted investigations have pursued homologous glycan-processing components or specific glycan structural classes.

The results of enlightened genetic studies, performed by experienced drosophilists with minds opened to new possibilities, have recently synergized with enhanced structural analysis, generating new routes for linking phenotypes to glycan expression [1–4]. In the hopes of catalyzing further associations between glycan function and

**Handbook of Glycomics**
Copyright © 2009 by Elsevier Inc. All rights of reproduction in any form reserved.

structure, we will summarize current knowledge regarding the diversity of N-linked, O-linked, and glycosphingolipid glycans in *Drosophila*. Within the constraints of this contribution it is not possible to adequately encompass the important and rapid advances in characterizing *Drosophila* glycosaminoglycan structure and function; excellent comprehensive reviews are available elsewhere [5–10]. Where appropriate, relevant functional, biosynthetic, and genetic studies will be highlighted. It will be seen that a simple rule emerges: *Drosophila* elaborates its own complexity, but in restricted contexts, and only as is needed to meet specific developmental or functional requirements.

## N-Linked Glycoprotein Glycans

The first studies of the N-linked glycans of *Drosophila* revealed a predominance of high- and pauci-mannose glycans [11–14]. The maximum extent of N-linked glycan processing was, therefore, suggested to be limited to Golgi-localized mannosidase activities. The capacity of this pathway was determined to be sufficient to trim almost half of the total pool of N-linked glycans to trimannosyl core structures ($Man_3GlcNAc_2$, M3N2). Core fucosylation was also detected as a prevalent modification to the M3N2 structure [4,15]. Similar profiles were detected in cultured cells of *Drosophila* (S2) and other insects (Sf9), further solidifying the impression that complex glycans were absent and probably not needed for normal development in the Arthropoda [16–18]. Seminal discoveries have gradually loosened this restriction, leading to a broader appreciation of N-linked glycan diversity in *Drosophila* (Figure 13.1). For instance, insect cells engineered to express major components of vertebrate glycosylation pathways accommodate these changes well, indicating the lack of a fundamental toxicity or cellular impediment to the generation of complex glycans in insects [17,19,20]. Completion of the *Drosophila* genome sequence has also proven to be a major advance for the field, revealing the presence of many candidate processing enzymes and glycosyltransferase genes.

Building on the availability of the genome sequence, a growing number of enzymes have been targeted for mutagenesis or have been expressed for direct biochemical characterization. Mutational and biochemical characterization of the *Drosophila* GlcNAcT-1 homolog revealed that *Drosophila* not only possesses this activity but requires it for normal development and adult behavior [21,22]. *Drosophila* adults that lack GlcNAcT-1 have reduced viability and altered locomotion. Perhaps as an underlying cause of their behavioral deficits, GlcNAcT-1 mutant adult brains exhibit altered morphology, in which bilaterally paired and normally separated brain lobes are instead fused at the midline to form a continuous neural structure. Interestingly, this fused-lobes phenotype was originally described in another mutation, appropriately named "fused-lobes" (*fdl*) [23–26]. The *fdl* gene encodes a hexosaminidase that specifically removes the GlcNAc residue added by GlcNAcT-1, thereby blocking progression toward the formation of hybrid or complex glycans and explaining the predominance of pauci-mannose glycans in insects (Figure 13.2).

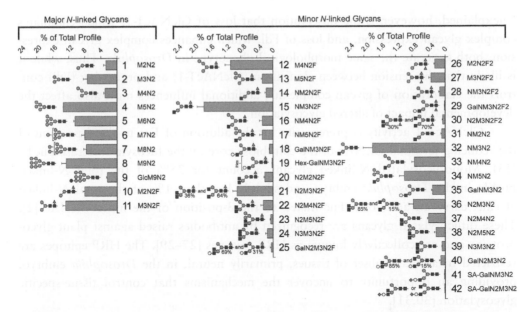

**FIGURE 13.1**   The *N*-linked glycan profile of wild-type *Drosophila* embryos. *N*-linked glycans were released from *Drosophila* embryo proteins by PNGAseA digestion. The total glycan profile was detected, characterized, and quantified by HPLC, exoglycosidase digestion, and mass spectrometry [2]. The prevalence of each major and minor glycan is expressed as a percent of the total pool of detected glycans. The representations of oligosaccharides in this and all other figures are in accordance with the current guidelines proposed by the Consortium for Functional Glycomics.

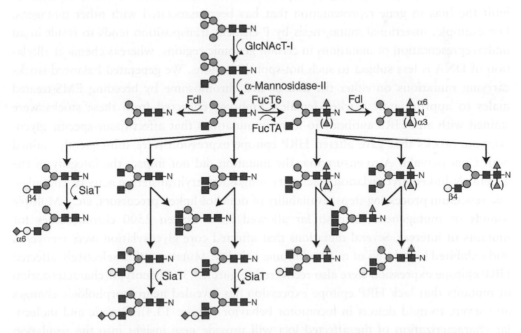

**FIGURE 13.2**   Probable processing and biosynthetic pathways for generating *N*-linked glycan diversity in *Drosophila*. *N*-linked glycan complexity is limited by the Fdl hexosaminidase and expanded by expression of uncharacterized branching GlcNAcT activities, undefined GalT activities, and a single SiaT activity. Two fucosyltransferases modify the chitobiose core. FucT6 adds Fuc to the 6-position and FucTA adds Fuc to the 3-position of the internal GlcNAc residue. A subset of the 6-linked, monofucosylated glycans also carry a second Fuc at the 3-position (see Figure 13.1, Structures 26–30). These glycans are known as HRP-epitopes.

Unexplained, however, is the observation that loss of GlcNAcT-1, which eliminates complex glycan expression, and loss of Fdl, which enhances complex glycan expression, both generate the same morphological phenotype in *Drosophila*. Thus, while it is likely that the tension between elongation (GlcNAcT-1) and processing (Fdl) controls the elaboration of glycan complexity, additional influences must also affect the developmental impact of altered glycosylation.

GlcNAcT-1 activity is prerequisite for the addition of Fuc to the 6-position of the reducing terminal GlcNAc of the chitobiose core by the fucosyltransferase FucT6 [15]. Monofucosylated *N*-linked glycans account for 25% of the total *N*-linked profile in the *Drosophila* embryo and slightly less than 1% of the total *N*-linked glycans carry an additional Fuc linked to the 3-position of the inner GlcNAc [2]. These difucosylated glycans are recognized by antibodies raised against plant glycoproteins and are collectively known as HRP epitopes [27–29]. The HRP epitopes are restricted to a small subset of tissues, primarily neural, in the *Drosophila* embryo, providing an opportunity to uncover the mechanisms that control tissue-specific glycosylation [30,31].

A particularly useful route to identify mechanisms that regulate glycosylation would take advantage of the genetic approaches available in *Drosophila*. We present here preliminary results from a random chemical mutagenesis screen designed to identify genes that affect HRP epitope expression. We chose to use chemical mutagenesis with the DNA alkylating reagent ethylmethane sulfonate (EMS, 25 mM) in order to limit the bias in gene representation that has been associated with other mutagens. For example, insertional mutagenesis by P-element transposition tends to result in an underrepresentation of mutations in certain genomic regions, whereas chemical alkylation of DNA is less subject to such hot-spot distortions. We generated balanced stocks carrying mutations on either the 2nd or 3rd chromosome by breeding EMS-treated males to appropriately marked females. Embryos collected from these stocks were stained with anti-HRP antibody to identify mutations that affect tissue-specific glycosylation. Stocks that gave altered HRP epitope expression were subsequently stained with Concanavalin A to ensure that the mutation did not impact the integrity of the general *N*-linked glycosylation machinery (oligosaccharyltransferase activity, endoplasmic reticulum processing steps, availability of dolichol-linked precursors, etc.). Multiple rounds of mutagenesis have so far allowed us to screen 2500 chromosomes for mutants of interest. Several mutations that affected core glycosylation were recovered and exhibited high levels of mid-embryonic lethality. Mutations that selectively affected HRP epitope expression were also recovered (Figure 13.3). Phenotypic characterization of mutants that lack HRP epitope expression has revealed neuromorphologic changes and severe-to-mild defects in locomotor behavior (Figure 13.4). Genetic and molecular characterization of the affected loci will provide new insight into the regulation of tissue-specific glycosylation.

In addition to core fucosylation, GlcNAcT-1 activity is also required for branching and elongation at the non-reducing termini of complex or hybrid glycans, which together account for 12% of the total glycan profile in the *Drosophila* embryo [2,4].

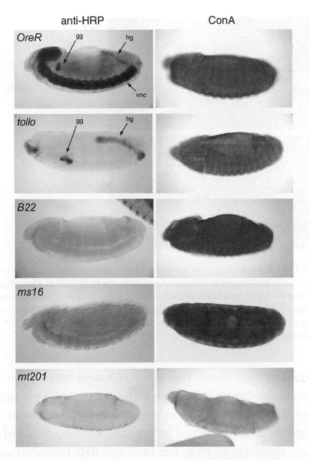

**FIGURE 13.3** Genetic screen for mutations that affect tissue-specific glycosylation. Wild-type *Drosophila* embryos (*OreR*) express difucosylated *N*-linked glycans in a restricted set of tissues (*vnc*, ventral nerve cord; *gg*, garland gland; *hg*, hindgut). Examples of mutations that alter HRP-epitope expression without affecting the general glycosylation machinery have been previously identified (*tollo*) or generated by EMS-mutagenesis (*B22*, *ms16*). The integrity of the general glycosylation machinery in these mutants is revealed by staining with Concanavalin A, reporting the distribution of high- or pauci-mannose glycans. EMS-mutagenesis also produces mutations that affect both HRP-epitope expression and general glycosylation (*mt201*). The blue color in the *tollo* panel results from lacZ activity staining, which is carried on the mutant chromosome and used as a convenient marker of embryo genotype. (See color plate 20.)

Genomic sequence and glycan structural analysis supports the existence of branching by endogenous GlcNAcT-2 and GlcNAcT-4 enzymes [2,4,32]. Extension by the addition of Gal is also detected but the responsible enzymes have yet to be identified. In fact, the best candidate GalT enzymes are more efficient at transferring GalNAc than Gal in vitro and may act primarily on glycosphingolipid substrates (see below) [33–36]. Capping with sialic acid has been detected on two very minor LacNAc-terminated glycans, reflecting the activity of an identified sialyltransferase, dSiaT [37]. The low prevalence of the sialylated glycans is consistent with the highly restricted expression of the dSiaT enzyme, which does not appear until very late in embryogenesis and is restricted to an exquisitely small number of neurons during larval, pupal, and adult stages, emphasizing the importance and difficulty of identifying the proteins that carry sialylated *N*-linked glycans [4,38].

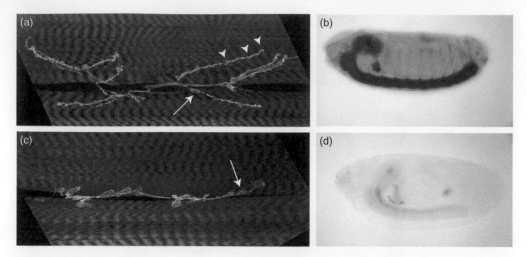

**FIGURE 13.4** Decreased HRP-epitope expression is associated with altered neuromuscular junction morphology. (a,c) The neuromuscular junctions (green) at two muscles (red) in the body wall of a wild-type (a) or EMS-mutant (c) larva are shown. In the wild-type, small (arrowhead) and large (arrow) nerve terminal specializations (boutons) are detected. In the mutant, small boutons are missing and the overall complexity of junctional branching is reduced. (b,d) Embryos of the same genotypes as shown in (a) and (c) stained with anti-HRP antibody (b, wild-type; d, EMS-mutant). (See color plate 21.)

Major unanswered questions regarding N-linked glycosylation in *Drosophila* remain to be addressed. Many key enzyme activities are still only sequence homologs and have not been fully characterized as active enzymes in vitro or in vivo. In some cases, substrate specificities determined in vitro are not corroborated by the detection of expected glycan structures in vivo. For instance, dSiaT exhibits an in vitro preference for LacdiNAc acceptors but the only sialylated glycans detected in *Drosophila* embryos possess subterminal LacNAc. Nucleotide sugar precursor flux is also uncharacterized in *Drosophila*. In particular, all the expected biosynthetic components for generating CMP-NeuAc are yet to be identified [39]. Furthermore, Golgi organization and trafficking dynamics are only partially characterized in *Drosophila* [40,41]. As was the case for understanding vertebrate N-linked glycosylation, the development of a broad range of compartment-specific markers in *Drosophila* would provide valuable tools for dissecting the underlying cellular architecture that supports N-linked glycosylation.

## O-Linked Glycoprotein Glycans

The characterization of glycans in O-linkage to animal glycoproteins had its beginnings in the study of mucins, the family of vertebrate secretory proteins that generally bear high densities of Ser/Thr-linked glycans. Pioneering biochemical characterization of *Drosophila* O-linked glycans revealed that the predominant structures are identical to the simplest mucin-type O-linked glycans of vertebrates, namely Tn-antigen (GalNAc) and the core 1 disaccharide known as T-antigen (Galβ3GalNAc) [4,18,42–47]. Lectin and antibody staining subsequently revealed intriguing subtleties in the tissue distribution and subcellular localization of these and other glycans [48–51]. Genomic survey and homology cloning has identified a family of 12 putative polypeptide GalNAc transferase genes (pgant family) that exhibit varying degrees of specificity and overlap in their tissue and cell type expression [52–54]. So far, mutants for two of these enzymes have been generated (pgant35a and pgant3), yielding cell polarity, embryo patterning, cell adhesion, and viability phenotypes [55–57]. Candidate approaches have also identified putative core 1 galactosyltransferase genes, one of which (C1GalTA) is required for normal development of the nervous system; mutant embryos display an elongated nerve cord and deformed brain hemispheres [58,59].

Other advances in *Drosophila* O-linked glycosylation have arisen from the characterization of mutants generated through various random genetic screens and from inspired, insightful connections drawn between vertebrate and *Drosophila* development. Foremost among these advances was the recognition that signaling through the Notch receptor required O-fucosylation of Ser/Thr residues located in appropriate EGF repeats in the Notch extracellular domain [60–62]. Depending on the context and the experimental approach, expression of the relevant O-fucosyl transferase (OFUT1) is either permissive for Notch signaling and/or necessary for trafficking of Notch to the cell surface [63]. Subsequent extension of the O-Fuc residue by the addition of GlcNAc is catalyzed by a specific β3GlcNAcTase known as Fringe [64–67]. Fringe (*fng*) activity shifts the ligand-binding preference of the Notch receptor, thereby generating specific patterns of cell-fate determination in the embryo and imaginal tissues. Elegant genetic approaches have probed and refined the respective biological functions of OFUT1 and Fng but direct demonstration of the respective glycosylation products has only been reported from cultured cells that overexpress the enzymes.

Recently, comprehensive characterization of the total O-linked glycan profile of developing *Drosophila* tissues has added a new layer of complexity to the O-Fuc and Fringe paradigm (Figure 13.5). Unbiased, sensitive, mass spectrometric characterization of the O-linked glycans released from glycoproteins extracted from *Drosophila* embryos and imaginal discs failed to detect O-Fuc or GlcNAcβ3Fuc [3]. Instead, a branched trisaccharide containing the GlcNAcβ3Fuc structure was detected. The branching residue, a GlcA linked β4 to Fuc, is novel and a relevant glucuronyl transferase is yet to be identified. The trisaccharide is detected in a tissue pattern consistent with a role in Notch signaling and also in a manner that is partially dependent on Fng. Loss of Fng does not completely eliminate the glycan, suggesting the presence of other Fng-like activities [68]. Furthermore, this glucuronylated O-Fuc structure is not a minor glycan in the embryo, indicating that the glycan will probably be found on other proteins in addition to Notch [3]. Therefore, the clear genetic and biochemical demonstrations of the importance of O-Fuc and Fng modifications for Notch signaling in *Drosophila* need to be evaluated in the context of a broader range of detected glycan structures and potential protein substrates.

Glucuronylation of O-linked glycans is not restricted to O-Fuc structures in *Drosophila*. GlcA was also detected on the core 1 disaccharide, both as a linear extension onto the terminal Gal residue and also as a branching residue linked to the

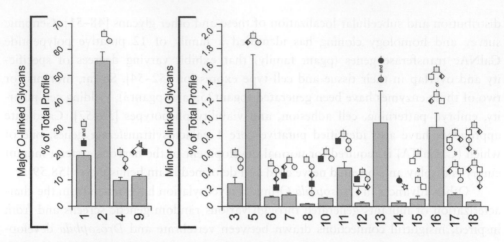

**FIGURE 13.5** The O-linked glycan profile of wild-type *Drosophila* embryos. O-linked glycans were released from *Drosophila* embryo proteins by reductive beta-elimination. The total glycan profile was detected, characterized, and quantified by HPLC, exoglycosidase digestion, and mass spectrometry [3]. The prevalence of each major and minor glycan is expressed as a percentage of the total pool of detected glycans.

internal GalNAc [3]. The linear form of the trisaccharide was first detected in glycans prepared from cultured *Drosophila* cells (S2) but the less prevalent branched glycan is, so far, unique to the intact organism [16]. GlcA is also incorporated into other extended O-linked glycans, including core 2 type structures. Other O-linked glycans with parallels in vertebrate glycobiology are also found in *Drosophila,* including O-GlcNAc, O-Man (without further extension), and O-Glc (with and without Xyl extension), and O-Fuc on thrombospondin repeats. Enzymes capable of transferring these monosaccharides (OGT, POMT1/POMT2 or Twisted/Rotated Abdomen, Rumi/POGLUT, OFUT-2) have also been described [69–76]. In contrast, sialylated O-linked glycans have not been detected in *Drosophila* tissues, despite the demonstrated existence of the dSiaT enzyme that modifies N-linked glycans.

The full diversity of O-linked glycans indicates that many glycosyltransferase activities remain to be identified in *Drosophila*. These missing enzyme activities include GlcA addition to Gal, GalNAc, and Fuc, the addition of HexNAc to GalNAc, and the addition of GlcNAc to GalNAc and GlcA (Figure 13.6). Some of these activities are evident in the biosynthesis of glycosaminoglycans and glycosphingolipids in *Drosophila*. However, in the context of core 1 and core 2 O-linked glycans, these activities are novel and targeted genetic analysis would provide new opportunities for dissecting O-linked glycan function.

## Glycosphingolipid Glycans

Insect glycosphingolipid biochemistry has a rich history of novel discoveries and rigorous structural characterization. The pioneering studies of Wiegandt, Sugita, Hori, and others demonstrated that arthropods build their ceramide-based glycolipids from a Manβ4Glcβ-Cer core, designated mactosylceramide, in comparison to the Galβ4Glcβ-Cer (lactosylceramide) of vertebrates [77–81]. Extension beyond the mactosylceramide

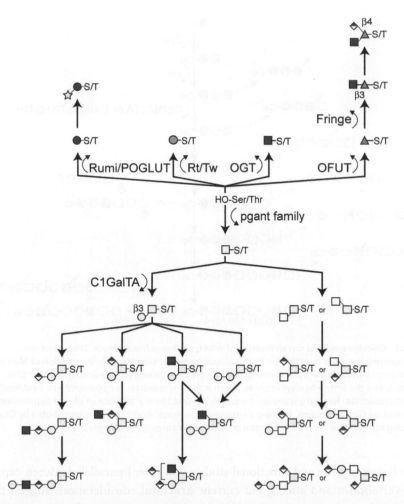

**FIGURE 13.6** Probable biosynthetic pathways for generating *O*-linked glycan diversity in *Drosophila*. Biosynthesis of each of the *O*-linked glycan families in *Drosophila* is initiated by distinct enzymatic activities. Rumi/POGLUT adds Glc, Rotated/Twisted (Rt/Tw) coordinately add Man, *O*-GlcNAc Transferase (OGT) adds GlcNAc, OFUT-1 and OFUT-2 add Fuc to distinct subsets of peptide substrates, and members of the polypeptide GalNAc transferase family (pgant) initiate mucin-type *O*-glycosylation. Subsequent extension steps are essentially uncharacterized, except for the addition of GlcNAc to Fuc by Fringe and the synthesis of the T-antigen by core 1 galactosyltransferases (C1GalTA). The broad distribution of GlcA residues on several types of core structures emphasizes that novel biosynthetic activities remain to be characterized.

core in the arthro-series glycosphingolipids is characterized by high levels of HexNAc residues, by the terminal addition of GlcA, and by modification of GlcNAc residues with 6-linked phosphoethanolamine [82–85] (Figure 13.7). The addition of capping GlcA residues has elicited comparisons to the sialic acid containing ganglioside glycosphingolipids of vertebrates. However, beyond the fact that GlcA and sialic acid are both acidic monosaccharides, very little existing data support such a parallel. Vertebrate sialic acids are frequently found as branching residues on glycosphingolipid glycan backbones and can also be covered by additional sialic acids in α2–8 linkage. In contrast, GlcA has only been found as a non-reducing terminal, capping residue linked to subterminal Gal in the arthro-series glycosphingolipids and GlcA dimers are unknown.

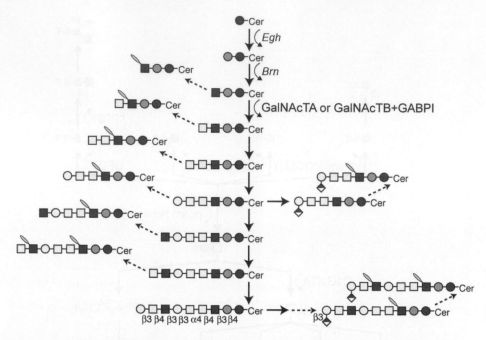

**FIGURE 13.7**   Glycosphingolipid glycan diversity of wild-type *Drosophila* embryos. The arthro-series glycosphingolipids are built by extension from a glucosylceramide core through the addition of Man by the Egghead mannosyltransferase (*Egh*), forming mactosylceramide (Manβ4Glcβ-ceramide). Brainiac (*Brn*), a GlcNAc transferase, forms the arthro-triaosylceramide, which is the first substrate for post-synthetic modification by phosphoethanolamine. Neutral glycans are frequently modified by the addition of phosphoethanolamine (orange ovals) on GlcNAc residues, forming zwitterionic structures. Acidic charge is contributed by GlcA linked to non-reducing terminal Gal residues on neutral or zwitterionic cores. (See color plate 22.)

Perhaps future genetic and functional studies will reveal parallels between capping residues in vertebrates and insects, but current structural considerations suggest that GlcA addition to the arthro-series and sialic acid addition to the ganglio-series glycosphingolipids are unrelated.

Genetic studies have contributed valuable insight into the function of *Drosophila* glycosphingolipids. Random mutagenesis screens produced neural development mutants called *egghead* and *brainiac,* so-named due to their phenotypic expansion of neural tissue [86–88]. In both mutants, over-proliferation of neural cells occurs at the expense of other cell lineages, indicating altered cell fate choices in the embryo. Molecular cloning and biochemical analysis revealed that *egghead* encodes a mannosyltransferase responsible for synthesis of the mactosylceramide core glycosphingolipid and *brainiac* encodes a GlcNAc transferase that extends this core. Although a molecular mechanism has not been elucidated, both *egghead* and *brainiac* interact with mutations in *notch*, indicating a central role for glycosphingolipids in cell signaling essential for normal development [89,90].

The triaosylceramide synthesized by Brainiac and the mactosylceramide synthesized by Egghead are relatively minor contributors to the full diversity of glycosphingolipids extracted from *Drosophila* embryos. Elongation effectively consumes the pool of mactosyl and triaosylceramide precursors to generate extended glycosphingolipids.

It remains to be determined whether arthro-series glycans elongated beyond the triao-sylceramide possess specific structural information that imparts function. However, genetic approaches indicate that extended glycosphingolipid glycans are important for normal development [91]. A search for *Drosophila* homologs of vertebrate galactosyltransferases identified a small number of candidates [33,35]. Biochemical analysis of recombinant forms of two of these enzymes indicated that one, designated GalNAcTA, preferentially transfers GalNAc, not Gal, to GlcNAc acceptors. The other candidate, designated GalNAcTB, was inactive. Subsequently, however, an ingenious expression-cloning screen identified an interacting protein (designated GABPI) with similarity to the DHHC acyltransferase family, which imparts GalNAcT activity to GalNAcTB [92]. GABPI appears to function by piloting GalNAcTB into Golgi membrane domains where it gains access to relevant acceptor substrates. Mutations in GalNAcTA generate defective locomotor behavior but have only a minor impact on glycosphingolipid composition [34,93]. On the other hand, mutations in GalNAcTB generate partially penetrant defects in epithelial morphogenesis but significantly alter glycosphingolipid composition [34,89]. Therefore, while these two enzymes exhibit partial redundancy, they must also contribute uniquely to the synthesis of specific structures in restricted cellular contexts. Currently, the only candidate GalT enzyme with demonstrated activity toward GalNAc-terminated glycosphingolipids is the enzyme previously identified as the *Drosophila O*-linked core 1 GalT (C1GalTA) [59]. The extent to which this enzyme serves double-duty in vivo for synthesis of core 1 disaccharide and for glycosphingolipid extension remains to be quantified.

The final levels of complexity found in the arthro-series glycosphingolipid glycans are terminal glucuronylation of Gal and phosphoethanolamine addition to GlcNAc residues. The relevant enzymes remain to be determined for both of these processes, although candidate GlcAT enzymes have been annotated within the genome and partially characterized as recombinantly expressed proteins [94]. The biosynthetic machinery responsible for phosphoethanolamine addition is a greater mystery. It is not known whether this post-synthetic modification is capable of modulating functional interactions between *Drosophila* glycosphingolipid glycans and other lipid or protein effectors, but it is noteworthy that the addition of one phosphoethanolamine to a neutral glycosphingolipid generates a zwitterionic species [81]. Further additions of phosphoethanolamine onto zwitterionic or acidic cores produce glycosphingolipid species with increasingly distinct biophysical characteristics. The fully and partially charged moieties (GlcA and phosphoethanolamine, respectively) on extended glycosphingolipid glycans may significantly impact the partition behavior of these lipids within the plane of the membrane and may also modulate their activities as molecular targets for cell recognition or protein interaction.

## Summary and Perspective

High-resolution probes for glycan expression are gradually becoming available for assessing the specificity of glycosylation in *Drosophila*. Structural analysis clearly demonstrates that *N*- and *O*-linked profiles are dominated by minimally processed glycans. For *N*-linked glycans, high- and pauci-mannose glycans predominate, reflecting the competition between elongation and removal of the GlcNAc added by GlcNAcT-1. Genetic mutations can shift this dynamic in specific tissues and the elaboration of glycan complexity can be restricted to very small numbers of cells, providing opportunities to decipher important regulatory mechanisms. Similarly, *O*-linked glycosylation is dominated by the relatively simple T- and Tn-antigen glycans. However, the detection of novel *O*-linked structures indicates that the functional ramifications of *O*-linked glycan diversity are not completely understood. *Drosophila* glycosphingolipid structures are markedly different than those of vertebrate species but, like their vertebrate counterparts, they modulate important cell signaling events during development. Further analysis of the importance of these glycans for *Drosophila* development may also establish new paradigms for investigating vertebrate glycosphingolipid function. Generally, *Drosophila* expresses minor, highly processed glycans in restricted cellular contexts at exquisitely sensitive choice points during development. This strict regulation of complex glycan expression presents analytic challenges but also opens new opportunities for functional studies.

It is almost cliché that the tools available for genetic analysis of function make *Drosophila* a superior experimental system for many purposes. Fortunately, in relation to glycobiology, this cliché is supported by almost two decades of insightful genetics and biochemistry. Advances in cell signaling, developmental patterning, cell fate selection, axon pathfinding, and the formation of epithelial polarity have arisen from understanding the contribution of glycans to *Drosophila* development and cellular function. For the most part, these advances have been driven by random mutagenesis strategies and the subsequent realization that mutagenized genes impact glycan synthesis or processing. That such non-biased approaches for gene discovery have begun to reveal the functional importance of glycans provides further impetus for expanding on-going efforts to enhance the structural analysis of minor glycans and to design new, targeted genetic strategies. Understanding the mechanisms that limit the complexity of glycosylation in this simple organism may reveal that the regulation of glycosylation in more complex organisms obeys a simple set of common rules.

# References

1. Ten Hagen KG, Zhang L, Tian E, Zhang Y. Glycobiology on the fly: developmental and mechanistic insights from *Drosophila*. *Glycobiology* 2009;**19**:102–11.

2. Aoki K, Perlman M, Lim J-M, Cantu R, Wells L, Tiemeyer M. Dynamic developmental elaboration of *N*-linked glycan complexity in the *Drosophila melanogaster* embryo. *J Biol Chem* 2007;**282**:9127–42.

3. Aoki K, Porterfield M, Lee SS, Dong B, Nguyen K, McGlamry KH, Tiemeyer M. The diversity of *O*-linked glycans expressed during *Drosophila melanogaster* development reflects stage- and tissue-specific requirements for cell signaling. *J Biol Chem* 2008;**283**:30385–400.

4. North SJ, Koles K, Hembd C, Morris HR, Dell A, Panin VM, Haslam SM. Glycomic studies of *Drosophila melanogaster* embryos. *Glycoconj J* 2006;**23**:345–54.

5. Beckett K, Franch-Marro X, Vincent JP. Glypican-mediated endocytosis of Hedgehog has opposite effects in flies and mice. *Trends Cell Biol* 2008;**18**:360–3.

6. Esko JD, Selleck SB. Order out of chaos: assembly of ligand binding sites in heparan sulfate. *Annu Rev Biochem* 2002;**71**:435–71.

7. Kramer KL, Yost HJ. Heparan sulfate core proteins in cell-cell signaling. *Annu Rev Genet* 2003;**37**:461–84.

8. Lander AD, Selleck SB. The elusive functions of proteoglycans: in vivo veritas. *J Cell Biol* 2000;**148**:227–32.

9. Lin X. Functions of heparan sulfate proteoglycans in cell signaling during development. *Development* 2004;**131**:6009–21.

10. Selleck SB. Genetic dissection of proteoglycan function in *Drosophila* and C. elegans. *Semin Cell Dev Biol* 2001;**12**:127–34.

11. Marchal I, Mir AM, Kmiecik D, Verbert A, Cacan R. Use of inhibitors to characterize intermediates in the processing of *N*-glycans synthesized by insect cells: a metabolic study with Sf9 cell line. *Glycobiology* 1999;**9**:645–54.

12. Parker GF, Williams PJ, Butters TD, Roberts DR. Detection of lipid-linked precursor oligosaccharide of *N*-linked protein glycosylation in *Drosophila melanogaster*. *FEBS Letters* 1991;**290**:58–60.

13. Seppo A, Tiemeyer M. Function and structure of *Drosophila* glycans. *Glycobiology* 2000;**10**:751–60.

14. Williams PJ, Wormald MR, Dwek RA, Rademacher TW, Parker GF, Roberts DR. Characterization of oligosaccharides from *Drosophila melanogaster* glycoproteins. *Bochimica et Biophysica Acta* 1991;**1075**:146–53.

15. Paschinger K, Staudacher E, Stemmer U, Fabini G, Wilson IB. Fucosyltransferase substrate specificity and the order of fucosylation in invertebrates. *Glycobiology* 2004;**15**:463–74.

16. Breloy I, Schwientek T, Lehr S, Hanisch F-G. Glucuronic acid can extend O-linked core 1 glycans, but it contributes only weakly to the negative surface charge of *Drosophila melanogaster* Schneider-2 cells. *FEBS Letters* 2008;**582**:1593–8.

17. Hollister J, Grabenhorst E, Nimtz M, Conradt H, Jarvis DL. Engineering the protein N-glycosylation pathway in insect cells for production of biantenary, complex *N*-glycans. *Biochemistry* 2002;**17**:15093–104.

18. Schwientek T, Mandel U, Roth U, Muller S, Hanisch F-G. A serial lectin approach to the mucin-type *O*-glycoproteome of *Drosophila melanogaster* S2 cells. *Proteomics* 2007;**7**:3264–77.

19. Hollister JR, Shaper JH, Jarvis DL. Stable expression of mammalian beta 1,4-galactosyltransferase extends the *N*-glycosylation pathway in insect cells. *Glycobiology* 1998;**8**:473–80.

20. Jarvis DL, Aumiller JJ. Novel baculovirus expression vectors that provide sialylation of recombinant glycoproteins in lepidopteran insect cells. *J Virol* 2001;**75**:6223–7.

21. Sarkar M, Leventis PA, Silvescu CI, Reinhold VN, Schachter H, Boulianne GL. Null mutations in *Drosophila* N-acetylglucosaminyltransferase I produce defects in locomotion and a reduced life span. *J Biol Chem* 2006;**281**:12776–85.

22. Sarkar M, Schachter H. Cloning and expression of *Drosophila melanogaster* UDP-GlcNAc:alpha-3-D-mannoside beta1,2-*N*-acetylglucosaminyltransferase I. *Biological Chemistry* 2001;**382**:209–17.

23. Geisler C, Aumiller JJ, Jarvis DL. A fused lobes gene encodes the processing beta-N-acetylglucosamine in Sf9 cells. *J Biol Chem* 2008;**283**:11330–9.

24. Altmann F, Schwihla H, Staudacher E, Glössl J, Marz L. Insect cells contain an unusual, membrane-bound beta-N-acetylglucosaminidase probably involved in the processing of protein N-glycans. *J Biol Chem* 1995;**270**:17344–9.

25. Aumiller JJ, Hollister JR, Jarvis DL. Molecular cloning and functional characterization of beta-N-acetylglucosaminidase genes from Sf9 cells. *Protein Expression and Purification* 2006;**47**:571–90.

26. Leonard R, Rendic D, Rabouille C, Wilson IB, Preat T, Altmann F. The *Drosophila* fused lobes gene encodes an N-acetylglucosaminidase involved in N-glycan processing. *J Biol Chem* 2006;**281**:4867–75.

27. Fabini G, Freilinger A, Altmann F, Wilson IB. Identification of core alpha 1,3 fucosylated glycans and cloning of the requisite fucosyltransferase cDNA from *Drosophila melanogaster*. Potential basis of the neural anti-horseradish peroxidase epitope. *J Biol Chem* 2001;**276**(30):28058–67.

28. Jan LY, Jan YN. Antibodies to horseradish peroxidase as specific neuronal markers in *Drosophila* and grasshopper embryos. *Proc Natl Acad Sci USA* 1982;**79**:2700–4.

29. Snow PM, Patel NH, Harrelson AL, Goodman CS. Neural-specific carbohydrate moiety shared by many surface glycoproteins in *Drosophila* and grasshopper embryos. *J Neuroscience* 1987;**7**:4137–44.

30. Seppo A, Matani P, Sharrow M, Tiemeyer M. Induction of neuron-specific glycosylation by Tollo/Toll-8, a *Drosophila* Toll-like receptor expressed in non-neural cells. *Development* 2003;**130**:1439–48.

31. Katz F, Moats W, Jan YN. A carbohydrate epitope expressed uniquely on the cell surface of *Drosophila* neurons is altered in the mutant *nac* (neurally altered carbohydrate). *EMBO J* 1988;**7**:3471–7.

32. Tsitilou SG, Grammenoudi S. Evidence for alternative splicing and developmental regulation of the *Drosophila melanogaster* Mgat2 (N-acetylglucosaminyltransferase II) gene. *Biochem Biophys Res Commun* 2003;**312**:1372–6.

33. Sasaki N, Yoshida H, Fuwa TJ, Kinoshita-Toyoda A, Toyoda H, Hirabayashi Y, Ishida H, Ueda R, Nishihara S. *Drosophila* beta 1,4-N-acetylgalactosaminyltransferase-A synthesizes the LacdiNAc structures on several glycoproteins and glycosphingolipids. *Biochem Biophys Res Commun* 2007;**354**:522–7.

34. Stolz A, Haines N, Pich A, Irvine KD, Hokke CH, Deelder AM, Gerardy-Schahn R, Wuhrer M, Bakker H. Distinct contributions of beta 4GalNAcTA and beta 4GalNAcTB to *Drosophila* glycosphingolipid biosynthesis. *Glycoconj J* 2008;**25**:167–75.

35. Haines N, Irvine KD. Functional analysis of *Drosophila* beta1,4-N-acetylgalactosaminyltransferases. *Glycobiology* 2005;**15**:335–46.

36. Vadaie N, Jarvis DL. Molecular cloning and functional characterization of a Lepidopteran insect beta4-N-acetylgalactosaminyltransferase with broad substrate specificity, a functional role in glycoprotein biosynthesis, and a potential functional role in glycolipid biosynthesis. *J Biol Chem* 2004;**279**:33501–18.

37. Koles K, Irvine KD, Panin VM. Functional characterization of *Drosophila* sialyltransferase. *J Biol Chem* 2004;**279**:4346–57.

38. Koles K, Lim J-M, Aoki K, Porterfield M, Tiemeyer M, Wells L, Panin V. Identification of N-glycosylated proteins from the central nervous system of *Drosophila melanogaster*. *Glycobiology* 2007;**17**:1388–403.

39. Koles K, Repnikova E, Pavlova G, Korochkin LI, Panin VM. Sialylation in protostomes: a perspective from *Drosophila* genetics and biochemistry. *Glycoconj J* 2009;**26**(3):313–24.

40. Bard F, Casano L, Mallabiabarrena A, Wallace E, Saito K, Kitayama H, Guizzunti G, Hu Y, Wendler F, Dasgupta R, Perrimon N, Malhotra V. Functional genomics reveals genes involved in protein secretion and Golgi organization. *Nature* 2006;**439**:604–7.

41. Yano H, Yamamoto-Hino M, Abe M, Kuwahara R, Haraguchi S, Kusaka I, Awano W, Kinoshita-Toyoda A, Toyoda H, Goto S. Distinct functional unit of the Golgi complex in *Drosophila* cells. *Proc Natl Acad Sci USA* 2005;**102**:13467–72.

42. Kramerov AA, Arbatsky NP, Rozovsky RM, Mickhaleva EA, Polesskaya OO, Gvozdev VA, Shibaev FN. Mucin-type glycoprotein from *Drosophila melanogaster* embryonic cells: characterization of carbohydrate component. *FEBS Letters* 1996;**378**:213–18.

43. Kramerov AA, Mikhaleva EA, Rozovsky YM, Pochechueva TV, Baikova NA, Arsenjeva EL, Gvozdev VA. Insect mucin-type glycoprotein: immunodetection of the O-glycosylated epitope in *Drosophila melanogaster* cells and tissues. *Insect Biochem Mol Biol* 1997;**27**:513–21.

44. Theopold U, Dorian C, Schmidt O. Changes in glycosylation during *Drosophila* development. The influence of ecdysone on hemomucin isoforms. *Insect Biochem Mol Biol* 2001;**31**:189–97.

45. Lopez M, Tetaert D, Juliant S, Gazon M, Cerutti M, Verbert A, Delanoy P. O-glycosylation potential of lepidopteran insect cell lines. *Biochim Biophys Acta* 1999;**1427**:49–61.

46. Thomsen DR, Post LE, Elhammer AP. Structure of O-glycosidically linked oligosaccharides synthesized by the insect cell line Sf9. *J Cell Biochem* 1990;**43**:67–79.

47. Uttenweiler-Joseph S, Moniatte M, Lambert J, Van Dorsselaer A, Bulet P. A matrix-assisted laser desorption ionization time-of-flight mass spectrometry approach to identify the origin of the glycan heterogeneity of diptericin, an O-glycosylated antibacterial peptide from insects. *Anal Biochem* 1997;**247**:366–75.

48. Callaerts P, Vulsteke V, Peumans W, De Loof A. Lectin binding sites during *Drosophila* embryogenesis. Development Genes and Evolution 1995;**204**:229–43.

49. D'Amico P, Jacobs JR. Lectin histochemistry of the *Drosophila* embryo. *Tissue & Cell* 1995;**27**:23–30.

50. Fredieu JR, Mahowald AP. Glycoconjugate expression during *Drosophila* embryogenesis. *Acta Anat* 1994;**149**:89–99.

51. Tian E, Ten Hagen G. O-linked glycan expression during *Drosophila* development. *Glycobiology* 2007;**17**:820–7.

52. Ten Hagen KG, Fritz TA, Tabak LA. All in the family: the UDP-GalNAc:polypeptide N-acetylgalactosaminyltransferases. *Glycobiology* 2003;**13**:1R–16R.

53. Ten Hagen KG, Tran DT, Gerken TA, Stein DS, Zhang Z. Functional characterization and expression analysis of members of the UDP-GalNAc:polypeptide N-acetylgalactosaminyltransferase family from *Drosophila melanogaster*. *J Biol Chem* 2003;**278**:35039–48.

54. Tian E, Ten Hagen K. Expression of the UDP-GalNAc:polypeptide N-acetylgalactosaminyltransferase family is spatially and temporally regulated during *Drosophila* development. *Glycobiology* 2006;**16**:83–95.

55. Zhang L, Zhang Y, Ten Hagen KG. A mucin-type O-glycosyltransferase modulates cell adhesion during *Drosophila* development. *J Biol Chem* 2008;**283**:34076–86.

56. Ten Hagen KG, Tran DT. A UDP-GalNAc:polypeptide N-acetylgalactosaminyltransferase is essential for viabilty in *Drosophila melanogaster*. *J Biol Chem* 2002;**277**:22616–22.

57. Tian E, Ten Hagen KG. A UDP-GalNAc:polypeptide N-acetylgalactosaminyltransferase is required for epithelial tube formation. *J Biol Chem* 2007;**282**:606–14.

58. Lin YR, Reddy BV, Irvine KD. Requirement for a core 1 galactosyltransferase in the *Drosophila* nervous system. *Dev Dyn* 2008;**237**:3703–14.

59. Muller R, Hulsmeier AJ, Altmann F, Ten Hagen K, Tiemeyer M, Hennet T. Characterization of mucin-type core-1 beta1–3 galactosyltransferase homologous enzymes in *Drosophila melanogaster*. *FEBS Journal* 2005;**272**:4295–305.

60. Okajima T, Irvine KD. Regulation of notch signaling by o-linked fucose. *Cell* 2002;**111**:893–904.

61. Panin VM, Shao L, Lei L, Moloney DJ, Irvine KD, Haltiwanger RS. Notch ligands are substrates for protein O-fucosyltransferase-1 and Fringe. *J Biol Chem* 2002;**277**:29945–52.

62. Stanley P. Regulation of Notch signaling by glycosylation. *Current Opinion in Structural Biology* 2007;**17**:530–5.

63. Okajima T, Reddy B, Matsuda T, Irvine KD. Contributions of chaperone and glycosyltransferase activities of O-fucosyltransferase 1 to Notch signaling. *BMC Biology* 2008;**6**:1.

64. Haines N, Irvine KD. Glycosylation regulates Notch signaling. *Nat Rev Mol Cell Biol* 2003;**4**:786–97.

65. Haltiwanger RS, Stanley P. Modulation of receptor signaling by glycosylation: fringe is an O-fucose-beta1,3-N-acetylglucosaminyltransferase. *Biochim Biophys Acta* 2002;**1573**:328–35.

66. Moloney DJ, Panin VM, Johnston SH, Chen J, Shao L, Wilson R, Wang Y, Stanley P, Irvine KD, Haltiwanger RS, Vogt TF. Fringe is a glycosyltransferase that modifies Notch. *Nature* 2000;**406**:357–8.

67. Xu A, Haines N, Dlugosz M, Rana NA, Takeuchi H, Haltiwanger RS, Irvine KD. In vitro reconstitution of the modulation of *Drosophila* Notch-ligand binding by Fringe. *J Biol Chem* 2007;**282**:35153–62.

68. Correia T, Popayannopoulos V, Panin V, Woronoff P, Jiang J, Vogt TF, Irvine KD. Molecular genetic analysis of the glycosyltransferase Fringe in *Drosophila*. *Proc Natl Acad Sci USA* 2003;**100**:6404–9.

69. Acar M, Jafar-Nejad H, Takeuchi H, Rajan A, Ibrani D, Rana NA, Pan H, Haltiwanger RS, Bellen HJ. Rumi is a CAP10 domain glycosyltransferase that modifies Notch and is required for Notch signaling. *Cell* 2008;**132**:247–58.

70. Ichimiya T, Manya H, Ohmae Y, Yoshida H, Takahashi K, Ueda R, Endo T, Nishihara S. The twisted abdomen phenotype of Drosphila POMT1 and POMT2 mutants coincides with their heterophilic protein O-mannosyltransferase activity. *J Biol Chem* 2004;**279**:42638–47.

71. Ishimizu T, Sano K, Uchida T, Teshima H, Omichi K, Hojo H, Nakahara Y, Hase S. Purification and substrate specificity of UDP-D-xylose:beta-D-glucoside alpha-1,3-D-xylosyltransferase involved in the biosynthesis of the Xyl alpha1–3Xyl alpha1–3Glc beta1-O-Ser on epidermal growth factor-like domains. *J Biochemistry* 2007;**141**:593–600.

72. Kelly WG, Hart GW. Glycosylation of chromosomal proteins: localization of O-linked N-acetylglucosamine in *Drosophila* chromatin. *Cell* 1989;**57**:243–51.

73. Kreppel LK, Bloomberg MA, Hart GW. Dynamic glycosylation of nuclear and cytosolic proteins. Cloning and characterization of a unique O-GlcNAc transferase with multiple tetratricopeptide repeats. *J Biol Chem* 1997;**272**:9308–15.

74. Lyalin D, Koles K, Roosendaal SD, Repnikova E, Van Wechel L, Panin VM. The twisted gene encodes *Drosophila* protein O-mannosyltransferase 2 and genetically interacts with the rotated abdomen gene encoding *Drosophila* protein O-mannosyltransferase 1. *Genetics* 2006;**172**:343–53.

75. Martin-Blanco E, Garcia-Bellido A. Mutations in the rotated abdomen locus affect muscle development and reveal an intrinsic asymmetry in *Drosophila*. *Proc Natl Acad Sci USA* 1996;**93**:6048–52.

76. Shao L, Luo Y, Moloney DJ, Haltiwanger RS. O-glycosylation of EGF repeats: Identification and initial characterization of a UDP-glucose:protein O-glucosyltransferase. *Glycobiology* 2002;**12**:763–70.

77. Sugita M, Iwasaki Y, Hori T. Studies on glycosphingolipids of larvae of the green-bottle fly, Lucillia caesar. II. Isolatiion and structural studies of three glycosphingolipids with novel sugar sequences. *J Biochemistry* 1982;**92**:881–7.

78. Sugita M, Nishida M, Hori T. Studies on glycosphingolipids of larvae of the green-bottle fly, *Lucilia caesar*, I. Isolation and characterization of glycosphingolipids having novel sugar sequences. *J Biochemistry* 1982;**92**:327–34.

79. Breidbach O, Dennis RD, Keller M, Wiegandt H. Evidence for the expression of a glucuronic acid-containing epitope in the central nervous system of two insects (*Calliphora vicina,* Diptera; *Tenebrio molitor,* Coleoptera). *Neuroscience Letters* 1990;**109**:265–70.

80. Briedbach O, Dennis R, Marx J, Görlach C, Wiegandt H, Wegerhoff R. Insect glial cells show differential expression of a glycolipid-derived glucuronic acid-containing epitope throughout neurogenesis: detection during postembryogenesis and regeneration in the central nervous system of *Tenebrio molitor*. *Neuroscience Letters* 1992;**147**:5–8.

81. Wiegandt H. Insect glycolipids. *Biochimica et Biophysica Acta* 1992;**1123**:117–26.

82. Sugita M, Inagaki F, Naito H, Hori T. Studies on glycosphingolipids in larvae of the green-bottle fly, *Lucilia caesar*: two neutral glycosphingolipids having large straight oligosaccharide chains with eight and nine sugars. *J Biochemistry* 1990;**107**:899–903.

83. Itonori S, Nishizawa M, Suzuki M, Inagaki F, Hori T, Sugita M. Polar glycosphingolipids in insect: chemical studies of glycosphingolipid series containing 2′-aminoethylphosphoryl-6-N-acetylglucosamine as a polar group from larvae of the green-bottle fly, *Lucilia caesar*. *J Biochemistry* 1991;**110**:479–85.

84. Sugita M, Itonori S, Inagaki F, Hori T. Characterization of two glucuronic acid-containing glycosphingolipids in larvae of the green-bottle fly, *Lucilia caesar*. *J Biol Chem* 1989;**264**:15028–33.

85. Seppo A, Moreland M, Schweingruber H, Tiemeyer M. Zwitterionic and acidic glycolipids of the *Drosophila melanogaster* embryo. *Eur J Biochem* 2000;**267**:3549–58.

86. Muller R, Altmann F, Zhou D, Hennet T. The *Drosophila melanogaster* brainiac protein is a glycolipid-specific beta 1,3N-acetylglucosaminyltransferase. *J Biol Chem* 2002;**277**:32417–20.

87. Schwientek T, Keck B, Levery SB, Jensen MA, Pedersen JW, Wandall HH, Stroud M, Cohen SM, Amado M, Clausen H. The *Drosophila* gene brainiac encodes a glycosyltransferase putatively involved in glycosphingolipid synthesis. *J Biol Chem* 2002;**277**:32421–9.

88. Wandall HH, Pedersen JW, Park C, Levery SB, Pizette S, Cohen SM, Schwientek T, Clausen H. *Drosophila* egghead encodes a beta 1,4-mannosyltransferase predicted to form the immediate precursor glycosphingolipid substrate for brainiac. *J Biol Chem* 2003;**278**:1411–14.

89. Chen YW, Pedersen JW, Wandall HH, Levery SB, Pizette S, Clausen H, Cohen SM. Glycosphingolipids with extended sugar chains have specialized functions in development and behavior of *Drosophila*. *Dev Biol* 2007;**306**:736–49.

90. Goode S, Melnick M, Chou TB, Perrimon N. The neurogenic genes egghead and brainiac define a novel signaling pathway essential for epithelial morphogenesis during *Drosophila* oogenesis. *Development* 1996;**122**:3863–79.

91. Wandall HH, Pizette S, Pedersen JW, Eichert H, Levery SB, Mandel U, Cohen SM, Clausen H. Egghead and brainiac are essential for glycosphingolipid biosynthesis in vivo. *J Biol Chem* 2005;**280**:4858–63.

92. Johswich A, Kraft B, Wuhrer M, Berger M, Deelder AM, Hokke CH, Gerardy-Schahn R, Bakker H. Golgi targeting of *Drosophila melanogaster* beta4GalNAcTB requires a DHHC protein family-related protein as a pilot. *J Cell Biol* 2009;**184**:173–83.

93. Haines N, Stewart BA. Functional roles for beta1,4-N-acetylgalactosaminyltransferase-A in *Drosophila* larval neurons and muscles. *Genetics* 2007;**175**:671–9.

94. Kim BT, Tsuchida K, Lincecum J, Kitagawa H, Bernfield M, Sugahara K. Identification and characterization of three *Drosophila melanogaster* glucuronyltransferases responsible for the synthesis of the conserved glycosaminoglycan-protein linkage region of proteoglycans. Two novel homologs exhibit broad specificity toward oligosaccharides from proteoglycans, glycoproteins, and glycosphingolipids. *J Biol Chem* 2003;**278**:9116–24.

# 14
# Malaria

Eugene Davidson

*DEPARTMENT OF CHEMISTRY, GEORGETOWN UNIVERSITY,
WASHINGTON, D.C., USA*

## Introduction

Malaria has a yearly case incidence in excess of 200 million; although the mortality rate seems relatively low, approximately 1%, this disease still accounts for about two million deaths yearly [1–3]. Actual figures are difficult to obtain due to the lack of medical infrastructure in the less-developed areas where the disease is prevalent, but it is important to note that most of the deaths occur in children under the age of 12. There is no vaccine and relatively little research on new drugs to contain this problem.

The causative agent of malaria is a protozoan parasite, species *Plasmodium*. Four species infect humans: *P. falciparum* (the most prevalent and responsible for 90+% of the mortality), *P. vivax*, *P. ovale*, and *P. malariae*. The latter two are far less common. Interestingly, *P. falciparum* is responsible for most deaths but recovery leaves the patient free of parasite, whereas *P. vivax* and the others have the capacity to go "cryptic," with documented recrudescences occurring as long as 20 years after the initial exposure. The tropism of the parasite is quite remarkable. Many mammals can contract malaria but those plasmodial species that infect humans will not infect routine laboratory animals (mice, rats, guinea-pigs) or even closely related primates (rhesus, chimpanzee, gorilla). Those mammals may still get malaria but the causative parasite does not infect humans. *P. falciparum* will infect the owl monkey (*Aotus* species) but the infection does not always take (in general, the animal needs to be splenectomized) and the disease pathology is often very different from that seen in the human host. The lack of a convenient animal model and the difficulty in growing the parasite in vitro (of the four species that infect humans, only the erythrocytic stage of *P. falciparum* can be grown in vitro) are contributing factors to the overall dearth of progress in the global management of this disease.

Malaria is transmitted by an insect vector—in the case of humans, the *Anopheles* mosquito. Disease control that targets the vector (insecticide, bed nets) has had some limited success but is not a long-term solution. The magnitude of the problem is underscored by the reintroduction of DDT in several areas of sub-Saharan Africa.

**Handbook of Glycomics**
Copyright © 2009 by Elsevier Inc. All rights of reproduction in any form reserved.

## Life Cycle

We can visualize the life cycle as beginning with the bite of an infected mosquito [3]. The female *Anopheles* relies on a blood meal to feed the eggs she is carrying. The heat-seeking insect lands on a human target and, prior to feeding, salivates onto the skin. This action has two purposes—the saliva contains both an anticoagulant and a topical anesthetic. This serves to facilitate blood flow and leave the donor largely unaware of what is happening. In an infected mosquito, the parasite is resident in the salivary gland and enters the circulation of the human host during feeding. Approximately 25 parasites (sporozoites) are transmitted [4,5]—if they do not enter a blood vessel directly, they can migrate through the dermis and enter a capillary within a few minutes. Once in the circulation, the parasites are rapidly cleared (about 15 minutes) via uptake into liver cells—initially into Kupffer cells but eventually into hepatocytes (first recognition event [6,7]). Once inside the liver cells, the parasites metamorphose into a new form (merozoites) and multiply. This process is pathologically silent and takes 7–10 days. At the end of this period, the infected liver cells rupture and up to several thousand merozoites are released into the circulation. This initiates the blood stage of the infection.

The released merozoite identifies, attaches to and invades the erythrocyte (second recognition event). This process takes between 5 and 30 minutes (failure to invade results in spontaneous lysis of the merozoite within an hour [8,9]. Once inside the erythrocyte, the merozoite multiplies (almost always producing a multiple of four—12, 16 ... up to 32) over the next 48 hours. Following the replication phase, the erythrocyte ruptures and the released merozoites reinvade fresh cells. The classic disease symptoms (fever, malaise) are associated with merozoite release. During a subsequent red cell cycle, a further transition occurs leading to the production of male and female gametocytes—this is likely stress related but the exact triggers are not known [10]. The gametocytes are acquired by a naïve mosquito during feeding. Once inside the mosquito, the gametocytes can mate and the resulting ookinete sequesters in the mosquito midgut (third recognition event). Development eventually produces sporozoites that migrate to the salivary gland to initiate the next cycle (fourth recognition event).

Additional features concern disease pathology and host response. The primary causes of death are overwhelming anemia and cerebral hemorrhage (fifth recognition event). It has also been known for some time that morbidity and mortality are more serious in pregnant women than age-matched cohorts (sixth recognition event).

The human immune system does not efficiently respond to this infection. Little or no cell-mediated reaction is seen and antibody production is sparse and short-lived. An individual who recovers from a *P. falciparum* infection has no significant immune memory and can be reinfected a few months later. There is little or no correlation between antibody levels to any malaria protein and disease protection. Lastly, it has long been known that individuals living in malaria endemic regions acquire resistance to disease pathology but still are able to function as carriers (i.e., they become infected, the parasite develops through to gametocytes, but the patient is not clinically ill).

Although infected erythrocytes are cleared with some efficiency by the spleen, they also have the capacity to avoid this fate by adherence to vascular endothelial cells. This function when combined with self-adherence (rosetting) can lead to capillary occlusion and subsequent hemorrhage (fifth recognition event).

# Glycobiology of *P. falciparum*

## Overview

As with eukaryotic organisms generally, the expectation is that some/most of the exo-directed surface proteins of plasmodia would be glycosylated. This is true for even simple unicellular organisms such as yeast as well as for other parasitic protozoa. However, in the case of plasmodia, the current evidence indicates that no significant glycosylation of surface proteins occurs in either the sporozoite or the merozoite—neither N- nor O-linked (less information is available regarding gametocyte surface proteins). This is not a function of the amino acid sequences of typical parasite proteins, nor of the inability of the parasite to synthesize nucleotide-linked saccharides. In fact, most of the exo-directed proteins of the parasite are embedded in the plasma membrane with a glycosylphosphatidylinositol anchor that characteristically contains glucosamine and mannose. Hence the parasite is able to synthesize both UDP-GlcNAc and GDP-Man but other intermediates of the classical N-glycosylation pathway appear to be absent.

During the erythrocytic cycle, the parasite goes through several morphologically distinct stages—rings, trophozoites, and schizonts—prior to the formation of mature merozoites. Although structures tentatively identified as Golgi have been seen, they are distinct only in trophozoites, and clearly different from the classical stacked membrane architecture associated with other eukaryotic cells [11]. The mature merozoite does not have such a structure. Nevertheless, it is clear that protein export does take place during the red cell stage—50 or more such products have been identified but none is known to be glycosylated. Most become embedded in the erythrocyte membrane where they serve functions needed for parasite growth and survival (e.g., transport of nutrients from the circulation to the multiplying parasite).

The interactions that lead to cellular recognition in both the human and mosquito, and erythrocyte binding all appear to involve saccharides on the host cells and, at least in the case of red cell invasion, differ among the various parasite species. It is noteworthy that an organism that does little or nothing in the way of N- or O-glycosylation retains dependency on saccharide-based recognition for essential functions.

# Experimental Results

## The Sporozoite (first recognition event)

A characteristic feature of the sporozoite is the presence of a major surface protein of approximately 45 000 molecular weight. This molecule, termed circumsporozoite protein

(CSP), is present in all species, distributed uniformly over the sporozoite surface and is embedded by means of a glycosylphosphatidylinositol (GPI) anchor [12–15].

A characteristic of the sporozoite stage of the parasite is that once it enters the bloodstream, it is cleared into hepatocytes extremely rapidly. Entry into the hepatocytes is mediated by passage through Kupffer cells once the parasite has reached the liver sinusoids [16–19]. This binding/invasion is strongly suggestive of a specific receptor–ligand interaction involving one or more elements on the sporozoite surface and exo-directed molecules on the liver cells. It should be appreciated that surface binding and invasion represent two different biological events with the latter likely more complex. In addition, the binding event must be sufficiently robust as to overcome the recoil effect anticipated after collision between a moving body (sporozoite) and a stationary target (liver cell).

Initial studies were carried out with recombinant circumsporozoite protein (CSP) although subsequent work indicated a role for a protein designated TRAP (thrombospondin-related anonymous protein) [20–23]. In both of these proteins, there is amino acid sequence homology to other proteins known to bind to heparin [24–26].

Direct interaction of CSP with glycosaminoglycan and other polyaninoic carbohydrates was measured by chromatography using Sepharose-immobilized saccharides and recombinant CSP [27]. The results showed relatively strong binding to heparin, fucoidan, and dextran sulfate (high molecular weight, approximately 500 000) but weak affinity for chondroitin 4- or 6-sulfate. In addition, binding of the recombinant protein to a hepatocyte cell in culture was diminished in cells deficient in the synthesis of heparan sulfate [28–30]. In this same study, sulfatide was found to be an effective ligand for sporozoites but other glycolipids, including gangliosides, were not; the likelihood is that this resulted from interaction with micelles but this was not documented. These data suggest that a charge-based interaction has a role in sporozoite attachment to hepatocytes but give no indication of any specific structural array that may be required. A key control showed that the protein core of the heparan sulfate proteoglycan had no effect.

In parallel, it was shown that lipoprotein remnants that are known to bind to heparan sulfate can decrease the ability of sporozoites to infect hepatocytes [31,32]. In addition, anionic polysaccharides such as those shown in the affinity studies can reduce infectivity in vivo [33–35].

In subsequent work, it was shown that both heparan sulfate and chondroitin sulfate on Kupffer cells can bind CSP whereas the latter is the primary ligand on the hepatocyte surface [35].

An apparently contradictory study showed that removal of cell surface heparan sulfate did not significantly diminish the ability of sporozoites to invade liver cells [36]. This work was extended, however, to demonstrate the importance of the charge-based interaction under dynamic conditions (active flow, as opposed to a simple over-lay of cells) [37]. In this latter work, the invasion assay was modified so that cellular interactions occurred in a flow chamber that could mimic conditions in vivo. Data showed that heparin was an effective inhibitor of sporozoite binding to hepatocytes whereas cytochalasin had no effect. Chemical removal of either O-linked or N-linked sulfate groups sharply diminished the inhibitory action of heparin. In addition, inhibition of glycosaminoglycan chain sulfation by chlorate resulted in decreased sporozoite binding in a dose-dependent manner. Interestingly, heparin in solution had relatively little effect on sporozoite invasion in these studies.

Note that although heparin is not present on the surface of hepatocytes, the nature of the heparan sulfate associated with those cells is more closely related to heparin than that present on other cell types (i.e., high sulfate content) [38]. It had already been established that CSP is more effectively bound by heparan sulfate domains of high sulfate content, analogous to those more frequently found in heparin [39–41].

With respect to TRAP, this protein is not surface immobilized as is CSP and its explicit role in binding/invasion is not completely defined. Nevertheless, it does bind to both heparin and heparan sulfate [42–44].

The tentative conclusion is that the CSP present uniformly on the sporozoite surface is probably the initial site of interaction with liver cells. This is likely charge-based since charge interactions are stronger than those involving secondary forces. Multiple sites on both the sporozoite and the liver cells provide the energy needed to overcome any recoil effect and could explain data from solution studies with single chains of any glycosaminoglycan. The apparent relationship between charge density and the requirement for sulfation supports this interpretation.

Given that the invasion of one cell by another is a highly complex process that likely involves both membrane and cytoskeletal components, many individual molecules are likely to be essential. However, a common facet must be a binding event that is sufficiently robust to allow contact to mature into entry. The initial interaction between the liver cell and the parasite surface may well be charge-based since those forces will be stronger at any distance than hydrogen-bond or Van der Waals attractions. It is also true that the specificity of these events argues for something more explicit than two regions of opposite charge. To date, that level of structural detail has not been resolved for sporozoite interactions (contrast with the anti-thrombin site of heparin).

The three possible uronide sugars and five possible hexosamines allow for a large variety of tetra- or pentasaccharides in the heparan sulfate structure. Binding sites in proteins for linear saccharide chains frequently encompass four or five sugars allowing for recognition of a specific sequence in the polysaccharide. It has been established that the heparan sulfate chains on the surface of hepatocytes are more highly charged (higher degree of sulfation) than those present on other endothelial cells and that a specific role exists for the iduronosyl moieties.

The multiple potential interaction sites on both the parasite and the hepatocyte serve to overcome the tendency of the two cells to move apart after collision (the sporozoite is in motion in the circulation and would normally recoil from a stationary target). Molecular details regarding the mechanism of hepatocyte invasion and the possible involvement of other surface molecules remain unknown.

## Erythrocyte Invasion (Second Recognition Event)

As with the sporozoite, the blood stage parasite must enter an erythrocyte in a short time or it will spontaneously lyse. When released into the bloodstream from the liver cell, the merozoite will encounter a red cell within a minute or less. This interaction must be arrested so as to allow attachment; the subsequent steps that allow for invasion involve reorientation of the merozoite so that the apical end is perpendicular to the erythrocyte surface, and changes in the red cell cytoskeleton so as to permit entry. The initial binding is critical, however, and many parasite proteins involved in the entry phase may not play a role in attachment.

More than 20 years ago, it was established that binding of a major merozoite surface protein (MSP-1) to erythrocytes could be inhibited by glycophorin A [45]. In addition, erythrocytes lacking glycophorin A (Ena-) showed a marked reduction in merozoite invasion capacity [46,47]. Extensions of these studies showed that this diminished invasion property was also exhibited by MkMk erythrocytes (neither glycophorin A nor glycophorin B), Tn erythrocytes (lack the terminal sialic acid as well as galactose on glycophorin), or erythrocytes treated with neuraminidase or trypsin (cleaves the sialyl-rich domain from glycophorin [48–51]. More recent work showed that a monoclonal antibody directed against the glycosylated domain of glycophorin (and requiring sialic acid for recognition) could completely block merozoite binding to the red cell [52–54].

Glycophorin A is a transmembrane erythrocyte protein with 15 O-glycosylation sites, and one N-glycosylation site [55]. It is somewhat unusual in that the variation in glycosylation observed in many surface proteins is not observed. Instead, all of the 15 O-linked sites bear the identical tetrasaccharide (Figure 14.1)

In addition, clustering of these sites occurs so that a single stretch of six amino acids (Ser-Ser-Ser-Thr-Thr-Thr) contains six glycosyl residues, each with two sialyl groups. This provides a microdomain with a net charge of −12 and is a likely first interaction site between the merozoite and the erythrocyte. This suggestion is supported

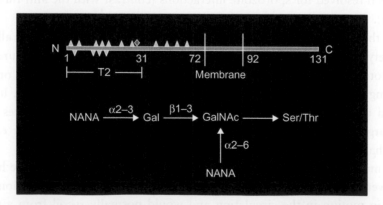

**FIGURE 14.1** Representation of glycophorin A. The triangles represent the tetrasaccharide structure shown. The diamond is the single N-linked saccharide unit. T2 is a peptide released by trypsin; note that this contains 11 of the 15 tetrasaccharides including the cluster at amino acids 10–15, the likely interaction site for the parasite.

by the ability of polyanions to inhibit merozoite invasion of the red cell [32,56,57]. As with the sporozoite, this effect is more pronounced with those polymers that have a high degree of sulfation (heparin) and is negligible with chondroitin sulfate. It has also been reported that the single *N*-linked saccharide unit is important in recognition of glycophorin C, another erythrocyte protein implicated in the binding/recogntion events.

Examination of MSP-1 for potential interaction sites with the erythrocyte led to the identification of a lysine-rich segment within that protein (115 amino acids, 27 lysines; P115) that bound specifically to erythrocytes and whose binding was dependent on sialic acid [58]. The likely role of this domain in attachment of the merozoite to the erythrocyte is indicated by the ability of recombinant P115 to bind to erythrocytes with specificity over and above that expected on a simple charge basis, and to inhibit merozoite invasion. Antibodies to P115 were also effective in blocking merozoite invasion.

Thus, both the liver and blood stages of the parasite appear dependent on cell surface saccharide recognition to complete their biological cycle. These events are similar in that both involve anionic host cell saccharides, although it is clear that some level of specificity must exist.

## Development in the Vector—Third and Fourth Recognition Events

Following uptake of gametocytes produced in the human host, the sexual forms mate in the mosquito to produce an ookinete. This vegetative body must complete its maturation cycle to produce several thousand sporozoites that in turn need to access the salivary gland of the insect. This is accomplished, in part, by the ookinete traversing the midgut barrier, and forming an oocyst that arrests at the basal lamina of the midgut. Once immobilized, sporozoites are formed, and when released, can access the salivary glands through the hemolymph of the mosquito. Adhesion of the oocyst and recognition/invasion of the salivary glands are processes that involve specific interactions.

There are relatively few data available on the binding of the oocyst to the basal lamina of the midgut. Given the general properties of gut epithelia, it is reasonable to infer that saccharides may well be involved in this event but no specific structures have been identified. Note that this is a technically difficult experimental system, especially since the externally directed components of the mosquito midgut have not been characterized.

It has been reported that the binding of ookinetes to midgut epithelial cells was largely abolished by pretreatment of the insect cells with periodate. In addition, *N*-acetylneuraminic acid competed with the ookinetes for binding but a variety of neutral sugars did not. However, since no sialic acid could be detected in the target cells (it has never been clearly established that insects are capable of sialic acid synthesis), it was concluded that some other carbohydrate may be involved. No effort was made in this work to examine polyanionic saccharides nor was the nature of the periodate-sensitive group identified.

Following release from the oocyst, sporozoites migrate, via the hemolymph, to the mosquito salivary gland and become resident there (via invasion) to await transfer to the mammalian host. The interaction between the sporozoite and the salivary gland cell is likely mediated by a surface receptor on the latter that can initially serve to tether the parasite and allow invasion. A number of candidates have been suggested to be involved, several of carbohydrate nature [59–63].

One of the initial studies showed that antibodies directed against mosquito salivary glands could block invasion but the specificity of these was not defined [64]. In the same work, a panel of lectins was examined for inhibitory effects. Most active were wheat germ agglutinin (WGA) and its succinylated derivative; *Pisum sativum* and soybean lectin were less so and a number of others, including concanavalin A and *Dolichos biflorus*, had no effect. WGA has known specificity for both GlcNAc and sialic acid and presumably is interacting with GlcNAc residues in this case. Whether this represents an actual receptor domain or is the result of a steric interaction amplified by proximity is not known.

More recent work has established that both heparan sulfate and chondroitin sulfate are present in both mosquito midgut and salivary glands. The insect heparan sulfate binds recombinant CSP and it has been suggested that this interaction is involved in the salivary gland invasion. It has also been reported that the salivary gland transcriptome is rich in the mRNA that codes for a syndecan core protein [66].

As with hepatocyte or erythrocyte invasion, charge-based interactions that allow for cellular attachment appear to have a significant role in parasite biology in the insect stage. These are most likely mediated by anionic saccharides with heparan sulfate playing a predominant role. It is also probable that this may not be the only mechanism at work. A recent study has indicated that at least one salivary gland protein has a motif that codes for tyrosine sulfation, that as many as 20 tyrosines may be involved, and that this protein may also be involved in the sporozoite/salivary gland events [63,67].

## Disease Pathology (Fifth Recognition Event)

The major determinants of malaria pathology in humans are cerebral involvement (capillary occlusion and hemorrhage) and massive anemia. The latter is related to the accelerated destruction of infected red cells in the spleen (recognition of surface alterations that may include passive attachment of malarial proteins to uninfected cells). The former derives from a property of infected erythrocytes. During the blood stage, the parasite synthesizes and exports a number of proteins to the erythrocyte outer membrane. Some are known to serve essential transport functions but the role of others remains unknown. One clear characteristic of the infected erythrocyte is its ability to bind to the external surface of capillary endothelial cells. In addition, the infected erythrocyte can bind other red cells (infected or not) to form aggregates (rosettes) that can block blood flow in small vessels [68–73]. This last property is key to the cerebral complications that can result in death.

A number of cell surface proteins and saccharides have been implicated in this binding event. The apparently diverse nature of this interaction has focused effort on the parasite product(s) responsible.

Among the proteins exported to the erythrocyte surface by the parasite is Pfemp1, a highly variable protein of 200 000–350 000 molecular weight that has been identified as a major participant in parasitized red cell adherence and resetting [74–76]. Fifty or more genes apparently code for this protein; the mechanism by which one or another of these genes is expressed is not known but the switching phenomenon is a relatively common strategy employed by microorganisms (*Neisseria*, *Borrelia*, trypanosomes) to evade host immune responses. This variability also allows parasitized erythrocytes to bind to a variety of cell surface targets including CD36, ICAM-1, ELAM-1, VCAM-1, heparan sulfate, and chondroitin-4-sulfate [77–86]. To date, the exact substructures involved in binding to these receptors have not been identified (certainly not for the entire gene family) and the possible role of specific saccharide sequences/moieties on the receptor remains undefined. It has been established, however, that the Pfemp1 genes code for sequences that have homology to proteins known to interact with the Duffy antigen (Duffy binding domain, DBL) or glycophorin. It should be noted that the DBL motif has also been shown to interact with glycosaminoglycan chains [74].

Not surprisingly, it has been known for some time that heparin is able to disrupt rosettes and can interfere with parasitized red cell adherence [87–91]. This effect, however, varies depending on the parasite strain and showed some dependence on the presence of *N*-sulfate groups. These data are consistent with the variable nature of the Pfemp1 gene product. Chemically prepared polyanionic saccharides such as dextran sulfate were also inhibitory as was fucoidan. Polysaccharides with lower charge density such as keratan sulfate or hyaluronic acid had no effect. It was initially reported that cleavage of cell surface heparan sulfate chains had no effect on the binding of parasitized red blood cells. However, a subsequent study showed that heparan sulfate was likely involved in this phenomenon and that the prior data resulted from the choice of a heparinase that was inefficient with the cell strain used [92].

A recent study has helped to clarify this complex problem. This work established that mature erythrocytes do contain a heparan sulfate proteoglycan at a level of about 2000 copies per cell [93]. This is much less than band 3 (one million copies) but comparable to the sodium/potassium ATPase (1–300 copies). The heparan sulfate chains isolated from erythrocytes bound to an affinity column made with DBL1, and this binding was abolished by treatment with heparinase.

In summary, the binding of parasitized erythrocytes to endothelial cells or other erythrocytes is mediated by Pfemp1, a parasite-encoded protein that is exported to and embedded in the red cell surface. The ligand(s) for this protein include heparan sulfate and a variety of endothelial cell surface molecules. The ability of highly charged saccharides such as heparin or dextran sulfate to interfere with this process is suggestive of a charge-basis for the interaction that involves heparan sulfate. As with sporozoite and merozoite invasion, the problem that must be solved is the arrest of the moving cell. The multiple surface receptors on the target facilitate this and the subsequent events that may utilize other surface molecules.

## Malaria in Pregnancy (Sixth Recognition Event)

It has been known for some time that infected pregnant women, especially primagravidae, experience more serious malaria pathology than their age-matched counterparts [94–97]. This is sufficiently severe as to be a major cause of stillbirth in the affected population. A key observation is that infected erythrocytes adhere to the placenta while they complete merozoite development. This avoids clearance in the spleen and leads to higher parasitemia with increased risk of cerebral complications.

As with the other cellular binding events discussed, this interaction relies on a placental glycosaminoglycan chain. It was initially reported that this molecule was chondroitin sulfate; a set of observations that included enzymatic digestion and inhibition studies provided the data. Definitive work on this problem was carried out by Gowda and colleagues [98–107]. A key finding from this laboratory was that the placenta produces an unusual chondroitin sulfate with a low sulfate content, analogous to that found in the cornea. In addition, they established that hyaluronic acid is not involved in placental binding contrary to earlier published work [100,107]. The binding site for the placenta was defined as a dodecasaccharide that contains two sulfated and four non-sulfated GalNAc residues [98]. The proteoglycan appears to be a product of the fetal tissues and may be increased in quantity as a result of parasite infection.

A recent study has identified a specific region of Pfemp1 that is the likely interaction site [108]. The interactive sequence is characterized by a flexible loop that in the presence of sulfate transforms to an organized, tight structure. This is associated with carbohydrate binding, contains one highly conserved lysine residue and two other basic amino acids sites (Lys, Arg) that show low variability among the different Pfemp1 genes.

# Direct Glycosylation by the Parasite

Protein modification by the addition of glycosyl residues is considered the most common form of posttranslational change. Originally thought to involve just asparagine (N-linked) or serine/threonine (O-linked) with defined linkage regions, many other types are now known. Early attempts to identify these modifications in *P. falciparum* provided some conflicting data, largely as a result of the experimental systems used [109–116]. In none of these studies was any protein target identified nor was the nature of the saccharides known. The following discussion reviews this work in chronological order to clarify the basis of current thinking.

## Experimental Conditions and Assays

Unlike many unicellular organisms, *P. falciparum* cannot be grown freely in culture. Of all of the developmental stages of the malaria parasite, as well as the different species of plasmodia, only the blood stage (merozoite) of *P. falciparum* can be maintained in vitro. Such cultures require fresh (outdated blood will not suffice) human erythrocytes, human serum, and a specialized incubator with reduced oxygen. Even under these conditions, cultures rarely last more than a few weeks. While red cells are considered not to perform protein synthesis or modification, material obtained from blood bank or other sources will contain a small percentage of reticulocytes (0.5–1.0%) that may be metabolically active with regard to protein. Passive pre-incubation of such cells for up to 5 days is necessary to eliminate this background problem. Since preparation of scale quantities of parasite proteins sufficient for direct chemical identification is not feasible, work has been carried out with radiolabeled sugar precursors. This also presents the problem of non-enzymatic glycosylation (as in the formation of hemoglobin A1c), trivial in a quantitative sense but able to provide labeled product. Most studies have used glucosamine as the precursor but at least one used glucose [116]. Since the latter sugar can serve as a precursor for several amino acids, it is not the best choice. Conclusions based only on lectin interactions require caveats regarding the presence of glycolipid anchors and general non-specific binding, virtually impossible to control for. Likewise, the use of glycosylation inhibitors such as tunicamycin is only meaningful if it is shown that, at the concentrations employed, other parasite functions such as protein synthesis are not affected [117].

Initial reports of parasite protein glycosylation were based on labeling/lectin studies. These concluded that O-linked glycosylation may be present because the labeled "glycoproteins" were resistant to the action of N-glycanase. It had also been reported that sialic acid was not detectable in the parasite nor was the parasite capable of its synthesis [118]. Other reports of O-glycosylation have not been confirmed and it should be appreciated that erythrocytes contain significant quantities of O-linked GlcNAc. In addition, GalNAc has not been detected in the parasite. Although O-glycanase was claimed to release a mixed set of oligosaccharides from parasite proteins, no compositional analysis was performed on released material [117]. This result is further clouded by the known specificity of this enzyme for Galβ1–3GalNAc-Ser/Thr.

A more detailed study employed tunicamycin and concluded that N-linked glycoproteins are involved in intraerythrocytic maturation of the parasite [116]. However, this work utilized glucose as a radiolabeled precursor, abnormally high concentrations of tunicamycin (as well as a longer incubation time), and did not consider the presence of glycolipid anchors [118].

Independently, it had been reported that the blood stage parasite does not contain N-linked glycoproteins and that the overall biosynthetic pathway for this process was absent in the merozoite [119].

Well-controlled studies indicated that a small proportion of incorporated label may be present in N-linked saccharides but no individual protein could be identified that contained such sugars [120]. Since GPI anchors were readily identified and since there is only one such per protein that is modified, even a single N-linked unit would have been detected. Finally, the apparent lack of such glycosylation is not due to the lack of consensus glycosylation sequences since it has been shown that expression of a fragment of MSP1 in mammalian cells results in N-glycosylation at the several sites available [121].

## Conclusion

The major carbohydrate present on parasite proteins appears to be that in the glycolipid anchors. No parasite protein has been identified that contains either N-linked or O-linked sugars; neither N-acetylgalactosamine nor sialic acid can be detected in the parasite. Glycosylation consensus sequences in *P. falciparum* can be functional in other backgrounds (only established for a limited number of proteins).

## Glycosylphosphatidylinositol (GPI) Anchors

The presence of GPI anchors in plasmodia was initially reported some 20 years ago [122]. It is noteworthy that their association with disease pathology was also recognized but that relatively little has been done to follow-up these observations.

In terms of overall saccharide incorporation into macromolecules, GPI anchors represent the major route. There is nothing unusual about the pathway itself and several of the classic intermediates have been identified. This biosynthesis requires both UDP-GlcNAc and GDP-Man but alternative utilization of these in the dolichol pathway is meager, at best. The structure of the parasite anchor has been determined by a combination of degradative studies and mass spectrometry [123] (Figure 14.2).

Although this is very similar to that normally found in humans, the differences, perhaps especially in the lipid moieties, are sufficient to make this molecule an effective hapten as well as a strong signaling molecule. It has been shown that individuals living in malaria endemic areas mount an immune response to the parasite GPI anchor [124–126]. The level of anti-GPI antibodies also correlates with increasing resistance to disease pathology, a characteristic of that environment. Note that this does not prevent infection/transmission but still represents a potential target that could reduce morbidity/mortality.

## Overall Summary

The malaria parasite uses host saccharides in both the insect vector and humans for recognition/binding events. These are usually anionic in nature and involve heparan sulfate, chondroitin sulfate, and sialyl residues. Specific domains on parasite surface proteins have been identified that are involved in these phenomena and confirm the charge-based nature of the primary interactions. Direct glycosylation of parasite

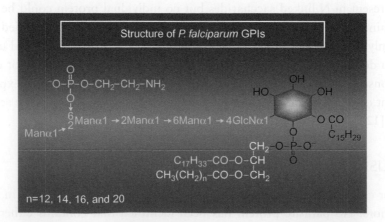

FIGURE 14.2 Structure of *Plasmodium falciparum* glycosylphosphatidylinositols.

proteins is primarily through the attachment of glycosylphosphatidylinositol anchors, a predominant mechanism for surface immobilization of parasite proteins. The anchors have a Man4 structure and are immunogenic in humans. Explicit identification of any protein of the parasite that contains *N*- or *O*-linked saccharides has not been accomplished although a small amount of high mannose structures may be present. The parasite does not contain sialic acid or *N*-acetylgalactosamine; intermediates in the dolichol pathway beyond UDP-GlcNAc and GDP-Man are not detectable.

# References

1. Sachs J, Malany P. The economic and social burden of malaria. *Nature* 2002;**415**:680–5.

2. Snow RW, Craig M, Deichmann U, Marsh K. Estimating mortality, morbidity and disability due to malaria among Africa's non-pregnant population. *Bull World Health Organ* 1999;**77**:624–40.

3. Sherman I. *A brief history of malaria and discovery of the parasite's life cycle, Malaria: Parasite Biology, Pathogenesis and Protection*. ASM Press: Washington, D.C.; 1998, p. 3–10.

4. Vanderberg JP. *Plasmodium berghei*: quantitation of sporozoites injected by mosquitoes feeding on a rodent host. *Exp Parasitol* 1977;**42**:169–81.

5. Ponnudurai TA, Lensen HW, Vangemert GJA, Bolmer MG, Meuwissen JHET. Feeding behavior and sporozoite ejection by infected *Anopheles stephensi*. *Trans Roy Soc Trop Med Hyg* 1991;**85**:175–80.

6. Sinden RE, Smith JE. The role of the Kupffer cell in the infection of rodents by sporozoites of *Plasmodium*: uptake of sporozoites by perfused liver and the establishment of infection in vivo. *Acta Trop* 1982;**39**:11–27.

7. Pradel G, Frevert U. Malaria sporozoites actively enter and pass through Kupffer cells prior to hepatocyte invasion. *Hepatology* 2001;**33**:1154–65.

8. Dvorak JA, Miller LH, Whitehouse WC, Shiroishi T. Invasion of erythrocytes by malaria merozoites. *Science* 1975 Feb 28;**187**(4178):748–50.

9. Johnson JG, Epstein N, Shiroishi T, Miller LH. Factors affecting the ability of *Plasmodium knowlesi* merozoites to attach to and invade erythrocytes. *Parasitology* 1980;**80**:539–50.

10. Silvestrini F, Alano P, Williams JL. Commitment to the production of male and female gametocytes in the human malaria parasite. *Parasitology* 2000;**121**:465–71.

11. Bannister LH, Hopkins JM, Fowler RE, Krishna S, Mitchell GH. A brief illustrated guide to the ultrastructure of *Plasmodium falciparum* asexual blood stages. *Parasitol Today* 2000;**16**:427–33.

12. Haldar K, Henderson CL, Cross GAM. Identification of the parasite transferrin receptor of *Plasmodium falciparum*-infected erythrocytes and its acylation via 1,2-diacyl-sn-glycerol. *Proc Natl Acad Sci USA* 1986;**83**:8565–9.

13. Schofield L, Hackett F. Signal transduction in host cells by a glycosylphosphatidylinositol toxin of malaria parasites. *J Exp Med* 1993;**177**:145–53.

14. Braun Breton C, Rosenberry TL, Pereira de Silva LH. Glycolipid anchorage of *Plasmodium falciparum* surface proteins. *Res Immunol* 1990;**141**:743–55.

15. Gerold P, Dieckmann-Schuppert A, Schwarz RT. Glycosylphosphatidylinositols synthesized by asexual erythrocytic stages of the malarial parasite, *Plasmodium falciparum*. Candidates for plasmodial

glycosylphosphatidylinositol membrane anchor precursors and pathogenicity factors. *J Biol Chem* 1994;**269**:2597–606.

16. Meis JFGM, Verhave JP, Jap PHK, Meuwissen JHET. An ultrastructural study on the role of Kupffer cells in the process of infection by *Plasmodium berghei* sporozoites in rats. *Parasitology* 1983;**86**:231–42.

17. Pradel G, Frevert U. Malaria sporozoites actively enter and passage through Kupffer cells prior to hepatocyte invasion. *Hepatology* 2001;**33**:1154–65.

18. Shin SCJ, Vanderberg JP, Terzakis JA. Direct infection of hepatocytes by sporozoites of *Plasmodium berghei*. *J Protozool* 1982;**29**:448–54.

19. Sinden RE, Smith JE. The role of the Kupffer cell in the infection of rodents by sporozoites of *Plasmodium*: uptake of sporozoites by perfused liver and the establishment of infection in vivo. *Acta Trop* 1982;**39**:11–27.

20. Sultan AA, Thathy V, Frevert U, Robson KJH, Crisanti A, Nussenzweug V, Nussenzweig R, Menard R. TRAP is necessary for gliding motility and infectivity of plasmodium sporozoites. *Cell* 1997;**90**:511–22.

21. Kappe ST, Bruderer T, Gantt S, Fujioka H, Nussenzweig V, Menard R. Conservation of a gliding motility and cell invasion machinery in apicomplexan parasites. *J Cell Biol* 1999;**147**:937–44.

22. Wengelnik K, Spaccapelo R, Naitza S, Robson KJH, Jansae CJ, Bistoni F, Waters AP, Crisanti A. The A-domain and the thrombospondin-related motif of *Plasmodium falciparum* TRAP are implicated in the invasion process of mosquito salivary glands. *EMBO J* 1999;**18**:5195–204.

23. Matuschewski K, Ross J, Brown SM, Kaiser K, Nussenzweig V, Kappe SHI. Infectivity-associated changes in the Transcriptional Repertoire of the Malaria Parasite Sporozoite Stage. *J Biol Chem* 2000;**277**: 41948–53.

24. Ying P, Shakibaei M, Patankar MS, Clavijo P, Beavis RC, Clark GF, Frevert U. The malaria circumsporozoite protein: interaction of the conserved regions I and II-plus with heparin-like oligosaccharides in heparan sulfate. *Exp Parasitol* 1997;**85**:168–82.

25. Rathore D, McCutchan TF, Garboczi DN, Toida T, Hernaiz MJ, LeBrun LA, et al. Direct measurement of the interactions of glycosaminoglycans and a heparin decasaccharide with the malaria circumsporozoite protein. *Biochemistry* 2001;**40**:11518–24.

26. Holt GD, Krivan HC, Gasic GJ, Ginsburg V. Antistasin, an inhibitor of coagulation and metastasis, binds to sulfatide (Gal(3-SO4) beta 1–1Cer) and has a sequence homology with other proteins that bind sulfated glycoconjugates. *J Biol Chem* 1989;**264**:12138–40.

27. Pancake SJ, Holt GD, Mellouk S, Hoffman SL. Malaria sporozoites and circumsporozoite proteins bind specifically to sulfated glycoconjugates. *J Cell Biol* 1992;**117**:1351–7.

28. Sinnis P, Clavijo P, Fenyo D, Chait BT, Cerami C, Nussenzweig V. Structural and functional properties of region II-plus of the malaria circumsporozoite protein. *J Exp Med* 1994;**180**:297–306.

29. Gantt SM, Clavijo P, Bai X, Esko JD, Sinnis P. Cell adhesion to a motif shared by the malaria circumsporozoite protein and thrombospondin is mediated by its glycosaminoglycan-binding region and not by CSVTCG. *J Biol Chem* 1997;**272**:19205–13.

30. Ying P, Shakibaei M, Manish M, Patankar S, Clavijo P, Beavis RC, Clark GF, Frevert U. The malaria circumsporozoite protein: interaction of the conserved regions I and II-plus with heparin-like oligosaccharides in heparan sulfate. *Exp Parasitol* 1997;**85**:168–82.

31. Sinnis P, Willnow TE, Briones MRS, Herz J, Nussenzweig V. Remnant lipoproteins inhibit malaria sporozoite invasion of hepatocytes. *J Exp Med* 1996;**184**:945–54.

32. Nussenzweig V. Malaria sporozoites and chylomicron remnants compete for binding sites in the liver. *Behring Inst Mitt* 1997;**99**:85–9.

33. Kulane A, Ekre H-P, Perlmann P, Rombo L, Mats Wahlgren M, Wahlin B. Effect of different fractions of heparin on Plasmodium falciparum merozoite invasion of red blood cells in vitro. *Am J Trop Med Hyg* 1992;**46**:589–94.

34. Clark DL, Su S, Davidson EA. Saccharide anions as inhibitors of the malaria parasite. *Glycoconj J* 1997;**14**:473–9.

35. Pradel G, Garapaty S, Frevert U. Proteoglycans mediate malaria sporozoite targeting to the liver. *Mol Microbiol* 2002;**45**:637–51.

36. Frevert U, Sinnis P, Esko JD, Nussenzweig V. Cell surface glycosaminoglycans are not obligatory for *Plasmodium berghei* sporozoite invasion in vitro. *Mol Biochem Parasitol* 1996;**76**:257–66.

37. Pinzon-Ortiz C, Friedman J, Esko JD, Sinnis P. The binding of the circumsporozoite protein to cell surface heparan sulfate is required for *Plasmodium* attachment to target cells. *J Biol Chem* 2001;**276**:26784–91.

38. Rathore D, McCutchan TF, Garboczi DN, Toida T, Hernáiz M, LeBrun LA, Lang SC, Linhardt RJ. Direct measurement of the interactions of glycosaminoglycans and a heparin decasaccharide with the malaria circumsporozoite. *Protein Biochemistry* 2001;**40**:11518–24.

39. Lyon M, Denkin JA, Gallagher JT. Liver heparan sulfate structure. A novel molecular design. *J Biol Chem* 1994;**269**:11208–15.

40. Soroka CJ, Farquhar MG. Characterization of a novel heparan sulfate proteoglycan found in the extracellular matrix of liver sinusoids and basement membrane. *J Cell Biology* 1991;**113**:1231–41.

41. Ying P, Shakibaei M, Patankar MS, Clavijo P, Beavis RC, Clark D and Frevert U. The malaria circumsporozoite protein: interaction of the conserved regions I and II-plus with heparin-like oligosaccharides in heparan sulfate. *Exp Parasitol* 1997;**85**:168–82.

42. Ancsin JB, Kisilevsky R. A binding site for highly sulphated heparan sulphate is identified at the N-terminus of the circumsporozoite protein: significance for malarial sporozoite attachment to hepatocytes. *J Biol Chem* 2004;**279**:21824–32.

43. McCormick CJ, Tuckwell DS, Crisanti A, Humphries MJ, Hollingdale MR. Identification of heparin as a ligand for the A-domain of Plasmodium falciparum thrombospondin-related adhesion protein. *Mol Biochem Parasitol* 1999;**100**(1):111–24.

44. Robson KJH, Frevert U, Reckmann I, Cowan G, Beier J, Scragg IG, Takehara K, Bishop DHL, Pradel G, Sinden R, Saccheo S, Muller HM, Crisanti A. Interaction of sporozoite surface proteins with heparin. *EMBO J* 1995;**14**:3883–94.

45. Perkins M. Inhibitory effects of erythrocyte membrane proteins on the in vitro invasion of the human malarial parasite (*Plasmodium falciparum*) into its host cell. *J Cell Biol* 1981;**90**:563–7.

46. Pasvol G, Wainscoat JS, Weatherall DJ. Erythrocytes deficient in glycophorin resist invasion by the malarial parasite *Plasmodium falciparum*. *Nature* 1982;**297**:64–7.

47. Miller LH, Haynes JD, McAuliffe FM, Shiroishi T, Durocher JR, McGinniss. Evidence for differences in erythrocytic surface receptors for the malarial parasites *Plasmodium falciparum* and *Plasmodium knowlesi*. *J Exp Med* 1977;**146**:277–81.

48. Perkins M. Binding of glycophorins to *Plasmodium falciparum* merozoites. *Mol Biochem Parasitol* 1984;**10**:67–74.

49. Davidson EA, Perkins M. Receptor binding domains of glycophorin A for *Plasmodium falciparum* surface proteins. *Ind J Biochem Biophys* 1988;**25**:90–4.

50. Perkins M, Rocco LJ. Sialic acid-dependent binding of *Plasmodium falciparum* merozoite surface antigen Pf200 to human erythrocytes. *J Immunol* 1988;**141**:3190–6.

51. Pasvol G, Jungery M, Weatherall DJ, Parsons SF, Anstee DJ, Tanner MJA. Glycophorin as a possible receptor for *Plasmodium falciparum*. *The Lancet* 1982;**320**:947–50.

52. Hadley TJ, Erkmen Z, Kaufman BM, Futrovsky S, McGinniss H, Graves P, Sadoff JC, Miller LH. Factors influencing invasion of erythrocytes by malaria parasites: the effects of an N-acetylglucosamine neo-glycoprotein and an anti-glycophorin A antibody. *Am J Trop Med Hyg* 1986;**35**:898–907.

53. Pasvol G, Chasis JA, Mohandas N, Anstee SJ, Tanner MJA, Mery AH. Inhibition of malarial parasite invasion by monoclonal antibodies against glycophorin A correlates with reduction in red cell deformability. *Blood* 1989;**74**:1836–44.

54. Su S, Sanadi AR, Ifon E, Davidson EA. A monoclonal antibody capable of blocking the binding of Pf200 (MSA-1) to human erythrocytes and inhibiting the invasion of *Plasmodium falciparum* merozoites into human erythrocytes. *J Immunol* 1993;**151**:2309–17.

55. Tomita T, Marchesi VT. Amino acid sequence and oligosaccharide attachment sites of human erythrocyte glycophorin. *Proc Natl Acad Sci USA* 1975;**251**:2964–9.

56. Xiao L, Yang C, Patterson PS, Udhayakumar V, Lal AA. Sulfated polyanions inhibit invasion of erythrocytes by plasmodial merozoites and cytoadherence of endothelial cells to parasitized erythrocytes. *Infection and Immunity* 1996;**64**:1373–8.

57. Dalton JP, Hudson D, Adams JH, Miller LH. Blocking of the receptor-mediated invasion of erythrocytes by *Plasmodium knowlesi* with sulfated polysaccharides and glycosaminoglycans. *Eur J Biochem* 1991;**195**:789–94.

58. Nikodem D, Davidson EA. Identification of a novel antigenic domain of *Plasmodium falciparum* merozoite surface protein-1 that specifically binds to human erythrocytes and inhibits parasite invasion, in vitro. *Mol Biochem Parasitol* 2000;**108**:79–91.

59. Zieler H, Nawrocki JP, Shahabuddin M. *Plasmodium gallinaceum* ookinetes adhere specifically to the midgut epithelium of *Aedes aegypti* by interaction with a carbohydrate ligand. *J Exp Biol* 1999;**202**:485–95.

60. Siden-Kiamos I, Louis C. Interactions between malaria parasites and their mosquito hosts in the midgut. *Insect Biochem Mol Biol* 2004;**34**:6685–709.

61. Brennan JDG, Kent M, Dhar R, Fujioka H, Kumar N. Anopheles gambiae salivary gland proteins as putative targets for blocking transmission of malaria parasites. *Proc Natl Acad Sci USA* 2000;**97**:13859–64.

62. Myung M, Marshall P, Sinnis P. The *Plasmodium* circumsporozoite protein is involved in mosquito salivary gland invasion by sporozoites. *Mol Biochem Parasitol* 2004;**133**:53–9.

63. Korochkina S, Barreau C, Pradel G, Jeffery E, Li J, Natarajan R, Shabanowitz J, Hunt D, Frevert U, Vernick KD. A mosquito-specific protein family includes candidate receptors for malaria sporozoite invasion of salivary glands. *Cellular Microbiology* 2006;**8**:163–75.

64. Barreau C, Touray M, Pimenta PF, Miller LH, Vernick KD. *Plasmodium gallinaceum*: sporozoite invasion of *Aedes aegypti* salivary glands is inhibited by anti-gland antibodies and by lectins. *Exp Parasitol* 1995;**81**:332–43.

65. Sinnis P, Coppi A, Toida T, Toyada H, Kinoshita-Toyada A, Xie J, Kemp MM, Linhardt RJ. Mosquito heparan sulfate and its potential role in malaria infection and transmission. *J Biol Chem* 2007;**282**:25376–84.

66. Valenzuela JG, Francischetti IMB, Pham VM, Garfield MK, Ribeiro JMC. Exploring the salivary gland transcriptome and proteome of the *Anopheles stephensi* mosquito. *Insect Biochemistry and Molecular Biology* 2003;**33**:717–32.

67. Choe H, Moore MJ, Owens CM, Wright PL, Vasilieva N, Li W, Singh AP, Shakri R, Chitnis CE, Farzan M. Sulphated tyrosines mediate association of chemokines and Plasmodium vivax Duffy binding protein with the Duffy antigen/receptor for chemokines (DARC). *Molecular Microbiology* 2005;**55**:1413–22.

68. Bignami B, Bastianeli G. Observations of estivo-autumnal malaria. *Reforma medica* 1889;**6**:1334–5.

69. Miller LH. Distribution of mature trophozoites and schizonts of *Plasmodium falciparum* in the organs of *Aotus trivergatus*, the night monkey. *Am J Trop Med Hyg* 1969;**18**:860–5.

70. Aley SB, Sherwood JA, Howard RJ. Knob-positive and knob-negative Plasmodium falciparum differ in expression of a strain-specific malarial antigen on the surface of erythrocytes. *J Exp Med* 1984;**160**:1585–90.

71. Leech JH, Barnwell JW, Miller LH, Howard RJ. Identification of a strain-specific malarial antigen exposed on the surface of *Plasmodium falciparum* infected erythrocytes. *J Exp Med* 1984;**159**:1567–75.

72. Miller LH, Good MF, Milon G. Malaria pathogenesis. *Science* 1994;**264**:1878–83.

73. Patnaik JK, Das BS, Mishra SK, Mohanty S, Satpathy SK, Mohanty D. Vascular clogging, mononuclear cell migration and enhanced vascular permeability in the pathogenesis of human cerebral malaria. *Am J Trop Med Hyg* 1994;**51**:842–7.

74. Baruch DI, Pasioske BL, Singh HB, Bi X, Ma XC, Feldman M, Taraschi TF, Howard RJ. Cloning the *P. falciparum* gene encoding Pfemp1, a malarial variant antigen and adherence receptor on the surface of parasitized human erythrocytes. *Cell* 1995;**82**:77–87.

75. Su X, Heatwole VM, Wertheimer SP, Guinet F, Herrfeldt F, Peterson DS, Ravetch J, Wellems TE. The large diverse gene family var encodes proteins involved in cytoadherence and antigenic variation of *Plasmodium falciparum*-infected erythrocytes. *Cell* 1995;**82**:89–100.

76. Craig A, Scherf A. Molecules on the surface of *Plasmodium falciparum*-infected erythrocytes and their role in malaria pathogenesis and immune evasion. *Mol Biochem Parasitol* 2001;**115**:129–43.

77. Barnwell JW, Ockenhouse CF, Knowles DM. Monoclonal antibody OKM5 inhibits the in vitro binding of *Plasmodium falciparum* infected erythrocytes to monocytes, endothelial and CD32 melanoma cells. *J Immunol* 1985;**135**:3494–7.

78. Barnwell JW, Asch AS, Nachman RL, Yamaya M, Aikawa M, Ingravallo P. A human 88Kd glycoprotein (CD36) functions in vitro as a receptor for a cytoadherence ligand on *Plasmodium falciparum*-infected erythrocytes. *J Clin Invest* 1989;**84**:765–72.

79. Treutiger CJ, Heddini A, Fernandez V, Muller WA, Wahlgren M. PECAM1/CD31, an endothelial receptor for binding *Plasmodium falciparum*-infected erythrocytes. *Nat Med* 1997;**3**:1405–8.

80. Newbold C, Warn P, Black G, Berendt A, Craig A, Snow B, Msobo M, Peshu N, Marsh K. Receptor-specific adhesion and clinical disease in *Plasmodium falciparum*. *Am J Trop Med Hyg* 1997;**57**:389–98.

81. Berendt AR, Simmons DL, Tansey J, Newbold CI, Marsh K. Intercellular adhesion molecule-1 is an endothelial cell adhesion receptor for *Plasmodium falciparum*. *Nature* 1989;**341**:57–9.

82. Berendt AR, McDowall A, Craig AG, Bates PA, Sternberg MJE, Marsh K, Newbold CI, Hogg N. The binding site on ICAM-1 for *Plasmodium falciparum*-infected erythrocytes overlaps, but is distinct from, the LFA-1-binding site. *Cell* 1992;**68**:71–81.

83. Baruch DI, Gormley JA, Ma C, Howard RJ. *Plasmodium falciparum*-erythrocyte membrane protein 1 is a parasitized erythrocyte receptor for adherence to CD36, thrombospondin and intracellular adhesion molecule 1. *Proc Natl Acad Sci USA* 1996;**93**:3497–502.

84. Flick K, Scholander C, Chen Q, Fernandez V, Pouvelle B, Gysin J, Wahlgren M. Role of nonimmune IgG bound to PfEMP1 in placental malaria. *Science* 2001;**293**:2098–100.

85. Scholander C, Treutiger CJ, Hultenby K, Wahlgren M. Novel fibrillar structure confers adhesive property to malaria-infected erythrocytes. *Nat Med* 1996;**2**:204–8.

86. Ockenhouse CF, Tegoshi T, Maeno Y, Benjamin C, Ho M, Kan KE, Thway Y, Win K, Aikawa M, Lobb RR. Human vascular cell adhesion receptors for *Plasmodium falciparum*-infected erythrocytes: Roles for endothelial leukocyte adhesion molecule 1 and vascular cell adhesion molecule 1. *J Exp Med* 1992;**176**:1183–9.

87. Barragan A, Spillmann D, Kremsner PG, Wahlgren M, Carlson J. *Plasmodium falciparum*: molecular background of strain specific rosette disruption by glycosaminoglycans and sulfated glycoconjugates. *Exp Parasitol* 1999;**91**:133–43.

88. Carlson J, Ekre H-P, Helmby H, Gysin G, Greenwood BM, Wahlgren M. Disruption of *Plasmodium falciparum* erythrocyte rosettes by standard heparin and heparin devoid of anticoagulant activity. *Am J Trop Med Hyg* 1992;**46**:595–602.

89. Chen Q, Barragan A, Fernandez V, Sundström A, Schlichtherle M, Sahlén A, Carlson J, Datta S and Wahlgren M. Identification of *Plasmodium falciparum* erythrocyte membrane protein 1 (PfEMP1) as the rosetting ligand of the malaria parasite *P. falciparum*. *J Exp Med* 1998;**187**:15–23.

90. Maccarana M, Sakura Y, Tawada A, Yoshida K, Lindahl U. Domain structure of heparan sulfate from bovine organs. *J Biol Chem* 1996;**273**:12960–6.

91. Barragan A, Fernandez V, Chen Q, von Euler A, Wahlgren M, Spillmann D. The Duffy-binding-like domain 1 of *Plasmodium falciparum* erythrocyte membrane protein 1 (Pfemp1) is a heparan sulfate ligand that requires 12-mers for binding. *Blood* 2000;**95**:3594–9.

92. Vogt AM, Barragan A, Chen Q, Jironde F, Spillmann D, Wahlgren M. Heparan sulfate on endothelial cells mediates the binding of *Plasmodium falciparum*-infected erythrocytes via the DBL1a domain of Pfemp1. *Blood* 2003;**101**:2405–11.

93. Vogt AM, Winter G, Wahlgren M, Spillmann D. Heparan sulphate identified on human erythrocytes: a *Plasmodium falciparum* receptor. *Biochem J* 2004;**381**:593–7.

94. McGregor IA, Wilson ME, Billewicz WZ. Malaria infection in of the placenta in The Gambia, West Africa: its incidence and relationship to stillbirth, birth weight and placental weight. *Trans Roy Soc Trop Med Hyg* 1983;**77**:232–44.

95. Menendez C, Ordi J, Ismail MR, Ventura PJ, Kahigawa E, Font F, Alonso PL. The impact of placental malaria on gestational age and birth weight. *J Infect Dis* 2000;**181**:1740–5.

96. Brabin BJ, Romagosa C, Abdelgalil S, Menendez C, Verhoeff FH, McGready R, Fletcher KA, Owens S, D'Alessandro U, Nosten F, Fischer PR, Ordi J. The sick placenta–the role of malaria. *Placenta* 2004;**25**:359–78.

97. Beeson JG, Duffy PE. The immunology and pathogenesis of malaria during pregnancy. *Curr Top Microbiol Immunol* 2005;**297**:187–227.

98. Fried M, Duffy PE. Adherence of *Plasmodium falciparum* to chondroitin sulfate A in the human placenta. *Science* 1996;**272**:1502–4.

99. Alkhalil A, Achur RN, Valiyaveettil M, Ockenhouse CF, Gowda DC. Structural requirements for the adherence of *Plasmodium falciparum*-infected erythrocytes to chondroitin sulfate proteoglycans of human placenta. *J Biol Chem* 2000;**275**:40357–64.

100. Achur RN, Valiyaveettil M, Alkhalil A, Ockenhouse CF, Gowda DC. Characterization of proteoglycans of human placenta and identification of unique chondroitin sulfate proteoglycans of the intervillous spaces that mediate the adherence of *Plasmodium falciparum*-infected erythrocytes to the placenta. *J Biol Chem* 2000;**275**:40344–56.

101. Valiyaveettil M, Achur RN, Alkhalil A, Ockenhouse CF, Gowda DC. *Plasmodium falciparum* cytoadherence to human placenta: evaluation of hyaluronic acid and chondroitin 4-sulfate for binding of infected erythrocytes. *Exp Parasitol* 2001;**99**:57–65.

102. Achur RN, Valiyaveettil M, Gowda DC. The low sulfated chondroitin sulfate proteoglycans of human placenta have sulfate group-clustered domains that can efficiently bind *Plasmodium falciparum*-infected erythrocytes. *J Biol Chem* 2003;**278**:11705–13.

103. Muthusamy A, Achur RN, Valiyaveettil M, Gowda DC. Plasmodium falciparum: adherence of the parasite-infected erythrocytes to chondroitin sulfate proteoglycans bearing structurally distinct chondroitin sulfate chains. *Exp Parasitol* 2004;**107**:183–8.

104. Muthusamy A, Achur RN, Bhavanandan VP, Fouda GG, Taylor DW, Gowda DC. *Plasmodium falciparum*-infected erythrocytes adhere both in the intervillous space and on the villous surface of human placenta by binding to the low-sulfated chondroitin sulfate proteoglycan receptor. *Am J Pathol* 2004;**164**:2013–25.

105. Valiyaveettil M, Achur RN, Muthusamy A, Gowda DC. Chondroitin sulfate proteoglycans of the endothelia of human umbilical vein and arteries and assessment for the adherence of *Plasmodium falciparum*-infected erythrocytes. *Mol Biochem Parasitol* 2004;**134**:115–26.

106. Achur RN, Muthusamy A, Madhunapantula SV, Bhavanandan VP, Seudieu C, Gowda DC. Chondroitin sulfate proteoglycans of bovine cornea: structural characterization and assessment for the adherence of *Plasmodium falciparum*-infected erythrocytes. *Biochim Biophys Acta* 2004;**1701**:109–19.

107. Muthusamy A, Achur RN, Valiyaveettil M, Botti JJ, Taylor DW, Leke RF, Gowda DC. Chondroitin sulfate proteoglycan but not hyaluronic acid is the receptor for the adherence of *Plasmodium falciparum*-infected erythrocytes in human placenta, and infected red blood cell adherence up-regulates the receptor expression. *Am J Pathol* 2007;**170**:1989–2000.

108. Higgins MK. The structure of a chondroitin sulfate-binding domain important in placental malaria. *J Biol Chem* 2008;**283**:21842–6.

109. Heidrich H-G, Strych W, Prehm P. Spontaneously released *Plasmodium falciparum* merozoites from cell cultures possess glycoproteins. *Z parasitenkd* 1984;**70**:747–51.

110. Fenton B, Clark JT, Wilson CS, McBride JS, Walliker D. Polymorphism of a 35–48 kDa *Plasmodium falciparum* merozoite surface antigen. *Mol Biochem Parasitol* 1989;**34**:79–86.

111. Vermeulen AN, van Deursen J, Brakenhoff RH, Lensen THW, Ponnudurai T, Meuwissen JHE. Characterization of *Plasmodium falciparum* sexual stage antigens and their biosynthesis in synchronised gametocyte cultures. *Mol Biochem Parasitol* 1986;**20**:155–63.

112. Ramasamy R. Studies on glycoproteins in the human malaria parasite *P. falciparum*—lectin binding properties and the possible carbohydrate–protein linkage. *Immunol Cell Biol* 1987;**65**:147–52.

113. Jakobsen PH, Theander TG, Jensen JB, Mølbak K, Jepsen S. Soluble *Plasmodium falciparum* antigens contain carbohydrate moieties important for immune reactivity. *J Clin Microbiol* 1987;**25**:2075–9.

114. Dayal-Drager R, Hoessli DC, Decrind C, Del Guidice G, Lambert P-H, ud-Din N. Presence of O-glycosylated glycoproteins in the *Plasmodium falciparum* parasite. *Carbohydr Res* 1991;**209**:c5–c8.

115. Dieckmann-Schuppert A, Bause E, Schwarz RT. Studies on O-glycans of *Plasmodium falciparum*-infected human erythrocytes. Evidence for O-glcNAc and O-glcNAc transferase in malaria parasites. *Eur J Biochem* 1993;**216**:779–88.

116. Kimura EA, Couto AS, Peres VJ, Casal OL, Katzin AM. N-linked glycoproteins are related to schizogony of the intraerythrocytic stage in *Plasmodium falciparum*. *J Biol Chem* 1996;**271**:14452–61.

117. Khan AH, Hoessli DC, Rahman A, Davidson EA, UdDin N. 2-amino-2-deoxy-D-mannose: Concurrent inhibition and incorporation of 2-amino-2-deoxy-D-glucose into malarial glycoproteins of *Plasmodium falciparum*. *Nat Prod Letters* 1997;**10**:17–24.

118. Schauer R, Wember M, Howard RJ. Malaria parasites do not contain or synthesize sialic acids. *Z Physiol Chem* 1984;**365**:185–94.

119. Dieckmann-Schuppert A, Bender S, Odenthal-Schnitter D, Bause E, Schwarz RT. Apparent lack of N-glycosylation in the asexual intraerythrocytic stage of *Plasmodium falciparum*. *Eur J Biochem* 1992;**205**:815–25.

120. Gowda DC, Gupta P, Davidson EA. Glycosylphosphatidylinositol anchors represent the major carbohydrate modification in proteins of intraerythrocytic stage *Plasmodium falciparum*. *J Biol Chem* 1997;**272**:6428–39.

121. Yang S, Nikodem D, Davidson EA, Gowda DC. Glycosylation and proteolytic processing of 70 kDA C-terminal recombinant polypeptides of *Plasmodium falciparum* merozoite surface protein 1 expressed in mammalian cells. *Glycobiology* 1999;**9**:1347–56.

122. Haldar K, Henderson CL, Cross GAM. Identification of the parasite transferrin receptor of *Plasmodium falciparum*-infected erythrocytes and its acylation via 1,2-diacyl-sn-glycerol. *Proc Natl Acad Sci USA* 1986;**83**:8565–9.

123. Naik RS, Branch OH, Woods AS, Vijaykumar M, Perkins DJ, Nahlen BL, Lal AA, Cotter RJ, Costello CE, Ockenhouse CF, Davidson EA, Gowda DC. Glycosylphosphatidylinositol anchors of *Plasmodium falciparum:* molecular characterization and naturally elicited antibody response that may provide immunity to malaria pathogenesis. *J Exp Med* 2000;**192**:1563–76.

124. deSouza JB, Todd J, Krishnegowda G, Gowda DC, Kwiatkowski D, Riley EM. Prevalence and boosting of antibodies to *Plasmodium falciparum* glycosylphosphatidylinositols and evaluation of their association with protection from mild and severe clinical malaria. *Infect Immunity* 2002;**70**:5045–51.

125. Boutlis CS, Gowda DC, Naik RS, Maguire GP, Mgone CS, Bockarie MJ, Lagog M, Ibam E, Lorry K, Anstey NM. Antibodies to *Plasmodium falciparum* glycosylphosphatidylinositols: inverse association with tolerance of parasitemia in Papua New Guinean children and adults. *Infect Immunity* 2002;**70**:5052–7.

126. Naik RS, Krishnegowda G, Ockenhouse CF, Gowda DC. Naturally elicited antibodies to glycosylphosphatidylinositols (GPIs) of *Plasmodium falciparum* require intact GPI structures for binding and are directed primarily against the conserved glycan moiety. *Infect Immunity* 2006;**74**:1412–15.

# 15

# Glycomics in Unraveling Glycan-Driven Immune Responses by Parasitic Helminths

Irma van Die[a] and Richard D. Cummings[b]

[a]DEPARTMENT OF MOLECULAR CELL BIOLOGY & IMMUNOLOGY,
VU UNIVERSITY MEDICAL CENTER, AMSTERDAM,
THE NETHERLANDS
[b]DEPARTMENT OF BIOCHEMISTRY, EMORY UNIVERSITY SCHOOL
OF MEDICINE, ATLANTA, USA

## Introduction

Infections with parasitic helminths are a major cause for human suffering and death. Worldwide, more than 4 billion humans are infected, with the majority of the life-threatening worm infections occurring in (sub)tropical areas. In the Western world parasite infections are becoming an increasing concern due to a growing consciousness that worm parasites can be transmitted from wild or domestic animals to humans (zoonoses), and the import of "tropical" diseases within infected immigrants or travellers coming from endemic areas. Whereas drugs are available in the Western world to treat human and veterinary parasitic helminth infections, these drugs are not generally affordable to the people in tropical areas who suffer most. In addition, reinfection and/or emerging resistance to the drugs are major problems not easily solved. Despite intensive research, no vaccines are available yet to prevent any of the known helminth infections. Therefore, there is an imminent need for novel approaches to develop effective and cheap diagnosis and treatment of these infections, both in humans and domestic animals. Increased understanding of the host immunological responses to these worms will allow a more rational design of novel drugs and may even open up the way to immunotherapy. In addition, understanding the immunological determinants may enhance the diagnostic methods to allow focused drug treatment.

**Handbook of Glycomics**
Copyright © 2009 by Elsevier Inc. All rights of reproduction in any form reserved.

Current diagnostics for most helminth infections rely on indirect and awkward techniques, such as visual observation of eggs in stool samples.

In a search for immunogenic structures that may be effective vaccine and diagnostic components, an enormous research effort has been focused on identification of protein antigens of the major helminth species, so far without much success. As an alternative strategy, several groups turned to characterization of antigen glycans of helminths. Glycans, covalently linked to proteins and lipids, are abundantly present on the surface of helminths and within their excretory/secretory products. Thus, they are easily recognized by the host immune system, and may contribute to mechanisms of host protection. It has been shown that the humoral immune response against several worm parasites is dominated by anti-glycan IgM and IgG responses. Second, helminth glycans have been associated with the capacity of many worms to modulate the host immune response to secure their survival. Whereas the immunomodulatory properties of the helminths may compromise efforts to induce protective immunity by vaccination, they may have potential for treatment of other chronic inflammatory diseases. To enable application of helminth glycans as prophylactic or therapeutic drugs, researchers have used glycomic approaches to define structures of the helminth glycans that induce the biological activity, and understand their mode of action. Currently, many helminth glycan structures have been identified, and there are also many reports indicating a role for helminth glycans of unknown identity to induce immunological responses. In the next decade it will be important to comprehensively link biological activities of the helminth glycans to defined glycan structures.

In this chapter we will provide an overview of the glycan structures that are synthesized by selected helminth species belonging to the *Platyhelminthes* and *Nematoda* phyla, and their contribution to induction or modulation of host immune responses. In addition we will review novel glycomic approaches that will facilitate the development of applications of helminth glycans in diagnosis and treatment of helminth infections, and possibly other inflammatory diseases.

## Glycan Structures in Helminths

Advances in mass spectrometry (MS) in the last decade allowed structural characterization of very small amounts of glycans which resulted in an enormous progress in the identification of helminth glycans, even from different developmental stages [1,2]. Several excellent reviews have been published recently that give an overview of these structural data [3–5]. We will only shortly summarize the main glycan structures and differences between species to illustrate functional aspects of the helminth glycans.

Helminths are multicellular eukaryotes that have a glycosylation potential resembling mammalian glycosylation in some ways. They synthesize *N*- and *O*-glycans on surface and secreted glycoproteins, as well as glycolipids and polysaccharides such as polylactosaminoglycans. However, important differences between the glycosylation of mammals and helminths are found in some core structures and in the terminal modifications of the glycans. Whereas in mammals terminal glycans are typically built up on LacNAc (Galβ1–4GlcNAc) units, many helminths express instead or in addition to the LacNAc unit, the LacdiNAc (GalNAcβ1–4GlcNAc) unit and/or a chitobiosyl (GlcNAcβ1–4GlcNAc) unit [6,7] (Figures 15.1 and 15.2). Most helminths synthesize glycosphingolipids (GSLs) with the Galβ1–4Glc-Cer or Galβ-Cer core structures that are commonly found in mammalian GSLs. Remarkably, schistosomes synthesize GalNAcβ1–4Glc-Cer core structures, also called the schisto-core [8] (Figure 15.1). It has been consistently observed that worms generate glycans that lack terminal sialic acid residues such as found in mammals. Instead, the glycans display other modifications, consisting, for example, of high levels of fucose residues, as observed in schistosomes. Many helminths express a number of common, similar glycan antigens, resulting in cross-reactivity between helminth species in serology of infected hosts (Figures 15.3 and 15.4). Glycan antigens may also be shared with the hosts of these helminths (Figure 15.4). In addition, the worms express modifications that are not found in the host (foreign glycan antigens) and which are helminth-specific, or found in a limited amount of helminths. A striking example is the presence of tyvelose in *T. spiralis* (Figure 15.5A) [9], which thus far is only found in this nematode, providing a useful diagnostic marker for detection of this parasite in infected hosts. The fact

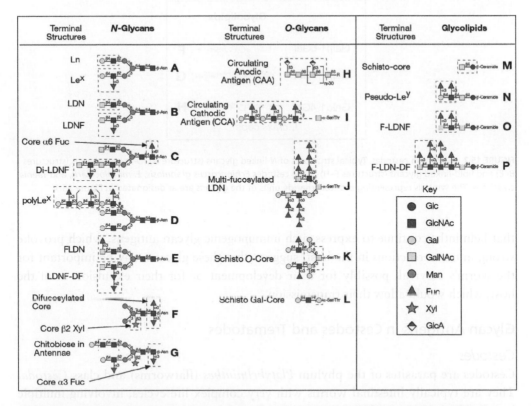

**FIGURE 15.1** Glycans in *Schistosoma* spp. Typical structures of schistosoma *N*-linked glycans (structures A–G), *O*-linked glycans (structures H, I) and lipid-linked glycans (structures M–P), with the short names indicated. The symbols representing monosaccharide units of the glycans are indicated, and are as proposed by the Consortium for Functional Glycomics (http://www.functionalglycomics.org).

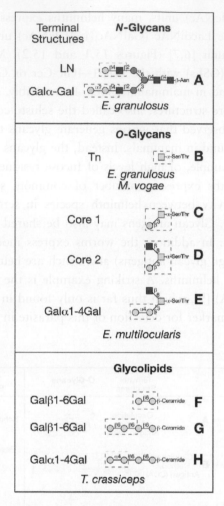

**FIGURE 15.2** Glycans in cestodes. Typical structures of *N*-linked glycans (structure A), *O*-linked glycans (structures B–E) and lipid-linked glycans (structures F–H) in the cestodes *Echinococcus granulosis, E. multilocularis*, and *Taenia crassiceps*. The symbols representing monosaccharide units of the glycans are as designated in Figure 15.1.

that helminths continue to express such immunogenic glycan antigens, which provoke strong immune reactions in the host, suggests that these glycans may be important for the worm's survival, possibly for their development or for their expulsion from the host, which would allow their transmission.

## Glycan Antigens in Cestodes and Trematodes

### Cestodes

Cestodes are parasites of the phylum *Platyhelminthes* (flatworms) and class *Cestoda*. They are typically intestinal worms with very complex life cycles, involving multiple hosts with varying specificity. Tapeworms, except for *Echinococcus* species, can grow to huge lengths, up to 20 m. Humans can be definitive hosts, such as the case with several fish tapeworms, or *Taenia solium*, the causative agent of an infection of the

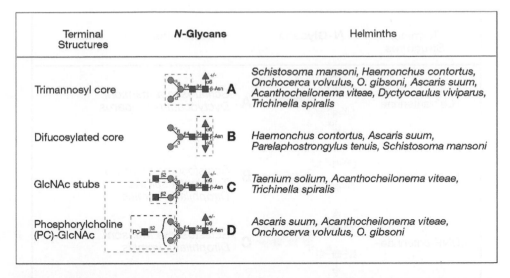

| Terminal Structures | *N*-Glycans | Helminths |
|---|---|---|
| Trimannosyl core | **A** | *Schistosoma mansoni, Haemonchus contortus, Onchocerca volvulus, O. gibsoni, Ascaris suum, Acanthocheilonema viteae, Dyctyocaulus viviparus, Trichinella spiralis* |
| Difucosylated core | **B** | *Haemonchus contortus, Ascaris suum, Parelaphostrongylus tenuis, Schistosoma mansoni* |
| GlcNAc stubs | **C** | *Taenium solium, Acanthocheilonema viteae, Trichinella spiralis* |
| Phosphorylcholine (PC)-GlcNAc | **D** | *Ascaris suum, Acanthocheilonema viteae, Onchocerva volvulus, O. gibsoni* |

**FIGURE 15.3**  Common non-mammalian *N*-linked glycans found in multiple helminth species. Multiple helminth species commonly express *N*-linked glycans that are not found in mammals as far as known. These include truncated *N*-glycans (structures A, C), and *N*-glycans with core $\alpha$(1–3)-fucose (structure B) or phosphorylcholine (PC)-modified *N*-acetylglucosamine (GlcNAc).

central nervous system called neurocysticercosis. Humans, however, can also function as intermediate hosts such as with *Echinococcus granulosis* that mostly uses dogs as definitive hosts. In either case, humans can be infected by eating raw meat, or contact with parasite-bearing animals.

There is limited information about glycosylation in cestodes. Several reports show the presence of mucins carrying the glycan antigens termed Tn (GalNAc$\alpha$1-Ser/Thr) (Figure 15.2B) and sialyl-Tn (Sia$\alpha$2–6GalNAc$\alpha$1-Ser/Thr). In *Echinococcus granulosus* and *Mesocestoides vogae*, the presence of the Tn antigen has been reported, along with a ppGalNAc-T (UDP-*N*-acetyl-d-galactosamine:polypeptide *N*-acetylgalactosaminyltransferase) involved in synthesis of the Tn antigen as a first step in *O*-glycosylation [10–13]. In *Mesocestoides vogae* both Tn and sialyl-Tn antigens were detected in in vitro cultured parasites, which makes it unlikely that they were host-derived. As a side note, it is important in studies of parasite-derived glycans to be aware of potential contamination from host-derived materials, such as blood cells or blood glycoproteins. Tn antigens, which in humans have been shown to be tumor-associated antigens, appear to be a common "helminth" antigen found across the phyla boundaries. In addition to the above examples, Tn antigens have also been observed in the cestodes *Taenia hydatigena* and *Mesocestoides corti*, the trematodes *Fasciola hepatica* and *S. mansoni*, as well as in nematodes (*C. elegans, Nippostrongylus brasiliensis*, and *Toxocara canis*) [10,14,15]. In *Echinococcus multilocularis*, a zoonosis causing alveolar echinococcosis, a mucin-type glycoprotein (Em2(G11)) of the laminated layer encapsulating the metacestode has been characterized by MS. The most abundant *O*-glycoforms of the Em2(G11) mucin appeared to be the common *O*-glycan core 1 (Gal$\beta$1–3GalNAc) and core 2 (GlcNAc$\beta$1–6[Gal$\beta$1–3]GalNAc) structures (Figure 15.2C,D). In addition

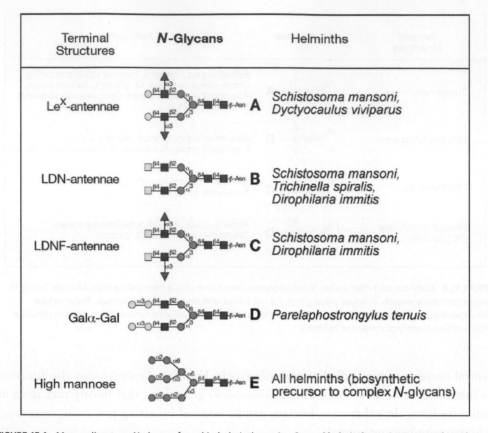

| Terminal Structures | *N*-Glycans | Helminths |
|---|---|---|

**FIGURE 15.4** Mammalian-type *N*-glycans found in helminth species. Several helminth species commonly and abundantly express *N*-linked glycans that closely resemble glycans that can be found in low amounts in mammals. These include glycans that contain the terminal antigens Le$^x$, LacdiNAc, or LDNF (structures A–C). Also the Galα1–3Gal terminal antigen (structure D) of helminths infecting non-primates belongs to the host-like glycans, since non-primates in contrast to humans commonly contain the Galα1–3Gal terminal antigen. Many helminths commonly express high-mannose or oligomannose type *N*-glycans, which are biosynthesis precursors of complex-type *N*-glycans that are, however seldom, found in mammals as final glycan structures.

a Galα1–4Gal moiety was demonstrated on these O-glycans, which might have a capping function (Figure 15.2E) [16].

Such Gal repeats do not occur only within O-glycans, but appear to be motifs that frequently occur in many different cestode glycoconjugates. The major *N*-glycans from the cyst membrane of *E. granulosus* were found to having complex-type antennae that terminated with GalαGalβ1–4 moieties (Figure 15.2A) [17]. In addition, *N*-glycans isolated from the fox tapeworm *Taenia crassiceps* metacestode extracts have been shown to contain complex *N*-glycans with a terminal structure consisting of Fucα1–3GlcNAc [18]. In glycosphingolipids isolated from *T. crassiceps,* both Galα1–4Gal and Galβ1–6Gal sequences have been demonstrated (Figure 15.2F,G,H) [19]. In *E. multilocularis* the neutral glycosphingolipids of the metacestodes contained a di-, a tri-, and a tetra-galactosyl-ceramide having also internal Galβ1–6Gal linkages that were found to be immunogenic in humans [20].

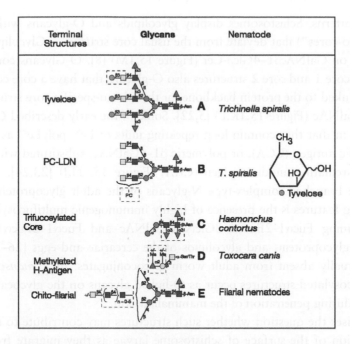

**FIGURE 15.5** Typical species-specific nematode glycans. Examples of glycans that can be regarded as species-specific, although it should be noted that elucidation of more helminth glycan structures in the future may discover additional species carrying these glycans.

## Schistosomes

Among the trematodes, many glycans from *Schistosoma* spp. are structurally characterized as recently reviewed extensively [3,5,21], and we refer to these reviews for structural details. Schistosomes have a complex life cycle involving an intermediate snail host. Infection occurs by contact with water that harbors free-swimming cercariae released from infected snails. The cercariae penetrate through the skin in their vertebrate hosts and transform within hours to schistosomula that migrate through the lymph, heart, lung, and liver, meanwhile maturing to adult worms. Male and female worms pair at 4–5 weeks post infection and produce thousands of eggs per day. The major human schistosome species *S. mansoni*, *S. japonicum*, and *S. haematobium*, each reside in a different location within the vascular system of their vertebrate hosts. The major pathology in schistosomiasis occurs as a consequence of egg deposition in host's tissues, which induces the induction of a granulomatous response. This granulomatous response, which is associated with a strong Th2 type cellular immune response, sequesters eggs and limits tissue damage.

Schistosomes generate a large array of glycan antigens in *O*- and *N*-glycans of glycoproteins and in glycolipids, which include the Le$^x$ antigen, polyLe$^x$, LacdiNAc, fucosylated LacdiNAc (GalNAc$\beta$1–4[Fuc$\alpha$1–3]GlcNAc, LDNF and Fuc$\alpha$1–3GalNAc$\beta$1–4GlcNAc, FLDN), difucosylated LacdiNAc (GalNAc$\beta$1–4[Fuc$\alpha$1–2Fuc$\alpha$1–3]GlcNAc, or multi-fucosylated LacdiNAc or chitobiose units, core $\alpha$1–3Fucose and core $\beta$1–2-xylose (Figure 15.1). These glycan antigens have restricted patterns and stage-dependent expression patterns. Schistosomes display glycolipids and O-glycans with core structures ("schisto-cores") that deviate from the usual core structures. Glycolipids are commonly based on GalNAcβ1–4Glcβ-Cer (Figure 15.1M) [8]. O-Glycans contain next to the common core 1 and core 2 structures also O-glycans that have a core consisting of a Gal residue linked to the protein backbone, or the schisto-specific core structure Galβ1–3(Galβ1–6)GalNAc (Figure 15.1K,L) [5,22]. Some of the early described O-glycans are very unusual in that they contain long repeating units of Le$^x$ (polyLe$^x$) as in the circulating cathodic antigen (CCA), or polymeric β1–6GalNAc, substituted with β1–3GlcA, such as in circulating anodic antigen (CAA) (Figure 15.1H,I) [23,24]. PolyLe$^x$ also occurs in the branched complex-type N-glycans of the adult glycoproteins [25]. One of the striking features is the presence of highly immunogenic multifucosylated glycans (those containing Fucα1–2Fucα1–3GlcNAc/GalNAc and Fucα1–2Fucα1–2Fuc linkages) within glycoproteins and glycolipids of the cercariae and eggs [26–29], whereas these are virtually absent from adult worm glycoconjugates. In S. mansoni cercariae, such multifucosylated structures occur as O-linked glycans on the glycocalyx, which is rapidly shed during penetration of the mammalian skin.

This raises the question whether such structures may contribute to the mechanical stabilization of the surface of schistosome larvae as they migrate from the snail intermediate host into their definitive mammalian host as suggested [27], or may facilitate the invasion process. Most likely, these multifucosylated antigens, which are rapidly shed during penetration of the mammalian host, may be potent immunological modulators [30,31].

Glycomic analysis of schistosome glycans by MS includes overall profiling of the major glycans within a life stage, as well as detailed and site-specific analysis of the glycans of a single protein [21]. Both glycomic approaches potentially provide important leads to study the function of the respective glycans. An overall analysis of the major glycans of different developmental stages provides us with information about stage-specific glycosylation. For example, expression of the pseudo-Le$^Y$ antigen is highly specific and only occurs in glycosphingolipids of schistosomal cercariae [32], the Le$^x$ antigen is found in all stages of the parasite, and egg and cercarial stages share many highly fucosylated terminal structures that are hardly expressed in the adult worms which may be of functional significance.

Remarkably, in addition to parasite-specific glycan antigens S. mansoni expresses definitive host-like glycan antigens, such as Le$^x$, as well as the intermediate host-like glycan antigens FLDN and β1–2Xyl (Table 15.1). Such mammalian-type N-glycans are found in many helminths (Figure 15.4A–E). These observations may support the concept of glycan-mediated molecular mimicry [33], possibly as a strategy to compromise the immune system in order to survive. In humans, the foreign, parasite-specific glycans appear to be highly immunogenic, whereas several self glycan antigens show interaction with C-type lectins of dendritic cells to modulate immune function [34,35] (Figure 15.7). However, C-type lectins can also act as pathogen receptors by recognizing pathogen-specific glycan structures. For example, the C-type lectin DC-SIGN recognizes in addition to the glycan antigens Le$^x$ and LDNF, which are

**Table 15.1  Presence of *S. mansoni* glycan antigens in the intermediate snail host *Biomphalaria glabrata* or the vertebrate human host**

| Glycan antigen *S. mansoni* | Snail host *B. glabrata* | Human host | Ref. |
|---|---|---|---|
| LDN | − | + | 7,157 |
| LDNF | − | + | 157 |
| Le$^x$ | +/− | + | 158 |
| FLDN | + | − | 159 |
| Core β1–2Xyl | + | − | 159 |
| LDN-DF | − | − | 157–159 |
| Tn | | + | 15 |

shared by schistosomes and humans [36], also the schistosome-specific glycan antigen pseudo-Le$^Y$ [37].

## Fasciola hepatica

The trematode *F. hepatica* causes fasciolosis, a chronic disease in humans and domestic animals such as sheep [38]. The disease is a zoonosis that can be transmitted from both domestic and wild animal host reservoirs to humans. Similar to schistosomes, the *F. hepatica* life cycle involves an intermediate snail host that produces the infective stage, which are encysted metacercariae in the case of *F. hepatica*. After infection of the mammalian host by ingestion of metacercariae, the flukes migrate from the digestive tract to the liver, where they enter the bile ducts and start egg production [38].

Despite its importance as a pathogen for humans and domestic animals, there is a virtual lack of structural information on *F. hepatica* protein-linked glycans. Early studies show a considerable cross-reactivity of infection sera between *F. hepatica* and *S. mansoni*, which most likely is caused by the presence of similar (fucosylated) glycan epitopes on glycoproteins and glycolipids [39–41]. It is believed that some of these—still unknown—common glycans may be responsible for the observed cross-protection that has been demonstrated by heterologous challenges with *S. mansoni* cercariae and *F. hepatica* metacercariae, respectively [39]. Recent data suggest the presence of truncated *O*-glycans including the Tn antigen in *F. hepatica*, as indicated by lectin-blotting and the presence of a ppGalNAcTase activity in adult flukes [42]. The occurrence of this antigen within many helminths may also contribute to observed serum cross-reactivity.

The glycolipids of *F. hepatica* have been structurally characterized in more detail by matrix-assisted laser desorption/ionization-time-of-flight (MALDI-TOF) and electrospray ionization mass spectrometry, in combination with enzymatic sequencing and linkage analysis [43–45]. In addition to both glucosyl- and galactosylceramide, several mammalian-type (iso)globotriaosylceramides, as well as Forssman antigen (GalNAc(α1–3)GalNAc(β1–3/4)Gal(α1–4/3)Gal(β1–4)Glc-ceramide), were found. Whereas the ceramide composition of the first group suggests these glycolipids are of parasite origin, the ceramide composition of the Forssman antigen, and its localization in the gut, indicate that this glycolipid is probably host derived. In addition several parasite-type glycolipids have been found in *F. hepatica*. Similar to cestodes, *F. hepatica* produces neutral glycolipids carrying terminal Gal(β1–6)Gal and Gal(α1–4)Gal epitopes (Figure 15.6C,D). The occurrence of such glycolipids and corresponding antibodies in *F. hepatica* as well as parasitic cestode infection sera may contribute to the described serological cross-reactivity observed in these infections [46,47]. Glycolipids carrying Galα1–3Gal moieties are also commonly found in nematodes (Figure 15.6B), whereas the acidic GlcNAcα1-HPO$_3$-6Gal-ceramide may be specific for *Fasciola* (Figure 15.6E) [45].

## Glycan Antigens in Nematodes

Nematodes are a large phylum consisting of roundworms that are covered with a non-cellular cuticle. Whereas most nematodes are free-living, a few cause diseases of great importance to humans, domestic animals, and plants. Glycans are abundantly present at the boundary of nematodes with their environment. A carbohydrate-rich surface coat covers the cuticle that may protect the nematodes from their environment, including the immune system in the parasitic members of this phylum. In addition, most nematodes possess an active excretory secretory (ES) system, by which they release substances that are often glycosylated, into their environment. From a glycobiological view, several nematode families are well characterized. We will review here examples of the glycan structures derived from a variety of nematodes belonging to different (super)families. A hallmark of nematodes is the modification of many glycans with phosphorylcholine (PC), resulting in highly antigenic structures [48,49]. PC-modified glycans are assumed to contribute to a long-term persistence of parasites within their host.

### Trichinella spiralis

*Trichinella spiralis* (order Trichuridae) is one of the clinically important nematode infections leading to the disease generally referred to as trichinosis. Most important

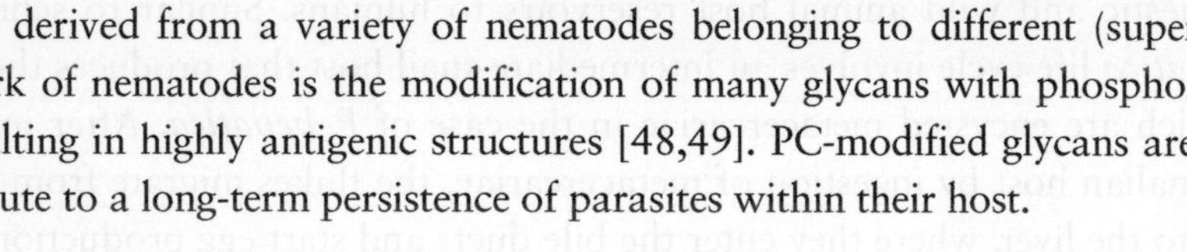

| Terminal Structures | Glycolipids from *Fasciola hepatica* | | Terminal Structures | Glycolipids from *Ascaris suum* | |
|---|---|---|---|---|---|
| Lactosyl-Ceramide | | A | Galα1-3GalNAc | | F |
| Galα1-3Gal | | B | PC and PE modifications | | G |
| Galα1-4Gal | | C | Blood group B moiety | | H |
| Galβ1-6Gal | | D | Sulfated Gal-Cer | | I |
| GlcNAc-phospho | | E | Phosphoinositol-glycosphingolipid | | J |

**FIGURE 15.6** Glycolipids from *Fasciola hepatica* and *Ascaris suum*. Whereas little is known about protein-linked glycans of the trematode *F. hepatica* and the nematode *A. suum*, their glycolipids have been studied quite well. Glycolipids of *F. hepatica* (structures A–E) include glycolipids that carry terminal Galβ1–6Gal and Galα1–4Gal moieties similar to those found in cestodes. The glycolipids from *A. suum* (structures F–J) include glycolipids carrying glycans with mammalian-type antigens (bloodgroup B antigen) and nematode-type modifications such as PC or PE groups.

for the life cycle of *T. spiralis*, however, is its parasitism of domestic and wild animals including pigs and rats. Humans are mostly infected by eating infected pork that is not properly cooked. An unusual property of *T. spiralis* is that the parasite's life cycle is completed within the same animal, with larval stages and adults living in different organs. Exceptional for a multicellular pathogen, *T. spiralis* lives intracellularly and can—like a virus—redirect host cell activities to its own benefit.

Early evidence shows that parasite glycans play an important role in survival of *T. spiralis*. Antibodies recognizing highly immunodominant 3,6-dideoxy-d-*arabino*-hexose (tyvelose)-bearing glycans on surface proteins of *T. spiralis* L1 larvae have protective potential, possibly by preventing the parasite to adhere to their intestinal epithelial niche, leading to expulsion of the parasites [50–57]. Detailed structural analysis of these glycans identified them as mainly tri- and tetraantennary *N*-linked glycans composed of LDNF antennae capped with a β3-linked tyvelose moiety (Figure 15.5A) [9,58,59]. Remarkably, whereas the anti-tyvelose antibodies are mostly generated late in infection, consistent with a role in preventing reinfection, [50], antibodies recognizing PC-modified glycans are generated very early in infection and do not seem to have a protective role [60]. The major *T. spiralis* *N*-glycans carrying PC modifications consist of multi-antennary LacdiNAc structures, with a PC linked to either the GlcNAc or GalNAc of the LacdiNAc moiety (Figure 15.5B). In addition, more common high-mannose *N*-glycans, as well as complex-type *N*-glycans with terminal GlcNAc stubs or LacdiNAc moieties with or without a core α6-fucose have been demonstrated in *T. spiralis* (Figure 15.3D) [61,62].

## Haemonchus contortus and Dictyocaulus viviparous

*Haemonchus contortus* and *Dictyocaulus viviparous* are veterinary nematodes of great economic importance. *H. contortus* is the most common and economically important parasite in small ruminants [63]. Sheep can develop a strong natural and long lasting immunity to *Haemonchus* and other trichostrongylids after natural exposure or experimental infections, indicating that the development of a vaccine may be feasible. Immunization with a native membrane protein from the intestinal cells of the parasite, designated H11, has been shown to be an effective vaccine antigen in a variety of sheep [64]. The finding that a significant portion of the antibody response to H11 appeared carbohydrate-specific led to the structural characterization of the *N*-glycans expressed on this protein. The major *N*-linked glycans identified in H11 contain up to three fucose residues attached to their chitobiose cores [65]. The fucose residues are found at the 3- and/or 6-positions of the proximal GlcNAc and at the 3-position of the distal GlcNAc (Figure 15.5C). The Fucα1–3GlcNAc moiety at the proximal GlcNAc of the core (core α3-fucose) is a highly antigenic epitope common within several helminth species as well as plant and insect glycoproteins and accounts for the serum cross-reactivity, in particular of IgE, between these organisms [66–69]. The fucose at the 3-position of the distal GlcNAc has not been identified in other species so far, and is not detected in L3 larvae, showing that this unusual modification is stage-specific [70].

Immunization with native *H. contortus* ES products also confers protection against this helminth to sheep. In these vaccination experiments high levels of anti-glycan antibodies were detected after immunization, whereas anti-peptide responses were low. Remarkably, protection was significantly associated with a strong IgG response against LDNF glycan antigens, suggesting that this glycan antigen may contribute to the protective response [71]. Western blot analysis showed that LDNF antigens are present as protein-linked antigens, but no data are available that establish whether the LDNF moieties are *N*- or *O*-linked.

*D. viviparous* is a lungworm that infects cattle, buffalo, and some deer species and is the etiologic agent of bovine parasitic bronchitis in both temperate and tropical regions worldwide. A remarkable feature of the life cycle of *D. viviparous* is that the L3 larvae, which are excreted in the feces of infected cattle, make use of a fungus to spread. The larvae invade sporangiophores of the fungus *Pilobolus* that grows on the feces, and when the mature sporangiophore bursts the larvae together with the spores fly for several meters out of the feces so that they can be ingested by grazing cattle [72].

The glycoproteins of adult *D. viviparous* contain high-mannose, truncated oligomannose and complex-type *N*-glycans. Remarkably, major components of the complex-type glycans are Le$^x$ moieties on bi-, tri-, and tetra-antennary *N*-glycans, which are also found within schistosomal glycans (Figure 15.4A) [73]. *D. viviparous* may also contain PC-substituted GlcNAc stubs (see Figure 15.3B), as is suggested by the binding of the PC-specific antibody TEPC-15 and WGA lectin to *N*-glycans of GP300, a major immunodominant glycoprotein of this parasite [74].

## Ascaridida

The ascaridids worms are among the largest intestinal nematodes. *Ascaris lumbricoides* and *Ascaris suum*, the first a human parasite and the second a parasite of pigs, are remarkably similar, suggesting a common ancestor that moved from pigs to humans or the other way around [75]. The disease caused by *Ascaris*, ascariasis, is characterized by low mortality but high morbidity, and is a serious problem especially in developing countries.

Whereas there is essentially no information about glycan structures of *A. lumbricoides*, those of *A. suum* have been described quite well. *Ascaris suum* has relatively simple *N*-glycans. Around 80% of the *N*-glycans are pauci- or oligomannosidic. Fucosylation of *N*-glycans of *A. suum* appears to be restricted to the *N*-glycan core structure. Truncated α3/α6 difucosylated core structures and short PC-GlcNAc-modified oligomannose structures have been found, similarly as in several other helminths (Figure 15.3A,B,D). In addition, some evidence indicates the presence of a hybrid PC-containing *N*-glycan [76].

Within the acidic glycolipid fraction of *A. suum* a 3-sulfogalactosylcerebroside (HSO$_3$-3Galβ-ceramide) [77], and an unusual phosphoinositol-glycosphingolipid (Galα1Ins-P-1-ceramide) have been demonstrated (Figure 15.6I,J) [77,78]. The sulfated galactosylcerebrosides may be of immunological interest, since they are good ligands for selectin receptors, suggesting a possible function in blocking selectin-dependent inflammatory responses in vivo [79].

The main structures identified within neutral glycosphingolipids are Galα1–3GalNAcβ1–4GlcNAcβ1–3Manβ1–4Glcβ-ceramide and precursors of this structure (Figure 15.6F) [80]. Zwitterionic glycosphingolipids are reported to be modified with PC and phosphorylethanolamine (PE) (Figure 15.6G). These consist of Galα1–3GalNAcβ1–4[PC6]GlcNAcβ1–3Manβ1–4Glcβ-ceramide with or without a PE group 6-linked to Man, and the corresponding truncated tri- and tetrasaccharides. In addition, extended structures were demonstrated that revealed novel structural moieties such as disubstituted αGal carrying two β-linked Gal residues, which were found to be partly further modified. Several modifications by fucose have been shown, such as on the PC-substituted GlcNAc, or within a blood group B moiety (Galα1–3(Fucα1–2)Gal) (Figure 15.6H) [81]. Some serological studies suggest that *Ascaris lumbricoides* may similarly express antigens belonging to the ABO blood group determinants [82]. The presence of PC is a major cause for cross-reactivity of IgG and IgM antibodies in infection sera between *A. suum* with *Litomosoides carinii* and *Nippostrongylus brasiliensis* [83], and with *T. canis* ES products (E. Pinelli, RIVM, the Netherlands, personal communication), thus predicting the presence of PC within glycoconjugates of these species.

*Toxocara canis* is a cosmopolitan ascaridid of dogs, which as a zoonosis can infect humans. In humans the larvae cannot complete their life cycle, and start a random wandering through the body which may result in disease, so-called visceral larva migrans. Whereas infections with several helminths such as schistosoma are thought to protect against allergy [84], evidence from epidemiological studies suggests that infection with *Toxocara* worms stimulates atopic diseases [85]. Larval parasites secrete glycoproteins with O-linked methylated oligosaccharide structures similar to the mammalian blood group H antigen, but bearing O-methylated substitutions on the terminal fucose and subterminal galactose residues (2-O-Me-Fuc(α1–2)-4-O-Me-Gal(β1–3)GalNAc-R and 2-O-Me-Fucα1–2Galβ1–3GalNAc-R) (Figure 15.5D) [86]. By contrast, *Toxocara cati* O-glycans consist predominantly of di-O-methylated trisaccharide. Not surprisingly, these unusual oligosaccharides are highly antigenic, being recognized by antibodies from human toxocariasis patients and murine antibodies, generated to *T. canis* infection [87].

## Filarial Nematodes

Filarial nematodes are tissue-dwelling parasites that all employ arthropods as intermediate hosts. They are well known to express a variety of PC-containing products that are reported to have strong immunomodulating properties [88,89]. The N-glycan structures of the major excretory secretory glycoprotein (ES-62) of the rodent filarial nematode *Acanthocheilonema viteae* have been identified as high mannose type structures, trimmed structures consisting of the trimannosyl core, and complex structures with and without core fucosylation carrying between one and four N-acetylglucosamine residues (GlcNAc stubs, see Figure 15.3D) partly modified with PC (Figure 15.3E).

Human *Onchocerca volvulus* infection results in the disease onchocerciasis, or river blindness. Sera of human patients were found to recognize zwitterionic glycolipids of *O. volvulus* and to cross-react with those of other parasitic nematodes (*A. suum*, *Setaria digitata*, and *Litomosoides sigmodontis*) [90]. Three glycolipid components of

*O. volvulus* were structurally identified as respectively PC-6GlcNAcβ1–3Manβ1–4Glc-Cer, GalNAcβ1–4(PC-6)GlcNAcβ1–3Manβ1–4Glc-Cer, and Galα1–3GalNAcβ1–4(PC-6)GlcNAcβ1–3Manβ1–4Glc-Cer, structures also found in *A. suum* (see Figure 15.6F, and precursors of F). Also *N*-glycans of *O. volvulus* and *O. gibsoni* appear to contain PC, similar to those found in ES-62 of *A. vitae* (Figure 15.3E) [91]. Remarkably, filarial nematodes also contain *N*-linked glycans, the antennae of which are composed of chito-oligomers (Figure 15.5E) [92].

The heart worm *Dirophilaria immitis* parasitizes the heart and pulmonary artery of dogs and other mammals throughout the world. Although no detailed analysis of glycan structures of *D. immitis* is available, biochemical and serological evidence indicates the presence of *N*-glycans with LacdiNAc, and possibly LDNF and core α1–3-fucose moieties (Figure 15.4B,C) [69,93].

## Recognition of Parasite Glycans by Collectins of the Innate Immune System

Recognition of glycans on pathogens, including helminths, by collectins contributes to the first line of host defense acting before the appearance of antigen-specific responses. Collectins can stimulate phagocytosis by recognizing carbohydrates on the surface of pathogens, regulate cytokine release by immune cells, and activate complement via the classical pathway. One of the best studied collectins is the mannan-binding lectin (MBL), which recognizes heavily mannosylated glycoconjugates. MBL can bind and activate the complement cascade through the "lectin pathway" in response *to S. mansoni* [94] and possibly *T. spiralis* [95]. In addition, a recent study using a MBL-A knockout mice shows that MBL plays a role in acquiring resistance to the filarial nematode *Brugia malayi* [96]. In in vitro studies, binding of pulmonary surfactant protein (SP)-D to FLDN moieties has been demonstrated, suggesting a possible role for this collectin in immunity towards lung stages of *Schistosoma mansoni* [97].

## Parasite Glycans are Key Players in Helminth-Induced Immune Regulation

Despite the pathology observed, most chronic helminth infections are relatively asymptomatic, which can be ascribed to the capacity of the worms to regulate the degree of inflammatory responses in their hosts. Helminth infections are typically characterized by attenuated Th1 responses, and induction of Th2 and regulatory T cell populations which favor survival of parasites in their host [98]. Recently these properties have raised an enormous interest focused on exploitation of helminths or their products to reduce excessive inflammation associated with several immune-mediated diseases. Helminth infections or helminth products have been shown to protect animals from diseases such as experimental colitis, reactive airway disease, experimental autoimmune encephalomyelitis, and type 1 diabetes [99–102]. Most remarkably, clinical trials show that exposure to helminths can reduce disease activity in patients with ulcerative colitis or Crohn's disease [99].

The ability of the host to generate a specific adaptive immune response to pathogens, including helminths, critically depends on the initial recognition of the pathogens by antigen-presenting cells of the innate immune system [103]. Dendritic cells (DC) play a key role herein by searching for invading pathogens using a repertoire of cell surface receptors including Toll-like receptors (TLRs) and C-type lectins (CLRs) (Figure 15.7). Recognition of a pathogen, which typically involves multiple receptor–ligand interactions, relays information about the interacting pathogen through intracellular signaling cascades, leading to DC maturation, the release of inflammatory cytokines, and the induction of a pathogen-specific T cell repertoire including Th1, Th2, Th17, and regulatory T cells [103–105].

Whereas the network of host–parasite communication is enormously complex, involving a wide variety of immune cells and molecules, there is increasing evidence supporting a role for helminth glycans in regulation of the host immune response towards Th2 and regulatory responses [106–111]. The molecular mechanisms underlying these specific glycan-mediated Th2-biased responses are incompletely understood but it is assumed that DCs play a crucial role. By expressing an extended repertoire of CLRs on their cell surface, DCs are well equipped to recognize a broad variety of glycan molecules on self-antigens or pathogens, thereby controlling immune responses either directly or via cross-talk with TLRs (Figure 15.7) [112].

It is clear that the immunomodulating properties of helminths appear to be more a general than a specific property. Apparently, evolution has driven individual helminth species, which stand far apart phylogenetically, to acquire similar strategies to

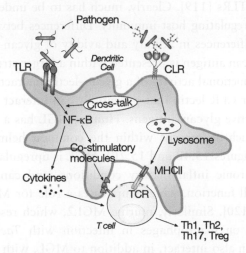

**FIGURE 15.7** Dendritic cells (DC) capture pathogens and control T cell responses. Immature DCs express different receptors (CLRs, TLRs) on their surface to capture pathogens. C-type lectin receptors (CLRs) are efficient in internalization of pathogens into lysosomes where they are degraded. Antigens derived from the pathogens are presented on MHC class II molecules. Upon maturation of the DC through specific signaling through Toll-like receptors (TLRs), MHC class II molecules are transported to the cell surface where they interact with the T cell receptor (TCR) to activate T cells. Cross-talk between TLRs and CLRs at the cell surface or at the level of intracellular signaling modulate the DC phenotype to induce pathogen-specific Th1, Th2, Th17, or Treg cells.

survive in their respective host species. One of the strategies employed may be the expression of "self" glycan antigens similar to or resembling host glycans. Parasitic cestodes, nematodes, and trematodes express similar glycan antigens on their surface or in their secreted products as deduced from the extensive cross-reactivity between infection sera observed between these helminths [113] (Figures 15.3 and 15.4). Several such cross-reactive "self" glycans (see Figure 15.4) have been found to target host lectins and may modulate immune responses. We showed that human DCs internalize *S. mansoni* soluble egg antigens (SEA) via interaction with the CLRs macrophage galactose lectin (MGL), the mannose receptor (MR), and DC-SIGN, thereby inducing a DC phenotype capable of skewing towards Th2 responses [35].

DC-SIGN is a C-type lectin receptor (CLR) on human DCs that strongly recognizes the glycan antigens $Le^x$ and LDNF, as well as high-mannose glycans [114]. It should be noted that these glycan antigens are not frequently found in mammals, but they appear as common glycan antigens on many different helminths (see Figure 15.4). DC-SIGN also recognizes the schistosoma-specific glycan Pseudo-$Le^Y$ via another binding mode [37]. Binding of pathogens to DC-SIGN can modulate Toll-like receptor signaling via Raf-1 kinase, which in turn may lead to acetylation of the NF-κB subunit p65 [115], or activation of MEK [116], dependent on the pathogen encountered. Signaling through DC-SIGN thus may contribute to a general Th2 or "tolerizing" capacity of helminths similar as has been shown for several other pathogens [117]. Remarkably, it has been reported that in mice, a $Le^x$-containing synthetic conjugate triggered maturation of DCs into a Th2–inducing phenotype via a TLR-4-dependent mechanism [118], whereas another study shows the capacity of $Le^x$ to modulate host responses in a Th1 direction via NF-κB p65, interferon gamma (IFN-γ), and macrophage TLRs [119]. Clearly, much has to be understood about the role that glycans play in regulating host immunity. Differences between human and mice receptors, but also differences in affinity and avidity of glycan–lectin binding, which may vary when a glycan antigen is presented within a larger structure, may be of crucial importance for functional activity of a glycan–lectin interaction.

MGL is another CLR lectin that potentially can interact with many helminths expressing cross-reactive glycan antigens. Human MGL has a strong preference for terminal GalNAc, such as present within the common helminth glycan antigens LacdiNAc and Tn (Figures 15.4B and 15.2B). MGL is upregulated on human tolerogenic DCs under chronic inflammatory conditions, and can induce downregulation of effector T cell function, which implicates a role for MGL in the control of adaptive immunity [120]. Similarly, murine MGL2, which resembles human MGL, appears upregulated on macrophages in infection with *Taenia crassiceps* [121]. LacdiNAc glycans can also interact, in addition to MGL, with galectin 3 [122]. In a murine model, LacdiNAc antigens showed the capacity to induce the Th2-associated formation of liver granulomas [123], possibly via interaction with the soluble lectin galectin-3 which appeared to be highly upregulated in granulomas. The finding that galectin-3 knockout mice show a decreased granuloma formation supports the in vitro observations [124].

Thus, the expression of glycan antigens resembling "self" antigens is an interesting property that allows the parasite to target CLRs on host cells. In some cases this allows the induction of a tolerizing DC phenotype [117], a mechanism by which pathogens may guide the immune system to create a safer environment for them by limiting immune attack. It should be noted however that interaction of pathogens with C-type lectins can also enhance a proinflammatory Th1 response [125,126] (Van Stijn and Van Die, unpublished observations), showing the complexity of host–pathogen interactions as mediated by multiple receptors and soluble factors.

Another example of a cross-reactive helminth glycan modification with highly immunomodulatory properties is the phosphorylcholine (PC) moiety linked to *N*-glycans or glycolipids expressed especially by a variety of nematodes. ES-62, which is a PC-containing glycoprotein secreted by the rodent filarial nematode *A. viteae*, is able to inhibit antigen receptor-stimulated proliferation of B and T lymphocytes in vitro and in vivo in a PC-dependent manner [88,127]. No lectins have been found to bind PC-linked glycans. The mechanism of action of PC appears to include the uncoupling of the antigen receptors from crucial intracellular signaling events that drive proliferation of the lymphocytes. Recently, the PC moiety of ES-62 has been shown essential to exert anti-inflammatory action in murine collagen-induced arthritis and in human rheumatoid arthritis-derived synovial tissue cultures [89]. Also ES-62 appears to direct the maturation of DCs towards a Th2-inducing phenotype by suppressing the production of proinflammatory cytokines and inducing the induction of anti-inflammatory cytokines. Remarkably, although PC appears to desensitize B and T cells, high anti-PC antibody levels are detected in many nematode-infected hosts [128]. In the host, such anti-PC antibodies may bind to and inactivate platelet-activating factor (PAF), a proinflammatory mediator of the host, reflecting another possible mechanism by which PC may limit inflammatory host responses [74]. This latter putative mechanism is supported by the observation that PAF receptor-deficient mice showed decreased inflammation and enhanced worm survival upon infection with the nematode *Strongyloides venezuelensis* [129].

## Secretion of Lectins by Parasitic Helminths

Helminths not only express a tremendous variety of glycans that can target host lectins, but they also secrete lectins that may bind host glycans. Although functional studies are restricted, it is hypothesized that secretion of such lectins that resemble host lectins may allow binding to host glycans and contribute to immune modulation or escape mechanisms, or may enable the parasites to settle in specific niches within its host. Many invertebrates produce galectins, among them several helminth parasites including *Onchocerca volvulus*, *Teladorsagia circumcincta*, *Haemonchus contortus*, and *Trichostrongylus colubriformis* [130]. Recently it was demonstrated that *H. contortus* L3 larvae secrete a mixture of galectins with in vitro chemoattractant properties for ovine bone marrow eosinophils, a property which may mimic those of the mammalian eosinophil chemokine galectin-9 [131].

In pathogenic nematodes, expression of C-type lectins has been demonstrated that show more similarities to mammalian C-type lectins than to those from *C. elegans*, suggesting an adaptation to host glycans [132,133]. The *Toxocara canis* lectin Tc-CTL-1 shows a high similarity to DC-SIGN and MBP-A, and appears to bind to glycans expressing *N*-acetylgalactosamine (GalNAc) and mannose (Man) residues. An interesting feature of this lectin is its broader binding specificity compared with mammalian C-type lectins that typically show either Gal/GalNAc or Man/Fuc/GlcNAc binding [134]. Other nematode C-type lectins show sequence similarities to distinct host C-type lectins, such as from *Necator americanus* (similarity to P-selectin) and *Ascaris suum* (similarity to DEC-205) [132]. The hookworm parasite *Ancylostoma ceylanicum* expresses a GlcNAc-binding CLR, which is suggested to be a male-specific CLR with a function in hookworm reproductive physiology [135].

Recently the identification has been reported of a novel CLR from the marine nematode *Laxus oneistus* cuticle, which was shown to be involved in acquisition of its bacterial symbionts [136]. Surprisingly, this nematode CLR is not only structurally similar to DC-SIGN, but shows also functional similarities as was demonstrated by its ability to inhibit binding of DC-SIGN to HIV-1 gp120, thereby inhibiting DC-SIGN-mediated HIV-1 transmission from DC to T cells [137]. Whereas this latter property can hardly be regarded as a function for this marine nematode CLR, it may reflect a high conservation in the mechanisms of host–microbe interactions throughout the animal kingdom [136].

## Anti-Glycan Humoral Responses

Humoral immune responses to many parasitic helminths are directed primarily to glycan determinants expressed on cell surface and secreted glycoconjugates. In schistosome infection of humans, primates, and rodents, all the major glycan antigens elicit generation of all isotypes of immunoglobulin—IgA, IgM, IgG, and IgE—and all subclasses of IgG, including IgG1 as reviewed in [138]. Antibodies to carbohydrate antigens effectively kill schistosomula in vitro either through complement-mediated pathways or by inducing the release of toxic granule proteins and reactive oxygen intermediates by activated leukocytes, and such antibodies may contribute to the phenomenon of concomitant immunity [139]. Concomitant immunity may be a general phenomenon among parasitic helminth infections, since many helminths share glycan antigens (see Figures 15.3 and 15.4). The discovery of anti-glycan antibodies to "self" glycan antigens, such as Le$^x$, was at first surprising. However, with the structural characterization of schistosome glycan structures it became clear that the mode of Le$^x$ presentation by the parasite is substantially different from the host, which may allow specific immune recognition. Still, immune responses to "foreign" glycan antigens such as LDN-DF in schistosomes may be much stronger than toward Le$^x$ [31,140].

The presence of anti-glycan antibodies have been demonstrated in many other helminth infections, or immunizations with helminth products. Examples are antibodies in serum of human toxocariasis patients recognizing the di-*O*-methylated H-type glycan antigens of *Toxocara canis* [87], and antibodies in *Trichinella*-infected hosts which recognize the tyvelose-containing moiety within the TSL-1 glycan antigens [141]. Also *Haemonchus contortus* triggers strong anti-glycan humoral responses, as demonstrated in lambs immunized with *H. contortus* ES antigen [71]. Thus, helminths in general appear to generate strong anti-glycan antibody responses that may be employed in serum diagnostics and/or vaccination approaches.

## Diagnostic Approaches Employing Helminth Glycans

The high abundance of different "foreign" glycan structures present, especially on secreted products of the helminths, and the dominant antibody responses generated in infected hosts, provide great opportunities for the development of glycan-based diagnostics. Especially the unique parasite species-specific glycan antigens such as the above described examples may provide powerful immuno-diagnostical targets. An elegant approach is the use of anti-glycan monoclonal antibodies that can detect unique, parasite-specific glycan-containing antigens secreted by the parasite in host fluids such as urine or serum. This approach is being used in the diagnosis of schistosomiasis, for which numerous different anti-glycan monoclonal antibodies (mAbs) have been generated and characterized [28,138,142]. mAbs specific for schistosoma circulating anodic (CAA) or (CCA) cathodic antigens have been of special importance over the years, and even presented as an easy-to-use strip assay, as described in [143–145]. The recent developments in sensitive and fast mass spectrometry methods may allow development of novel assays for detection of schistosomiasis. Recently, the detection of a series of egg-derived fuco-oligosaccharides in urine of schistosomiasis patients has been reported, as an initial step towards a new approach for diagnosis of schistosomiasis [146].

## Conclusions and Future Prospects

Glycomic approaches to identify the types of glycans made by human and animal parasites have been highly successful, but many glycans remain undefined since no complete description of the glycome of any parasite has been obtained. There is a tremendous interest in defining the antigenic glycans generated by parasites and in defining the enzymes, especially glycosyltransferases and glycosidases, in parasites that may be unique drug targets for therapies. Identification of major glycan antigens may lead to novel vaccine candidates, since glycans are normally abundant on the surfaces of many different parasites and represent targets of both humoral and cellular immunity. The unique glycan antigens of parasites and immune responses to them also represent targets of screening for diagnosis and prognosis in infection. There is a great need to synthesize parasite glycans to use as antigens in microarray or ELISA-based approaches. Such glycan antigens can be synthesized chemically, as has been reported, for example, for fuco-oligosaccharides of *Schistosoma* [147], the unique mono- or dimethylated *Toxocara* glycan antigens (Figure 15.5D) [148], or structures that contain the unique Galβ1–6Galβ sequences occurring in metacestodes of *Echinococcus multilocularis*

[149]. Alternatively, enzymatic approaches may be used, such as reported for synthesis of the unique *Schistosoma* LDN-DF structure [28]. For the latter method, the availability of recombinant parasite-type glycosyltransferases would be extremely valuable. The cloning of only a few of such enzymes have been reported from *C. elegans*, snail, or plant sources [150–153].

There is special interest in using glycan microarrays [154] for diagnostic purposes. Such glycan microarrays have been explored for *Schistosoma* sp. in which multiple glycans are covalently attached to glass slides and interrogated with antibodies, in an elegant approach coupling the glycomics and immunobiology of glycans in this helminth infection [4]. More studies in this direction are needed to incorporate multiple glycan antigens in a glycan profile that could be made specific for a given helminth species (or other pathogens), as earlier initiated for bacterial polysaccharides [155] and as envisaged in the recent pathogen arrays being developed within the Consortium for Functional Glycomics (http://www.functionalglycomics.org) [156]. It is hoped that these modern approaches of glycomics, chemical/enzymatic synthesis, and new methods of generating glycan–protein conjugates for immunization, could lead to novel glycan-based diagnostics and vaccines for parasitic infections.

## Acknowledgments

This work was supported by the Technology Foundation STW to I.v.D and NIH Grant 2R56AI047214 to R.D.C.

# References

1. Haslam SM, Morris HR, Dell A. Mass spectrometric strategies: providing structural clues for helminth glycoproteins. *Trends Parasitol* 2001;**17**:231–5.

2. Geyer H, Geyer R. Strategies for analysis of glycoprotein glycosylation. *Biochim Biophys Acta* 2006;**1764**:1853–69.

3. Cummings RD, Turco S. Parasitic Infections. In: Varki A, Cummings RD, Esko JD, Freeze HH, Stanley P, Bertozzi CR, Hart GW, Etzler ME, editors. *Essentials of Glycobiology*. Woodbury, NY: Cold Spring Harbor Laboratory Press; 2009, p. 553–66.

4. Hokke CH, Fitzpatrick JM, Hoffmann KF. Integrating transcriptome, proteome and glycome analyses of Schistosoma biology. *Trends Parasitol* 2007;**23**:165–74.

5. Jang-Lee J, Curwen RS, Ashton PD, Tissot B, Mathieson W, Panico M, Dell A, Wilson RA, Haslam SM. Glycomics analysis of *Schistosoma mansoni* egg and cercarial secretions. *Mol Cell Proteomics* 2007;**6**:1485–99.

6. Van den Eijnden DH, Neeleman AP, Bakker H, Van Die I. Novel pathways in complex-type oligosaccharide synthesis. New vistas opened by studies in invertebrates. *Adv Exp Med Biol* 1998;**435**:3–7.

7. Van den Eijnden DH, Neeleman AP, Van der Knaap WP, Bakker H, Agterberg M, Van Die I. Novel glycosylation routes for glycoproteins: the lacdiNAc pathway. *Biochem Soc Trans* 1995;**23**:175–9.

8. Makaaru CK, Damian RT, Smith DF, Cummings RD. The human blood fluke *Schistosoma mansoni* synthesizes a novel type of glycosphingolipid. *J Biol Chem* 1992;**267**:2251–7.

9. Reason AJ, Ellis LA, Appleton JA, Wisnewski N, Grieve RB, McNeil M, Wassom DL, Morris HR, Dell A. Novel tyvelose-containing tri- and tetra-antennary N-glycans in the immunodominant antigens of the intracellular parasite Trichinella spiralis. *Glycobiology* 1994;**4**:593–603.

10. Casaravilla C, Freire T, Malgor R, Medeiros A, Osinaga E, Carmona C. Mucin-type O-glycosylation in helminth parasites from major taxonomic groups: evidence for widespread distribution of the Tn antigen (GalNAc-Ser/Thr) and identification of UDP-GalNAc:polypeptide N-acetylgalactosaminyltransferase activity. *J Parasitol* 2003;**89**:709–14.

11. Alvarez Errico D, Medeiros A, Miguez M, Casaravilla C, Malgor R, Carmona C, Nieto A, Osinaga E. O-glycosylation in *Echinococcus granulosus:* identification and characterization of the carcinoma-associated Tn antigen. *Exp Parasitol* 2001;**98**:100–9.

12. Medeiros A, Chiribao ML, Ubillos L, Festari MF, Saldana J, Robello C, Dominguez L, Calvete JJ, Osinaga E. Mucin-type O-glycosylation in *Mesocestoides vogae* (syn. corti). *Int J Parasitol* 2008;**38**:265–76.

13. Freire T, Fernandez C, Chalar C, Maizels RM, Alzari P, Osinaga E, Robello C. Characterization of a UDP-N-acetyl-D-galactosamine:polypeptide N-acetylgalactosaminyltransferase with an unusual lectin domain from the platyhelminth parasite Echinococcus granulosus. *Biochem J* 2004;**382**:501–10.

14. Hagen FK, Nehrke K. cDNA cloning and expression of a family of UDP-N-acetyl-D-galactosamine: polypeptide N-acetylgalactosaminyltransferase sequence homologs from *Caenorhabditis elegans*. *J Biol Chem* 1998;**273**:8268–77.

15. Nyame K, Cummings RD, Damian RT. Characterization of the N- and O-linked oligosaccharides in glycoproteins synthesized by *Schistosoma mansoni* schistosomula. *J Parasitol* 1988;**74**:562–72.

16. Hulsmeier AJ, Gehrig PM, Geyer R, Sack R, Gottstein B, Deplazes P, Kohler P. A major *Echinococcus multilocularis* antigen is a mucin-type glycoprotein. *J Biol Chem* 2002;**277**:5742–8.

17. Khoo KH, Nieto A, Morris HR, Dell A. Structural characterization of the N-glycans from *Echinococcus granulosus* hydatid cyst membrane and protoscoleces. *Mol Biochem Parasitol* 1997;**86**:237–48.

18. Lee JJ, Dissanayake S, Panico M, Morris HR, Dell A, Haslam SM. Mass spectrometric characterisation of *Taenia crassiceps* metacestode N-glycans. *Mol Biochem Parasitol* 2005;**143**:245–9.

19. Dennis RD, Baumeister S, Geyer R, Peter-Katalinic J, Hartmann R, Egge H, Geyer E, Wiegandt H. Glycosphingolipids in cestodes. Chemical structures of ceramide monosaccharide, disaccharide, trisaccharide and tetrasaccharide from metacestodes of the fox tapeworm, *Taenia crassiceps* (Cestoda: Cyclophyllidea). *Eur J Biochem* 1992;**207**:1053–62.

20. Persat F, Bouhours JF, Mojon M, Petavy AF. Glycosphingolipids with Gal β1–6Gal sequences in metacestodes of the parasite *Echinococcus multilocularis*. *J Biol Chem* 1992;**267**:8764–9.

21. Hokke CH, Deelder AM, Hoffmann KF, Wuhrer M. Glycomics-driven discoveries in schistosome research. *Exp Parasitol* 2007;**117**(3):275–83.

22. Huang HH, Tsai PL, Khoo KH. Selective expression of different fucosylated epitopes on two distinct sets of *Schistosoma mansoni* cercarial O-glycans: identification of a novel core type and Lewis x structure. *Glycobiology* 2001;**11**:395–406.

23. Bergwerff AA, van Dam GJ, Rotmans JP, Deelder AM, Kamerling JP, Vliegenthart JF. The immunologically reactive part of immunopurified circulating anodic antigen from *Schistosoma mansoni* is a threonine-linked polysaccharide consisting of →6)-(β-D-GlcpA-(1→3))-β-D-GalpNAc-(1→ repeating units. *J Biol Chem* 1994;**269**:31510–17.

24. Van Dam GJ, Bergwerff AA, Thomas-Oates JE, Rotmans JP, Kamerling JP, Vliegenthart JF, Deelder AM. The immunologically reactive O-linked polysaccharide chains derived from circulating cathodic antigen

isolated from the human blood fluke *Schistosoma mansoni* have Lewis x as repeating unit. *Eur J Biochem* 1994;**225**:467–82.

25. Srivatsan J, Smith DF, Cummings RD. The human blood fluke Schistosoma mansoni synthesizes glycoproteins containing the Lewis x antigen. *J Biol Chem* 1992;**267**:20196–203.

26. Khoo KH, Chatterjee D, Caulfield JP, Morris HR, Dell A. Structural mapping of the glycans from the egg glycoproteins of *Schistosoma mansoni* and *Schistosoma japonicum:* identification of novel core structures and terminal sequences. *Glycobiology* 1997;**7**:663–77.

27. Khoo KH, Sarda S, Xu X, Caulfield JP, McNeil MR, Homans SW, Morris HR, Dell A. A unique multifucosylated -3GalNAc β1–4GlcNAc β1–3Gal α 1- motif constitutes the repeating unit of the complex O-glycans derived from the cercarial glycocalyx of *Schistosoma mansoni*. *J Biol Chem* 1995;**270**:17114–23.

28. Van Remoortere A, Hokke CH, van Dam GJ, van Die I, Deelder AM, van den Eijnden DH. Various stages of schistosoma express Lewis(x), LacdiNAc, GalNAcβ1–4 (Fucα1–3)GlcNAc and GalNAcβ1–4(Fucα1–2Fucα1–3)GlcNAc carbohydrate epitopes: detection with monoclonal antibodies that are characterized by enzymatically synthesized neoglycoproteins. *Glycobiology* 2000;**10**:601–9.

29. Robijn ML, Koeleman CA, Wuhrer M, Royle L, Geyer R, Dwek RA, Rudd PM, Deelder AM, Hokke CH. Targeted identification of a unique glycan epitope of *Schistosoma mansoni* egg antigens using a diagnostic antibody. *Mol Biochem Parasitol* 2007;**151**:148–61.

30. Van der Kleij D, Van Remoortere A, Schuitemaker JH, Kapsenberg ML, Deelder AM, Tielens AG, Hokke CH, Yazdanbakhsh M. Triggering of innate immune responses by schistosome egg glycolipids and their carbohydrate epitope GalNAc β1–4(Fucα1–2Fucα1–3)GlcNAc. *J Infect Dis* 2002;**185**:531–9.

31. van Remoortere A, van Dam GJ, Hokke CH, van den Eijnden DH, van Die I, Deelder AM. Profiles of immunoglobulin M (IgM) and IgG antibodies against defined carbohydrate epitopes in sera of Schistosoma-infected individuals determined by surface plasmon resonance. *Infect Immun* 2001;**69**:2396–401.

32. Wuhrer M, Dennis RD, Doenhoff MJ, Lochnit G, Geyer R. *Schistosoma mansoni* cercarial glycolipids are dominated by Lewis x and pseudo-Lewis y structures. *Glycobiology* 2000;**10**:89–101.

33. Damian RT. Parasite immune evasion and exploitation: reflections and projections. *Parasitology* 1997;**115**(Suppl):S169–75.

34. van Die I, Cummings RD. Glycans modulate immune responses in helminth infections and allergy. *Chem Immunol Allergy* 2006;**90**:91–112.

35. van Liempt E, van Vliet SJ, Engering A, Garcia Vallejo JJ, Bank CM, Sanchez-Hernandez M, van Kooyk Y, van Die I. *Schistosoma mansoni* soluble egg antigens are internalized by human dendritic cells through multiple C-type lectins and suppress TLR-induced dendritic cell activation. *Mol Immunol* 2007;**44**:2605–15.

36. van Die I, van Vliet SJ, Nyame AK, Cummings RD, Bank CM, Appelmelk B, Geijtenbeek TB, van Kooyk Y. The dendritic cell-specific C-type lectin DC-SIGN is a receptor for *Schistosoma mansoni* egg antigens and recognizes the glycan antigen Lewis x. *Glycobiology* 2003;**13**:471–8.

37. Meyer S, van Liempt E, Imberty A, van Kooyk Y, Geyer H, Geyer R, van Die I. DC-SIGN mediates binding of dendritic cells to authentic pseudo-Lewis Y glycolipids of *Schistosoma mansoni* cercariae, the first parasite-specific ligand of DC-SIGN. *J Biol Chem* 2005;**280**:37349–59.

38. Mas-Coma MS, Esteban JG, Bargues MD. Epidemiology of human fascioliasis: a review and proposed new classification. *Bull World Health Organ* 1999;**77**:340–6.

39. Hillyer GV. Immunity of schistosomes using heterologous trematode antigens--a review. *Vet Parasitol* 1984;**14**:263–83.

40. Aronstein WS, Lewis SA, Norden AP, Dalton JP, Strand M. Molecular identity of a major antigen of *Schistosoma mansoni* which cross-reacts with Trichinella spiralis and Fasciola hepatica. *Parasitology* 1986;**92**(Pt 1):133–51.

41. Abdul-Salam F, Mansour MH. Identification and localization of a schistosome-associated fucosyllactose determinant expressed by *Fasciola hepatica. Comp Immunol Microbiol Infect Dis* 2000;**23**:99–111.

42. Freire T, Casaravilla C, Carmona C, Osinaga E. Mucin-type O-glycosylation in *Fasciola hepatica:* characterisation of carcinoma-associated Tn and sialyl-Tn antigens and evaluation of UDP-GalNAc: polypeptide N-acetylgalactosaminyltransferase activity. *Int J Parasitol* 2003;**33**:47–56.

43. Wuhrer M, Berkefeld C, Dennis RD, Idris MA, Geyer R. The liver flukes *Fasciola gigantica* and *Fasciola hepatica* express the leucocyte cluster of differentiation marker CD77 (globotriaosylceramide) in their tegument. *Biol Chem* 2001;**382**:195–207.

44. Wuhrer M, Grimm C, Dennis RD, Idris MA, Geyer R. The parasitic trematode *Fasciola hepatica* exhibits mammalian-type glycolipids as well as Gal(β1–6)Gal-terminating glycolipids that account for cestode serological cross-reactivity. *Glycobiology* 2004;**14**:115–26.

45. Wuhrer M, Grimm C, Zahringer U, Dennis RD, Berkefeld CM, Idris MA, Geyer R. A novel GlcNAcα1-HP03–6Gal(1–1)ceramide antigen and alkylated inositol-phosphoglycerolipids expressed by the liver fluke *Fasciola hepatica. Glycobiology* 2003;**13**:129–37.

46. Bossaert K, Farnir F, Leclipteux T, Protz M, Lonneux JF, Losson B. Humoral immune response in calves to single-dose, trickle and challenge infections with *Fasciola hepatica. Vet Parasitol* 2000;**87**:103–23.

47. Schantz PM, Ortiz-Valqui RE, Lumbreras H. Nonspecific reactions with the intradermal test for hydatidosis in persons with other helminth infections. *Am J Trop Med Hyg* 1975;**24**:849–52.

48. Lochnit G, Dennis RD, Geyer R. Phosphorylcholine substituents in nematodes: structures, occurrence and biological implications. *Biol Chem* 2000;**381**:839–47.

49. Pery P, Petit A, Poulain J, Luffau G. Phosphorylcholine-bearing components in homogenates of nematodes. *Eur J Immunol* 1974;**4**:637–9.

50. Appleton JA, Romaris F. A pivotal role for glycans at the interface between *Trichinella spiralis* and its host. *Vet Parasitol* 2001;**101**:249–60.

51. Appleton JA, Schain LR, McGregor DD. Rapid expulsion of *Trichinella spiralis* in suckling rats: mediation by monoclonal antibodies. *Immunology* 1988;**65**:487–92.

52. Denkers EY, Hayes CE, Wassom DL. *Trichinella spiralis:* influence of an immunodominant, carbohydrate-associated determinant on the host antibody response repertoire. *Exp Parasitol* 1991;**72**:403–10.

53. Denkers EY, Wassom DL, Hayes CE. Characterization of *Trichinella spiralis* antigens sharing an immunodominant, carbohydrate-associated determinant distinct from phosphorylcholine. *Mol Biochem Parasitol* 1990;**41**:241–9.

54. Ellis LA, Reason AJ, Morris HR, Dell A, Iglesias R, Ubeira FM, Appleton JA. Glycans as targets for monoclonal antibodies that protect rats against *Trichinella spiralis. Glycobiology* 1994;**4**:585–92.

55. McVay CS, Bracken P, Gagliardo LF, Appleton J. Antibodies to tyvelose exhibit multiple modes of interference with the epithelial niche of *Trichinella spiralis. Infect Immun* 2000;**68**:1912–18.

56. McVay CS, Tsung A, Appleton J. Participation of parasite surface glycoproteins in antibody-mediated protection of epithelial cells against *Trichinella spiralis. Infect Immun* 1998;**66**:1941–5.

57. Otubu OE, Carlisle-Nowak MS, McGregor DD, Jacobson RH, Appleton JA. *Trichinella spiralis:* the effect of specific antibody on muscle larvae in the small intestines of weaned rats. *Exp Parasitol* 1993;**76**:394–400.

58. Wisnewski N, McNeil M, Grieve RB, Wassom DL. Characterization of novel fucosyl- and tyvelosyl-containing glycoconjugates from *Trichinella spiralis* muscle stage larvae. *Mol Biochem Parasitol* 1993;**61**:25–35.

59. Ellis LA, McVay CS, Probert MA, Zhang J, Bundle DR, Appleton JA. Terminal bβ-linked tyvelose creates unique epitopes in *Trichinella spiralis* glycan antigens. *Glycobiology* 1997;**7**:383–90.

60. Peters PJ, Gagliardo LF, Sabin EA, Betchen AB, Ghosh K, Oblak JB, Appleton JA. Dominance of immunoglobulin G2c in the antiphosphorylcholine response of rats infected with *Trichinella spiralis*. *Infect Immun* 1999;**67**:4661–7.

61. Morelle W, Haslam SM, Morris HR, Dell A. Characterization of the N-linked glycans of adult *Trichinella spiralis*. *Mol Biochem Parasitol* 2000;**109**:171–7.

62. Morelle W, Haslam SM, Olivier V, Appleton JA, Morris HR, Dell A. Phosphorylcholine-containing N-glycans of *Trichinella spiralis:* identification of multiantennary lacdiNAc structures. *Glycobiology* 2000;**10**:941–50.

63. Getachew T, Dorchies P, Jacquiet P. Trends and challenges in the effective and sustainable control of *Haemonchus contortus* infection in sheep. Review. *Parasite* 2007;**14**:3–14.

64. Munn EA, Smith TS, Graham M, Tavernor AS, Greenwood CA. The potential value of integral membrane proteins in the vaccination of lambs against *Haemonchus contortus*. *Int J Parasitol* 1993;**23**:261–9.

65. Haslam SM, Coles GC, Munn EA, Smith TS, Smith HF, Morris HR, Dell A. *Haemonchus contortus* glycoproteins contain N-linked oligosaccharides with novel highly fucosylated core structures. *J Biol Chem* 1996;**271**:30561–70.

66. Kubelka V, Altmann F, Staudacher E, Tretter V, Marz L, Hard K, Kamerling JP, Vliegenthart JF. Primary structures of the N-linked carbohydrate chains from honeybee venom phospholipase A2. *Eur J Biochem* 1993;**213**:1193–204.

67. Lerouge P, Cabanes-Macheteau M, Rayon C, Fischette-Laine AC, Gomord V, Faye L. N-glycoprotein biosynthesis in plants: recent developments and future trends. *Plant Mol Biol* 1998;**38**:31–48.

68. van Ree R, Cabanes-Macheteau M, Akkerdaas J, Milazzo JP, Loutelier-Bourhis C, Rayon C, Villalba M, Koppelman S, Aalberse R, Rodriguez R, Faye L, Lerouge P. Beta(1,2)-xylose and alpha(1,3)-fucose residues have a strong contribution in IgE binding to plant glycoallergens. *J Biol Chem* 2000;**275**:11451–8.

69. van Die I, Gomord V, Kooyman FN, van den Berg TK, Cummings RD, Vervelde L. Core α1–3fucose is a common modification of N-glycans in parasitic helminths and constitutes an important epitope for IgE from *Haemonchus contortus* infected sheep. *FEBS Lett* 1999;**463**:189–93.

70. Haslam SM, Coles GC, Reason AJ, Morris HR, Dell A. The novel core fucosylation of *Haemonchus contortus* N-glycans is stage specific. *Mol Biochem Parasitol* 1998;**93**:143–7.

71. Vervelde L, Bakker N, Kooyman FN, Cornelissen AW, Bank CM, Nyame AK, Cummings RD, van Die I. Vaccination-induced protection of lambs against the parasitic nematode *Haemonchus contortus* correlates with high IgG antibody responses to the LDNF glycan antigen. *Glycobiology* 2003;**13**:795–804.

72. Jorgensen RJ, Ronne H, Helsted C, Iskander AR. Spread of infective *Dictyocaulus viviparus* larvae in pasture and to grazing cattle: experimental evidence of the role of Pilobolus fungi. *Vet Parasitol* 1982;**10**:331–9.

73. Haslam SM, Coles GC, Morris HR, Dell A. Structural characterization of the N-glycans of *Dictyocaulus viviparus:* discovery of the Lewis(x) structure in a nematode. *Glycobiology* 2000;**10**:223–9.

74. Kooyman FN, de Vries E, Ploeger HW, van Putten JP. Antibodies elicited by the bovine lungworm, *Dictyocaulus viviparus,* cross-react with platelet-activating factor. *Infect Immun* 2007;**75**:4456–62.

75. Sprent JF. Anatomical distinction between human and pig strains of Ascaris. *Nature* 1952;**170**:627–8.

76. Poltl G, Kerner D, Paschinger K, Wilson IB. N-glycans of the porcine nematode parasite Ascaris suum are modified with phosphorylcholine and core fucose residues. *Febs J* 2007;**274**:714–26.

77. Lochnit G, Nispel S, Dennis RD, Geyer R. Structural analysis and immunohistochemical localization of two acidic glycosphingolipids from the porcine, parasitic nematode *Ascaris suum. Glycobiol* 1998;**8**:891–9.

78. Sugita M, Mizunoma T, Aoki K, Dulaney JT, Inagaki F, Suzuki M, Suzuki A, Ichikawa S, Kushida K, Ohta S, Kurimoto A. Structural characterization of a novel glycoinositolphospholipid from the parasitic nematode, *Ascaris suum. Biochim Biophys Acta* 1996;**1302**:185–92.

79. Marinier Jr. A, Martel A, Banville J, Bachand C, Remillard R, Lapointe P, Turmel B, Menard M, Harte WE, Wright JJ, Todderud G, Tramposch KM, Bajorath J, Hollenbaugh D, Aruffo A. Sulfated galactocerebrosides as potential antiinflammatory agents. *J Med Chem* 1997;**40**:3234–47.

80. Lochnit G, Dennis RD, Zahringer U, Geyer R. Structural analysis of neutral glycosphingolipids from *Ascaris suum* adults (Nematoda:Ascaridida). *Glycoconj J* 1997;**14**:389–99.

81. Friedl CH, Lochnit G, Zahringer U, Bahr U, Geyer R. Structural elucidation of zwitterionic carbohydrates derived from glycosphingolipids of the porcine parasitic nematode *Ascaris suum. Biochem J* 2003;**369**:89–102.

82. Ponce-Leon P, Foresto P, Valverde J. *Ascaris lumbricoides:* heterogeneity in ABO epitopes expression. *Invest Clin* 2006;**47**:385–93.

83. Dennis RD, Baumeister S, Smuda C, Lochnit C, Waider T, Geyer E. Initiation of chemical studies on the immunoreactive glycolipids of adult *Ascaris suum. Parasitology* 1995;**110**(Pt 5):611–23.

84. Smits HH, Yazdanbakhsh M. Chronic helminth infections modulate allergen-specific immune responses: Protection against development of allergic disorders? *Ann Med* 2007;**39**:428–39.

85. Pinelli E, Dormans J, van Die I. Toxocara and Asthma. In: Holland C, Smith H, editors. *Toxocara: The enigmatic parasite.* U.K.: CABI, Publishers; 2006, p. 42–57.

86. Khoo KH, Maizels RM, Page AP, Taylor GW, Rendell NB, Dell A. Characterization of nematode glycoproteins: the major O-glycans of Toxocara excretory-secretory antigens are O-methylated trisaccharides. *Glycobiology* 1991;**1**:163–71.

87. Schabussova I, Amer H, van Die I, Kosma P, Maizels RM. O-methylated glycans from Toxocara are specific targets for antibody binding in human and animal infections. *Int J Parasitol* 2007;**37**:97–109.

88. Goodridge HS, McGuiness S, Houston KM, Egan CA, Al-Riyami L, Alcocer MJ, Harnett MM, Harnett W. Phosphorylcholine mimics the effects of ES-62 on macrophages and dendritic cells. *Parasite Immunol* 2007;**29**:127–37.

89. Harnett MM, Kean DE, Boitelle A, McGuiness S, Thalhamer T, Steiger CN, Egan C, Al-Riyami L, Alcocer MJ, Houston KM, Gracie JA, McInnes IB, Harnett W. The phosphorylcholine moiety of the filarial nematode immunomodulator ES-62 is responsible for its anti-inflammatory action in arthritis. *Ann Rheum Dis* 2008;**67**:518–23.

90. Wuhrer M, Rickhoff S, Dennis RD, Lochnit G, Soboslay PT, Baumeister S, Geyer R. Phosphocholine-containing, zwitterionic glycosphingolipids of adult *Onchocerca volvulus* as highly conserved antigenic structures of parasitic nematodes. *Biochem J* 2000;**348**(Pt 2):417–23.

91. Houston KM, Harnett W. Structure and synthesis of nematode phosphorylcholine-containing glycoconjugates. *Parasitology* 2004;**129**:655–61.

92. Haslam SM, Houston KM, Harnett W, Reason AJ, Morris HR, Dell A. Structural studies of N-glycans of filarial parasites. Conservation of phosphorylcholine-substituted glycans among species and discovery of novel chito-oligomers. *J Biol Chem* 1999;**274**:20953–60.

93. Kang S, Cummings RD, McCall JW. Characterization of the N-linked oligosaccharides in glycoproteins synthesized by microfilariae of *Dirofilaria immitis*. *J Parasitol* 1993;**79**:815–28.

94. Klabunde J, Berger J, Jensenius JC, Klinkert MQ, Zelck UE, Kremsner PG, Kun JF. *Schistosoma mansoni*: adhesion of mannan-binding lectin to surface glycoproteins of cercariae and adult worms. *Exp Parasitol* 2000;**95**:231–9.

95. Gruden-Movsesijan A, Petrovic M, Sofronic-Milosavljevic L. Interaction of mannan-binding lectin with *Trichinella spiralis* glycoproteins, a possible innate immune mechanism. *Parasite Immunol* 2003;**25**:545–52.

96. Carter T, Sumiya M, Reilly K, Ahmed R, Sobieszczuk P, Summerfield JA, Lawrence RA. Mannose-binding lectin A-deficient mice have abrogated antigen-specific IgM responses and increased susceptibility to a nematode infection. *J Immunol* 2007;**178**:5116–23.

97. van de Wetering JK, van Remoortere A, Vaandrager AB, Batenburg JJ, van Golde LM, Hokke CH, van Hellemond JJ. Surfactant protein D binding to terminal alpha1–3-linked fucose residues and to *Schistosoma mansoni*. *Am J Respir Cell Mol Biol* 2004;**31**:565–72.

98. Maizels RM, Balic A, Gomez-Escobar N, Nair M, Taylor MD, Allen JE. Helminth parasites--masters of regulation. *Immunol Rev* 2004;**201**:89–116.

99. Elliott DE, Summers RW, Weinstock JV. Helminths as governors of immune-mediated inflammation. *Int J Parasitol* 2007;**37**:457–64.

100. Zheng X, Hu X, Zhou G, Lu Z, Qiu W, Bao J, Dai Y. Soluble egg antigen from *Schistosoma japonicum* modulates the progression of chronic progressive experimental autoimmune encephalomyelitis via Th2-shift response. *J Neuroimmunol* 2008;**194**:107–14.

101. Ruyssers NE, De Winter BY, De Man JG, Loukas A, Herman AG, Pelckmans PA, Moreels TG. Worms and the treatment of inflammatory bowel disease: are molecules the answer? *Clin Dev Immunol* 2008;**2008**:567314.

102. Cooke A, Tonks P, Jones FM, O'Shea H, Hutchings P, Fulford AJ, Dunne DW. Infection with *Schistosoma mansoni* prevents insulin dependent diabetes mellitus in non-obese diabetic mice. *Parasite Immunol* 1999;**21**:169–76.  .

103. Medzhitov R. Recognition of microorganisms and activation of the immune response. *Nature* 2007;**449**:819–26.

104. Kapsenberg ML. Dendritic-cell control of pathogen-driven T-cell polarization. *Nat Rev Immunol* 2003;**3**:984–93.

105. Jin D, Zhang L, Zheng J, Zhao Y. The inflammatory Th 17 subset in immunity against self and non-self antigens. *Autoimmunity* 2008;**41**:154–62.

106. Okano Jr. M, Satoskar AR, Nishizaki K, Abe M, Harn DA. Induction of Th2 responses and IgE is largely due to carbohydrates functioning as adjuvants on *Schistosoma mansoni* egg antigens. *J Immunol* 1999;**163**:6712–17.

107. Tawill S, Le Goff L, Ali F, Blaxter M, Allen JE. Both free-living and parasitic nematodes induce a characteristic Th2 response that is dependent on the presence of intact glycans. *Infect Immun* 2004;**72**:398–407.

108. Gomez-Garcia L, Rivera-Montoya I, Rodriguez-Sosa M, Terrazas LI. Carbohydrate components of *Taenia crassiceps* metacestodes display Th2-adjuvant and anti-inflammatory properties when co-injected with bystander antigen. *Parasitol Res* 2006;**99**:440–8.

109. Hokke CH, Yazdanbakhsh M. Schistosome glycans and innate immunity. *Parasite Immunol* 2005;**27**:257–64.

110. Okano Jr. M, Satoskar AR, Nishizaki K, Harn DA Lacto-N-fucopentaose III found on *Schistosoma mansoni* egg antigens functions as adjuvant for proteins by inducing Th2-type response. *J Immunol* 2001;**167**:442–50.

111. Faveeuw C, Mallevaey T, Paschinger K, Wilson IB, Fontaine J, Mollicone R, Oriol R, Altmann F, Lerouge P, Capron M, Trottein F. Schistosome N-glycans containing core α3-fucose and core β2-xylose epitopes are strong inducers of Th2 responses in mice. *Eur J Immunol* 2003;**33**:1271–81.

112. van Vliet SJ, Dunnen JD, Gringhuis SI, Geijtenbeek TB, van Kooyk Y. Innate signaling and regulation of dendritic cell immunity. *Curr Opin Immunol* 2007;**19**(4):435–40.

113. Ishida MM, Rubinsky-Elefant G, Ferreira AW, Hoshino-Shimizu S, Vaz AJ. Helminth antigens (Taenia solium, Taenia crassiceps, Toxocara canis, Schistosoma mansoni and Echinococcus granulosus) and cross-reactivities in human infections and immunized animals. *Acta Trop* 2003;**89**:73–84.

114. van Liempt E, Bank CM, Mehta P, Garcia-Vallejo JJ, Kawar ZS, Geyer R, Alvarez RA, Cummings RD, Kooyk Y, van Die I. Specificity of DC-SIGN for mannose- and fucose-containing glycans. *FEBS Lett* 2006;**580**:6123–31.

115. Gringhuis SI, den Dunnen J, Litjens M, van Het Hof B, van Kooyk Y, Geijtenbeek TB. C-type lectin DC-SIGN modulates Toll-like receptor signaling via Raf-1 kinase-dependent acetylation of transcription factor NF-kappaB. *Immunity* 2007;**26**:605–16.

116. Hovius JW, de Jong MA, den Dunnen J, Litjens M, Fikrig E, van der Poll T, Gringhuis SI, Geijtenbeek TB. Salp15 binding to DC-SIGN inhibits cytokine expression by impairing both nucleosome remodeling and mRNA stabilization. *PLoS Pathog* 2008;**4**:e31.

117. van Kooyk Y, Engering A, Lekkerkerker AN, Ludwig IS, Geijtenbeek TB. Pathogens use carbohydrates to escape immunity induced by dendritic cells. *Curr Opin Immunol* 2004;**16**:488–93.

118. Thomas PG, Carter MR, Atochina O, Da'Dara AA, Piskorska D, McGuire E, Harn DA. Maturation of dendritic cell 2 phenotype by a helminth glycan uses a Toll-like receptor 4-dependent mechanism. *J Immunol* 2003;**171**:5837–41.

119. Dissanayake S, Shahin A. Induction of interferon-gamma by *Taenia crassiceps* glycans and Lewis sugars in naive BALB/c spleen and peritoneal exudate cells. *Mol Immunol* 2007;**44**:1623–30.

120. van Vliet SJ, Saeland E, van Kooyk Y. Sweet preferences of MGL: carbohydrate specificity and function. *Trends Immunol* 2008;**29**:83–90.

121. Raes G, Brys L, Dahal BK, Brandt J, Grooten J, Brombacher F, Vanham G, Noel W, Dogaert P, Boonefaes T, Kindt A, Van den Bergh R, Leenen PJ, De Baetselier P, Ghassabeh GH. Macrophage galactose-type C-type lectins as novel markers for alternatively activated macrophages elicited by parasitic infections and allergic airway inflammation. *J Leukoc Biol* 2005;**77**:321–7.

122. van den Berg TK, Honing H, Franke N, van Remoortere A, Schiphorst WE, Liu FT, Deelder AM, Cummings RD, Hokke CH, van Die I. LacdiNAc-glycans constitute a parasite pattern for galectin-3-mediated immune recognition. *J Immunol* 2004;**173**:1902–7.

123. Van de Vijver KK, Deelder AM, Jacobs W, Van Marck EA, Hokke CH. LacdiNAc- and LacNAc-containing glycans induce granulomas in an in vivo model for schistosome egg-induced hepatic granuloma formation. *Glycobiology* 2006;**16**:237–43.

124. Breuilh L, Vanhoutte F, Fontaine J, van Stijn CM, Tillie-Leblond I, Capron M, Faveeuw C, Jouault T, van Die I, Gosset P, Trottein F. Galectin-3 modulates immune and inflammatory responses during helminthic infection: impact of galectin-3 deficiency on the functions of dendritic cells. *Infect Immun* 2007;**75**:5148–57.

125. Gantner BN, Simmons RM, Canavera SJ, Akira S, Underhill DM. Collaborative induction of inflammatory responses by dectin-1 and Toll-like receptor 2. *J Exp Med* 2003;**197**:1107–17.

126. Steeghs L, van Vliet SJ, Uronen-Hansson H, van Mourik A, Engering A, Sanchez-Hernandez M, Klein N, Callard R, van Putten JP, van der Ley P, van Kooyk Y, van de Winkel JG. *Neisseria meningitidis* expressing lgtB lipopolysaccharide targets DC-SIGN and modulates dendritic cell function. *Cell Microbiol* 2006;**8**:316–25.

127. Harnett W, Harnett MM. Modulation of the host immune system by phosphorylcholine-containing glycoproteins secreted by parasitic filarial nematodes. *Biochim Biophys Acta* 2001;**1539**:7–15.

128. van Riet E, Wuhrer M, Wahyuni S, Retra K, Deelder AM, Tielens AG, van der Kleij D, Yazdanbakhsh M. Antibody responses to Ascaris-derived proteins and glycolipids: the role of phosphorylcholine. *Parasite Immunol* 2006;**28**:363–71.

129. Negrao-Correa D, Souza DG, Pinho V, Barsante MM, Souza AL, Teixeira MM. Platelet-activating factor receptor deficiency delays elimination of adult worms but reduces fecundity in Strongyloides venezuelensis-infected mice. *Infect Immun* 2004;**72**:1135–42.

130. Young AR, Meeusen EN. Galectins in parasite infection and allergic inflammation. *Glycoconj J* 2004;**19**:601–6.

131. Turner DG, Wildblood LA, Inglis NF, Jones DG. Characterization of a galectin-like activity from the parasitic nematode, *Haemonchus contortus*, which modulates ovine eosinophil migration in vitro. *Vet Immunol Immunopathol* 2008;**122**:138–45.

132. Loukas A, Maizels RM. Helminth C-type lectins and host-parasite interactions. *Parasitol Today* 2000;**16**:333–9.

133. Mallo GV, Kurz CL, Couillault C, Pujol N, Granjeaud S, Kohara Y, Ewbank JJ. Inducible antibacterial defense system in *C. elegans*. *Curr Biol* 2002;**12**:1209–14.

134. Drickamer K. C-type lectin-like domains. *Curr Opin Struct Biol* 1999;**9**:585–90.

135. Brown AC, Harrison LM, Kapulkin W, Jones BF, Sinha A, Savage A, Villalon N, Cappello M. Molecular cloning and characterization of a C-type lectin from *Ancylostoma ceylanicum*: evidence for a role in hookworm reproductive physiology. *Mol Biochem Parasitol* 2007;**151**:141–7.

136. Bulgheresi S, Schabussova I, Chen T, Mullin NP, Maizels RM, Ott JA. A new C-type lectin similar to the human immunoreceptor DC-SIGN mediates symbiont acquisition by a marine nematode. *Appl Environ Microbiol* 2006;**72**:2950–6.

137. Nabatov AA, de Jong MA, de Witte L, Bulgheresi S, Geijtenbeek TB. C-type lectin Mermaid inhibits dendritic cell mediated HIV-1 transmission to CD4+ T cells. *Virology* 2008;**378**(2):323–8.

138. Nyame AK, Kawar ZS, Cummings RD. Antigenic glycans in parasitic infections: implications for vaccines and diagnostics. *Arch Biochem Biophys* 2004;**426**:182–200.

139. Clegg JA, Smithers SR, Terry RJ. Concomitant immunity and host antigens associated with schistosomiasis. *Int J Parasitol* 1971;**1**:43–9.

140. van Remoortere A, Vermeer HJ, van Roon AM, Langermans JA, Thomas AW, Wilson RA, van die I, van den Eijnden DH, Agoston K, Kerekgyarto J, Vliegenthart JF, Kamerling JP, van dam GJ, Hokke

CH, Deelder AM. Dominant antibody responses to Fucα1–3GalNAc and Fucα1–2Fucα1–3GlcNAc containing carbohydrate epitopes in Pan troglodytes vaccinated and infected with *Schistosoma mansoni. Exp Parasitol* 2003;**105**:219–25.

141. Bruschi F, Moretti A, Wassom D, Piergili Fioretti D. The use of a synthetic antigen for the serological diagnosis of human trichinellosis. *Parasite* 2001;**8**:S141–3.

142. Deelder AM, van Dam GJ, Kornelis D, Fillie YE, van Zeyl RJ. Schistosoma: analysis of monoclonal antibodies reactive with the circulating antigens CAA and CCA. *Parasitology* 1996;**112**(Pt 1):21–35.

143. van Dam GJ, Wichers JH, Ferreira TM, Ghati D, van Amerongen A, Deelder AM. Diagnosis of schistosomiasis by reagent strip test for detection of circulating cathodic antigen. *J Clin Microbiol* 2004;**42**:5458–61.

144. van Lieshout L, Polderman AM, Deelder AM. Immunodiagnosis of schistosomiasis by determination of the circulating antigens CAA and CCA, in particular in individuals with recent or light infections. *Acta Trop* 2000;**77**:69–80.

145. Midzi N, Butterworth AE, Mduluza T, Munyati S, Deelder AM, van Dam GJ. Use of circulating cathodic antigen strips for the diagnosis of urinary schistosomiasis. *Trans R Soc Trop Med Hyg* 2009;**103**:45–51.

146. Robijn ML, Planken J, Kornelis D, Hokke CH, Deelder AM. Mass spectrometric detection of urinary oligosaccharides as markers of *Schistosoma mansoni* infection. *Trans R Soc Trop Med Hyg* 2008;**102**:79–83.

147. van Roon AM, Aguilera B, Cuenca F, van Remoortere A, van der Marel GA, Deelder AM, Overkleeft HS, Hokke CH. Synthesis and antibody-binding studies of a series of parasite fuco-oligosaccharides. *Bioorg Med Chem* 2005;**13**:3553–64.

148. Amer H, Hofinger A, Kosma P. Synthesis of neoglycoproteins containing O-methylated trisaccharides related to excretory/secretory antigens of Toxocara larvae. *Carbohydr Res* 2003;**338**:35–45.

149. Yamamura T, Hada N, Kaburaki A, Yamano K, Takeda T. Synthetic studies on glycosphingolipids from Protostomia phyla: total syntheses of glycosphingolipids from the parasite, *Echinococcus multilocularis. Carbohydr Res* 2004;**339**:2749–59.

150. Bakker H, Schijlen E, de Vries T, Schiphorst WE, Jordi W, Lommen A, Bosch D, van Die I. Plant members of the α1,3/4-fucosyltransferase gene family encode an α1,4-fucosyltransferase, potentially involved in Lewis(a) biosynthesis, and two core α1,3-fucosyltransferases. *FEBS Lett* 2001;**507**:307–12.

151. Kawar ZS, Van Die I, Cummings RD. Molecular cloning and enzymatic characterization of a UDP-GalNAc:GlcNAcβ-R β1,4-N-acetylgalactosaminyltransferase from *Caenorhabditis elegans. J Biol Chem* 2002;**277**:34924–32.

152. Bakker H, Agterberg M, Van Tetering A, Koeleman CA, Van den Eijnden DH, Van Die I. A *Lymnaea stagnalis* gene, with sequence similarity to that of mammalian β1,4-galactosyltransferases, encodes a novel UDP-GlcNAc:GlcNAc β-R β1,4-N-acetylglucosaminyltransferase. *J Biol Chem* 1994;**269**:30326–33.

153. Zheng Q, Van Die I, Cummings RD. Molecular cloning and characterization of a novel α1,2-fucosyltransferase (CE2FT-1) from *Caenorhabditis elegans. J Biol Chem* 2002;**277**:39823–32.

154. Blixt O, Head S, Mondala T, Scanlan C, Huflejt ME, Alvarez R, Bryan MC, Fazio F, Calarese D, Stevens J, Razi N, Stevens DJ, Skehel JJ, van Die I, Burton DR, Wilson IA, Cummings R, Bovin N, Wong CH, Paulson JC. Printed covalent glycan array for ligand profiling of diverse glycan binding proteins. *Proc Natl Acad Sci USA* 2004;**101**:17033–8.

155. Wang D, Liu S, Trummer BJ, Deng C, Wang A. Carbohydrate microarrays for the recognition of cross-reactive molecular markers of microbes and host cells. *Nat Biotechnol* 2002;**20**:275–81.

156. Blixt O, Hoffmann J, Svenson S, Norberg T. Pathogen specific carbohydrate antigen microarrays: a chip for detection of Salmonella O-antigen specific antibodies. *Glycoconj J* 2008;**25**:27–36.

157. Nyame AK, Yoshino TP, Cummings RD. Differential expression of LacdiNAc, fucosylated LacdiNAc, and Lewis x glycan antigens in intramolluscan stages of *Schistosoma mansoni*. *J Parasitol* 2002;**88**:890–7.

158. Lehr T, Beuerlein K, Doenhoff MJ, Grevelding CG, Geyer R. Localization of carbohydrate determinants common to *Biomphalaria glabrata* as well as to sporocysts and miracidia of *Schistosoma mansoni*. *Parasitology* 2008;**135**:1–12.

159. Lehr T, Geyer H, Maass K, Doenhoff MJ, Geyer R. Structural characterization of N-glycans from the freshwater snail *Biomphalaria glabrata* cross-reacting with *Schistosoma mansoni* glycoconjugates. *Glycobiology* 2007;**17**:82–103.

SECTION
VI

# Disease Glycomes

# 16
# Cancer Glycomics

J. Michael Pierce

*THE UNIVERSITY OF GEORGIA CANCER CENTER,
COMPLEX CARBOHYDRATE RESEARCH CENTER, AND
DEPARTMENT OF BIOCHEMISTRY AND MOLECULAR BIOLOGY,
THE UNIVERSITY OF GEORGIA, ATHENS, GA, USA*

## Introduction

As discussed in chapter 5 on "Glycotranscriptomics", our best estimates of the total number of "glyco-related genes" in the human genome is 750–800. Of these, there are around 210 glycosyltransferase genes, with about half of those sequences representing glycosyltransferases whose activities have been demonstrated or whose sequences show high identity to known enzymes. Furthermore, there are another 150 or so genes that encode enzymes or acceptor proteins whose expression levels can directly alter the complement of specific glycan structures produced by a particular human cell. Each of these glyco-genes is, no doubt, subject to many types of expression regulation, although we know the details of only a handful of them. Plus, in addition to the glycosyltransferases, the expression of other glyco-related genes can clearly affect glycan expression; for example, the endoplasmic reticulum and Golgi-processing glycosidases and sugar nucleotide transporters.

If we think of the expression of glycan structures simply as "readouts" of changes in the expression of genes during oncogenic transformation and progression, and in light of the many alterations of intracellular signaling that occur as cells show loss of proliferative, adhesive, and apoptotic control, it would be logical to predict that the glycan readout of a cancer cell should show altered, or aberrant, changes. These glycans are covalently bound to proteins and glycosphingolipids that are, in most cases, secreted or show residency on the outer cell surface from which they can be shed or released by hydrolytic activities, finding their way into the serum or into other extracellular fluids (Figure 16.1). A useful term to describe these glycans is as specific "glycosignatures" on particular proteins and lipids.

We now know of many examples of altered or aberrant expression of glycans on specific glycoconjugates during oncogenic transformation and, in some cases, cancer progression [1,2]. There are examples of the same protein being expressed and secreted

**Handbook of Glycomics**
Copyright © 2009 by Elsevier Inc. All rights of reproduction in any form reserved.

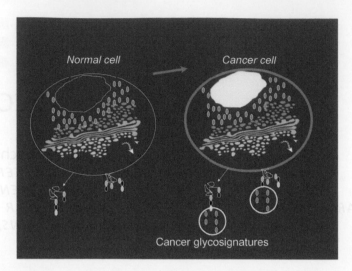

**FIGURE 16.1**   Glycosignatures on cell surface glycoproteins and those released or secreted from cells are altered after oncogenic transformation.

both before and after oncogenic transformation but showing aberrant glycosylation—an altered glycosignature—when synthesized in the cancer cell. These cell surface glycoconjugates are, therefore, logical candidates for specific markers of cancer cells at various stages of progression to which chemotherapeutic agents could be directed. Moreover, cell-released glycoconjugates are potential markers in serum and other fluids. The precise glycosignatures of these potential markers, therefore, must be identified, and in the case of a particular glycoprotein of interest, the identification would include its glycan structures as well as the site(s) on the protein where specific glycans are expressed. Of course, we know of many examples where changes in glycan structure of a cell-associated glycoprotein cause dramatic functional changes, and several examples of these will be noted.

It is also interesting to note that nine of the cancer markers in most common clinical use are either glycoproteins or, in one case, a glycan itself. These glycoproteins include chorionic gonadotropin (hCG), the mucin epitopes CA27–29 and CA15–3, prostatic-specific antigen (PSA), alpha-fetoprotein, CA125, carcinoembryonic antigen (CEA), and thyroglobulin [3]. The glycan epitope is CA19–9, the sialyl Lewis a glycan [4]. In general, these markers are used to confirm possible diagnoses and to monitor course of disease after treatment, since their specificities and/or sensitivities are not high enough to be used routinely as diagnostic markers. PSA is commonly used in conjunction with the digital rectal exam to screen for prostatic carcinoma, although its sensitivity has recently been determined to be 21%, while its specificity is 94%, where normal values were defined as less than 4.0 ng/mL [5]. For those unaccustomed to biomarker terminology, sensitivity relates to the number of false negatives a test gives: 21% sensitivity means there is a 79% false negative rate; specificity relates to a test's false positives: a 94% specificity means that 6% of the test's positive indications are false. As discussed below, there is evidence that a focus on the glycans expressed on some of the cancer marker glycoproteins can increase their specificity as cancer markers [1].

The US Food and Drug Administration has approved as Class II an instrument developed by Wako Pharmaceuticals to test for the ratio of alpha-fetoprotein with and without core $\alpha$1,6 fucosylation as an aid to diagnose hepatocellular carcinoma in at-risk populations [6]. An assay that determines the glycosignature of a particular glycoprotein cancer marker, alpha-fetoprotein, is approved for clinical use in the USA, which should assist the approval of tests for the glycosignatures of additional cancer markers.

Perhaps the earliest observations of glycan changes in cancerous tissue were those of Masamune at Tohoku Medical School in Sendai, Japan, in the early 1950s, who documented changes in the ABO blood group carbohydrates in cases of gastric cancer [7,8]. It is interesting to note that later experiments followed up on these original studies, showing a positive correlation between loss of blood group A and a more positive survival rate of patients with several cancers [9–11]. Interestingly, a similar association of loss of glycans with blood group A reactivity has more recently been documented for small cell lung cancer [12].

The discovery that lectins showed differential binding to oncogenically transformed cells relative to non-diseased cells was a critical step on the path to show that aberrant glycosylation is a hallmark of oncogenesis. Interestingly, Aub et al. [13] were the first to describe a differential agglutination of cancer cells. The agglutinin was actually a contaminant of a wheat germ lipase preparation that was being tested for its effects on mouse lymphoma cells vs. thymocytes. The agglutination of the lymphoma cells did not correlate with specific activity of the lipase, was not found in pancreatic lipase preparations, and survived heat treatment that inactivated the wheat germ lipase. A subsequent study showed the differential agglutination of the lymphoma cells was due to a lectin, wheat germ agglutinin (WGA) [14,15]. A 1972 review in *Science* by Sharon and Lis summarized the current knowledge concerning lectins and their glycan-binding specificities, highlighting their abilities to distinguish some oncogenically transformed cells and foreshadowing their current use to separate glycoproteins and glycans as a means to identify potential cancer markers [16].

Alterations in glycosphingolipid glycan expression were some of the first well-characterized glycan structures shown to be affected in cancer cells and tissues, since many of this type of glycoconjugate could be resolved using thin-layer chromatography after chloroform:methanol:water extraction of cancer cells or tissues [17–21]. Subsequently, numerous examples of changes in glycosphingolipid expression have been documented for a variety of oncogenically transformed cells and tissues, including human cancers, reviewed by Hakomori [2]. These glycosphingolipids clearly regulate plasma membrane receptor signaling and changes in their expression in cancer cells may have important effects on intracellular signaling [22]. Moreover, tumor cell-specific gangliosides are now the subject of cancer vaccine development [23].

Detailed studies of the glycosylation of proteins, glycoproteomics, has lagged behind that of the glycosphingolipids because of the complexity of glycan structures that can be found even on a single glycoprotein, plus the difficulty of isolating the glycoproteins themselves and solving the structures of the glycans that are released from them. One of first observations of glycoprotein glycan expression changes in oncogenically transformed cells was that of Robbins [24,25], Glick and colleagues [26–29], who noted alterations of glycopeptide and glycan size after oncogenic transformation of cultured cells by using metabolic radiolabeling of glycans with glucosamine or fucose. More detailed structural characterization of transformed cell glycoprotein glycans followed when Kobata's laboratory [30] showed that one of the primary effects of viral transformation on fibroblast glycan structures was an increase in a class of branched N-linked structures with a β1,6-linked N-acetylglucosamine (GlcNAc) attached to the α1,6-linked mannose (Man) [31]. The enzymatic basis for this increased expression of N-linked β1,6 branches was subsequently shown to be due to increased activity of N-acetylglucosaminyltransferase-V (GnT-V), the enzyme that synthesizes this branch [32]. These findings were confirmed using complementary methodologies [33,34]. Purification of GnT-V over a million-fold from rat kidney [35] allowed the subsequent isolation of a cDNA encoding this enzyme [36]. Utilizing this sequence, it was possible to demonstrate that viral transformation causes an upregulated *ras-raf-ets* signaling pathway that results in increased GnT-V transcription and subsequent increased expression of the N-linked β1,6-branched glycans [37]. A cDNA encoding the human GnT-V was isolated by Taniguchi's laboratory [38]. Transcriptional regulation of GnT-V by *ets*-1 signaling was demonstrated by this group in bile duct carcinoma cells [39]. The upregulation of GnT-V (now known also as GnT-Va) caused by oncogene activation results in promotion of cell motility and invasiveness due to changes in cell–cell and cell–matrix adhesion, reviewed recently [40], although many glycoproteins show altered expression of GnT-Va products. Experiments to exploit these changes in glycan expression on specific glycoproteins in order to develop potential tumor markers are discussed below. Other glycosyltransferases have been shown to be responsive to oncogene signaling and may have even more pronounced effects on cell adhesion than GnT-Va. For example, hypersialylation of β1 integrins, observed in colon adenocarcinoma, may contribute to cancer progression [41]. In addition, tumor suppressor [42] and metastasis suppressor expression can alter glycosylation and glycosyltransferase activity [43].

Another type of glycan change observed frequently in some types of tumors is that of the expression of sialyl Lewis x and sialyl Lewis a, a change which, in several instances, has been inversely associated with patient survival (discussed in [44]). The hypothesis connecting the expression of these glycans and decreased survival suggests that they can serve as ligands for the family of glycan-binding proteins known as the selectins, which are known to mediate interactions of cells in circulation with endothelial cells. These interactions of tumor cells and activated or induced endothelial cells via selectin and tumor cell–glycan may then serve to allow extravasation of tumor cells into tissues to develop metastatic foci [45]. The mechanisms by which tumor cells show aberrant expression of the Lewis glycans are, however, as yet unclear [46]. Nonetheless, research is focused on inhibiting these interactions with various compounds, including heparins [47].

Outstanding reviews by Kobata and Amano [1], as well as Fuster and Esko [44], have discussed the alteration of protein glycosylation in malignant cells as it relates to immunotherapy and diagnosis of tumors, and to glycans and the enzymes that synthesize them as potential cancer therapeutic targets, with special attention to glycosaminoglycans. In addition, the chapter on "Glycosylation changes in cancer," by Varki, Kannagi, and Toole, in the *Essentials of Glycobiology*, provides comprehensive background and discussion [48]. This chapter will emphasize recent studies that have focused on using newly developed analytical technologies to characterize in detail altered glycan and glycoconjugate expression in cancer cells, tissues, and fluids such as serum from patients with cancer. In particular, emphasis will be placed on studies that seek to exploit these aberrant glycan changes to develop potential tumor markers that could be used to direct chemotherapeutic agents or serve as diagnostic or prognostic markers of cancer or cancer risk.

## Analysis of Glycans on Specific Glycoproteins

Several studies, beginning with those from Kobata's laboratory, have documented structural glycan structural changes in hCG secreted into urine from normal pregnant women and those with hydatidiform mole compared to those with choriocarcinoma [49,50]. These changes involve increases in $\beta$1,4-branched *N*-linked glycans with and without fucose. In addition, increased core 2 branching of *O*-linked structures, at Ser127 or Ser132 were observed in the urine of choriocarcinoma patients [50]. Glycoforms of hCG detected in urine can be utilized to distinguish the presence of benign hydatidiform mole from cancerous choriocarcinoma.

Carcinoembryonic antigen (CEA) is in use as a serum marker of colorectal cancer, primarily to follow disease course. Non-diseased, non-colorectal epithelial cells can also produce gene products similar to CEA, which, unfortunately, decreases CEA's specificity as a cancer screening marker. Yamashita and Kobata have extensively studied the glycans expressed on CEA and glycoproteins that cross-react with CEA antibodies (NCA-2, NFA-2) and noted several important differences [1,51]. Only CEA expressed any high-mannose containing glycans, and its sialic acid residues were Sia$\alpha$2,6Gal$\beta$1,4GlcNAc. By contrast the sialic acids linkages of NCA-2 and NFA-2 were identified as Sia$\alpha$2,3Gal$\beta$1,4GlcNAc. In addition, CEA contained 6-sulfated galactose residues in the sequence, Gal$\beta$1,4(HSO$_3^-$-6)GlcNAc, while the related glycoproteins contained 6-sulfated galactose residues linked $\beta$1,4 to GlcNAc: Gal$\beta$1,3(HSO$_3^-$-6)GlcNAc. If antibodies were available to distinguish the CEA-specific glycan structures, it is conceivable that assays could be developed that would measure the colorectal carcinoma-specific forms of CEA and be used for early diagnosis of this disease.

One of the first glycoproteins in cancer cells whose aberrant glycosylation was directly shown to have functional consequences was matriptase, a serine protease released from many types of cancers [52,53]. The methodology used, described first by Taniguchi and collaborators [54], was to overexpress glycosyltransferases in certain cell types to identify specific glycoprotein acceptors that can lead to identifying altered function by aberrant glycosylation. Using gastric cancer cells, a secreted protease with gelatinase activity was identified. The activity of this protease was significantly increased when cells were transfected to overexpress GnT-V. The increased branching, which was demonstrated by L-PHA binding, resulted not in increased protein expression of released matripase, but in sustained activity. This effect on matripase activity was subsequently shown to result from the branched glycan at Asn772, which caused the matripase itself to be resistant to proteolytic degradation. Gastric cancer cells expressing a GnT-V cDNA or, separately, a cDNA encoding matripase both showed increased lymph node metastasis in an athymic mouse model. Coexpression of GnT-V and matripase activity was shown to parallel one another when 132 cases of various types of thyroid cancer were examined. The highest correlation was found in papillary carcinoma (<1 cm), suggesting that GnT-V and matripase coexpression is required in the early stage of papillary carcinoma progression, rather than in late stages [55].

Barrabes et al. [56] studied the N-linked glycans of serum ribonuclease I, a potential marker of pancreatic cancer. The major difference was an increase in core fucosylation of the sialylated biantennary glycans in the cancer sera. Interestingly, endothelial cells had been shown to release ribonuclease I into culture fluid. Analysis of the glycans of ribonuclease I released from three types of cultured endothelial cells showed similarity to those found in the cancer patients' sera, suggesting that a major source of ribonuclease I in the sera was secreted from endothelial cells, and not the pancreas.

As mentioned earlier, the mucin glycoprotein CA125 is used as a clinical serum marker for ovarian carcinoma. An extensive study of the glycans expressed on CA125 released into the culture media of OVCAR-3 ovarian tumor cells was recently reported [57]. The results show that this glycoprotein contains both high mannose and highly branched N-linked glycans. In addition, it contains both type 1 and type 2 O-linked glycans. Reducing the circulating levels of CA125 is associated with increased survival of patients with ovarian carcinoma [58]; therefore, the various glycans associated with CA125 may function in immunosuppression.

The majority of recent glycomic analysis of cancer cells, tissues, and fluids has been focused on identifying potential glycoprotein or glycoprotein-derived markers. As discussed in various chapters of this *Handbook*, attempts at a comparative analysis of the *total* glycoproteome based on characterization of *total* glycoproteins and glycopeptides derived from cancer and non-diseased tissues is presently not possible because of sample complexity and enormous ranges in the levels of expression of glycoproteins in these types of analytes. It is feasible at the present time, however, to use a focused glycoproteomic approach to separate classes of glycoproteins or glycopeptides, and then drill deeply into their amino acid sequences and glycan structures. Moreover, it is also feasible to analyze in increasing detail glycans, glycopeptides, and glycoproteins derived from fluids (e.g., serum), which do not require prior solubilization using detergents that are not easily compatible with subsequent mass spectrometric analysis.

## Comparative Glycoproteomics of Serum to Identify Potential Tumor Markers

The identification of potential tumor marker glycoproteins from cancer patient sera has been difficult because of the extreme complexity of the proteins and glycoproteins present in serum, coupled with the fact that over 99% (by weight) of the total proteins present in sera are irrelevant to the discovery of tumor markers [59]. The simplest strategy to identify potential cancer biomarkers in serum is to release glycans from the total glycoprotein fraction from cancer patient sera and perform a comparative analysis with matched control sera. To reduce the contributions of the glycans from the predominant serum glycoproteins (e.g., IgG, haptoglobin), some investigators perform fractionation prior to glycan release using immobilized antibodies to the predominant glycoproteins to remove them from the analysis. Although a great deal of information is lost when glycans are separated from the glycoproteins that express them, several investigators have generated results that suggest strongly that there are reproducible differences in the types and amounts of glycans released from cancer serum glycoproteins compared to control serum glycoproteins.

Since N-glycans can be released quantitatively from a total glycoprotein or glycopeptide fraction, many investigators have used this methodology successfully to compare N-linked glycans in sera from patients with various types of cancer and benign diseases. Block and Mehta have utilized this approach to identify an exciting candidate for a marker for hepatocellular carcinoma (HCC) [60–62]. The initial studies utilized a woodchuck model for HCC, releasing serum glycoprotein N-glycans, tagging them with fluorescent 2-aminopyridine, and separating them using normal-phase high-performance liquid chromatography (HPLC). IgGs were removed prior to glycan release using an immunoaffinity column. The conclusion from these studies was that there was a significant increase in N-linked core $\alpha$1,6-fucosylation of glycoproteins in the sera of animals with HCC. Several plant lectins were known to bind to core fucosylated glycans, and these were used to identify serum glycoproteins after two-dimensional isoelectric focusing/polyacrylamide gel electrophresis (2D IEF/PAGE) that were increased in the HCC model. A glycoprotein that showed increased core fucosylation in HCC woodchuck sera was identified as GP73 (Golgi protein 73).

An initial experiment quantified relative amounts of GP73 in human patient sera with HCC and non-diseased patients, plus sera from patients with hepatitis B infections with and without active disease, and from patients with colorectal carcinoma. The results from this study showed a consistent, significant increase (average of 30-fold) in serum GP73 levels of patients with HCC compared with controls. One individual with HCC, however, showed low serum levels of GP73, similar to non-diseased patients. This particular sample of GP73 showed hyper core-fucosylation, as did all GP73 in HCC patients, suggesting that an assay focusing on measuring this glycoform could lead to a highly specific and sensitive assay for HCC. Additional glycoproteins (19) in HCC sera were identified and shown in a later study to also show increased levels of core fucosylation [60], suggesting that increased core fucosylation of liver glycoproteins is likely a common occurrence in HCC. After examining total GP73 levels in sera, but not the specific core-fucosylated glycoform, in 352 patients with HCC and those with cirrhosis, GP73 levels showed a sensitivity of 69% and a specificity of 75%, with an area under the receiver operating curve (ROC) of 0.79. This value compares to an ROC of 0.61 for alpha-fetoprotein for the same samples, suggesting that total GP73 is a better marker to distinguish HCC from cirrhosis than alpha-fetoprotein [33].

A recent study by Drake et al. [63] utilized lectin precipitation to show core fucosylated GP73 and hemopexin demonstrated a relatively high sensitivity and specificity in a small number of patients to distinguish HCC and cirrhosis. The convincing conclusion from these studies is that potential tumor markers with increased specificity and sensitivity can be discovered and developed by focusing on specific glycoprotein marker glycosignatures.

Other laboratories had noted the increase in core fucosylation of liver proteins in HCC serum [64–73], mainly alpha-fetoprotein. As mentioned above, a commercial assay to determine the relative ratio of core fucosylated versus non-core fucosylated alpha-fetoprotein is in use. In another study, the core fucosylation status of haptoglobin was characterized in HCC and other cancer patients' sera using *Aleuria aurantia* lectin (AAL) binding. Sera from 5 out of 23 HCC cases were shown to exhibit core fucosylated haptoglobin, while 29 out of 49 pancreatic cancer patients' sera showed this positive glycoform [74]. Elevated total haptoglobin levels themselves have been shown to be a useful diagnostic for HCC versus other chronic liver diseases, and SNA and LCA binding to serum haptoglobin was used to characterize glycoforms by 2D IEF/PAGE. The results showed increased acidic glycoforms (sialylated?) that were also fucosylated [75]. Initial experiments suggested that elevated Fut8, the single fucosyltransferase in humans responsible for core fucosylation, could be at least partly responsible for the increased levels of the $\alpha$1,6 core fucose on N-linked glycans on alpha-fetoprotein [76]. A more recent study, however, suggested that increased expression of the Golgi GDP-transporter could be responsible for the increased core fucosylation seen in secreted liver proteins from patients with HCC, by contrast to those with cirrhosis [77]. This result could also explain the increased core fucosylation of GP73 in HCC patient sera. Core fucosylation of N-glycans has also been suggested to be a putative signal for the secretion of glycoproteins into the liver bile ducts, since mice that lack expression of Fut8 exhibit low levels of secreted glycoproteins in their bile [78]. Increased core fucosylation in HCC livers could divert increased amounts of secreted proteins, such as alpha-fetoprotein, into the serum.

Interestingly, haptoglobin glycoforms have been studied as potential serum markers for other cancers, including pancreatic cancer and prostatic cancer. Haptoglobin from pancreatic carcinoma patients showed increases in a site-specific glycoform, Asn211, expressing a Lewis x (non-core fucosylated) structure, when analyzed by liquid chromatography coupled to mass spectrometry with electrospray ionization (LC-ESI-MS). A difucosylated structure (Lewis y) at this site was seen only in HCC sera [74,79]. Increased levels of haptoglobin have also been documented in sera from patients with prostatic carcinoma. Lewis x glycans were identified on two N-glycan sites, 207 and 211, but not in haptoglobin found in control subject sera [80]. These studies suggest that increased expression of haptoglobin and some of its glycoforms is relevant to marker discovery in cancers other than HCC. A recent study by Miyoshi and Nakano [81] demonstrated that core fucosylated haptoglobin is significantly increased in 66% of pancreatic cancer patient sera (57 of 87), the highest percentage of the five types of cancer studied. Increased core fucosylation was seen on biantennary *N*-linked glycans on sites 2 and 4 and on triantennary glycans at all four sites on the haptoglobin polypeptide when sera from patients with pancreatic cancer were compared to that from normal subjects and those with chronic pancreatitis. Co-culture of pancreatic carcinoma with a human hepatoma cell line resulted in production of core-fucosylated haptoglobin, suggesting the intriguing possibility that the pancreatic cancer cells can release a soluble factor that can induce the fucosylated haptoglobin expression in the hepatoma cells.

## Glycans Released from Serum Glycoproteins

Novotny's laboratory has focused on the analysis of *N*-linked glycans isolated from total serum glycoproteins and found in several cases that a collection of these glycans show statistically significant changes (both increased and decreased) that can likely be used to predict the presence of a specific carcinoma. Serum glycoproteins were subjected to reduction, alkylation, and trypsinization, followed by peptide *N*-glycosidase F (PNGase F) release, purification, and permethylation. Permethylation neutralizes the negative charge of sialic acids, allowing MS analysis in positive-ion mode, increases sensitivity over 100-fold, and reduces the variation in sensitivity seen with underivatized glycans, as detailed elsewhere in this *Handbook*. Permethylated *N*-glycans were then analyzed by matrix-assisted laser desorption/ionization-tandem time-of-flight-mass spectrometry (MALDI-TOF/TOF-MS). Sera from 24 patients with prostatic carcinoma were analyzed and compared with 10 non-diseased males. The results showed 12 glycans that were significantly changed statistically (10 increased and 2 decreased) using both ROC (AUC) and ANOVA *P*-test analyses [82]. Some structures (e.g., disialylated-triantennary with and without core fucosylation) showed large increases in the cancer sera with no statistical overlap with levels found in control sera. Only 10μL of sera was necessary for analysis, and permethylation was accomplished using a capillary procedure [83]. A related study focused on sera from 203 total patients with HCC, those with chronic liver disease, and controls analyzing permethylated *N*-linked glycans, as well as peptides [84]. The results showed that the top three peptides showed an average sensitivity and specificity of 81% and 88%, respectively; the top three glycans showed a sensitivity and specificity of 91% and 95%, respectively. Combining these peptides and glycans yielded an overall sensitivity of 94% and specificity of 90%. An initial study of sera from patients with HCC and cirrhosis was recently done using ion mobility MS to separate permethylated *N*-linked glycans. The results from this study showed that a principal component analysis (PCA) of the "drift times" for two glycans, mono- and disialylated biantennary structures, showed significant cluster differences between the two groups [85].

A recent study by Novotny's group extended studies of *N*-linked glycan serum analysis to patients with breast cancer. Sera from 82 patients with breast cancer ranging from Stage I to Stage IV and 27 control women were analyzed. Six *N*-glycan species were shown to be highly accurate predictors of breast cancer, including those women with Stage I cancer [86]. Five of these glycans showed AUC values of >0.98 and ANOVA *P*-values of <0.01, with many individual *P*-values of <0.002.

These glycans consisted of tri- and tetra-antennary glycans with core fucose (4/5) and with outer-branched fucose (2/5). Some of these same structures were observed in culture supernatants of breast carcinoma cells compared with a "normal" breast epithelial line. Thus, these results were consistent with some changes in N-linked glycosylation known to occur in breast cancer tissues [87], although in the case of the glycans identified in cancer serum, it is clearly not known whether the glycans are attached to proteins that have been secreted or released from the carcinoma itself, or are from glycoproteins produced in other cell types whose release was triggered by the carcinoma.

This group has recently studied the glycans released from six cultured breast cell lines, one of which was a "normal" breast epithelial line, two of which were invasive breast carcinoma lines, and three of which were non-invasive breast cancer lines [88]. The analytical methods and statistical analyses developed and applied for analysis of released N- and O-linked glycans were clearly state-of-the-art. Lysed cells were subjected to centrifugation at $100K \times g$ and separated into "cytosolic" and membrane-bound fractions. Both the N-linked and O-linked glycans released from glycoproteins in the cytosolic fractions, when analyzed by PCA, allowed the cancer cell lines to be segregated from the "normal" three lines. The N-linked cytosolic glycoprotein glycans showed statistically significant (ANOVA) increased high mannose structures in the cancer cells compared with controls, as well as a decrease in hybrid structures and increases in tetra-antennary glycans ± fucose. Increased high-mannose glycans were found released from the membrane-bound glycoproteins, in particular, $GlcNAc_2Man_5$ structures. By contrast, and somewhat surprising in light of the study discussed above, no statistically significant differences in highly branched N-linked glycans from membrane-bound glycoproteins were identified. A significant difference was found in fucosylated O-linked glycans from glycoproteins found in the cytosolic fractions of invasive cells when compared with control cells. In addition, fucosylated O-linked structures in this fraction also showed increases between the invasive and non-invasive cancer cells, as well as a marked decrease of sialylated structures. It may well be that "normal" breast cells cultured for many passages on plastic may go through a selection such that their glycans more resemble oncogenically transformed cells that underwent in vivo selection processes to form tumors.

Reinhold's laboratory has reported a detailed analysis and comparison of N-linked glycans from two metastatic astrocytoma cell lines and one non-metastatic cell line, all prepared from a spontaneous mouse astrocytoma that was passaged intracranially through two hosts [89]. A control astrocyte cell line was purchased from ATCC. Glycans were released by N-glycanase, reduced, permethylated, and analyzed by NSI-MS$^n$ using an LTQ instrument, including state-of-the-art CID (collision-induced dissociation) fragmentation. The results showed one isomeric N-linked structure that was identified in the three cancer cell lines, compared to the astrocyte control line, and one structure that was found only in the non-metastatic line. Analysis was accomplished using GlySpy, software that includes tools for data manipulation and a glycan topology assignment algorithm.

The Rudd-Dwek groups have developed and applied glycoproteomic strategies to identify serum glycoproteins whose glycans show alterations in sera taken from patients with several types of cancers. Methodologies employed by this group involve N-glycan release from serum or fluid glycoproteins, tagging the released glycans with fluorescent 2-aminobenzamide, separation by HPLC, and subsequent analysis using MALDI-TOF and negative ion nanospray mass spectrometry [90]. The glycan structures of prostate specific antigen (PSA) were studied in serum and seminal fluid from patients with prostatic carcinoma [91]. Results showed that in a small number of samples, PSA showed an increase of N-linked sialyl Lewis x structures, and in one case, a decrease of α2,3-linked sialic acid. These results suggest that the glycans of PSA might be useful to distinguish elevated PSA values caused by the presence of carcinoma versus elevated PSA caused by other factors. No evaluation of PSA from patients with benign prostatic hyperplasia (BPH) was made, however, which is the benign condition that can also result in elevated PSA. A subsequent study of sera from ovarian cancer patients showed an increase of N-linked non-galactosylated biantennary glycans with sialyl Lewis x outer fucose and also core fucose. Two-dimensional IEF/PAGE was performed, and this glycan change was found on three glycoproteins, haptoglobin (β-chain), α1-acid glycoprotein, and α$_1$-antichymotrypsin, all acute-phase proteins synthesized in the liver. IgG heavy chains in ovarian cancer sera also showed twice the level of this glycan, compared with that in control sera [92]. Changes in the N-linked glycans on serum proteins were also seen in patients with the inflammatory diseases sepsis and acute pancreatitis, including increased non-core fucosylation, increases in tetrasialylated structures, and degree of branching [93]. Taken together, these studies show that the inflammatory response yields defined changes in N-linked glycans of acute phase proteins and IgG, and these glycoproteins likely constitute a majority of the glycan changes seen in serum in the presence of cancer. Interestingly, there is speculation that these changes in liver acute phase protein glycosylation by the presence of some cancers may lead to increased survival of the cancer cells [94].

Recently, this research group has published its strategy to identify potential glycan markers on serum glycoproteins found in breast cancer patients [95]. N-Linked structures were first released and tagged, and separated by HPLC. Glycans were characterized by exoglycosidase digestions and MALDI-TOF mass spectrometry, revealing an increase in trisialylated, tri-antennary glycans with α1,3-linked fucose in the breast cancer sera. In addition, an unusual monogalactosylated tri-antennary structure with α1,3-linked fucose showed a twofold increase in the sera from breast cancer patients. Moreover, monitoring 10 patients for several months suggested a positive association between increased expression of this structure on serum glycoproteins and disease progression, although the sample size in this initial experiment was very low. In this study, the unusual glycan structure mentioned above appeared to be a better indicator than the commonly used markers, CA15–3 and CEA. Subsequent glycoproteomic analysis showed that the major glycoproteins expressing this glycan are α1-acid glycoprotein, α1-antichymotrypsin, and haptoglobin. Each of these proteins, when used as reporters for the mono-galactosylated glycan structure showed increased glycan expression before CA15–3 increases. This study underscores the promise of the measurement of specific glycan structures and serum glycoproteins that express them in developing new markers for tumor progression. The changes in serum N-linked glycan changes in cancer and inflammation have recently been summarized [96]. Large numbers of samples for serum/plasma N-glycan determination can now be analyzed in a semi-automated fashion. Plasma glycoprotein glycans from over 1000 individuals were recently analyzed; age and gender interaction effects were statistically significant in many of the N-glycans groups analyzed [97].

## Use of Targeted Glycoproteomics to Identify Changes in Glycoprotein Expression in Cancer Serum

Hancock's laboratory has developed impressive methodologies to utilize lectin separations to enrich for serum glycoproteins for comparisons between cancer patient and control sera. This methodology uses multiple lectins in combination with affinity-separate glycoproteins, followed by nano-HPLC coupled to electrospray ionization-Fourier transform-linear trap quadrupole (ESI-FT-LTQ) mass spectrometry for identification of N-linked glycopeptides and peptides in serum [98]. This multi-lectin affinity chromatography, M-LAC, seeks to mimic the "glycoside cluster effect" through the use of a simple admixture of multiple lectins, shown to yield a tenfold increase in affinity constants for thyroglobulin over the use of single lectins [99]. In a study of sera from 10 breast cancer patients and 10 controls, a combination of immobilized concanavalin A (Con A), WGA, and Jacalin lectins was utilized, followed by trypsinization and protein identification by proteomics. Potential glycoproteins of biomarker interest were identified in the cancer sera, including apolipoprotein B and C III, ceruloplasmin, several proteases inhibitors, and prothrombin. These glycoproteins were individually identified by other proteomic studies, validating the strategy of serum analysis using multi-lectin separations. Recently, this methodology was used to study potential markers released by a human MCF-7 breast cancer cell line in a mouse xenograft system. The human cancer cells were surgically implanted into a fat pad of an immunocompromised mouse. Sera from the recipient mouse were then analyzed during the course of tumor growth, and proteomic identification of peptides used to distinguish mouse and human sequences [100]. As the tumor mass increased, several human tumor-derived glycoproteins linked to cell signaling, immune response, and transcriptional regulation were identified. These results clearly showed that the xenograft approach with increased sensitivity can be developed to identify very early glycoprotein markers secreted by tumors, which can then be searched for in human breast cancer patient sera. The multi-lectin separation and proteomic identification is clearly a significant contribution to glycoproteomics, and this strategy employed in the xenograft model will likely lead to the identification of potential serum markers of several types of cancer.

A related, lectin-based separation approach has been developed and applied by Lubman and collaborators to identify glycan differences in cancer patient serum and urine. In one study, sera from two patients with pancreatic carcinoma were processed and compared with sera from two non-diseased patients [101]. Sera were subjected to Con A chromatography, followed by reduction/alkylation, trypsinization, and subsequent re-chromatography of tryptic glycopeptides on Con A. Bound glyco-peptides were eluted and subjected to *N*-glycanase treatment, followed by peptide identification and glycan characterization using MALDI quadrupole ion trap TOF mass spectrometry. Also, glycosylation sites were mapped by nano-reverse phase LC-ESI-MS/MS. The cancer serum showed 44 "distinct" *N*-linked glycans compared to the controls, characterized by increased branching, fucosylation, and sialylation. Some structures were verified by fragmentation analysis. Interestingly, abundant complex branched *N*-linked glycans were identified using this methodology, although glycopeptides containing these structures have been shown to not be bound by Con A [102]. It is possible that the tryptic glycopeptides expressed multiple glycosylation sites per peptide which included at least one glycan that bound to Con A. A subsequent study by Lubman's laboratory focused on sera from colorectal carcinoma patients (5), those with adenoma (5), and non-diseased individuals (9) [103,104]. *N*-Linked glycoproteins were separated using Con A chromatography, separated by silica reverse-phase HPLC, blotted to nitrocellulose-coated slides, and probed with various lectins. Conclusions were that cancer serum glycoproteins had increased sialylation and fucosylation. Individual glycoproteins with these changes were identified in cancer serum: complement C3, histidine-rich glycoprotein, and kininogen-1. The ability of the analysis of these glycoproteins to distinguish sera from colorectal cancer patients from those with adenoma and "normal" controls was verified on an independent set of samples. A related study examined glycoproteins in urine of individuals with bladder cancer [104]. Con A chromatography was used to isolate *N*-linked glycoproteins from urine of individuals with bladder carcinoma and controls. Bound glycoproteins were subjected to shotgun proteomics and identification. The results showed that several glycoproteins were associated with bladder cancer, in particular $\alpha$1B-glycoprotein, which was detected in all cancer samples, but none of the controls. These promising results have identified a potential marker for bladder cancer using glycoproteomic separation and analysis.

L-PHA affinity chromatography was used in Ko's laboratory to identify potential biomarkers in sera from patients with colorectal carcinoma [105]. A ProteomPrep 20 column was first used to remove high-abundance proteins/glycoproteins, followed by denaturation and alkylation, L-PHA chromatography, elution with urea, and shotgun proteomics. Twenty-six proteins with *N*-linked sequons and more than one peptide coverage were identified. Three control and 10 sera from patients with Stage I–III cancers were studied. Two glycoproteins were identified with 100% specificity and 90% sensitivity: isoform long of diacylglycerol kinase zeta and JmjC domain-containing histone demethylation protein 2 A. This excellent study demonstrates the efficacy of using lectins that bind specific glycans known to be increased in a particular cancer type to identify potential serum markers.

Profiling of glycoproteins that contain *N*-linked structures by shotgun proteomics has been performed on lung adenocarcinoma pleural effusions [106], based on the methodology developed by Zhang and Ebersold [107,108]. In this method, glycoproteins are subjected to mild periodate oxidation which reacts with many glycan structures to generate aldehydes, which are then reacted with a support with attached hydrazine moieties. Underivatized peptides are separated from the immobilized glycoproteins by washing the support. The immobilized glycoproteins are then subjected to reduction and alkylation, followed by trypsinization. Unbound peptides are then separated from bound glycopeptides. As illustrated in the case of the lung adenocarcinoma effusions, the next step in this approach is to treat the support with PNGase F, releasing the peptides that were attached to the support through $N$-glycans. Five samples of effusate from patients with lung adenocarcinoma and that from non-malignant control patients were analyzed using this methodology, followed by LC/MS-MS peptide online identification. The results showed that 170 and 278 non-redundant proteins were identified with probabilities of >0.9 and >0.5, respectively. Identification of proteins in the range of microgram to nanogram per milliliter was observed. Several glycoproteins associated with cancer were identified, CA125, CD44, CD166, as well as LAMP-2, multimerin 2, and periostin. For several of these markers, immunoassays on tissues or effusates were performed, and the results showed a good correlation for the glycoproteins identified by MS analysis. The major limitation of this methodology, however, is that all information concerning glycan structures on particular glycoproteins is lost, making impossible comparisons of glycosylation changes between non-diseased and diseased conditions, for example.

## Comparison of $O$-Linked Glycans in Cancer Serum and on Cancer Cells

Lebrilla's group is one of the few laboratories to have studied $O$-linked glycans in cancer patients' serum, releasing them from glycoproteins by beta-elimination [109–113]. In one study of ovarian cancer, cultured cell line supernatants were first used to optimize the methodology for glycan release and characterization by MALDI-FT-MS analysis, followed by the analysis of sera from patients with ovarian cancer and attempts for further glycan characterization using infrared multiphoton dissociation (IRMPD). The results revealed several glycan species in the cell supernatants and patients' sera, but not in "normal" sera [109]. In addition, a series of unusual hexuronic acid-containing saccharides were identified in human serum, and several of these species were described as unique to ovarian cancer patients' sera. Further characterization of these saccharides has not as yet been reported, however. A more recent study utilized breast carcinoma cell line supernatants, sera from an implanted mouse breast cancer, and breast cancer patient sera to expand upon the results obtained in the ovarian cancer study [110]. Several $O$-linked glycans released from serum glycoproteins were found in several human breast cancer cell lines but not a "pre-cancerous" cell line or epithelial-like (MCF-10a) cell line. Many $O$-linked glycans were identified in sera from mice with implanted breast tumors taken from mice expressing the polyoma-middle T antigen at the time of implantation. Several $O$-linked glycan signatures began to appear after two weeks and were increased by eight weeks after implantation, suggesting that these signatures were the result of increased tumor growth or exposure. Moreover, six molecular ions seemed to be expressed predominantly in the samples from the four breast cancer patients and cultured breast carcinoma supernatants. These results were displayed using PCA and suggest that this approach may be able to identify *O*-linked glycan differences when non-diseased and cancer sera are compared.

A further study has applied this methodology to the analysis of ovarian cancer patient sera [111]. The results showed that 44 of 48 patient sera displayed the diagnostic *O*-linked glycan glycoprotein signals described above, while 23 of 24 control patients had no detectable signals. Sensitivity and specificity values of this assay were shown to be 92% and 96%, respectively, and the AUC of the ROC analysis for all 72 samples was 0.954. Thus, this analysis showed better discrimination than the use of the marker CA125, whose levels were tested in parallel. CA125 is the most common test for the presence of ovarian cancer. A more detailed study of glycoproteins from cultured ovarian cell lines and cancer sera showed that four glycoproteins were identified from pooled sera: two forms of apolipoprotein B100, fibronectin, and dimmunoglobulin A1. Mass spectrometric analysis of the *O*-linked glycans on these proteins demonstrated aberrant glycosylation when the results from non-diseased sera and ovarian cancer patient sera were compared [114].

Changes in the *O*-linked structures of glycoproteins in serum and other fluids, such as secreted ovarian cyst glycoproteins and lung pleural effusions, from patients with cancer have been noted for many years. The mucin family of glycoproteins has been extensively studied in cancer (reviewed in [115]), and new analytical technologies have been applied to determine site-specific glycosylation changes in various cancers. Hollingsworth's laboratory studied *O*-linked glycan expression in 77 gastric carcinomas and showed a high level of Tn and sialyl-Tn antigens, with lower expression of the T antigen [116]. Gastric carcinomas from patients that were homozygous for MUC1 large tandem repeat alleles showed higher expression of the Thomsen-Friedenreich T antigen, however. These and other in vitro studies indicated that polymorphism in the MUC1 tandem repeats can influence tumor antigen presentation, and that this type of analysis might be used to predict the possibility of more aggressive tumors in patients that have MUC1 large tandem repeats.

Storr et al. studied the *O*-linked glycosylation of secreted/shed MUC1 in the serum of a patient with advanced breast cancer [95]. The conclusions were that core 1 glycans (83%) were dominant and contained high degrees of sialylation, up to tri-sialylation; the (T) antigen made up 14% of the core 1 structures. Core 2 structures made up 17% of total structures. About 9% of total structures were polysialylated core 1. The percent of structures with the Tn antigen (Ser/Thr-*O*-GalNAc) was not reported.

A detailed investigation of the *O*-linked structures present on human ovarian cyst glycoproteins was recently performed using up-to-date mass spectrometry and CID fragmentation [117]. The results showed convincingly that most of the glycans were built on core 1 and core 2 glycans with extensions of type I (Gal-β1,3-GlcNAc) and type II (Gal-β1,4-GlcNAc) chains. Moreover, there were additional branches on the first Gal on the β1,3 arm that yielded structures that mimicked N-linked bi-antennary and tri-antennary structures. Fucose was also present in α1,3/4 linkages with and without sialylation to produce Lewis x/a and sialylated Lewis x/a structures. Sulfated residues were detected but not described. Abundant T and STn structures were also observed. The authors also suggested that there is a very dense glycosyl-ation pattern on the mucins, particularly of Le$^x$ and SLe$^z$ structures that might have function in tumor progression.

O-Linked structures on cancer glycoproteins clearly change during oncogenesis and, in some cases, appear to be able to regulate the metastatic potential of various carcinomas. The best example is the appearance of the Tn (T null) and SialylTn (STn) structures, which are Ser/Thr-O-GalNAc and Ser/Thr-O-GalNAc-α2,6-sialic acid, respectively. These structures, most often defined as lectin or antibody-binding epitopes, are commonly found on mucin and other glycoproteins after oncogenic transforma-tion. A recent publication from Cummings' laboratory described the mechanism that is responsible for the expression of these glycans. Rather than mutations in the T syn-thase, the β1,3-galactosyltransferase (T-synthase) that converts the Tn to T structure, Ju et al. [118] demonstrated somatic mutations in a specific chaperone, COSMC, a chap-erone whose expression is absolutely required for functional T synthase. These muta-tions were identified in colon carcinoma, melanoma, and cervical cancer. Expression of the Tn and STn antigens correlates with poor prognosis and metastatic potential, and their expression is present in the majority of human cancers. COSMC is encoded on the X chromosome; thus, loss of COSMC function requires only one mutation per cell. These observations suggest either a hyper-sensitivity of the COSMC gene to undergo mutations, and/or an in vivo selection pressure for carcinomas with these loss-of-func-tion mutations. In addition to the T synthase, downregulation of the β1,3-N-acetyl-glucosaminyltransferase 6 (core 3 synthase), which often occurs in colon and gastric carcinomas, results in an increase in invasiveness and metastasis, as assayed in a mouse model [119]. These results demonstrate that altered O-glycosylation is perhaps the most common alteration in tumor cell glycans, that these alterations have critical functional ramifications on tumor cell behavior, and that for as yet unknown reasons there is likely a selection for cells that have undergone these glycan alterations.

## Comparative Glycoproteomic Analysis of Cancer Cells and Tissues

Several studies have analyzed cell lines for changes in glycoproteins and glycans using glycoproteomic analyses. As proof of concept, one study was done using Chinese hamster cells (CHO) transfected with N-acetylglucosaminyltransferase III, which syn-thesizes the bisecting GlcNAc on N-linked complex glycans. N-Linked glycans were released and characterized by graphitized carbon LC/MS. Peaks with an additional HexNAc were seen in the transfected cells. Cell lystates before or after transfection were subjected to SDS-PAGE, followed by lectin blotting with E-PHA, which rec-ognizes specifically the bisected glycans, and extraction of the lectin-positive bands at 70–120 kDa. These samples were then subjected to 2D IEF/SDS-PAGE and lectin blotting. A major series of bands that were E-PHA positive were extracted and subjected to LC/MS-MS, identifying the α3 integrin subunit as a major carrier of bisected glycans in the transfected CHO cells [120].

Ko's laboratory developed methods to study gastric cancer and non-diseased tissues using 2D IEF/SDS-PAGE, followed by lectin blotting with either L-PHA (binds β1,6-branched N-linked glycans) or LCA (binds core α1,6 fucose), gel excision, and ESI-MS analysis. Six sets of either cancer or control tissues were extracted in 0.2 M NaCl and Tris buffer with protease inhibitors and precipitated with TCA. The pellets were solubilized in 5 M urea, 2 M thiourea, 4% CHAPS buffer, 1% DTT, and 2% IPG buffer prior to isoelectrofocusing. A MASCOT search of the data obtained from the gel excisions identified 29 proteins found only in cancer tissues and 7 proteins in the non-cancer tissues [121]. In a related study, this group focused on WiDr human colon cancer cells transfected with GnT-V to identify potential colorectal cancer (CRC) markers [122]. The secretome of these cells was separated first by LCA chromatography, followed by L-PHA separation. Analysis of transfected and control cells by 2D IEF/SDS-PAGE, followed by L-PHA blotting, revealed an apparent increase in stained proteins in the transfectants. Spots were picked and subjected to shotgun proteomics using ESI-MS analysis. Fifteen proteins were identified in total. An interesting protein, protein tyrosine phosphatase kappa (PTPκ), was identified in the transfected secretome that was bound by L-PHA. Evidence was presented that in the transfectant, higher levels of PTPκ reactive with L-PHA were found secreted. Altered glycosylation in the transfectant was postulated to regulate release of PTPκ and result in altered migration. Thus, a cell-released, potential marker for CRC was identified by this study.

Two additional studies by this group have focused on using L-PHA to identify potential markers for CRC. First, using GnT-V-transfected WiDr cells again, 30 proteins were identified as those increased in binding L-PHA. In particular, TIMP-1 (tissue inhibitor of metalloproteinase-1) was identified and shown to be a weaker inhibitor of matrix metalloproteinases -2 and -9 [123]. Next, immobilized L-PHA was used to identify glycoproteins in serum of patients with CRC compared with control serum [124]. Considerable optimization of the separation protocol led the investigators to develop a protocol whereby serum was immunodepleted of abundant glycoproteins using a ProteomPrep 20 plasma ID column, subjected to reduction and alkylation, and applied to L-PHA-biotin-avidin agarose. Bound glycoproteins were released by urea and subjected to shotgun proteomics using ESI-MS/MS analysis. Ten CRC patient sera and six control sera were analyzed. Twenty-six proteins were identified, each with a specificity of 100% for patient sera. Two proteins showed a sensitivity of 90%, isoform long of diacylglycerol kinase zeta, and JmjC domain-containing histone demethylation protein 2 A. An additional two proteins were identified with sensitivities of 80%, the long isoform of protocadherin alpha C2 precursor and brain-rescue-factor-1.

Lubman's laboratory recently studied glycoprotein expression differences between a precancerous (MCF10AT1) breast cell and a malignant, metastatic (MCF10CA1a) cultured breast cancer cell, utilizing membrane extractions, lectin affinity enrichment

(Con A mixed with WGA), followed by LC-MS/MS analysis [125]. Several glycoproteins were identified as specifically expressed in the malignant and metastatic cells: CD44, galectin-3-binding protein, syndecan-1, and γ-glutamyl hydrolyase. These glycoproteins have all been identified previously as potential markers for breast cancer, thereby showing that utilizing a mixed lectin affinity methodology to isolate and identify relevant glycoproteins for cancer marker discovery is quite promising.

The Lille group has applied glycoproteomics to the study of the glycoproteins of a human colon carcinoma cell line that can be induced to differentiate into enterocytes [126]. Membrane pellets were subjected to solubilization, followed by Con A lectin-affinity chromatography, SDS-PAGE, and MS analysis to identify proteins. Twelve membrane-bound proteins with N-linked sequons were identified, and this group included glycoproteins involved in adhesion and membrane transport. MAA and SNA lectin blots were performed to identify sialylated glycoproteins and those whose sialic acid linkages differed after differentiation. Most interestingly, glycan analysis of permethylated N-linked glycans by MALDI-MS/MS revealed a tenfold increase of an unusual glycan in the differentiated cells, a tri-antennary, non-galactosylated (GlcNAc-terminated) structure, which was confirmed by methylation analysis using GC-MS. These results point out that lectin affinity chromatography is also a very useful means to separate glycoproteins from non-glycosylated proteins prior to proteomic identification. Although there appears to be significant sialylation of glycoproteins in the differentiated cells, a large increase in a non-sialylated, non-glycosylated glycan was observed. Understanding the mechanism of the production of this glycan and testing for its presence in other cell types may support its use as a differentiation marker.

Abbott et al. used lectin affinity chromatography to identify potential markers in breast carcinoma tissue samples [127]. Focusing on surgical specimens from patients with invasive ductal breast carcinoma and adjacent, non-diseased tissue as controls, samples were first delipidated and then suspended in buffer with 0.1% NP-40 with burst sonication. Insoluble material was removed by centrifugation, and supernatants were exchanged into ammonium bicarbonate buffer and subjected to biotinylated L-PHA binding. L-PHA and bound complexes were removed from unbound material by the use of paramagnetic streptavidin beads, followed by elution with buffered 2 M urea/0.2 mM DTT. After carboxyamidomethylation and trypsinization, peptides were desalted and analyzed by LC-NSI-MS/MS. In addition, samples eluted from the L-PHA beads were subjected to N-glycanase treatment and released glycans were analyzed by NSI-LTQ-MS by the total ion mapping (TIM) method [128]. Increased β1,6 branching of N-linked glycans in breast cancer samples compared with non-diseased adjacent tissue was confirmed in four out of five breast cancer samples. Focusing on these four samples, 12 glycoproteins were common to all breast cancers but not observed in any non-diseased adjacent tissues. Two of these glycoproteins, periostin and haptoglobin-related protein, were verified to be present in the breast cancer samples and bound by L-PHA using commercially available antibodies. Studies to quantify the expression of these and other potential marker glycoproteins identified in this study in sera from patients with invasive ductal carcinoma are in progress.

The availability of methodology to analyze the glycotranscriptome of cells and tissues using nucleic acid probes immobilized on Affimetrix chips or a battery of qRT-PCR primers to transcripts from genes related to glycosylation, as detailed in this *Handbook*, has provided an approach to profile or quantify glycotranscript changes in cancer cells and tissues [129]. *N*-Linked glycan changes in epithelial ovarian cancer of both human and mouse tissues were recently studied [130]. The results for both human and mouse showed qualitative similarities in the expression of nine glycosyltransferase and glycosidase transcript levels which are active in the *N*-linked glycan biosynthetic pathway in both non-diseased and ovarian epithelial carcinoma ($n = 3$). The largest changes in the human cancer were seen in Mgat1, Mgat3, and Mgat5, with by far the greatest change in Mgat3 transcripts (18-fold increase). The transcript results were extended by blotting glycoproteins from the tissues with various lectins specific for glycans predicted to show expression changes. E-PHA, DSL, and AAL binding were elevated in the cancer samples compared to non-diseased ovary, showing increased expression of glycoproteins expressing these glycans. In order to identify potential ovarian carcinoma markers, these lectins are now being used to identify glycoproteins present in the human cancer specimens which are either not present in non-diseased ovary tissue or are expressed but do not bind the lectin in these control samples, using the same methodology described above [117].

Large increases in the activity of GnT-III, -IV, and -V were reported many years ago by Chen et al. [132], who compared the specific activities of these enzymes in 10 samples of pancreatic carcinoma and 9 non-diseased pancreas samples. Focusing on the activity of polycytidylate-specific RNase in sera from the subjects, more of this RNase activity was observed to not bind Con A in the sera of patients with pancreatic cancer than controls, consistent with increased branching of *N*-linked glycans on this reporter glycoprotein. These results suggest that an approach to those glycoproteomic experiments described above will be useful in exploiting aberrant glycosylation to identify potential markers of pancreatic carcinoma.

## Antibody Arrays to Detect Glycan Variation on Glycoprotein Markers

A technology that looks very promising to rapidly assess the glycosignatures on particular glycoproteins had been developed by Haab's laboratory using immobilized antibody arrays [132]. Basically, capture antibodies to glycoproteins of interest (CEA, MUC1, haptoglobin, etc.) are printed onto glass slides coated with a derivatization reagent such as nitrocellulose. Complex mixtures of glycoproteins, such as found in serum or urine, are then incubated with the slide and captured by the immobilized antibodies [133]. Tagged lectins or anti-glycan antibodies are then used to bind to the captured glycoproteins and microarray technologies used to measure binding and analyze data. This type of multiplexed analysis will prove to be very powerful in identifying patterns of glycosylation changes across many glycoproteins in serum that occur during early oncogenesis. These results will likely point to changes in glycosylation that can be reduced to analyses that can be performed in a clinical setting.

# Antibodies to Glycan Epitopes Expressed in Cancer Tissues

Antibodies in the sera of cancer patients bind to proteins and glycoproteins, as well as glycans [134–136]. In addition, several monoclonal antibodies that have been prepared against cancer cells and tissues have been shown to bind glycan epitopes; for example [137,138]. There are also reports of IgM antibodies that show both cancer tissue-binding specificity and epitope sensitivity to glycosidase treatment, implying that glycans are part of their epitopes or regulate their epitope's conformation [139,140]. Clearly, taken together these results are another demonstration that expression of specific cell surface glycan epitopes occurs during oncogenesis, and they suggest that a means for detection of circulating anti-cancer glycan antibodies could be exploited as potential diagnostic or prognostic indicators for disease course.

Yu's laboratory, for example, has shown recently that antibodies to several glycosphingolipids are found in sera of patients with various neural tumors [141], including anti-GD3 and anti-sulfoglucuronosyl paragloboside (HNK-1) antibodies. In addition, there is a report of serum antibodies against GD3 and O-acetyl GD3 in melanoma patients [142]. It would seem a very productive area to determine if any of the anti-glycosphingolipid antibodies can be detected at very early times during tumor formation for use as potential diagnostics.

The advent of technologies to array glycans on plates or glass slides is allowing hundreds of structures to be interrogated using sera or plasma; for example, [143–146], and Core H of the Consortium for Functional Glycomics: http://www.functionalglycomics.org/static/consortium/resources/resourcecorec1.shtml. Wang's laboratory, for example, used this type of technology to identify circulating antibodies to glycans expressed on infectious microbes, including *Bacillus anthracis* [147]. Several laboratories are applying this technology to identify glycans that are bound by natural antibodies in the sera of patients with various cancers, and initial results appear promising.

# Conclusion

New technologies are now being applied to identify specific glycan, glycopeptide, and glycoprotein differences found in cancer patient sera when compared to non-diseased, control sera. In a very few cases, differences have been found in sera from patients with early stage cancer; for example, Stage I breast cancer [86]. In all cases, however, results have to be used to work toward finding markers from the earliest stages of cancer. Moreover, non-diseased controls have to be expanded to include samples from a variety of cancers, benign diseases, as well as inflammatory conditions, in order to demonstrate the specificity of a particular marker. Analyses must also include double-blinded samples that are analyzed in the same manner as those used to demonstrate particular associations with cancer diagnosis. Detection of circulating antibodies to cancer-specific glycopeptides is clearly an exciting area of experimentation, particularly since these antibodies have the potential to detect very early stage disease and perhaps be used for screening purposes. With the possibility of utilizing actual tumor-derived glycans and glycopeptides immobilized in arrays on slides, the odds of detecting antibodies to cancer-specific glycopeptides would appear to significantly increase. Moreover, there is new promise in the generation of robust immune responses to glycopeptides [148]. Applying glycomic technologies for the identification of cancer-specific glycopeptides could, therefore, lead to the generation of efficacious anti-cancer vaccines.

Differences in glycoprotein glycoform expression in cancer patient sera have been identified in some cases, but the origin of the glycoproteins is not clear (i.e., are they released from the tumor itself, adjacent stromal tissue, or are they products of immune or inflammatory response?). One system used human breast carcinoma cells that were ectopically transplanted in an immunodeficient mouse and human glycoprotein marker sequences were identified during tumor progression [100]. It is clear, therefore, that released or secreted tumor glycoproteins can be detected in sera. There is also evidence, however, that pancreatic cancer cells, when co-cultured with hepatic cells, can cause production of specific glycoforms of haptoglobin from the hepatic cells [81]. The mechanisms that cause potential glycoform markers to be found in sera are clearly complex, and their elucidation will clearly expand our knowledge of the regulation of glycosylation.

Measurement of core fucosylated liver proteins in the sera of patients with HCC is clearly the foremost example of the application of glycoproteomics to identify a cancer marker and to develop an assay that will significantly aid in cancer diagnosis. The approval by the US Food and Drug Administration (FDA) of a cancer glycoprotein glycoform marker, core-fucosylated alpha-fetoprotein, represents a significant step forward for the identification and development of glycoform markers for other cancers. This approval should serve as stimulus for funding in this research area by both government and the private sector. A formidable challenge that impacts all glycomics marker discoveries is to develop reliable clinical assay systems for detection of specific cancer glycoprotein glycoform such that large numbers of samples can be assayed during clinical trials. Complicated sample preparation and time-consuming analytical procedures must be reduced and adapted so that analyses can be performed in a specialized clinical laboratory. Many of the potential glycoprotein markers of interest will likely be associated or aggregated with other protein components of serum, causing difficulties with standard types of ELISA assays, however. Ultimately, in order to achieve an assay with both high specificity and high sensitivity, it will likely be necessary to measure multiple markers in a "multiplex" type of format [149]. Lectins and a few antibodies are our only reagents to recognize specific glycan structures at the present time. A strong emphasis must now be placed on developing reagents that can recognize specific sugar structures (for example [150,151]) and, in conjunction with peptide-specific antibodies, be adapted to a clinical assay format. Many studies beginning over 30 years ago have shown that changes in glycan expression accompany oncogenic transformation, and in some cases these changes have been shown to have functional consequences. We now have the tools to identify glycan changes in cancer tissues and fluids and to connect these changes to specific glycoproteins and, in some cases, specific glycosylation sites. Developing assay systems to test the association of specific glycan or glycoprotein glycoform expression in sera with the presence of early stage cancer and putting these associations to test in large-scale trials are now major challenges for the field of cancer glycomics.

## Acknowledgments

I am grateful for comments and information provided by many of my colleagues, including Sen-itoh Hakomori, Akira Kobata, Naoyuki Taniguchi, Harry Schachter, and Nathan Sharon.

## References

1. Kobata A, Amano J. Altered glycosylation of proteins produced by malignant cells, and application for the diagnosis and immunotherapy of tumours. *Immunol Cell Biol* 2005;**83**(4):429–39.

2. Hakomori S. Glycosylation defining cancer malignancy: new wine in an old bottle. *Proc Natl Acad Sci USA* 2002;**99**(16):10231–3.

3. Hoefner DM. Serum tumor markers. Part I: Clinical utility. *MLO Med Lab Obs* 2005;**37**(12):20, 22, 24.

4. Magnani JL, Nilsson B, Brockhaus M, Zopf D, Steplewski Z, Koprowski H, Ginsburg V. A monoclonal antibody-defined antigen associated with gastrointestinal cancer is a ganglioside containing sialylated lacto-N-fucopentaose II. *J Biol Chem* 1982;**257**(23):14365–9.

5. Thompson IM, Ankerst DP, Chi C, Lucia MS, Goodman PJ, Crowley JJ, Parnes HL, Coltman CA Jr. Operating characteristics of prostate-specific antigen in men with an initial PSA level of 3.0 ng/ml or lower. *JAMA* 2005;**294**(1):66–70.

6. Sterling RK, Jeffers L, Gordon F, Sherman M, Venook AP, Reddy KR, Satomura S, Schwartz ME. Clinical utility of AFP-L3% measurement in North American patients with HCV-related cirrhosis. *Am J Gastroenterol* 2007;**102**(10):2196–205.

7. Masamune H, Yosizawa Z, Oh-Uti K, Matuda Y, Masudawa A. Biochemical studies on carbohydrates. CLVI. On the sugar components of the hexosamine-containing carbohydrates from gastric cancers, normal human gastric mucosa and human liver and of the glacial acetic acid-soluble proteins from those tissues as well as a metastasis in liver of gastric cancer. *Tohoku J Exp Med* 1952;**56**(1–2):37–42.

8. Masamune H, Yosizawa Z, Masukawa A. Comparison of the group carbohydrate of gastric cancer with the corresponding carbohydrate of gastric mucosa. *Tohoku J Exp Med* 1953;**58**(3–4):381–98.

9. Davidsohn I, Ni LY. Loss of isoantigens A, B, and H in carcinoma of the lung. *Am J Pathol* 1969;**57**(2):307–34.

10. Davidsohn I, Ni LY, Stejskal R. Tissue isoantigens A,B, and H in carcinoma of the stomach. *Arch Pathol* 1971;**92**(6):456–64.

11. Davidsohn I, Ni LY, Stejskal R. Tissue isoantigens A, B, and H in carcinoma of the pancreas. *Cancer Res* 1971;**31**(9):1244–50.

12. Lee JS, Ro JY, Sahin AA, Hong WK, Brown BW, Mountain CF, Hittelman WN. Expression of blood-group antigen A—a favorable prognostic factor in non-small-cell lung cancer. *N Engl J Med* 1991;**324**(16):1084–90.

13. Aub JC, Tieslau C, Lankester A. Reactions of normal and tumor cell surfaces to enzymes. I. Wheat-germ lipase and associated mucopolysaccharides. *Proc Natl Acad Sci USA* 1963;**50**:613–19.

14. Aub JC, Sanford BH, Wang LH. Reactions of normal and leukemic cell surfaces to a wheat germ agglutinin. *Proc Natl Acad Sci USA* 1965;**54**(2):400–2.

15. Aub JC, Sanford BH, Cote MN. Studies on reactivity of tumor and normal cells to a wheat germ agglutinin. *Proc Natl Acad Sci USA* 1965;**54**(2):396–9.

16. Sharon N, Lis H. Lectins: cell-agglutinating and sugar-specific proteins. *Science* 1972;**177**(53):949–59.

17. Hakomori SI, Murakami WT. Glycolipids of hamster fibroblasts and derived malignant-transformed cell lines. *Proc Natl Acad Sci USA* 1968;**59**(1):254–61.

18. Brady RO, Borek C, Bradley RM. Composition and synthesis of gangliosides in rat hepatocyte and hepatoma cell lines. *J Biol Chem* 1969;**244**(23):6552–4.

19. Mora PT, Brady RO, Bradley RM, McFarland VW. Gangliosides in DNA virus-transformed and spontaneously transformed tumorigenic mouse cell lines. *Proc Natl Acad Sci USA* 1969;**63**(4):1290–6.

20. Hakomori S. Cell density-dependent changes of glycolipid concentrations in fibroblasts, and loss of this response in virus-transformed cells. *Proc Natl Acad Sci USA* 1970;**67**(4):1741–7.

21. Hakomori S. Tumor malignancy defined by aberrant glycosylation and sphingo(glyco)lipid metabolism. *Cancer Res* 1996;**56**(23):5309–18.

22. Hakomori S. Bifunctional role of glycosphingolipids. Modulators for transmembrane signaling and mediators for cellular interactions. *J Biol Chem* 1990;**265**(31):18713–16.

23. Osorio M, Gracia E, Rodríguez E, Saurez G, Arango Mdel C, Noris E, Torriella A, Joan A, Gómez E, Anasagasti L, González JL, Melgares Mde L, Torres I, González J, Alonso D, Rengifo E, Carr A, Pérez R, Fernández LE. Heterophilic NeuGcGM3 ganglioside cancer vaccine in advanced melanoma patients: results of a Phase Ib/IIa study. *Cancer Biol Ther* 2008;**7**(4):488–95.

24. Wu HC, Meezan E, Black PH, Robbins PW, et al. Comparative studies on the carbohydrate-containing membrane components of normal and virus-transformed mouse fibroblasts. I. Glucosamine-labeling patterns in 3T3, spontaneously transformed 3T3, and SV-40-transformed 3T3 cells. *Biochemistry* 1969;**8**(6):2509–17.

25. Meezan E, Wu HC, Black PH, Robbins PW. Comparative studies on the carbohydrate-containing membrane components of normal and virus-transformed mouse fibroblasts. II. Separation of glycoproteins and glycopeptides by Sephadex chromatography. *Biochemistry* 1969;**8**(6):2518–24.

26. Buck CA, Glick MC, Warren L. A comparative study of glycoproteins from the surface of control and rous sarcoma virus transformed hamster cells. *Biochem* 1970;**9**:4567–76.

27. Buck CA, Glick MC, Warren L. Glycopeptides from the surface of control and virus-transformed cells. *Science* 1971;**172**(979):169–71.

28. Glick MC, Rabinowitz Z, Sachs L. Surface membrane glycopeptides correlated with tumorigenesis. *Biochemistry* 1973;**12**(24):4864–9.

29. Glick MC, Rabinowitz Z, Sachs L. Surface membrane glycopeptides which coincide with virus transformation and tumorigenesis. *J Virol* 1974;**13**(5):967–74.

30. Ogata SI, Muramatsu T, Kobata A. New structural characteristic of the large glycopeptides from transformed cells. *Nature* 1976;**259**(5544):580–2.

31. Yamashita K, Ohkura T, Tachibana Y, Takasaki S, Kobata A. Comparative study of the oligosaccharides released from baby hamster kidney cells and their polyoma transformant by hydrazinolysis. *J Biol Chem* 1984;**259**(17):10834–40.

32. Yamashita K, Tachibana Y, Ohkura T, Kobata A. Enzymatic basis for the structural changes of asparagine-linked sugar chains of membrane glycoproteins of baby hamster kidney cells induced by polyoma transformation. *J Biol Chem* 1985;**260**(7):3963–9.

33. Pierce M, Arango J. Rous sarcoma virus-transformed baby hamster kidney cells express higher levels of asparagine-linked tri- and tetraantennary glycopeptides containing [GlcNAc-b(1,6)Man -a(1,6)Man] and poly-N-acetyllactosamine sequences than baby hamster kidney cells. *J Biol Chem* 1986;**261**:10772–7.

34. Arango J, Pierce M. Comparison of N-acetylglucosaminyltransferase V activities in Rous sarcoma-transfered baby hamster kidneys (RS-BHK) and BHK cells. *J Cell Biochem* 1988;**37**:225–31.

35. Shoreibah M, Hindsgaul O, Pierce M. Purification and characterization of N-acetylglucosaminyltransferase V from rat kidney. *J Biol Chem* 1992;**267**:2920–7.

36. Shoreibah M, Perng GS, Adler B, Weinstein J, Basu R, Cupples R, Wen D, Browne JK, Buckhaults P, Fregien N, Pierce M. Isolation, characterization, and expression of a cDNA encoding N-acetylglucosaminyltransferase V. *J Biol Chem* 1993;**268**(21):15381–5.

37. Buckhaults P, Chen L, Fregien N, Pierce M. Transcriptional regulation of N-acetylglucosaminyltransferase V by the src oncogene. *J Biol Chem* 1997;**272**:19575–81.

38. Saito H, Nishikawa A, Gu J, Ihara Y, Soejima H, Wada Y, Sekiya C, Niikawa N, Taniguchi N. cDNA cloning and chromosomal mapping of human N-acetylglucosaminyltransferase V+. *Biochem Biophys Res Commun* 1994;**198**(1):318–27.

39. Kang R, Saito H, Ihara Y, Miyoshi E, Koyama N, Sheng Y, Taniguchi N. Transcriptional regulation of the N-acetylglucosaminyltransferase V gene in human bile duct carcinoma cells (HuCC-T1) is mediated by Ets-1. *J Biol Chem* 1996;**271**(43):26706–12.

40. Zhao YY, Takahashi M, Gu JG, Miyoshi E, Matsumoto A, Kitazume S, Taniguchi N. Functional roles of N-glycans in cell signaling and cell adhesion in cancer. *Cancer Sci* 2008;**99**(7):1304–10.

41. Seales EC, Jurado GA, Brunson BA, Wakefield JK, Frost AR, Bellis SL. Hypersialylation of beta1 integrins, observed in colon adenocarcinoma, may contribute to cancer progression by up-regulating cell motility. *Cancer Res* 2005;**65**(11):4645–52.

42. André S, Sanchez-Ruderisch H, Nakagawa H, Buchholz M, Kopitz J, Forberich P, Kemmner W, Böck C, Deguchi K, Detjen KM, Wiedenmann B, von Knebel Doeberitz M, Gress TM, Nishimura S, Rosewicz S, Gabius HJ. Tumor suppressor p16INK4a—modulator of glycomic profile and galectin-1 expression to increase susceptibility to carbohydrate-dependent induction of anoikis in pancreatic carcinoma cells. *FEBS J* 2007;**274**(13):3233–56.

43. Guo HB, Liu F, Zhao JH, Chen HL. Down-regulation of N-acetylglucosaminyltransferase V by tumorigenesis- or metastasis-suppressor gene and its relation to metastatic potential of human hepatocarcinoma cells. *J Cell Biochem* 2000;**79**(3):370–85.

44. Fuster MM, Esko JD. The sweet and sour of cancer: glycans as novel therapeutic targets. *Nat Rev Cancer* 2005;**5**(7):526–42.

45. Kim YJ, Borsig L, Varki NM, Varki A. P-selectin deficiency attenuates tumor growth and metastasis. *Proc Natl Acad Sci USA* 1998;**95**(16):9325–30.

46. Nakamori S, Nishihara S, Ikehara Y, Nagano H, Dono K, Sakon M, Narimatsu H, Monden M. Molecular mechanism involved in increased expression of sialyl Lewis antigens in ductal carcinoma of the pancreas. *J Exp Clin Cancer Res* 1999;**18**(3):425–32.

47. Stevenson JL, Choi SH, Varki A. Differential metastasis inhibition by clinically relevant levels of heparins—correlation with selectin inhibition, not antithrombotic activity. *Clin Cancer Res* 2005;**11**(19 Pt 1):7003–11.

48. Varki A, Kannagi R, Toole BP, et al. Glycosylation changes in cancer. In: Varki A, editor. *Essentials of Glycobiology*. Cold Spring Harbor, New York: Cold Spring Harbor Press; 2008, p. 617–32.

49. Endo T, Nishimura R, Kawano T, Mochizuki M, Kobata A. Structural differences found in the asparagine-linked sugar chains of human chorionic gonadotropins purified from the urine of patients with invasive mole and with choriocarcinoma. *Cancer Res* 1987;**47**(19):5242–5.

50. Valmu L, Alfthan H, Hotakainen K, Birken S, Stenman UH. Site-specific glycan analysis of human chorionic gonadotropin beta-subunit from malignancies and pregnancy by liquid chromatography–electrospray mass spectrometry. *Glycobiology* 2006;**16**(12):1207–18.

51. Yamashita K, Kobata A. Cancer cells and metastasis. In: Montreuil J, Vliegenthart JFG, Schacter H, editors. *Glycoproteins and Disease.* Elsevier Science B.V.; 1996, p. 229–39.

52. Ihara S, Miyoshi E, Ko JH, Murata K, Nakahara S, Honke K, Dickson RB, Lin CY, Taniguchi N. Prometastatic effect of N-acetylglucosaminyltransferase V is due to modification and stabilization of active matriptase by adding beta 1–6 GlcNAc branching. *J Biol Chem* 2002;**277**(19):16960–7.

53. Ihara S, Miyoshi E, Nakahara S, Sakiyama H, Ihara H, Akinaga A, Honke K, Dickson RB, Lin CY, Taniguchi N. Addition of beta1–6 GlcNAc branching to the oligosaccharide attached to Asn 772 in the serine protease domain of matriptase plays a pivotal role in its stability and resistance against trypsin. *Glycobiology* 2004;**14**(2):139–46.

54. Taniguchi N, Ekuni A, Ko JH, Miyoshi E, Ikeda Y, Ihara Y, Nishikawa A, Honke K, Takahashi M. A glycomic approach to the identification and characterization of glycoprotein function in cells transfected with glycosyltransferase genes. *Proteomics* 2001;**1**(2):239–47.

55. Ito Y, Akinaga A, Yamanaka K, Nakagawa T, Kondo A, Dickson RB, Lin CY, Miyauchi A, Taniguchi N, Miyoshi E. Co-expression of matriptase and N-acetylglucosaminyltransferase V in thyroid cancer tissues–its possible role in prolonged stability in vivo by aberrant glycosylation. *Glycobiology* 2006;**16**(5):368–74.

56. Barrabés S, Pagès-Pons L, Radcliffe CM, Tabarés G, Fort E, Royle L, Harvey DJ, Moenner M, Dwek RA, Rudd PM, De Llorens R, Peracaula R. Glycosylation of serum ribonuclease 1 indicates a major endothelial origin and reveals an increase in core fucosylation in pancreatic cancer. *Glycobiology* 2007;**17**(4):388–400.

57. Kui Wong N, Easton RL, Panico M, Sutton-Smith M, Morrison JC, Lattanzio FA, Morris HR, Clark GF, Dell A, Patankar MS. Characterization of the oligosaccharides associated with the human ovarian tumor marker CA125. *J Biol Chem* 2003;**278**(31):28619–34.

58. Wagner U, Köhler S, Reinartz S, Giffels P, Huober J, Renke K, Schlebusch H, Biersack HJ, Möbus V, Kreienberg R, Bauknecht T, Krebs D, Wallwiener D. Immunological consolidation of ovarian carcinoma recurrences with monoclonal anti-idiotype antibody ACA125: immune responses and survival in palliative treatment. See The biology behind: K. A. Foon and M. Bhattacharya-Chatterjee, Are solid tumor anti-idiotype vaccines ready for prime time? *Clin Cancer Res* 2001;**7**:1112–5. *Clin Cancer Res* 2001;**7**(5):1154–62.

59. Zhang H, Chan DW. Cancer biomarker discovery in plasma using a tissue-targeted proteomic approach. *Cancer Epidemiol Biomarkers Prev* 2007;**16**(10):1915–17.

60. Comunale MA, Lowman M, Long RE, Krakover J, Philip R, Seeholzer S, Evans AA, Hann HW, Block TM, Mehta AS. Proteomic analysis of serum associated fucosylated glycoproteins in the development of primary hepatocellular carcinoma. *J Proteome Res* 2006;**5**(2):308–15.

61. Marrero JA, Romano PR, Nikolaeva O, Steel L, Mehta A, Fimmel CJ, Comunale MA, D'Amelio A, Lok AS, Block TM. GP73, a resident Golgi glycoprotein, is a novel serum marker for hepatocellular carcinoma. *J Hepatol* 2005;**43**(6):1007–12.

62. Block TM, Comunale MA, Lowman M, Steel LF, Romano PR, Fimmel C, Tennant BC, London WT, Evans AA, Blumberg BS, Dwek RA, Mattu TS, Mehta AS. Use of targeted glycoproteomics to identify serum glycoproteins that correlate with liver cancer in woodchucks and humans. *Proc Natl Acad Sci USA* 2005;**102**(3):779–84.

63. Drake RR, Schwegler EE, Malik G, Diaz J, Block T, Mehta A, Semmes OJ. Lectin capture strategies combined with mass spectrometry for the discovery of serum glycoprotein biomarkers. *Mol Cell Proteomics* 2006;**5**(10):1957–67.

64. Taketa K, Endo Y, Sekiya C, Tanikawa K, Koji T, Taga H, Satomura S, Matsuura S, Kawai T, Hirai H. A collaborative study for the evaluation of lectin-reactive alpha-fetoproteins in early detection of hepatocellular carcinoma. *Cancer Res* 1993;**53**(22):5419–23.

65. Naitoh A, Aoyagi Y, Asakura H. Highly enhanced fucosylation of serum glycoproteins in patients with hepatocellular carcinoma. *J Gastroenterol Hepatol* 1999;**14**(5):436–45.

66. Aoyagi Y, Isokawa O, Suda T, Watanabe M, Suzuki Y, Asakura H. The fucosylation index of alpha-fetoprotein as a possible prognostic indicator for patients with hepatocellular carcinoma. *Cancer* 1998;**83**(10):2076–82.

67. Aoyagi Y, Suzuki Y, Igarashi K, Yokota T, Mori S, Suda T, Naitoh A, Isemura M, Asakura H. Highly enhanced fucosylation of alpha-fetoprotein in patients with germ cell tumor. *Cancer* 1993;**72**(2):615–18.

68. Breborowicz J, Mackiewicz A, Breborowicz D. Microheterogeneity of alpha-fetoprotein in patient serum as demonstrated by lectin affino-electrophoresis. *Scand J Immunol* 1981;**14**(1):15–20.

69. Taketa K, Sekiya C, Namiki M, Akamatsu K, Ohta Y, Endo Y, Kosaka K. Lectin-reactive profiles of alpha-fetoprotein characterizing hepatocellular carcinoma and related conditions. *Gastroenterology* 1990;**99**(2):508–18.

70. Aoyagi Y, Suzuki Y, Igarashi K, Saitoh A, Oguro M, Yokota T, Mori S, Nomoto M, Isemura M, Asakura H. The usefulness of simultaneous determinations of glucosaminylation and fucosylation indices of alpha-fetoprotein in the differential diagnosis of neoplastic diseases of the liver. *Cancer* 1991;**67**(9):2390–4.

71. Aoyagi Y, Suzuki Y, Igarashi K, Saitoh A, Oguro M, Yokota T, Mori S, Suda T, Isemura M, Asakura H. Carbohydrate structures of human alpha-fetoprotein of patients with hepatocellular carcinoma: presence of fucosylated and non-fucosylated triantennary glycans. *Br J Cancer* 1993;**67**(3):486–92.

72. Sato Y, Nakata K, Kato Y, Shima M, Ishii N, Koji T, Taketa K, Endo Y, Nagataki S. Early recognition of hepatocellular carcinoma based on altered profiles of alpha-fetoprotein. *N Engl J Med* 1993;**328**(25):1802–6.

73. Yamashita F, Tanaka M, Satomura S, Tanikawa K. Prognostic significance of *Lens culinaris* agglutinin A-reactive alpha-fetoprotein in small hepatocellular carcinomas. *Gastroenterology* 1996;**111**(4): 996–1001.

74. Okuyama N, Ide Y, Nakano M, Nakagawa T, Yamanaka K, Moriwaki K, Murata K, Ohigashi H, Yokoyama S, Eguchi H, Ishikawa O, Ito T, Kato M, Kasahara A, Kawano S, Gu J, Taniguchi N, Miyoshi E. Fucosylated haptoglobin is a novel marker for pancreatic cancer: a detailed analysis of the oligosaccharide structure and a possible mechanism for fucosylation. *Int J Cancer* 2006;**118**(11):2803–8.

75. Ang IL, Poon TC, Lai PB, Chan AT, Ngai SM, Hui AY, Johnson PJ, Sung JJ. Study of serum haptoglobin and its glycoforms in the diagnosis of hepatocellular carcinoma: a glycoproteomic approach. *J Proteome Res* 2006;**5**(10):2691–700.

76. Noda K, Miyoshi E, Uozumi N, Yanagidani S, Ikeda Y, Gao C, Suzuki K, Yoshihara H, Yoshikawa K, Kawano K, Hayashi N, Hori M, Taniguchi N. Gene expression of alpha1–6 fucosyltransferase in human hepatoma tissues: a possible implication for increased fucosylation of alpha-fetoprotein. *Hepatology* 1998;**28**(4):944–52.

77. Moriwaki K, Noda K, Nakagawa T, Asahi M, Yoshihara H, Taniguchi N, Hayashi N, Miyoshi E. A high expression of GDP-fucose transporter in hepatocellular carcinoma is a key factor for increases in fucosylation. *Glycobiology* 2007;**17**(12):1311–20.

78. Nakagawa T, Uozumi N, Nakano M, Mizuno-Horikawa Y, Okuyama N, Taguchi T, Gu J, Kondo A, Taniguchi N, Miyoshi E. Fucosylation of N-glycans regulates the secretion of hepatic glycoproteins into bile ducts. *J Biol Chem* 2006;**281**(40):29797–806.

79. Nakano M, Nakagawa T, Ito T, Kitada T, Hijioka T, Kasahara A, Tajiri M, Wada Y, Taniguchi N, Miyoshi E. Site-specific analysis of N-glycans on haptoglobin in sera of patients with pancreatic cancer: a novel approach for the development of tumor markers. *Int J Cancer* 2008;**122**(10):2301–9.

80. Fujimura T, Shinohara Y, Tissot B, Pang PC, Kurogochi M, Saito S, Arai Y, Sadilek M, Murayama K, Dell A, Nishimura S, Hakomori SI. Glycosylation status of haptoglobin in sera of patients with prostate cancer vs. benign prostate disease or normal subjects. *Int J Cancer* 2008;**122**(1):39–49.

81. Miyoshi E, Nakano M. Fucosylated haptoglobin is a novel marker for pancreatic cancer: detailed analyses of oligosaccharide structures. *Proteomics* 2008;**8**(16):3257–62.

82. Kyselova Z, Mechref Y, Al Bataineh MM, Dobrolecki LE, Hickey RJ, Vinson J, Sweeney CJ, Novotny MV. Alterations in the serum glycome due to metastatic prostate cancer. *J Proteome Res* 2007;**6**(5):1822–32.

83. Kang P, Mechref Y, Klouckova I, Novotny MV. Solid-phase permethylation of glycans for mass spectrometric analysis. *Rapid Commun Mass Spectrom* 2005;**19**(23):3421–8.

84. Ressom HW, Varghese RS, Goldman L, An Y, Loffredo CA, Abdel-Hamid M, Kyselova Z, Mechref Y, Novotny M, Drake SK, Goldman R. Analysis of MALDI-TOF mass spectrometry data for discovery of peptide and glycan biomarkers of hepatocellular carcinoma. *J Proteome Res* 2008;**7**(2):603–10.

85. Isailovic D, Kurulugama RT, Plasencia MD, Stokes ST, Kyselova Z, Goldman R, Mechref Y, Novotny MV, Clemmer DE. Profiling of human serum glycans associated with liver cancer and cirrhosis by IMS-MS. *J Proteome Res* 2008;**7**(3):1109–17.

86. Kyselova Z, Mechref Y, Kang P, Goetz JA, Dobrolecki LE, Sledge GW, Schnaper L, Hickey RJ, Malkas LH, Novotny MV. Breast cancer diagnosis and prognosis through quantitative measurements of serum glycan profiles. *Clin Chem* 2008;**54**(7):1166–75.

87. Fernandes B, Sagman U, Auger M, Demetrio M, Dennis JW. Beta 1–6 branched oligosaccharides as a marker of tumor progression in human breast and colon neoplasia. *Cancer Res* 1991;**51**:718–23.

88. Goetz JA, Mechref Y, Kang P, Jeng MH, Novotny MV. Glycomic profiling of invasive and non-invasive breast cancer cells. *Glycoconj J* 2009;**26**:117–31.

89. Prien JM, Huysentruyt LC, Ashline DJ, Lapadula AJ, Seyfried TN, Reinhold VN. Differentiating N-linked glycan structural isomers in metastatic and nonmetastatic tumor cells using sequential mass spectrometry. *Glycobiology* 2008;**18**(5):353–66.

90. Harvey DJ, Royle L, Radcliffe CM, Rudd PM, Dwek RA. Structural and quantitative analysis of N-linked glycans by matrix-assisted laser desorption ionization and negative ion nanospray mass spectrometry. *Anal Biochem* 2008;**376**(1):44–60.

91. Tabarés G, Radcliffe CM, Barrabés S, Ramírez M, Aleixandre RN, Hoesel W, Dwek RA, Rudd PM, Peracaula R, de Llorens R. Different glycan structures in prostate-specific antigen from prostate cancer sera in relation to seminal plasma PSA. *Glycobiology* 2006;**16**(2):132–45.

92. Saldova R, Royle L, Radcliffe CM, Abd Hamid UM, Evans R, Arnold JN, Banks RE, Hutson R, Harvey DJ, Antrobus R, Petrescu SM, Dwek RA, Rudd PM. Ovarian cancer is associated with changes in glycosylation in both acute-phase proteins and IgG. *Glycobiology* 2007;**17**(12):1344–56.

93. Gornik O, Royle L, Harvey DJ, Radcliffe CM, Saldova R, Dwek RA, Rudd P, Lauc G. Changes of serum glycans during sepsis and acute pancreatitis. *Glycobiology* 2007;**17**(12):1321–32.

94. Saldova R, Wormald MR, Dwek RA, Rudd PM. Glycosylation changes on serum glycoproteins in ovarian cancer may contribute to disease pathogenesis. *Disease Markers* 2008;**25**:1–14.

95. Storr SJ, Royle L, Chapman CJ, Hamid UM, Robertson JF, Murray A, Dwek RA, Rudd PM. The O-linked glycosylation of secretory/shed MUC1 from an advanced breast cancer patient's serum. *Glycobiology* 2008;**18**(6):456–62.

96. Arnold JN, Saldova R, Hamid UM, Rudd PM. Evaluation of the serum N-linked glycome for the diagnosis of cancer and chronic inflammation. *Proteomics* 2008;**8**(16):3284–93.

97. Knezevi A, Polasek O, Gornik O, Rudan I, Campbell H, Hayward C, Wright A, Kolcic I, O'Donoghue N, Bones J, Rudd PM, Lauc G. Variability, heritability and environmental determinants of human plasma N-glycome. *J Proteome Res* 2009;**8**(2):694–701.

98. Wang Y, Wu SL, Hancock WS. Approaches to the study of N-linked glycoproteins in human plasma using lectin affinity chromatography and nano-HPLC coupled to electrospray linear ion trap—Fourier transform mass spectrometry. *Glycobiology* 2006;**16**(6):514–23.

99. Ralin DW, Dultz SC, Silver JE, Travis JC, Kullolli M, Hancock WS, Hincapie M. Kinetic analysis of glycoprotein-lectin interactions by label free internal reflection. *Clin Proteomics* 2008;**4**:37–46.

100. Orazine CI, Hincapie M, Hancock WS, Hattersley M, Hanke JH. A proteomic analysis of the plasma glycoproteins of a MCF-7 mouse xenograft: a model system for the detection of tumor markers. *J Proteome Res* 2008;**7**(4):1542–54.

101. Zhao J, Qiu W, Simeone DM, Lubman DM. N-linked glycosylation profiling of pancreatic cancer serum using capillary liquid phase separation coupled with mass spectrometric analysis. *J Proteome Res* 2007;**6**(3):1126–38.

102. Merkle RK, Cummings RD. Lectin affinity chromatography of glycopeptides. *Methods Enzymol* 1987;**138**:232–59.

103. Qiu Y, Patwa TH, Xu L, Shedden K, Misek DE, Tuck M, Jin G, Ruffin MT, Turgeon DK, Synal S, Bresalier R, Marcon N, Brenner DE, Lubman DM. Plasma glycoprotein profiling for colorectal cancer biomarker identification by lectin glycoarray and lectin blot. *J Proteome Res* 2008;**7**(4):1693–703.

104. Kreunin P, Zhao J, Rosser C, Urquidi V, Lubman DM, Goodison S. Bladder cancer associated glycoprotein signatures revealed by urinary proteomic profiling. *J Proteome Res* 2007;**6**(7):2631–9.

105. Kim YS, Son OL, Lee JY, Kim SH, Oh S, Lee YS, Kim CH, Yoo JS, Lee JH, Miyoshi E, Taniguchi N, Hanash SM, Yoo HS, Ko JH. Lectin precipitation using phytohemagglutinin-L(4) coupled to avidin-agarose for serological biomarker discovery in colorectal cancer. *Proteomics* 2008;**8**(16):3229–35.

106. Soltermann A, Ossola R, Kilgus-Hawelski S, von Eckardstein A, Suter T, Aebersold R, Moch H. N-glycoprotein profiling of lung adenocarcinoma pleural effusions by shotgun proteomics. *Cancer* 2008;**114**(2):124–33.

107. Zhang H, Li XJ, Martin DB, Aebersold R. Identification and quantification of N-linked glycoproteins using hydrazide chemistry, stable isotope labeling and mass spectrometry. *Nat Biotechnol* 2003;**21**(6):660–6.

108. Zhang 2nd H, Yi EC, Li XJ, Mallick P, Kelly-Spratt KS, Masselon CD, Camp DG, Smith RD, Kemp CJ, Aebersold R. High throughput quantitative analysis of serum proteins using glycopeptide capture and liquid chromatography mass spectrometry. *Mol Cell Proteomics* 2005;**4**(2):144–55.

109. An HJ, Miyamoto S, Lancaster KS, Kirmiz C, Li B, Lam KS, Leiserowitz GS, Lebrilla CB. Profiling of glycans in serum for the discovery of potential biomarkers for ovarian cancer. *J Proteome Res* 2006;**5**(7):1626–35.

110. Kirmiz C, Li B, An HJ, Clowers BH, Chew HK, Lam KS, Ferrige A, Alecio R, Borowsky AD, Sulaimon S, Lebrilla CB, Miyamoto S. A serum glycomics approach to breast cancer biomarkers. *Mol Cell Proteomics* 2007;**6**(1):43–55.

111. Leiserowitz GS, Lebrilla C, Miyamoto S, An HJ, Duong H, Kirmiz C, Li B, Liu H, Lam KS. Glycomics analysis of serum: a potential new biomarker for ovarian cancer? *Int J Gynecol Cancer* 2008;**18**(3):470–5.

112. Ye J, Liu H, Kirmiz C, Lebrilla CB, Rocke DM. On the analysis of glycomics mass spectrometry data via the regularized area under the ROC curve. *BMC Bioinformatics* 2007;**8**:477.

113. Williams TI, Saggese DA, Toups KL, Frahm JL, An HJ, Li B, Lebrilla CB, Muddiman DC. Investigations with O-linked protein glycosylations by matrix-assisted laser desorption/ionization Fourier transform ion cyclotron resonance mass spectrometry. *J Mass Spectrom* 2008;**43**:1215–23.

114. Li B, An HJ, Kirmiz C, Lebrilla CB, Lam KS, Miyamoto S. Glycoproteomic analyses of ovarian cancer cell lines and sera from ovarian cancer patients show distinct glycosylation changes in individual proteins. *J Proteome Res* 2008;**7**(9):3776–88.

115. Hollingsworth MA, Swanson BJ. Mucins in cancer: protection and control of the cell surface. *Nat Rev Cancer* 2004;**4**(1):45–60.

116. Santos-Silva F, Fonseca A, Caffrey T, Carvalho F, Mesquita P, Reis C, Almeida R, David L, Hollingsworth MA. Thomsen-Friedenreich antigen expression in gastric carcinomas is associated with MUC1 mucin VNTR polymorphism. *Glycobiology* 2005;**15**(5):511–17.

117. Wu AM, Khoo KH, Yu SY, Yang Z, Kannagi R, Watkins WM. Glycomic mapping of pseudomucinous human ovarian cyst glycoproteins: identification of Lewis and sialyl Lewis glycotopes. *Proteomics* 2007;**7**(20):3699–717.

118. Ju T, Lanneau GS, Gautam T, Wang Y, Xia B, Stowell SR, Willard MT, Wang W, Xia JY, Zuna RE, Laszik Z, Benbrook DM, Hanigan MH, Cummings RD. Human tumor antigens Tn and sialyl Tn arise from mutations in Cosmc. *Cancer Res* 2008;**68**(6):1636–46.

119. Iwai T, Kudo T, Kawamoto R, Kubota T, Togayachi A, Hiruma T, Okada T, Kawamoto T, Morozumi K, Narimatsu H. Core 3 synthase is down-regulated in colon carcinoma and profoundly suppresses the metastatic potential of carcinoma cells. *Proc Natl Acad Sci USA* 2005;**102**(12):4572–7.

120. Hashii N, Kawasaki N, Itoh S, Hyuga M, Kawanishi T, Hayakawa T. Glycomic/glycoproteomic analysis by liquid chromatography/mass spectrometry: analysis of glycan structural alteration in cells. *Proteomics* 2005;**5**(18):4665–72.

121. Kim YS, Hwang SY, Oh S, Sohn H, Kang HY, Lee JH, Cho EW, Kim JY, Yoo JS, Kim NS, Kim CH, Miyoshi E, Taniguchi N, Ko JH. Identification of target proteins of N-acetylglucosaminyl-transferase V and fucosyltransferase 8 in human gastric tissues by glycomic approach. *Proteomics* 2004;**4**(11):3353–8.

122. Kim YS, Hwang SY, Oh S, Sohn H, Kang HY, Lee JH, Cho EW, Kim JY, Yoo JS, Kim NS, Kim CH, Miyoshi E, Taniguchi N, Ko JH. Identification of target proteins of N-acetylglucosaminyl transferase V in human colon cancer and implications of protein tyrosine phosphatase kappa in enhanced cancer cell migration. *Proteomics* 2006;**6**(4):1187–91.

123. Kim YS, Hwang SY, Kang HY, Sohn H, Oh S, Kim JY, Yoo JS, Kim YH, Kim CH, Jeon JH, Lee JM, Kang HA, Miyoshi E, Taniguchi N, Yoo HS, Ko JH. Functional proteomics study reveals that N-Acetylglucosaminyltransferase V reinforces the invasive/metastatic potential of colon cancer through aberrant glycosylation on tissue inhibitor of metalloproteinase-1. *Mol Cell Proteomics* 2008;**7**(1):1–14.

124. Kim YS, Son OL, Lee JY, Kim SH, Oh S, Lee YS, Kim CH, Yoo JS, Lee JH, Miyoshi E, Taniguchi N, Hanash SM, Yoo HS, Ko JH. Lectin precipitation using phytohemagglutinin-L(4) coupled to avidin-agarose for serological biomarker discovery in colorectal cancer. *Proteomics* 2008;**16**:3229–35.

125. Wang Y, Ao X, Vuong H, Konanur M, Miller FR, Goodison S, Lubman DM. Membrane glycoproteins associated with breast tumor cell progression identified by a lectin affinity approach. *J Proteome Res* 2008;**7**(10):4313–25.

126. Vercoutter-Edouart AS, Slomianny MC, Dekeyzer-Beseme O, Haeuw JF, Michalski JC. Glycoproteomics and glycomics investigation of membrane N-glycosylproteins from human colon carcinoma cells. *Proteomics* 2008;**8**(16):3236–56.

127. Abbott KL, Aoki K, Lim JM, Porterfield M, Johnson R, O'Regan RM, Wells L, Tiemeyer M, Pierce M. Targeted glycoproteomic identification of biomarkers for human breast carcinoma. *J Proteome Res* 2008;**7**(4):1470–80.

128. Aoki K, Perlman M, Lim JM, Cantu R, Wells L, Tiemeyer M. Dynamic developmental elaboration of N-linked glycan complexity in the *Drosophila melanogaster* embryo. *J Biol Chem* 2007;**282**(12):9127–42.

129. Nairn AV, York WS, Harris K, Hall EM, Pierce JM, Moremen KW. Regulation of glycan structures in animal tissues: transcript profiling of glycan-related genes. *J Biol Chem* 2008;**283**(25):17298–313.

130. Abbott KL, Nairn AV, Hall EM, Horton MB, McDonald JF, Moremen KW, Dinulescu DM, Pierce M. Focused glycomic analysis of the N-linked glycan biosynthetic pathway in ovarian cancer. *Proteomics* 2008;**8**:3210–20.

131. Nan BC, Shao DM, Chen HL, Huang Y, Gu JX, Zhang YB, Wu ZG. Alteration of N-acetylglucosaminyltransferases in pancreatic carcinoma. *Glycoconj J* 1998;**15**(10):1033–7.

132. Chen S, LaRoche T, Hamelinck D, Bergsma D, Brenner D, Simeone D, Brand RE, Haab BB. Multiplexed analysis of glycan variation on native proteins captured by antibody microarrays. *Nat Methods* 2007;**4**(5):437–44.

133. Shafer MW, Mangold L, Partin AW, Haab BB. Antibody array profiling reveals serum TSP-1 as a marker to distinguish benign from malignant prostatic disease. *Prostate* 2007;**67**(3):255–67.

134. Chapman C, Murray A, Chakrabarti J, Thorpe A, Woolston C, Sahin U, Barnes A, Robertson J. Autoantibodies in breast cancer: their use as an aid to early diagnosis. *Ann Oncol* 2007;**18**(5):868–73.

135. Graves CR, Robertson JF, Murray A, Price MR, Chapman CJ. Malignancy-induced autoimmunity to MUC1: initial antibody characterization. *J Pept Res* 2005;**66**(6):357–63.

136. Vollmers HP, Brandlein S. Natural antibodies and cancer. *J Autoimmun* 2007;**29**(4):295–302.

137. Krueger P, Nitz C, Moore J, Foster R, Gelber O, Gelber C. Monoclonal antibody identifies a distinctive epitope expressed by human multiple myeloma cells. *J Immunother* 2001;**24**(4):334–44.

138. Krueger P, Nitz C, Foster R, MacDonald C, Gelber O, Lalehzadeh G, Goodson R, Winter J, Gelber C. A new small cell lung cancer (SCLC)-specific marker discovered through antigenic subtraction of neuroblastoma cells. *Cancer Immunol Immunother* 2003;**52**(6):367–77.

139. Brandlein S, Pohle T, Ruoff N, Wozniak E, Müller-Hermelink HK, Vollmers HP. Natural IgM antibodies and immunosurveillance mechanisms against epithelial cancer cells in humans. *Cancer Res* 2003;**63**(22):7995–8005.

140. Rauschert N, Brändlein S, Holzinger E, Hensel F, Müller-Hermelink HK, Vollmers HP. A new tumor-specific variant of GRP78 as target for antibody-based therapy. *Lab Invest* 2008;**88**(4):375–86.

141. Ariga T, Suetake K, Nakane M, Kubota M, Usuki S, Kawashima I, Yu RK. Glycosphingolipid antigens in neural tumor cell lines and anti-glycosphingolipid antibodies in sera of patients with neural tumors. *Neurosignals* 2008;**16**(2–3):226–34.

142. Ravindranath MH, Morton DL, Irie RF. An epitope common to gangliosides O-acetyl-GD3 and GD3 recognized by antibodies in melanoma patients after active specific immunotherapy. *Cancer Res* 1989;**49**(14):3891–7.

143. Adams EW, Ratner DM, Bokesch HR, McMahon JB, O'Keefe BR, Seeberger PH. Oligosaccharide and glycoprotein microarrays as tools in HIV glycobiology; glycan-dependent gp120/protein interactions. *Chem Biol* 2004;**11**(6):875–81.

144. Carroll GT, et al. Photons to illuminate the universe of sugar diversity through bioarrays. *Glycoconj J* 2008;**25**(1):5–10.

145. Wang D, Liu S, Trummer BJ, Deng C, Wang A. Carbohydrate microarrays for the recognition of cross-reactive molecular markers of microbes and host cells. *Nat Biotechnol* 2002;**20**(3):275–81.

146. Stevens J, Blixt O, Paulson JC, Wilson IA. Glycan microarray technologies: tools to survey host specificity of influenza viruses. *Nat Rev Microbiol* 2006;**4**(11):857–64.

147. Wang D, Carroll GT, Turro NJ, Koberstein JT, Kovác P, Saksena R, Adamo R, Herzenberg LA, Herzenberg LA, Steinman L. Photogenerated glycan arrays identify immunogenic sugar moieties of *Bacillus anthracis* exosporium. *Proteomics* 2007;**7**(2):180–4.

148. Ingale S, Wolfert MA, Gaekwad J, Buskas T, Boons GJ. Robust immune responses elicited by a fully synthetic three-component vaccine. *Nat Chem Biol* 2007;**3**(10):663–7.

149. Nick AM, Sood AK. The ROC 'n' role of the multiplex assay for early detection of ovarian cancer. *Nat Clin Pract Oncol* 2008;**5**(10):568–9.

150. Masud MM, Kuwahara M, Ozaki H, Sawai H. Sialyllactose-binding modified DNA aptamer bearing additional functionality by SELEX. *Bioorg Med Chem* 2004;**12**(5):1111–20.

151. Li M, Lin N, Huang Z, Du L, Altier C, Fang H, Wang B. Selecting aptamers for a glycoprotein through the incorporation of the boronic acid moiety. *J Am Chem Soc* 2008;**130**(38):12636–8.

# 17

# Introduction to Human Glycosylation Disorders

Hudson H. Freeze and Erik Eklund
*BURNHAM INSTITUTE FOR MEDICAL RESEARCH,*
*SANFORD CHILDREN'S HEALTH RESEARCH CENTER,*
*LA JOLLA, CALIFORNIA, USA*

## Introduction

The last decade witnessed an explosive discovery of human diseases caused by altered glycan biosynthesis—and this process is only just starting. Finding those first 35+ diseases required collaboration between basic scientists and clinicians and future progress depends on enhancing that alliance as we become increasingly interdependent on cross-feeding within scientific disciplines and with medical specialties.

The purpose of this review is to introduce the known human disorders, provide current information on the nature of the defects, how they are diagnosed and (rarely) treated. We will suggest where there are gaps in our understanding of the basic mechanisms, what tools that are needed, and strategies that basic scientists might adopt to position themselves to address the unanswered questions. Many excellent reviews provide broad perspectives of these diseases [1–8].

## Brief History and Nomenclature

### What is a Glycosylation Disorder?

Genetic defects have now been identified in each of the known human glycosylation pathways (Table 17.1). Twenty-five of these affect the *N*-glycosylation pathway [9]; some of those affect other pathways as well. Even defining a "glycosylation disorder" is tricky. Clearly, mutations in glycosyltransferases, processing enzymes, chaperones, precursor biosynthetic enzymes, and nucleotide sugar transporters fully qualify. However, recent additions include proteins needed for Golgi homeostasis and intracellular protein trafficking [10,11]. Some favor including other disorders long known by other names, such as galactosemia and hereditary fructose intolerance, since one of their consequences is altered glycosylation of some proteins. We will not define

**Handbook of Glycomics**
Copyright © 2009 by Elsevier Inc. All rights of reproduction in any form reserved.

Table 17.1 Summary of glycosylation-associated genes relevant to human disorders.

| Gene | Defect step | Type of defect | Human disorder |
|------|-------------|----------------|----------------|
| *Asparagine (N)-linked glycosylation* | | | |
| PMM2 | Man-6-P⇌Man-1-P | Monosaccharide activation | Phosphomannomutase deficiency (CDG-Ia) |
| MPI | Fructose-6-P⇌Man-6-P | Monosaccharide activation | Phosphomannose isomerase deficiency (CDG-Ib) |
| DPAGT1 | Dol-P + UDP-GlcNAc→GlcNAc-PP-Dol | Glycosyltransferase | N-Acetyl glucosaminyltransferase I deficiency (CDG-Ij) |
| HMT1 | GlcNAc$_2$-PP-Dol→Man$_1$GlcNAc$_2$-PP-Dol | Glycosyltransferase | Mannosyltransferase I deficiency (CDG-Ik) |
| ALG2 | Man$_1$GlcNAc$_2$-PP-Dol→Man$_2$GlcNAc$_2$-PP-Dol | Glycosyltransferase | Mannosyltransferase II deficiency (CDG-Ii) |
| RFT-1 | Flips Man$_5$GlcNAc$_2$-PP-Dol to ER lumen | Substrate localization | RFT1 deficiency (CDG-In) |
| NOT56L | Man$_5$GlcNAc$_2$-PP-Dol→Man$_6$GlcNAc$_2$-PP-Dol | Glycosyltransferase | Mannosyltransferase VI deficiency (CDG-Id) |
| DIBD1 | Man$_6$GlcNAc$_2$-PP-Dol→Man$_7$GlcNAc$_2$-PP-Dol and Man$_8$GlcNAc$_2$-PP-Dol→Man$_9$GlcNAc$_2$-PP-Dol | Glycosyltransferase | Mannosyltransferase VII/IX deficiency (CDG-IL) |
| ALG12 | Man$_7$GlcNAc$_2$-PP-Dol→Man$_8$GlcNAc$_2$-PP-Dol | Glycosyltransferase | Mannosyltransferase VIII deficiency (CDG-Ig) |
| ALG6 | Man$_9$GlcNAc$_2$-PP-Dol→Glc$_1$Man$_9$GlcNAc$_2$-PP-Dol | Glycosyltransferase | Glucosyltransferase I deficiency (CDG-Ic) |
| ALG8 | Glc$_1$Man$_9$GlcNAc$_2$-PP-Dol→Glc$_2$Man$_9$GlcNAc$_2$-PP-Dol | Glycosyltransferase | Glucosyltransferase II deficiency (CDG-Ih) |
| N33/TUSC3 | Glc$_3$Man$_9$GlcNAc$_2$-PP-Dol→Glc$_3$Man$_9$GlcNAc$_2$-protein | Glycosyltransferase | Oligosaccharyltransferase deficiency |
| IAP | Glc$_3$Man$_9$GlcNAc$_2$-PP-Dol→Glc$_3$Man$_9$GlcNAc$_2$-protein | Glycosyltransferase | Oligosaccharyltransferase deficiency |
| GLS1 | Glc$_3$Man$_9$GlcNAc$_2$-protein→Glc$_2$Man$_9$GlcNAc$_2$-protein | Glycosidase | Glucosidase I deficiency (CDG-IIb) |
| MGAT2 | GlcNAc-Man$_3$GlcNAc$_2$-protein→GlcNAc$_2$-Man$_3$GlcNAc$_2$-protein | Glycosyltransferase | N-Acetyl glucosaminyltransferase II deficiency (CDG-IIa) |
| GNPTAB | Man-glycan-protein→GlcNAc-P-Man-glycan-protein | Glycosyltransferase | I-Cell disease/mucolipidosis II and III |
| *O-linked glycosylation* | | | |
| GALNT3 | N-Acetyl galactosaminylation of some O-glycans | Glycosyltransferase | Familial tumoral calcinosis |
| COSMC | Impaired synthesis of T-antigen (Galβ1,3GalNAcα-Ser/Thr) | Chaperone | Tn syndrome |
| POMT1 | O-mannosyl glycan synthesis | Glycosyltransferase | WWS, type II lissencephaly, LGMD2K |
| POMT2 | O-mannosyl glycan synthesis | Glycosyltransferase | WWS, type II lissencephaly |
| POMGnT1 | O-mannosyl glycan synthesis | Glycosyltransferase | MEB, type II lissencephaly |
| FKTN | Glycosylation of α-dystroglycan | Unknown | Fukuyama CMD, WWS, LGMD2M |
| FKRP | Glycosylation of α-dystroglycan | Unknown | MDC1C, WWS, MEB, LGMD2I |

| Gene | Pathway/step | Function | Disorder |
|---|---|---|---|
| *LARGE* | Glycosylation of α-dystroglycan | Putative glycosyltransferase | MDC1D, WWS |
| *GYLTL1B* | Glycosylation of α-dystroglycan | Putative glycosyltransferase | Non detected |
| *LFNG* | Extension of O-fucosylated structures | Glycosyltransferase | Spondylocostal dysostosis type 3 |
| *B3GALTL* | Extension of O-fucosylated structures | Glycosyltransferase | Peter's plus syndrome |
| *Lipid glycosylation* | | | |
| *SIAT9* | Synthesis of GM3 (Sia2,3Galβ1,4Glc-ceramide) | Glycosyltransferase | Amish infantile-onset epilepsy |
| *PIGA* | GPI-anchor biosynthesis | Glycosyltransferase | Marchiafava–Micheli syndrome/PNH |
| *PIGM* | GPI-anchor biosynthesis | Glycosyltransferase | A hepatic and portal veins thrombosis with epilepsy syndrome |
| *Glycosaminoglycan biosynthesis* | | | |
| *B4GALT7* | Link region biosynthesis | Glycosyltransferase | Progeria variant of Ehlers–Danlos syndrome |
| *EXT1* | Heparan sulfate biosynthesis | Glycosyltransferase | Hereditary multiple exostosis |
| *EXT2* | Heparan sulfate biosynthesis | Glycosyltransferase | Hereditary multiple exostosis |
| *Two or more glycosylation pathways* | | | |
| *DPM1* | Synthesis of Dol-P-Man | Monosaccharide activation | Dol-P-Man synthase deficiency (CDG-Ie) |
| *MPDU1* | Utilization of Dol-P-Man and Dol-P-Glc | Substrate utilization | MPDU1 deficiency (CDG-If) |
| *DK1* | Phosphorylation of Dol | Lipid anchor activation | DK1 deficiency (CDG-Im) |
| *B4GALT1* | Galactosylation of N- and O-linked oligosaccharides | Glycosyltransferase | B4GALT1 deficiency (CDG-IId) |
| *SLC35A1* | Sialic acid transport | Nucleotide sugar transport | SLC35A1 deficiency (CDG-IIf) |
| *SLC35C1* | Fucose transport | Nucleotide sugar transport | SLC35C1 deficiency (CDG-IIc) |
| *COG1* | Retrograde Golgi transport/tethering | Golgi function | COG1 deficiency (CDG-IIg) |
| *COG7* | Retrograde Golgi transport/tethering | Golgi function | COG7 deficiency (CDG-IIe) |
| *COG8* | Retrograde Golgi transport/tethering | Golgi function | COG8 deficiency (CDG-IIh) |
| *ATP6V0A2* | pH regulation of Golgi | Golgi function | Autosomal recessive cutis laxa type II |
| *DTDST* | Sulfate ion transport | Ion transporter | Diastrophic dysplasia, achondrogenesis type IB, and others[a] |
| *ATPSK2* | PAPS synthesis | Substrate activation | Spondyloepimetaphyseal dysplasia |
| *GNE* | UDP-GlcNAc epimerase/kinase for de novo synthesis of sialic acid | Monosaccharide biosynthesis | Hereditary inclusion body myopathy-II (IBM2). Adult-onset myopathy |

WWS, Walker–Warburg syndrome; LGMD, limb girdle muscular dystrophy; MEB, muscle–eye–brain syndrome; CMD, congenital muscular dystrophy; PNH, paroxysmal nocturnal hemoglobinuria.

[a] For example, neonatal osseous dysplasia I and autosomal recessive multiple epiphyseal dysplasia.

"glycosylation disorder" here, but rather will present science and clinical features in a flexible framework. Certainly, additional genes and their more precise annotations will continue to refine our perspectives of glycosylation and consequences of its alterations.

## The Discovery of Glycosylation Disorders

I-Cell disease (mucolipidosis II and III), which results in the mistargeting of many lysosomal enzymes, can arguably be considered the first N-glycosylation pathway disorder to be identified (Table 17.1). Around the same time as the molecular basis of glycosylation was determined, several patients were described who were later shown to have deficiencies in the glycosylation of several serum glycoproteins. The basis of the disorder was not known, but the clinical appearance defined a "syndrome" that led to the name "carbohydrate-deficient glycoprotein syndrome" or CDGS. A second clinically distinct type of patient was discovered to have altered glycosylation, and this led to a designation of CDGS-II. As additional disorders were identified, the acronym changed to CDG—congenital disorder of glycosylation [12]. To help clinicians unfamiliar with glycosylation categorize their patients, an N-glycosylation-centric system was proposed. Disorders were divided into two groups: I and II. Group I was defined by mutations in any gene that affected the biosynthesis and transfer of a dolichol-bound glycan chain to its acceptor protein. Group II covered mutations in any genes affecting N-glycan processing—or mutations in any other glycosylation pathway. The specific mutated genes defined the type in order of their discovery (e.g., CDG-Ia, CDG-IIa, etc.). This non-comprehensive and biased nomenclature, however, could not easily accommodate the rapid pace of discovery. Moreover, the designation was of little practical value to clinicians seeing patients.

A revised nomenclature has been proposed under which defects are defined by the name of the gene with (CDG) as an addendum. Thus, CDG-Ia can now be called PMM2 (CDG-Ia). Informal groupings are based on the type(s) of glycans affected [13].

# Glycobiologists, Clinicians, and Geneticists Collaborate in Patient Diagnosis

Discovery of glycosylation disorders required a close cooperation between glycobiologists, physicians, and geneticists. Most glycosylation disorders were initially identified by biochemical approaches. These included incomplete or altered glycan structures, reduced enzymatic activity, or abnormal lectin or antibody binding. Glycan structural abnormalities helped to focus the search on most likely defective genes. Glycobiologists provided an understanding of glycan function, analytical techniques, together with cell-based and animal model systems for functional analysis. Clinicians trained in a variety of specialties provided detailed descriptions of their patients and patient-derived materials (serum, cells, biopsies, and autopsies). These assets generated insights into the pathophysiological impact of impaired glycosylation in multiple organ systems. Biochemical analysis provided the clues needed to identify the defective genes, and routine sequencing of gDNA or cDNA identified potential pathological mutations that were confirmed by their failure to rescue yeast or mammalian cell lines defective in the human homolog. At the same time, glycosylation-deficient patient cell lines were shown to become normal when transfected with adenovirus or lentivirus containing the normal human allele.

Geneticists have also discovered glycosylation disorders by applying techniques such as SNP analysis [14], and mapping to study sets of families with specific clinical phenotypes [15,16]. This focuses the search on a small region of the chromosome, and if pre-warned that patients express altered glycosylation, the search is further refined to focus on genes suspected of contributing to glycosylation. Clearly, it is important to show that the putative mutated genes actually alter glycosylation in one or more pathways. Improved efficiency and lower costs of large-scale DNA sequencing technology mean that future glycosylation defects are more likely to be found via this approach than by the traditional biochemical methods. Simply knowing the name of the gene is essential, but it is only the first step, and biochemical confirmation of the lesion remains essential. Clinical, biochemical, and genetic approaches complement and instruct each other. All will continue to be important.

## Spectrum of Disorders

Table 17.1 lists the currently known human glycosylation disorders. All major endoplasmic reticulum (ER)/Golgi-based pathways are represented. Defects limited to N-glycan biosynthesis or those in multiple pathways including N-glycosylation are the most numerous at 26. Mutations that affect O-mannosyl based glycan synthesis on α-dystroglycan account for 6 defective genes. Defects in the glycosaminoglycan biosynthetic pathway affect the core linkage region, chain elongation, and sulfation. The clinical spectrum of patients among these disorders is extremely diverse and clinical features cannot be used to diagnose the specific gene.

## N-Glycan Defects

### Biochemical Diagnosis

Most of these defects were identified by altered glycosylation of serum transferrin (Tf) or other serum/plasma glycoproteins such as α1-antitrypsin (AAT) [17]. Tf contains two N-glycosylation sites, mostly occupied by bi-antennary, disialylated glycans producing tetrasialo-Tf. Some molecules have tri-antennary, trisialylated chains. Absence of one or both chains (type I disorders) yields di- or asialo-Tf, while alterations in glycan processing produce incompletely sialylated species with 0–3 sialic acids (type II disorders). Differences in sialylation are detected by using, for example, isoelectric focusing [18], kits using ion-exchange chromatography [19], capillary electrophoresis [20], high-performance liquid chromatography (HPLC) [21], sodium dodecyl sulfate polyacrylamide gel electrophoresis (SDS-PAGE) [22], surface enhanced laser desorption/ionization time-of-flight mass spectrometry (SELDI-TOF-MS) [23], and matrix-assisted laser desorption/ionization mass spectrometry (MALDI-MS) [24]. Electrospray ionization mass spectrometry (ESI-MS) can also be used [25]. The latter method uses immunocapture and elution into the mass spectrometer, and is the most informative because it distinguishes absence of the glycan chain from altered glycan structures [26]. This method also identified patients with a combination of incomplete processing and absence of entire chains [27]. Separation methods that rely on charge or charge/mass ratio require digestion with sialidase to confirm that differences are due to sialic acid content and not amino acid polymorphisms.

N-Glycosylation defects also can be demonstrated by analysis of other serum glycoproteins such as AAT and α1-antichymotrypsin [28,29]. In type I CDG defects there can be increased serum levels of the lysosomal enzyme aspartylglucosamini-dase [30]. LC-MS/MS analysis of Tf and AAT showed that loss of glycosylation site occupancy is site specific and correlates with the clinical severity of CDG-I patients; more severe patients have a greater proportion of unoccupied N-glycosylation sites [31]. In Tf, the first N-glycosylation site is preferentially occupied, while AAT shows a decreasing preference for the first, third, and second N-glycosylation sites [31]. Another study suggested that patients showing lower glycosylation site occupancy correlate with increased severity [32]. In rare instances heterozygous parents may show slight increase in unoccupied sites.

Impaired Tf glycosylation site occupancy also occurs in uncontrolled galacto-semia [33,34], uncontrolled hereditary fructose intolerance [34], prolonged heavy alcohol consumption [35], and altered N-glycan processing in some acquired liver disorders [36]. The physiological basis and mechanism leading to hypoglycosylation in these pathologies is not known. Accumulation of fructose-1-P was proposed to inhibit phosphomannose isomerase and limit availability of mannose-6-phosphate (Man-6-P), but this has not been substantiated [37].

Not all defects can be identified by Tf analysis because some do not alter Tf sialylation. This includes GLS1 deficiency (CDG-IIb), the first α1,2 glucosidase in N-glycan processing, and SLC35C1 (CDG-IIc), one of the two Golgi GDP-fucose transporters. The most recently discovered defect, a deficiency in TUSC3, the Ost3/Ost6 homolog subunit of the oligosaccharyltransferse (OST) complex, does not alter Tf glycosylation [38,39], reinforcing the concept that multiple OST complexes differ in protein substrate selection and specificity [40]. No diagnostic proteins have been identified for the TUSC3 defect.

## Nature of the Defects

Defects involving primarily the N-glycan biosynthetic pathways include monosaccharide substrate activation (PMM2, MPI), glycosyltransferases (ALG6, NOT561, ALG8, ALG12, ALG2, DPAGT1, DIB1, and MGAT2), a protein used to flip the lipid-linked oligosaccharide (LLO) glycan from the cytoplasmic to the luminal side of the ER (RFT1), processing glycosidase (GLS1), and oligosaccharyltransferase subunits (TUSC3 and IAP).

## Defects in Multiple Pathways

Other defects were previously included in the CDG nomenclature, and while many were detected by an abnormal N-glycosylation of an indicator protein, they are not limited to the N-glycosylation pathway. Indeed, it is likely that some clinical manifestations are due to abnormalities in other glycosylation pathways or unrelated to abnormal glycosylation.

DPM1, the catalytic subunit for Dol-P-mannose synthase, provides a substrate for N- and O-linked glycans, GPI anchors, and C-mannosylation [41]. Dolichol kinase (DK1, also called CDG-Im) deficiency could potentially impair these pathways also [42]. MPDU1 (CDG-If) is needed for proper delivery or orientation of both Dol-P-Man and Dol-P-Glc for these pathways [43–45]. B4GALT1 is needed for β1,4Gal addition to serum glycoproteins, but it can also be used for O-glycans. Both the CMP-sialic acid transporter (SLC35A1) and the GDP-fucose transporter (SLC35C1) service multiple pathways.

Defects in a conserved oligomeric Golgi (COG) protein complex disrupt several pathways by altering intracellular trafficking and distribution of glycosylation-relevant proteins (transferases and nucleotide sugar transporters) [46]. COG complex is composed of eight distinct subunits and defects have been found in three (COG1, 7, and 8) [10,11,47]; mutations in others will likely be identified. Many of the patients have poor bone growth and loose skin, a characteristic not typically seen in defects specific for the N-glycosylation pathway. Other patients with a loose skin phenotype, called cutis laxa type II, led to the identification of another glycosylation disorder caused by mutations in a vacuolar $H^+$ ATPase subunit ($V_0$ A2) [48]. Mutations in the gene alter protein trafficking, probably by disturbing the intra-Golgi or intravesicular pH balance needed for vesicle recycling.

Efficient glycosylation is arguably the Golgi's most important function, and the need to shuttle and localize nucleotide sugar transporters and recycling them into and within the Golgi opens a Pandora's Box of potential "glycosylation defects." Traditional biochemical and cell biological approaches are not well equipped to identify candidate genes. Genetic approaches are more robust.

## Therapy

There are few therapeutic options for glycosylation-deficient patients. MPI (phospho-mannose isomerase)-deficient patients can be treated with oral mannose supplements and most of their symptoms disappear within a few weeks [49,50]. Oral mannose is absorbed through the intestine, enters the bloodstream, and is carried into cells via facilitated diffusion transporters, but mannose-specific transporters have not been identified in humans. Once inside the cells, mannose is converted to Man-6-P, bypassing the defect in the predominant fructose-6-P→Man-6-P pathway. Children benefit from mannose, but there are no MPI-deficient adult patients who are currently taking mannose supplements, suggesting that the stress of childhood precipitates the disease symptoms [1].

Some patients with a defect in the GDP-fucose transporter can be treated effectively with oral fucose supplements [51,52]. These patients have elevated circulating leukocytes since they lack the sialyl Lewis x glycan needed for selectin-dependent rolling on the endothelium. Fucose immediately reduces the leukocyte number, improves the patients' growth, and decreases their susceptibility to infections [51].

There is no therapy available, however, for the hundreds of PMM2 (phosphomannomutase 2)-deficient (CDG-Ia) patients who make insufficient Man-1-P from Man-6-P. Oral mannose therapy was unsuccessful in these patients: it neither increased nor decreased serum protein glycosylation [53–55]. This result is not surprising since the defect is distal to the potential increase in Man-6-P. Recent studies using streptolysin-$O$ permeabilized cells cautioned that increased Man-6-P reduced the amount of $Glc_3Man_9GlcNAc_2$-PP-Dol precursor by an unknown mechanism [56]. This finding raised concerns that providing mannose to glycosylation-compromised PMM2-deficient patients might further decrease glycosylation by accumulating intracellular Man-6-P. This seems unlikely since MPI activity remains normal in PMM2-deficient patients and would catabolize Man-6-P. This is supported by studies in PMM2 or MPI patient cells that showed no increase in Man-6-P when exogenous mannose was increased from $10–500\,\mu M$ [57].

Another unsuccessful approach to treat PMM2 deficiency was providing membrane-permeable Man-1-P derivatives that generated substrate within the cell [58–60]. While they restored normal glycosylation and elevated intracellular GDP-mannose, their instability and cytotoxicity made their therapeutic use impractical [59]. Recent preliminary studies of a new generation of similar compounds claim a rescue of glycosylation along with lower toxicity [60]. Further studies are needed. Other approaches such as enzyme or gene replacement therapy have not been attempted.

## Animal Models of N-Glycosylation Disorders

Genetically deficient mice, flies, and worms have been used to model the known human glycosylation disorders with variable success. Only a few mouse models have been developed. Mutant flies and worms have been developed and zebrafish models may become a rewarding approach.

### Mouse Models

Many mouse mutants in the glycosylation pathways have been created. Only those related to the known human disorders will be mentioned here. In the N-glycosylation pathway, *Pmm2* ablation is lethal within the first few days of embryonic development [61]. *Mpi*-null mice die about e10.5. Since neither one produces a useful model, patient mutations are being engineered into these genes to create hypomorphic alleles with the hope that they survive and exhibit a disease-like phenotype. Both models are in progress. Organ or temporal-specific knockout of these genes would probably not model the diseases very well since they would likely result in ER stress-induced apoptosis of deficient cells. No mouse models are available for the LLO biosynthetic pathway.

Knockout of a GDP-fucose transporter (Slc35c1) reduced GDP-fucose import by ~95%. Mice appeared normal at birth, but developed profound growth retardation and most died within a few weeks [62,63]. Overall fucosylation was greatly decreased in tissues and individual cells and providing fucose to cells partially reversed the loss of fucose-specific AAL staining. No results were reported for fucose effects on the mice. Leukocyte rolling and adhesion were severely impaired due to lack of selectin ligands. Similar to patients, the null mice showed a fivefold increase in circulating neutrophils and two- to threefold increase in other leukocytes. Longer term survivors showed dialated alveoles and thin alveolar walls, similar to those seen in Fut8-null animals that die early in postnatal life [64]. The survival of the Slc35c1 mice points to the existence of another GDP-fucose transporter [62]. A follow-up study confirmed that homing of lymphocytes to lymph nodes was 1–2% of normal. However, trafficking to the spleen was normal, offering a possible explanation for the normal lymphocyte function in patients.

Mutations in MGAT-II cause CDG-IIa [65] that results in a nearly total absence of enzymatic activity. Mice lacking GlcNAcT-2 (Mgat2) fail to make complex *N*-glycans and are a surprisingly faithful model of CDG-IIa patients. Most of the mouse embryos developed normally, but then died shortly after birth. No progeny survived beyond two weeks when the null allele was expressed in a C57B16 background. However, if the mutation was crossed into a strain (ICR) with a different genetic background, approximately 2% of the mice survived for many months. The symptoms of these survivors were somewhat milder and more heterogeneous than those in the C57B16 background. The background gene(s) that permitted this limited survival were not identified, but it clearly showed that genetic background has a major impact on survival and symptoms.

## Caenorhabditis elegans

The advantages of *C. elegans* as a model system include a powerful genetic system to study glycans during development and the existence of primitive organ systems. The position and cell lineage of each of the 959 cells is precisely defined and small interfering RNA (siRNA) were first used for gene silencing. Its 3.5-day life cycle can produce oocytes for about 4 days. The nematode makes mostly paucimannose glycans; many are heavily fucosylated or contain phosphorylcholine, making them considerably different from the typical mammalian glycans [66–68]. Mutants and siRNA knockdowns have been made for a variety of glycosylation-related genes, but little is known about their expression or function. Organisms mutated in oligosaccharyl transferase subunit Dad-1 (defender against apoptotic cell death) are abnormal and RNAi knockdown of five other subunits also causes abnormalities [69]. The worm has three genes encoding GlcNAcT-1 with differential expression. Deletion of all three produces abnormal glycans, but no obvious change in phenotype. However, the glycans play a role in the interaction with pathogenic bacteria, suggesting that these *N*-glycans are components of the worm's innate immune system [70].

Worms may offer insights into genes that may be important in the search for new types of glycosylation disorders and the interaction of various components. For instance, *C. elegans* has U-shaped gonad arms and directed migration of distal tip cells (DTC) are required. A metalloprotease (ADAM), mig17, is required for the migration [71] and anther gene, mig23, controls the action of mig17 [72]. *Mig23* encodes a luminal nucleoside diphosphatase (NDP), which is needed to hydrolyze the NDPs that arise after the transfer of monosaccharides to acceptors in the Golgi. MIG-17 is heavily glycosylated and *N*-glycans are important for its full function. Its expression in mig23 mutants produced a less glycosylated product. MIG-23 appears necessary for the normal glycan maturation in MIG-17, but loss of MIG-23 would be expected to affect maturation of multiple proteins and other mutations that affect Golgi integrity should produce similar results.

Mutations in cogc-3 and cogc-1 also cause misdirected DTC migration similar to that seen in mig-17 mutants [73]. RNAi knockdown of any of the other COG subunits also produces DTC migration defects and they are required for the glycosylation and gonadal localization of MIG-17, but not for its secretion from muscle cells. These findings link the COG complex to functional ADAM proteases, and suggest that *C. elegans* will be a model to study COG function in animal development [74].

Another nucleotide sugar transporter study in *C. elegans* showed that a single transporter carried both UDP-GalNAc and UDP-GlcNAc [75], but transport of each substrate is simultaneous and independent of the other [76]. A mutated version lost UDP-GlcNAc transport but retained UDP-GalNAc transport.

A well established set of genes, *sqv1–sqv8,* define the squashed vulva phenotype and encode genes needed for precursor biosynthesis, transport, or assembly of chondroitin sulfate in the worm [77,78].

## Drosophila melanogaster

*Drosophila* also makes paucimannose *N*-glycans that are initially converted to GlcNAc-terminated complex types and further trimmed using a Golgi-localized *N*-acetylglucosaminidase. Suppression of this enzymatic activity leads to production of GlcNAc and/or Gal-terminated glycans [79]. In contrast to *C. elegans,* loss of Mgat1 has more dramatic defects. Homozygous deficient flies are viable, but have pronounced defects in adult locomotion. Mutant males are sterile and have reduced lifespans. Adult brain architecture is abnormal [80].

Phosphorylation of the extracellular domains of proteins can occur in the Golgi. A gene called Four-jointed was identified as the first Golgi-localized protein kinase that phosphorylates Ser or Thr residues on the extracellular domains of an atypical cadherin and its ligand, Dachsous. Phosphorylation regulates signaling through the receptor [81].

*Drosophila* mutants affected in nucleotide sugar delivery and disposal offer additional perspectives. A GDP-fucose transporter (*Gfr*) in *Drosophila* can complement the defective transporter in CDG-IIc cells [82], and the normal human gene complements the Gfr-null fly. Loss of Gfr in the fly decreases Notch signaling, mostly likely due to loss of O-fucose glycans [83].

The biosynthesis, function, and genetics of O-fucose-based glycans in *Drosophila* have been extensively studied and reviewed in detail [84,85]. This will not be covered here except to suggest that many of the effects on development seen in patients with mutations in genes that limit the amount of GDP-Man could impact fucosylation of Notch. This remains to be established and further work in O-fucose glycans of *Drosophila* is likely to receive considerable attention in the future.

### Zebrafish (*Danio rerio*)

Zebrafish is arguably the model system with the brightest future for studying N-glycosylation disorders. The major advantages include external fertilization, rapid development, translucent embryos, easily recognizable phenotypes, and ease of maintenance. Probably the most important feature though is the ability to knockdown gene expression using morpholinos. The limiting factor in trying to establish mouse models of the human N-glycosylation disorders is the embryonic lethality of knockouts. This led to knocking in patient mutations and hoping that they exhibit a postnatal phenotype. Use of heritable hypomorphic alleles and morpholinos circumvents the problem. Forward genetic analysis of the known N-glycosylation defects would provide a robust opportunity to explore the currently elusive phenotypes affecting skeletal, brain, and liver development. Likewise, reverse genetic screens probing the development of these systems is likely to turn up glycosylation-related genes.

## Congenital Muscular Dystrophies

Mutations in six different glycosylation-related genes cause a series of congenital muscular dystrophies (CMD) (also known as α-dystroglycanopathies). The defects all impact the synthesis of a series of O-mannose-based glycans primarily found on α-dystroglycan (α-DG), which is part of the dystrophin glycoprotein complex that assembles on the sarcolemma of skeletal muscle cells. The complex links the muscle cell cytoskeleton containing dystrophin to the extracellular matrix. The O-Man glycans in the mucin-like domain in α-DG bind to laminin in the extracellular matrix. Clinically these autosomal recessive diseases have distinctive muscle, eye, and brain phenotypes, but vary in their presentations and severities.

Biosynthesis of O-mannose glycans begins in the ER with POMT1 (protein-O-mannosyltransferase 1) and POMT2, forming a complex that uses Dol-P-Man to form the Man-O-Ser/Thr linkage. This linkage is extended by addition of β1,2GlcNAc using a pathway-specific GlcNAc transferase (POMGnT1). In α-DG, this disaccharide is usually extended with β1,4Gal and α2,3Sia residues. More complex branched glycans with glucuronic acid or sulfate can be found in brain, suggesting that this pathway is much more complex. Monoclonal antibodies IIH6 and VIA4–1 recognize an O-Man glycan epitope in α-DG, and loss or decreased binding of the antibodies is seen in these CMD patients (Table 17.1).

Analysis of synthetic α-DG peptides and others containing mucin-like domains showed that selected Thr residues (Thr351 and Thr414) are sequentially mannosylated [86]. The mucin regions of α-DG may contain both O-Man- and O-GalNAc-based glycans. Circular dichroism (CD) and nuclear magnetic resonance (NMR) approaches show that the GalNAc-based structures are more organized than those containing Man [87].

At least three other genes are involved in the synthesis of the O-Man glycans (i.e., Fukutin (FKTN), Fukutin-related protein (FKRP) and LARGE), and mutations in each can cause a CMD [88]. Each has glycosyltransferase-like domains (LARGE has two) and characteristic catalytic amino acid residues, but none have been shown to have glycosyltransferase activity in vitro [6]. FKTN was initially found to be deficient in patients with Fukuyama CMD [89]. It is thought to physically interact with POMGnT1 and regulate its activity. Some FKTN mutations in CMD patients cause FKTN to be retained in the ER, leading to POMGnT1 retention as well. Further, POMGnT1 is co-precipitated with FKTN and mice with knockin mutations in FKTN have a decreased POMGnT1 activity [90]. The function of FKRP remains obscure, but it is a medial Golgi resident with glycosyltransferase motifs. Its deficiency in tissue seems to elicit the unfolded protein response (UPR) [91]. The LARGE protein enhances glycosylation of α-DG, and is suggested to be a glycosyltransferase, although its substrates are unknown. In one report its similarity to UDP-glucose:glucosyltransferases (UGGTs) was stressed, which could implicate a role for LARGE in α-DG folding/quality control [92].

## Diagnosis

Enzymatic assays of POMT1, POMT2, and POMGnT1 using lymphoblasts from patients have been used diagnostically and later confirmed by identifying mutations in the genes [93,94]. Fibroblasts from patients can also be assayed [95]. However, Muntoni et al. have recently demonstrated that allelic mutations in any of the six known O-Man-related genes (POMT1, POMT2, POMGnT1, FKTN, FKRP, and LARGE) can result in a much wider spectrum of clinical conditions. The general view is that the phenotypic severity does not primarily suggest which gene is mutated, but how severely the mutation affects glycosylation. A recent report argues, however, that hypoglycosylation of α-DG does not consistently correlate with clinical phenotype in patients carrying mutations in FKTN and FKRP [96]. Systematic mutation analysis of these six known glycosylation-related genes in patients with an α-dystroglycanopathy identifies mutations in approximately 65%, suggesting that more genes have yet to be identified [97]. In a set of 43 patients with Walker–Warburg syndrome (WWS) from various geographical and ethnic origins, 40% were found to have mutations in POMT1, POMT2, FKTN, or FKRP, but no mutations were found in POMGnT1 or LARGE [98]. Moreover, in approximately 100 Australian patients with CMD, about 10% had reduced glycosylated α-DG based on glycan antibody immunoblots. Sequencing revealed mutations in one patient each for FKRP, FKTN, POMGnT1, and

POMT1 genes in patients with abnormal α-DG immunofluorescence and two patients had mutations in POMT2 [99].

## Models of Muscular Dystrophy

There are several animal models of CMD with hypoglycosylation of α-DG at hand, forwarding the borders of knowledge in these complex disorders. Several knockout and patient mutation knockin mouse models have been created. *Drosophila* and zebrafish also provide excellent model systems.

The first characterized glycosylation-related CMD mouse model is the spontaneous myodystrophy (*myd*) mouse with a deletion of exons 4–7 of the mouse ortholog of LARGE (Large). The phenotype includes abnormal gait, abnormal posturing when suspended by the tail, and small litters. Serum creatine kinase is elevated and muscle histology is typical of a progressive myopathy. Furthermore, there is sensorineural deafness, reduced lifespan and decreased reproductive fitness [100]. Another Large mouse mutant, enervated (Large(enr)), results from a transgene integration disrupting the Large gene expression. In addition to the Large (*myd*) phenotype, the mice show peripheral nerve abnormalities [101].

A mouse model of muscle–eye–brain (MEB) disease with a similar phenotype to the human disorder was created by gene trapping of the POMGnT1 [102]. In this model, the gene is completely silenced with a solid decrease in the glycosylation of α-DG, and multiple defects in the muscles, eyes, and the brain. FKTN-null mice have also been created, but succumb by e9.5 of gestation due to basement membrane fragility, thus its resemblance to Fukuyama CMD could not be established [103].

A zebrafish model of FKRP deficiency was created, showing several defects including alterations in somitic structure and muscle fiber organization, defects in developing neuronal structures and eye morphology. The phenotype correlates with a reduction in α-DG glycosylation and concomitantly reduced laminin binding [104].

The mutants of the *Drosophila* homologs of POMT1 and POMT2, PomT1 and PomT2, are called rotated abdomen (rt) and twisted (tw), respectively [105–107]. Initially rt was reported to cause a clockwise helical rotation of the body. Null animals are viable. Embryos show rotation of the whole larval body and helical staggering of cuticular patterns in abdominal segments of the adult [105]. Later [106,108] RNAi knockdown of either PomT1, PomT2, or both, showed defects in muscle attachment, altered muscle contraction, and changed muscle membrane resistance, analogous to the α-DG (dg) mutant [109]. Mutation of PomT1 also decreased efficacy of synaptic transmission and changed the subunit composition of the postsynaptic glutamate receptors at the neuromuscular junction, suggesting a possible cause of developmental delay in patients [107].

## Therapy

Alterations of the α-DG glycosylation pathway constitute a potential approach to the treatment of muscular dystrophies [6]. It was recently shown that the overexpression of LARGE in cells from CMD patients (WWS, MEB, and Fukuyama CMD) increased the glycosylation of α-DG and thereby restored its function. Interestingly, a homolog of LARGE, LARGE L, seems to induce α-DG glycosylation even better and it is a good candidate for gene transfer treatment of patients [110].

Another approach used is the overexpression of non-CMD related glycosyl-transferase, in order to increase α-DG glycosylation. In a mouse model of Duchenne muscular dystrophy, overexpression of the glycosyltransferase Galgt2 inhibited the muscular dystrophy but induced glycosylation of α-DG [111]. Also, in the mdx model of Duchenne muscular dystrophy, overexpression of this GalNAc transferase rescues the phenotype, possibly by recruiting utrophin to the sarcolemma [88,112]. Gene transfer-based therapies aimed at increasing the α-DG glycosylation, thus providing a possible future treatment of dystroglycanopathies.

## Adult Onset Muscular Dystrophy

All of the muscular dystrophies described so far in this chapter are evident in children. However, inclusion body myopathy type 2 (IBM2) shows its first onset in young adults. This autosomal recessive disorder occurs worldwide, but it is especially common among Persian Jews (1:1500) [113] and in Japan, where it is known as distal myopathy with rimmed vacuoles, or Nonaka myopathy [114–116]. The patients accumulate vacuoles containing β-amyloid, tau, and presenilin. These allelic disorders result from mutations in GNE, which encodes a bifunctional, two-domain enzyme UDP-GlcNAc 2-epimerase/N-acetylmannosamine kinase [117]. This enzyme catalyzes successive steps in the de novo pathway for sialic acid biosynthesis. In IBM2, mutations occur in various combinations in both domains with the most common (M712T) occurring in the kinase domain [118,119]. In vitro assays show moderately reduced enzyme activity (20–60%) in both domains and it remains controversial whether the mutations in patients cause defective sialylation [116,118,120–122]. Sialic acid is efficiently salvaged and some cells may be less reliant on de novo synthesis than others.

Gne-null mice show embryonic lethality; mice carrying the homozygous hypomorphic allele causing an M712T substitution are born, but die only a few days after birth [123]. They have no signs of myopathy. Instead, they show severe hematuria and proteinuria due to a defective glomerular basement membrane. The cause appears to be the undersialylation of the major sialoprotein in foot podocytes (podocalexin). Providing N-acetylmannosamine to the pups rescues some and increases sialylation of podocalexin. N-Acetylmannosamine may provide a therapy for IBM2 patients, assuming that the muscle defect is due to reduced sialylation.

Another mouse model, carrying a Gne mutation common in the Japanese population (D176V), develops a pathological muscle phenotype involving β-amyloid deposition that precedes the accumulation of inclusion bodies [124]. No therapeutic studies have been reported on this model, but they will likely emerge and would help further clarify whether undersialylation contributes to the pathology [125–127].

Dominant mutations in GNE also cause another disease called sialuria. These mutations cause loss of an allosteric binding site for CMP-sialic, resulting in the continuous secretion of sialic acid [128]. IBM2 patients do not have this phenotype.

# Defects in Glycosphingolipid and Glycosylphosphaditylinositol Biosynthesis

## Disorders in Glycosphingolipid Biosynthesis

Mutations in the gene *SIAT9*, encoding GM3 synthase, cause an autosomal recessive infantile-onset epilepsy syndrome, found in the Amish population [129]. The syndrome is characterized by epilepsy, developmental stagnation, and blindness, the latter due to optic nerve atrophy. The defect enzyme, a sialyltransferase, converts lactosyl-ceramide (Galβ1,4Glc-ceramide) to the ganglioside GM3 (Sia2,3Galβ1,4Glc-ceramide), a precursor of more complex gangliosides. Deficient GM3 synthase activity increases the level of non-sialylated glycolipids in plasma. Mice that lack GM3 synthase, in contrast to humans, do not have seizures or a shortened lifespan, whereas strains lacking both GM3 synthase and a GalNAc transferase, required for making other complex gangliosides, do develop seizures [130]. The absence of these more complex gangliosides suggests that their absence causes the seizures.

## Disorders of Glycosylphosphatidylinositol Anchor Biosynthesis

Paroxysmal nocturnal hemoglobinuria (Marchiafava–Micheli syndrome) is a non-malignant hematopoietic disorder characterized by complement-mediated hemolytic anemia, hemoglobinuria, and thrombotic events. It results from an acquired deficiency of glycosylphosphatidylinositol (GPI)-anchored proteins in the erythrocyte membrane, caused by deficiency of the GlcNAc transferase that initiates the GPI biosynthesis. More than 100 somatic mutations have been shown in its corresponding X-linked gene, *PIGA* [131]. The GPI-deficient hematopoietic stem cells expand clonally, which has been linked to apoptosis resistance [132], by decreased susceptibility to NK cell killing. This is possibly mediated by loss of stress-inducible UL16-binding proteins (ULBP1 and ULBP2), proteins that activate NK and T cells [133].

A novel syndrome characterized by thrombosis of the portal and hepatic veins, and development of persistent absence epilepsy, was found to be due to a deficiency in the first mannosyltransferase for GPI anchor synthesis [134]. This enzyme is encoded by the *PIGM* gene and exists as a complex with PIG-X, required for stability. Two unrelated patients were identified, with a point mutation disrupting the binding site of the transcription factor Sp1 in the promoter, thus reducing transcription of PIGM and severe GPI deficiency [135]. Interestingly, one of the patients responded to a very simple, mutation-targeted therapy [136]. As the mutation results in histone hypoacetylation at the promoter, *PIGM* transcription could be re-induced by providing butyrate, a histone deacetylase inhibitor, ceasing the seizures.

## Defects in O-GalNAc Glycan Biosynthesis

The only inherited defect in O-GalNAc glycosylation is a variant of familial tumoral calcinosis (FTC). FTC is a severe autosomal recessive condition that exhibits phosphatemia and massive calcium deposits in the skin and subcutaneous tissues [137]. The molecular explanation is mutations in the *GALNT3* gene, encoding an O-GalNAc transferase. This GalNAc transferase transfers GalNAc to FGF23, and mutations in FGF23 itself also cause FTC, suggesting a link between *GALNT3* and FGF23 function [138]. Absence of O-GalNAc glycans on FGF23 may alter its normal proteolytic cleavage, since other mutations that prevent FGF23 cleavage also result in increased phosphaturic activity [139].

A rare autoimmune disease called Tn syndrome is caused by somatic mutations in the X-linked gene *COSMC*, which encodes a highly specific chaperone needed for proper folding and normal activity of T-synthase, converting GalNAcα-Ser/Thr (Tn antigen) to Galβ1,3GalNAcα-Ser/Thr (T-antigen) [140]. Subpopulations of all types of blood cells carry the Tn antigen due to somatic mutations in a stem cell. Since most people have anti-Tn antibodies, recognition of this antigen in Tn syndrome patients leads to an autoimmune response that causes anemia, leukopenia, and thrombocytopenia [141]. *COSMC* mutations might be found in other Tn-related disorders such as IgA nephropathy [142], but other factors may also be important in causing this condition [143].

Loss-of-function mutations in COSMC in colon cancer and melanoma-derived cells lines lead to expression of both Tn and STn antigen and mutations in COSMC were found in human cervical cancer specimens that made Tn/STn antigens showing that somatic mutations in multiple tumors can alter global cell surface O-glycan expression [144].

### X-Linked Genes to Note

Other X-linked glycan biosynthetic genes that could cause glycosylation disorders include the UDP-galactose transporter [145] *ALG13*, which is required for addition of the second GlcNAc to LLO [146] and *OGT*, the O-GlcNAc transferase [147].

## Defects in O-Fucose-Linked Glycans

O-Fucose-linked glycans occur on proteins containing an appropriate consensus sequence in the context of epidermal growth factor (EGF) repeats or in thromospondin type I repeats (TSR). These repeats are small (~40–60 amino acids) and both have a series of six conserved cysteines forming three disulfide bonds. The Notch family of signaling receptors is the best known EGF repeat, containing molecules that are O-fucosylated proteins. Notch signaling is involved in many aspects of embryonic development [7,84].

The addition of O-fucose to EGF repeats is catalyzed by OPFUT1 which can then be extended by addition of β1,3GlcNAc and β1,4Gal and α2–3/6Sia. Mutations in the fucose-specific β1,3GlcNAc transferase called lunatic fringe (LFNG) causes spondylocostal dysostosis type 3 (SCD03), which causes vertebral anomalies. Mutations in other genes involved in the Notch signaling pathway cause similar abnormal dysostosis [148]. LFNG was considered a candidate gene based on the similar phenotype of a mouse null mutant in *Lfng* [149], and a homozygous mutation (F188L) was found in one child. No other patients have been reported.

The O-fucose glycans are found in TSR repeats, in thrombospondin 1, properdin, F-spondin, ADAMTS-13, and ADAMTSL-1 are initiated by OPFUT2 and elongated to Glcβ1,3Fuc disaccharide using a fucose-specific β3-glucosyltransferase. These proteins perform essential functions in embryonic development, tissue remodeling, angiogenesis, neurogenesis, and complement activation [150]. Defects in this enzyme cause an autosomal recessive disorder called Peter's plus syndrome in which patients have anterior eye-chamber abnormalities, short stature, and developmental delay. DNA array analysis initially showed a microdeletion in a patient, which then led to the identification of mutations in the β1,3-galactosyltransferase-like gene (B3GALTL) in 20 patients [151]. Clearly, the focus on O-fucose glycan biosynthesis in TSR repeats required the biochemical characterization of the gene product showing that it was a β1,3-glucosyltransferase, not a β1,3-galactosyltransferase [152,153], and that it was specific for this pathway. The requirement for biochemical verification of an incorrectly annotated gene product cannot be overstated. This theme will occur time and again: Biochemical approaches are required to validate difficult to assign homologies [154].

# Defects in Glycosaminoglycan Synthesis

Some defects in glycosaminoglycan (GAG) synthesis have been known for quite some time and many of these give rise to skeletal- and chondrodysplasias.

## Linkage Glycan Defects

Mutations in *B4GALT7* cause the progeria variant of Ehlers–Danlos syndrome. The gene product (β1,4-galactosyltransferase) [155] adds the first Gal residue to xylose in the linkage region of GAG chains [156] and only a few patients have been described [155,157–159]. A dermatan sulfate proteoglycan from one patient's fibroblasts contained only xylose [159].

## Defective Sulfation

Several autosomal recessive chondrodysplasias result from defective GAG sulfation. They primarily affect chondroitin sulfate synthesis, which is critical to bone and joint development.

In diastrophic dysplasia, achondrogenesis type IB, neonatal osseous dysplasia I, and autosomal recessive multiple epiphyseal dysplasia, the molecular etiology is mutations in the sulfate ion transporter, encoded by *DTDST* (also known as *SLC26A2*) [160]. The defect in diastrophic dysplasia was discovered by positional cloning with fine structure mapping. Later, the others were found to be allelic.

In spondyloepimetaphyseal dysplasia, the defective gene is ATPSK2, encoding the 3'-phosphoadenosine 5'-phosphosulfate (PAPS) synthase [161]. In a mouse model of this disease, there is reduced limb length and axial skeletal size. In spondyloepimetaphyseal dysplasia there is progressive reduction in the size of the columnar and hypertrophic zones in the epiphyseal growth plates [162].

Macular corneal dystrophy is caused by a deficiency in a tissue-specific sulfo-transferase (CHST6), corneal N-acetylglucosamine-6-sulfotransferase (GlcNAc6ST), which is responsible for the production of corneal keratan sulfate [163]. Unsulfated keratan chains are poorly soluble and their eventual precipitation disrupts the collagen network, leading to thinning and loss of transparency of the corneal stroma.

## Impaired Heparan Sulfate Synthesis: Hereditary Multiple Exostosis

The most common and well-studied GAG-related disorder is hereditary multiple exostosis (HME), caused by mutations in the synthetic machinery used for heparan sulfate (HS). It is one of the few autosomal dominant diseases and has a prevalence of about 1:50 000 [164]. More specifically, HME is caused by missense or frameshift mutations in either EXT1 or EXT2, which encode HS polymerases [165]. Patients with HME develop bony outgrowths at the growth plates of the long bones. Normally the growth plates contain well-ordered chondrocytes in various stages of development, embedded in a collagen–chondroitin sulfate-containing matrix [165], but in HME patients, the outgrowths consist of disorganized chondrocytes. These outgrowths are painful and must sometimes be surgically removed. In addition, 1–2% of patients also develop osteosarcomas [164]. In one report two patients carrying EXT1 mutations were also diagnosed with autism [166], and it may be a more frequent consequence of the disease.

In the HME population, about 60–70% of mutations occur in EXT1, and EXT2 account for the rest. These proteins probably form a complex in the Golgi that function as a polymerase, adding repeating GlcNAc and GlcA residues to the growing HS chain. Both enzymes have polymerase activity, but EXT1 is enzymatically more efficient in vitro. However, the partial loss of one allele of either gene appears sufficient to cause HME, which is unusual since most glycan biosynthetic enzymes are produced in excess capacity.

The disruption of HS synthesis, and the concurrent decrease in tissue HS probably leads to an abnormal distribution of HS-binding growth factors. These include several members of the FGF family, and morphogens, such as hedgehog (Hh), decapentaplegic (Dpp), and wingless (Wg) [167]. In Drosophila, loss of HS disrupts Hh, Wg, and Dpp pathways [168]. Embryonic lethality and failure to gastrulate occur in mice that are null for either Ext gene [169,170]. However, EXT2 heterozygous animals are viable. These develop a single visible exostosis on the ribs in about a third of the litters [170]. Hh signaling in these animals is normal since no difference was detected in the protein distribution based on immunohistochemistry, thus not explaining the phenotype. EXT1 heterozygotes also develop exostoses, but the penetrance is strain-dependent. This is also seen within different families who carry the same mutations, suggesting a profound gene modifier effect.

# Opportunities

Basic understanding of glycan biosynthesis and function can be applied to human glycosylation disorders. Here we chose to focus on N-linked glycans, but there are clearly opportunities in studying other pathways.

## Identification of Markers for *N*-Glycosylation Disorders

Plasma/serum glycoproteins, primarily Tf and sometimes AAT, have been invaluable for identifying glycosylation-deficient patients, but they have several shortcomings. Most notable is that many of the mutations affecting N-glycan processing and O-GalNAc glycans are not disease-specific. For instance, lack of sialic acid and galactose in multiple glycan classes indicate likely lesions in shared steps such as intracellular trafficking defects (COG components, vacuolar ATPase, and likely others), but they do not indicate the defect. Increasing the sensitivity of detecting altered glycans is unlikely to provide additional clues about the specific defect, unless disease-specific, altered molecules are found. Since geneticists, rather than glycosylation experts, will probably identify future defects, there will be less attention focused on the specific changes in glycosylation, and important clues and subtle glycan structural differences may be overlooked. Changes in glycosylation may be used as a generic categorization of new disorders. Moreover, demonstrated changes in N- or O-linked glycosylation are not necessarily the cause of the specific pathology, especially in trafficking defects such as COGs; alterations in extracellular matrix components such as GAGs or collagen may be the predominant alteration in such disorders. Determining the effects of these trafficking defects on heparan sulfate, chondroitin sulfate, or dermatan sulfate synthesis may provide new insights into the pathological outcomes and define new aspects of biosynthesis of multiple glycan classes.

The surprising finding that non-syndromic mental retardation patients with deficiencies in TUSC3 (oligosaccharyltransferase component OST3) have normal Tf means that additional N-glycosylation disorders must certainly have been missed by routine diagnosis [38,39]. Clearly, it will be important to identify specific marker proteins, perhaps in the brain or cerebrospinal fluid, which rely on OST3 for normal glycosylation. It has been pointed out that mutations in GRIK2, a heavily glycosylated subunit of the glutamate receptor, produce a similar mental retardation phenotype [171]. Routine access to appropriate patient diagnostic material may present an obstacle, however.

Another diagnostic shortcoming of Tf is that, on occasion, its glycosylation may spontaneously normalize even in patients with documented pathological PMM2 mutations [172]. This suggests a variable "window" of abnormal glycosylation that could be disease- and mutation-specific. Without knowing the defect, variable Tf glycosylation may confuse rather than instruct. How many patients fall outside the diagnostic window is unknown. Finding more consistent diagnostic markers should be explored in patients who have shown variable Tf results.

## New Tools to Assess Glycosylation Status in Cells

Fibroblasts from type I N-glycosylation-deficient patients may not show overall glycosylation deficiencies [56,173], but brief metabolic labeling with [2-3H] mannose usually indicates reduced synthesis of LLO or its transfer to protein [174–176]. One potential marker is an engineered form of DNaseI containing a single N-glycosylation sequon—NDS. In normal cells, the site is about 80% occupied, but several CDG-I lines show considerable less site occupancy [177]. DNaseI introduced in an adenoviral construct has been used to titrate the amount of mannose needed to normalize glycosylation in MPI-deficient cells and the amount of Man-1-P generating prodrugs to normalize glycosylation in PMM2-deficient cells.

# Possible Therapies for CDG-Ia?

Mannose and fucose have been the only therapies for patients with the N-glycosylation disorders CDG-Ib and CDG-IIc, respectively. These cases are extremely rare. More prevalent are the PMM2-deficient CDG-Ia patients, where mannose has no beneficial effect. Recently, it was proposed that mannose might be more effective if one could inhibit MPI, the enzyme that competes with the depleted PPM2 for Man-6-P. Since loss of MPI in CDG-Ib patients can be rescued by mannose, small molecule inhibitors might decrease the loss of Man-6-P to glycolysis and encourage its flux into the glycosylation pathway [178]. High-throughput screening of compounds in vitro with cell-based confirmation assays has already produced a few candidates and their characterization continues (Sharma, Ng, and Freeze, unpublished observations). Another high-throughput assay could search for PMM2 activators. Both of these approaches have drawbacks in that they may only be effective for selected mutants, but they have the advantage that small molecules can cross the blood–brain barrier. Enzyme replacement therapy, which was successful for treating a few lysosomal storage disorders [179], could be used regardless of the specific mutation, but the therapeutic proteins would not cross the blood–brain barrier. Another limitation here is that, while lysosomal enzyme replacement therapy exploits normal surface→ lysosome trafficking, the defective enzyme, PMM2, resides in the cytoplasm. Whether a sufficient amount of the endocytosed enzyme would escape vesicular transport into the lysosome is unknown. Chemically modified (guanidinylated) proteins that bind to cell surface heparan sulfate, however, can access the cytoplasmic compartment [180]. Another potential approach to reduce the impact of hypoglycosylation on key proteins was suggested by Shang et al. [181]. They pointed out that slowing the rate of protein synthesis might lessen the flux through the glycosylation-deficient pathway. The rationale for this approach is that ER stress induced by hypoglycosylation slows the rate of protein synthesis by phosphorylation of eIF2$\alpha$ via PERK. Attenuating protein synthesis also slowed the rate of LLO turnover in cells [182].

Challenges and opportunities exist in finding therapies for rare diseases [183], which are defined as those affecting <200000 individuals. Since therapies for these diseases do not usually generate high profits for the pharmaceutical industry, the US Congress and the Food and Drug Administration (FDA) have extended special considerations and inducements to encourage development and sales of orphan disease therapies. The success of the Orphan Drug Act of 1983 is clear in that 11% of all new drug approvals and 24% of all new molecular and biological entities approved by the FDA between 1996 and 2006 were for orphan disorders.

## Some Unanswered Questions

### Searching for New Genetic Defects in Glycosylation

How will new human glycosylation disorders be identified? The fruitful biochemical metabolic approach to identify new defects is unlikely to be the primary discovery tool in the future. While routine and inexpensive sequencing of an individual's complete genome is not a reality at this time, it will be in the near- to mid-term. When combined with the explosion of information on the frequency of variants, the "metabalome," and "protein interactome," the discovery of new glycosylation-related defects will almost certainly fall in the geneticists' camp. Methods such as homozygosity mapping are likely to dominate. This approach uses whole-genome genotyping of different consanguineous patients to identify potential disease loci and subsequent positional candidate genes. Data from high-density single nucleotide polymorphism (SNP) genotyping arrays rely on the assumption that unrelated patients from several consanguineous families have mutations in the same gene [184]. Similar approaches have been used to identify genes involved in the development of complex disorders such as autism [185].

As mentioned earlier, assessing the impact of a mutation that affects glycosylation will require a broad collaboration of basic scientists, clinicians, and geneticists. Glycobiology specialists can contribute by finding more specific glycosylation indicators and developing flexible model systems. Establishing a consortium of clinicians and scientists, analogous to the Euroglycanet of the recent past [186], could provide mutually beneficial expertise, momentum, and continuity for those interested in developing and maintaining an investment in medically relevant projects.

### Defects that Affect Multiple Portions of the *N*-Glycosylation Pathway

Several cases have been reported in which the glycosylation defect leads to the absence of entire *N*-glycans and also of individual monosaccharides from serum glycoproteins [27]. There are few cellular pathways that are expected to affect both portions of the *N*-glycan biosynthesis (chain addition and processing) [187]. One possible pathway would be alterations in proteins that control the distribution of phosphatidylinositol phosphates between the ER and Golgi. For instance, phosphatidylinositol-4-P has been shown to play a role in PI4P homeostasis [188]. The lipid is preferentially located in the Golgi during cell growth, but in quiescent cells it moves in the ER. SAC1, a PI4P-specific phosphatase, accumulates in the Golgi in quiescent cells, depleting the Golgi

PI4P, but the enzyme moves into the ER in growing cells thus allowing accumulation of PI4P in the Golgi [189]. In growing yeast, Sac1p binds to Dpm1p, the enzyme responsible for Dol-P-Man synthesis [190], but Sac1p moves to the Golgi following nutrient depletion. Sac1p also appears to be partially responsible for full-length LLO glycans. Recent studies using siRNA-mediated knockdown of SAC1 in HeLa cells alters both $N$- and $O$-linked glycans (unpublished observations). Deficiencies in other enzymes regulating membrane lipid homeostasis and trafficking could also alter the distribution of glycosylation elements in the Golgi and ER.

For glycosylation disorders that affect multiple pathways, an important issue is to determine which impaired pathway is responsible for specific phenotypes [9]. A good example of the complexity is the bone and skin abnormalities seen in some COG-deficient or vacuolar ATPase-deficient patients [48]. These clinical presentations would suggest deficiencies in extracellular matrix biosynthesis or assembly since these patients appear different from patients with defects restricted to $N$-glycan addition or modification [191]. A similar problem for disorders restricted to $N$-glycosylation is assessing which protein(s) are responsible for a specific phenotype and the mechanisms involved. Examples of successes are few. Loss of Fut8, which adds core fucose residues to $N$-glycans, in mice causes early postnatal lethality due to dysregulation of transforming growth factor β1 (TGFβ1) receptor activation. This leads to an emphysema-like condition due to excessive expression of matrix metalloproteinases along with a downregulation of matrix proteins such as elastin [64]. These mice also have increased serum levels of insulin-like growth factor (IGF)-binding protein-3 (IGFBP-3) because its endocytosis through low-density lipoprotein (LDL) receptor-related protein-1 (LRP-1) is impaired [192]. Another example is seen in mice lacking GlcNAcT-IVa [193]. They have reduced expression of glucose transporter 2 (Glut2) in pancreatic beta cells because the transporter is endocytosed more rapidly without its normal tri-antennary $N$-glycan. This impairs glucose-stimulated insulin secretion and metabolic abnormalities similar to type 2 diabetes [193].

## Conclusions

Glycobiologists have used biochemical and molecular tools to identify the mutated genes in a large number of inherited human diseases that affect glycosylation. Many more such genetic diseases will be identified using powerful genetic mapping tools. Their discoveries will create even more opportunities for scientists interested in glycan function to play a major role in determining the specific functions of the mutated genes. Developing vertebrate and non-vertebrate model systems containing one or more interacting hypomorphic alleles can be a major contribution to understanding the roles of glycans in human physiology and health. The development of sensitive and specific diagnostic indicators of selected disorders will help to insure that defects in the 1–2% of the human genes used for glycosylation are not overlooked in the clinic or the laboratory.

## Acknowledgments

This work was supported by grants to HHF from the National Institutes of Health (R01 DK55695), The Rocket Williams Fund, Mason's Hope, and the CDG Family Network.

# References

1. Eklund EA, Freeze HH. The congenital disorders of glycosylation: a multifaceted group of syndromes. *NeuroRx* 2006;**3**:254–63.

2. Freeze HH. Genetic defects in the human glycome. *Nat Rev Genet* 2006;**7**:537–51.

3. Grunewald S. Congenital disorders of glycosylation: rapidly enlarging group of (neuro)metabolic disorders. *Early Hum Dev* 2007;**83**:825–30.

4. Jaeken J, Matthijs G. Congenital disorders of glycosylation: a rapidly expanding disease family. *Annu Rev Genomics Hum Genet* 2007;**8**:261–78.

5. Leroy JG. Congenital disorders of N-glycosylation including diseases associated with O- as well as N-glycosylation defects. *Pediatr Res* 2006;**60**:643–56.

6. Muntoni F, Torelli S, Brockington M. Muscular dystrophies due to glycosylation defects. *Neurotherapeutics* 2008;**5**:627–32.

7. Rampal R, Luther KB, Haltiwanger RS. Notch signaling in normal and disease States: possible therapies related to glycosylation. *Curr Mol Med* 2007;**7**:427–45.

8. Wopereis S, Lefeber DJ, Morava E, Wevers RA. Mechanisms in protein O-glycan biosynthesis and clinical and molecular aspects of protein O-glycan biosynthesis defects: a review. *Clin Chem* 2006;**52**:574–600.

9. Schachter H, Freeze H. Glycosylation diseases: Quo vadis? *Biochim Biophys Acta* 2008 [Epub ahead of print].

10. Foulquier F. COG defects, birth and rise! *Biochim Biophys Acta* 2008 [Epub ahead of print].

11. Smith RD, Lupashin VV. Role of the conserved oligomeric Golgi (COG) complex in protein glycosylation. *Carbohydr Res* 2008;**343**:2024–31.

12. Aebi M, Helenius A, Schenk B, Barone R, Fiumara A, Berger EG, Hennet T, Imbach T, Stutz A, Bjursell C, Uller A, Wahlstrom JG, Briones P, Cardo E, Clayton P, Winchester B, Cormier-Dalre V, de Lonlay P, Cuer M, Dupre T, Seta N, de Koning T, Dorland L, de Loos F, Kupers L. Carbohydrate-deficient glycoprotein syndromes become congenital disorders of glycosylation: an updated nomenclature for CDG. First International Workshop on CDGS. *Glycoconj J* 1999;**16**:669–71.

13. Jaeken J, Hennet T, Freeze HH, Matthijs G. On the nomenclature of congenital disorders of glycosylation (CDG). *J Inherit Metab Dis* 2008;**31**:669–72.

14. Rhee H, Lee JS. MedRefSNP: a database of medically investigated SNPs. *Hum Mutat* 2009;**30**(Mar (3)):E460–6.

15. Losh M, Sullivan PF, Trembath D, Piven J. Current developments in the genetics of autism: from phenome to genome. *J Neuropathol Exp Neurol* 2008;**67**:829–37.

16. Psychiatric GWAS Consortium Steering Committee. A framework for interpreting genome-wide association studies of psychiatric disorders. *Mol Psychiatry* 2009;**14**:10–17.

17. Fang J, Peters V, Assmann B, Korner C, Hoffmann GF. Improvement of CDG diagnosis by combined examination of several glycoproteins. *J Inherit Metab Dis* 2004;**27**:581–90.

18. Stibler H, Jaeken J. Carbohydrate deficient serum transferrin in a new systemic hereditary syndrome. *Arch Dis Child* 1990;**65**:107–11.

19. Colome C, Ferrer I, Artuch R, Vilaseca MA, Pineda M, Briones P. Personal experience with the application of carbohydrate-deficient transferrin (CDT) assays to the detection of congenital disorders of glycosylation. *Clin Chem Lab Med* 2000;**38**:965–9.

20. Carchon HA, Chevigne R, Falmagne JB, Jaeken J. Diagnosis of congenital disorders of glycosylation by capillary zone electrophoresis of serum transferring. *Clin Chem* 2004;**50**:101–11.

21. Helander A, Bergstrom J, Freeze HH. Testing for congenital disorders of glycosylation by HPLC measurement of serum transferrin glycoforms. *Clin Chem* 2004;**50**:954–8.

22. Schwarz M, Thiel C, Lubbehusen J, Dorland B, de Koning T, von Figura K, Lehle L, Korner C. Deficiency of GDP-Man:GlcNAc2-PP-dolichol mannosyltransferase causes congenital disorder of glycosylation type Ik. *Am J Hum Genet* 2004;**74**:472–81.

23. Mills K, Mills P, Jackson M, Worthington V, Beesley C, Mann A, Clayton P, Grunewald S, Keir G, Young L, Langridge J, Mian N, Winchester B. Diagnosis of congenital disorders of glycosylation type-I using protein chip technology. *Proteomics* 2006;**6**:2295–304.

24. Wada Y. Mass spectrometry in the detection and diagnosis of congenital disorders of glycosylation. *Eur J Mass Spectrom (Chichester, Eng)* 2007;**13**:101–3.

25. Marklova E, Albahri Z. Screening and diagnosis of congenital disorders of glycosylation. *Clin Chim Acta* 2007;**385**:6–20.

26. O'Brien 3rd JF, Lacey JM, Bergen HR. Detection of hypo-N-glycosylation using mass spectrometry of transferring. *Curr Protoc Hum Genet Chapter* 2007;**17**. Unit 17 14.

27. Mandato C, Brive L, Miura Y, Davis JA, Di Cosmo N, Lucariello S, Pagliardini S, Seo NS, Parenti G, Vecchione R, Freeze HH, Vajro P. Cryptogenic liver disease in four children: a novel congenital disorder of glycosylation. *Pediatr Res* 2006;**59**:293–8.

28. Mills PB, Mills K, Johnson AW, Clayton PT, Winchester BG. Analysis by matrix assisted laser desorption/ionisation-time of flight mass spectrometry of the post-translational modifications of alpha 1-antitrypsin isoforms separated by two-dimensional polyacrylamide gel electrophoresis. *Proteomics* 2001;**1**:778–86.

29. Mills K, Mills PB, Clayton PT, Mian N, Johnson AW, Winchester BG. The underglycosylation of plasma alpha 1-antitrypsin in congenital disorders of glycosylation type I is not random. *Glycobiology* 2003;**13**:73–85.

30. Jackson M, Clayton P, Grunewald S, Keir G, Mills K, Mills P, Winchester B, Worthington V, Young E. Elevation of plasma aspartylglucosaminidase is a useful marker for the congenital disorders of glycosylation type I (CDG I). *J Inherit Metab Dis* 2005;**28**:1197–8.

31. Hulsmeier AJ, Paesold-Burda P, Hennet T. N-glycosylation site occupancy in serum glycoproteins using multiple reaction monitoring liquid chromatography-mass spectrometry. *Mol Cell Proteomics* 2007;**6**:2132–8.

32. Barone R, Sturiale L, Sofia V, Ignoto A, Fiumara A, Sorge G, Garozzo D, Zappia M. Clinical phenotype correlates to glycoprotein phenotype in a sib pair with CDG-Ia. *Am J Med Genet A* 2008;**146A**:2103–8.

33. Sturiale L, Barone R, Fiumara A, Perez M, Zaffanello M, Sorge G, Pavone L, Tortorelli S, O'Brien JF, Jaeken J, Garozzo D. Hypoglycosylation with increased fucosylation and branching of serum transferrin N-glycans in untreated galactosemia. *Glycobiology* 2005;**15**:1268–76.

34. Pronicka E, Adamowicz M, Kowalik A, Ploski R, Radomyska B, Rogaszewska M, Rokicki D, Sykut-Cegielska J. Elevated carbohydrate-deficient transferrin (CDT) and its normalization on dietary treatment as a useful biochemical test for hereditary fructose intolerance and galactosemia. *Pediatr Res* 2007;**62**:101–5.

35. Golka K, Wiese A. Carbohydrate-deficient transferrin (CDT)--a biomarker for long-term alcohol consumption. *J Toxicol Environ Health B Crit Rev* 2004;**7**:319–37.

36. Schulz BL, Laroy W, Callewaert N. Clinical laboratory testing in human medicine based on the detection of glycoconjugates. *Curr Mol Med* 2007;**7**:397–416.

37. Jaeken J, Pirard M, Adamowicz M, Pronicka E, van Schaftingen E. Inhibition of phosphomannose isomerase by fructose 1-phosphate: an explanation for defective N-glycosylation in hereditary fructose intolerance. *Pediatr Res* 1996;**40**:764–6.

38. Garshasbi M, Hadavi V, Habibi H, Kahrizi K, Kariminejad R, Behjati F, Tzschach A, Najmabadi H, Ropers HH, Kuss AW. A defect in the TUSC3 gene is associated with autosomal recessive mental retardation. *Am J Hum Genet* 2008;**82**:1158–64.

39. Molinari F, Foulquier F, Tarpey PS, Morelle W, Boissel S, Teague J, Edkins S, Futreal PA, Stratton MR, Turner G, Matthijs G, Gecz J, Munnich A, Colleaux L. Oligosaccharyltransferase-subunit mutations in nonsyndromic mental retardation. *Am J Hum Genet* 2008;**82**:1150–7.

40. Kelleher DJ, Gilmore R. An evolving view of the eukaryotic oligosaccharyltransferase. *Glycobiology* 2006;**16**:47R–62R.

41. Maeda Y, Kinoshita T. Dolichol-phosphate mannose synthase: structure, function and regulation. *Biochim Biophys Acta* 2008;**1780**:861–8.

42. Kranz C, Jungeblut C, Denecke J, Erlekotte A, Sohlbach C, Debus V, Kehl HG, Harms E, Reith A, Reichel S, Grobe H, Hammersen G, Schwarzer U, Marquardt T. A defect in dolichol phosphate biosynthesis causes a new inherited disorder with death in early infancy. *Am J Hum Genet* 2007;**80**:433–40.

43. Anand M, Rush JS, Ray S, Doucey MA, Weik J, Ware FE, Hofsteenge J, Waechter CJ, Lehrman MA. Requirement of the Lec35 gene for all known classes of monosaccharide-P-dolichol-dependent glycosyltransferase reactions in mammals. *Mol Biol Cell* 2001;**12**:487–501.

44. Kranz C, Denecke J, Lehrman MA, Ray S, Kienz P, Kreissel G, Sagi D, Peter-Katalinic J, Freeze HH, Schmid T, Jackowski-Dohrmann S, Harms E, Marquardt T. A mutation in the human MPDU1 gene causes congenital disorder of glycosylation type If (CDG-If). *J Clin Invest* 2001;**108**:1613–19.

45. Schenk B, Imbach T, Frank CG, Grubenmann CE, Raymond GV, Hurvitz H, Korn-Lubetzki I, Revel-Vik S, Raas-Rotschild A, Luder AS, Jaeken J, Berger EG, Matthijs G, Hennet T, Aebi M. MPDU1 mutations underlie a novel human congenital disorder of glycosylation, designated type If. *J Clin Invest* 2001;**108**:1687–95.

46. Wu X, Steet RA, Bohorov O, Bakker J, Newell J, Krieger M, Spaapen L, Kornfeld S, Freeze HH. Mutation of the COG complex subunit gene COG7 causes a lethal congenital disorder. *Nat Med* 2004;**10**:518–23.

47. Zeevaert R, Foulquier F, Jaeken J, Matthijs G. Deficiencies in subunits of the Conserved Oligomeric Golgi (COG) complex define a novel group of Congenital Disorders of Glycosylation. *Mol Genet Metab* 2008;**93**:15–21.

48. Kornak U, Reynders E, Dimopoulou A, van Reeuwijk J, Fischer B, Rajab A, Budde B, Nurnberg P, Foulquier F, Lefeber D, Urban Z, Gruenewald S, Annaert W, Brunner HG, van Bokhoven H, Wevers R, Morava E, Matthijs G, Van Maldergem L, Mundlos S. Impaired glycosylation and cutis laxa caused by mutations in the vesicular H+-ATPase subunit ATP6V0A2. *Nat Genet* 2008;**40**:32–4.

49. Freeze HH. Congenital disorders of glycosylation and the pediatric liver. *Semin Liver Dis* 2001;**21**:501–15.

50. de Lonlay P, Seta N. The clinical spectrum of phosphomannose isomerase deficiency, with an evaluation of mannose treatment for CDG-Ib. *Biochim Biophys Acta* 2008 [Epub ahead of print].

51. Marquardt T, Luhn K, Srikrishna G, Freeze HH, Harms E, Vestweber D. Correction of leukocyte adhesion deficiency type II with oral fucose. *Blood* 1999;**94**:3976–85.

52. Wild MK, Luhn K, Marquardt T, Vestweber D. Leukocyte adhesion deficiency II: therapy and genetic defect. *Cells Tissues Organs* 2002;**172**:161–73.

53. Mayatepek E, Schroder M, Kohlmuller D, Bieger WP, Nutzenadel W. Continuous mannose infusion in carbohydrate-deficient glycoprotein syndrome type I. *Acta Paediatr* 1997;**86**:1138–40.

54. Mayatepek E, Kohlmuller D. Mannose supplementation in carbohydrate-deficient glycoprotein syndrome type I and phosphomannomutase deficiency. *Eur J Pediatr* 1998;**157**:605–6.

55. Kjaergaard S, Kristiansson B, Stibler H, Freeze HH, Schwartz M, Martinsson T, Skovby F. Failure of short-term mannose therapy of patients with carbohydrate-deficient glycoprotein syndrome type 1A. *Acta Paediatr* 1998;**87**:884–8.

56. Gao N, Shang J, Lehrman MA. Analysis of glycosylation in CDG-Ia fibroblasts by fluorophore-assisted carbohydrate electrophoresis: implications for extracellular glucose and intracellular mannose 6-phosphate. *J Biol Chem* 2005;**280**:17901–9.

57. Higashidani A, Bode L, Nishikawa A, Freeze HH. Exogenous mannose does not raise steady state mannose-6-phosphate pools of normal or N-glycosylation-deficient human fibroblasts. *Mol Genet Metab* 2009;**96**(Apr (4)):268–72.

58. Rutschow S, Thiem J, Kranz C, Marquardt T. Membrane-permeant derivatives of mannose-1-phosphate. *Bioorg Med Chem* 2002;**10**:4043–9.

59. Eklund EA, Merbouh N, Ichikawa M, Nishikawa A, Clima JM, Dorman JA, Norberg T, Freeze HH. Hydrophobic Man-1-P derivatives correct abnormal glycosylation in Type I congenital disorder of glycosylation fibroblasts. *Glycobiology* 2005;**15**:1084–93.

60. Hardre R, Khaled A, Willemetz A, Dupre T, Moore S, Gravier-Pelletier C, Le Merrer Y. Mono, di and tri-mannopyranosyl phosphates as mannose-1-phosphate prodrugs for potential CDG-Ia therapy. *Bioorg Med Chem Lett* 2007;**17**:152–5.

61. Thiel C, Lubke T, Matthijs G, von Figura K, Korner C. Targeted disruption of the mouse phosphomannomutase 2 gene causes early embryonic lethality. *Mol Cell Biol* 2006;**26**:5615–20.

62. Hellbusch CC, Sperandio M, Frommhold D, Yakubenia S, Wild MK, Popovici D, Vestweber D, Grone HJ, von Figura K, Lubke T, Korner C. Golgi GDP-fucose transporter-deficient mice mimic congenital disorder of glycosylation IIc/leukocyte adhesion deficiency II. *J Biol Chem* 2007;**282**:10762–72.

63. Yakubenia S, Frommhold D, Scholch D, Hellbusch CC, Korner C, Petri B, Jones C, Ipe U, Bixel MG, Krempien R, Sperandio M, Wild MK. Leukocyte trafficking in a mouse model for leukocyte adhesion deficiency II/congenital disorder of glycosylation IIc. *Blood* 2008;**112**:1472–81.

64. Wang X, Inoue S, Gu J, Miyoshi E, Noda K, Li W, Mizuno-Horikawa Y, Nakano M, Asahi M, Takahashi M, Uozumi N, Ihara S, Lee SH, Ikeda Y, Yamaguchi Y, Aze Y, Tomiyama Y, Fujii J, Suzuki K, Kondo A, Shapiro SD, Lopez-Otin C, Kuwaki T, Okabe M, Honke K, Taniguchi N. Dysregulation of TGF-beta1 receptor activation leads to abnormal lung development and emphysema-like phenotype in core fucose-deficient mice. *Proc Natl Acad Sci USA* 2005;**102**:15791–6.

65. Wang Y, Schachter H, Marth JD. Mice with a homozygous deletion of the Mgat2 gene encoding UDP-N-acetylglucosamine:alpha-6-D-mannoside beta1,2-N-acetylglucosaminyltransferase II: a model for congenital disorder of glycosylation type IIa. *Biochim Biophys Acta* 2002;**1573**:301–11.

66. Cipollo JF, Costello CE, Hirschberg CB. The fine structure of Caenorhabditis elegans N-glycans. *J Biol Chem* 2002;**277**:49143–57.

67. Cipollo JF, Awad AM, Costello CE, Hirschberg CB. N-Glycans of Caenorhabditis elegans are specific to developmental stages. *J Biol Chem* 2005;**280**:26063–72.

68. Paschinger K, Gutternigg M, Rendic D, Wilson IB. The N-glycosylation pattern of *Caenorhabditis elegans*. *Carbohydr Res* 2008;**343**:2041–9.

69. Schachter H. Protein glycosylation lessons from *Caenorhabditis elegans*. *Curr Opin Struct Biol* 2004;**14**:607–16.

70. Shi H, Tan J, Schachter H. N-glycans are involved in the response of Caenorhabditis elegans to bacterial pathogens. *Methods Enzymol* 2006;**417**:359–89.

71. Nishiwaki K, Hisamoto N, Matsumoto K. A metalloprotease disintegrin that controls cell migration in Caenorhabditis elegans. *Science* 2000;**288**:2205–8.

72. Nishiwaki K, Kubota Y, Chigira Y, Roy SK, Suzuki M, Schvarzstein M, Jigami Y, Hisamoto N, Matsumoto K. An NDPase links ADAM protease glycosylation with organ morphogenesis in *C. elegans*. *Nat Cell Biol* 2004;**6**:31–7.

73. Kubota Y, Sano M, Goda S, Suzuki N, Nishiwaki K. The conserved oligomeric Golgi complex acts in organ morphogenesis via glycosylation of an ADAM protease in *C. elegans*. *Development* 2006;**133**:263–73.

74. Kubota Y, Nishiwaki K. *C. elegans* as a model system to study the function of the COG complex in animal development. *Biol Chem* 2006;**387**:1031–5.

75. Caffaro CE, Luhn K, Bakker H, Vestweber D, Samuelson J, Berninsone P, Hirschberg CB. A single Caenorhabditis elegans Golgi apparatus-type transporter of UDP-glucose, UDP-galactose, UDP-N-acetylglucosamine, and UDP-N-acetylgalactosamine. *Biochemistry* 2008;**47**:4337–44.

76. Caffaro CE, Hirschberg CB, Berninsone PM. Independent and simultaneous translocation of two substrates by a nucleotide sugar transporter. *Proc Natl Acad Sci USA* 2006;**103**:16176–81.

77. Cummings R, Doering T. Nematoda. In: Varki A, Cummings RD, Esko JD, Freeze HH, Stanley P, Bertozzi CR, Hart GW, Etzler ME, editors. *Essentials of Glycobiology*. 2nd edn. Cold Spring Harbor, NY: Cold Spring Harbor Laboratory Press; 2008, p. 333–45.

78. Herman T, Horvitz HR. Three proteins involved in Caenorhabditis elegans vulval invagination are similar to components of a glycosylation pathway. *Proc Natl Acad Sci USA* 1999;**96**:974–9.

79. Kim YK, Kim KR, Kang DG, Jang SY, Kim YH, Cha HJ. Suppression of β-N-acetylglucosaminidase in N-glycosylation pathway for complex glycoprotein formation in Drosophila S2 cells. *Glycobiology* 2009;**19**(3):301–8.

80. Sarkar M, Leventis PA, Silvescu CI, Reinhold VN, Schachter H, Boulianne GL. Null mutations in Drosophila N-acetylglucosaminyltransferase I produce defects in locomotion and a reduced life span. *J Biol Chem* 2006;**281**:12776–85.

81. Ishikawa HO, Takeuchi H, Haltiwanger RS, Irvine KD. Four-jointed is a Golgi kinase that phosphorylates a subset of cadherin domains. *Science* 2008;**321**:401–4.

82. Luhn K, Laskowska A, Pielage J, Klambt C, Ipe U, Vestweber D, Wild MK. Identification and molecular cloning of a functional GDP-fucose transporter in Drosophila melanogaster. *Exp Cell Res* 2004;**301**:242–50.

83. Ishikawa HO, Higashi S, Ayukawa T, Sasamura T, Kitagawa M, Harigaya K, Aoki K, Ishida N, Sanai Y, Matsuno K. Notch deficiency implicated in the pathogenesis of congenital disorder of glycosylation IIc. *Proc Natl Acad Sci U S A* 2005;**102**:18532–7.

84. Haltiwanger RS, Lowe JB. Role of glycosylation in development. *Annu Rev Biochem* 2004;**73**:491–537.

85. Stanley P. Regulation of Notch signaling by glycosylation. *Curr Opin Struct Biol* 2007;**17**:530–5.

86. Manya H, Suzuki T, Akasaka-Manya K, Ishida HK, Mizuno M, Suzuki Y, Inazu T, Dohmae N, Endo T. Regulation of mammalian protein O-mannosylation: preferential amino acid sequence for O-mannose modification. *J Biol Chem* 2007;**282**:20200–6.

87. Liu M, Borgert A, Barany G, Live D. Conformational consequences of protein glycosylation: preparation of O-mannosyl serine and threonine building blocks, and their incorporation into glycopeptide sequences derived from alpha-dystroglycan. *Biopolymers* 2008;**90**:358–68.

88. Martin PT. Congenital muscular dystrophies involving the O-mannose pathway. *Curr Mol Med* 2007;**7**:417–25.

89. Kobayashi K, Nakahori Y, Miyake M, Matsumura K, Kondo-Iida E, Nomura Y, Segawa M, Yoshioka M, Saito K, Osawa M, Hamano K, Sakakihara Y, Nonaka I, Nakagome Y, Kanazawa I, Nakamura Y, Tokunaga K, Toda T. An ancient retrotransposal insertion causes Fukuyama-type congenital muscular dystrophy. *Nature* 1998;**394**:388–92.

90. Xiong H, Kobayashi K, Tachikawa M, Manya H, Takeda S, Chiyonobu T, Fujikake N, Wang F, Nishimoto A, Morris GE, Nagai Y, Kanagawa M, Endo T, Toda T. Molecular interaction between fukutin and POMGnT1 in the glycosylation pathway of alpha-dystroglycan. *Biochem Biophys Res Commun* 2006;**350**:935–41.

91. Boito CA, Fanin M, Gavassini BF, Cenacchi G, Angelini C, Pegoraro E. Biochemical and ultrastructural evidence of endoplasmic reticulum stress in LGMD2I. *Virchows Arch* 2007;**451**:1047–55.

92. Patnaik SK, Stanley P. Mouse large can modify complex N- and mucin O-glycans on alpha-dystroglycan to induce laminin binding. *J Biol Chem* 2005;**280**:20851–9.

93. Endo T, Manya H. O-mannosylation in mammalian cells. *Methods Mol Biol* 2006;**347**:43–56.

94. Manya H, Bouchet C, Yanagisawa A, Vuillaumier-Barrot S, Quijano-Roy S, Suzuki Y, Maugenre S, Richard P, Inazu T, Merlini L, Romero NB, Leturcq F, Bezier I, Topaloglu H, Estournet B, Seta N, Endo T, Guicheney P. Protein O-mannosyltransferase activities in lymphoblasts from patients with alpha-dystroglycanopathies. *Neuromuscul Disord* 2008;**18**:45–51.

95. Vajsar J, Zhang W, Dobyns WB, Biggar D, Holden KR, Hawkins C, Ray P, Olney AH, Burson CM, Srivastava AK, Schachter H. Carriers and patients with muscle-eye-brain disease can be rapidly diagnosed by enzymatic analysis of fibroblasts and lymphoblasts. *Neuromuscul Disord* 2006;**16**:132–6.

96. Jimenez-Mallebrera C, Torelli S, Feng L, Kim J, Godfrey C, Clement E, Mein R, Abbs S, Brown SC, Campbell KP, Kroger S, Talim B, Topaloglu H, Quinlivan R, Roper H, Childs AM, Kinali M, Sewry CA, Muntoni F. A comparative study of alpha-dystroglycan glycosylation in dystroglycanopathies suggests that the hypoglycosylation of alpha-dystroglycan does not consistently correlate with clinical severity. *Brain Pathol* 2008 [Epub ahead of print].

97. Muntoni F, Brockington M, Godfrey C, Ackroyd M, Robb S, Manzur A, Kinali M, Mercuri E, Kaluarachchi M, Feng L, Jimenez-Mallebrera C, Clement E, Torelli S, Sewry CA, Brown SC. Muscular dystrophies due to defective glycosylation of dystroglycan. *Acta Myol* 2007;**26**:129–35.

98. Manzini MC, Gleason D, Chang BS, Hill RS, Barry BJ, Partlow JN, Poduri A, Currier S, Galvin-Parton P, Shapiro LR, Schmidt K, Davis JG, Basel-Vanagaite L, Seidahmed MZ, Salih MA, Dobyns WB, Walsh CA. Ethnically diverse causes of Walker-Warburg syndrome (WWS): FCMD mutations are a more common cause of WWS outside of the Middle East. *Hum Mutat* 2008;**29**:E231–41.

99. Peat RA, Smith JM, Compton AG, Baker NL, Pace RA, Burkin DJ, Kaufman SJ, Lamande SR, North KN. Diagnosis and etiology of congenital muscular dystrophy. *Neurology* 2008;**71**:312–21.

100. Grewal PK, Holzfeind PJ, Bittner RE, Hewitt JE. Mutant glycosyltransferase and altered glycosylation of alpha-dystroglycan in the myodystrophy mouse. *Nat Genet* 2001;**28**:151–4.

101. Levedakou EN, Popko B. Rewiring enervated: thinking LARGEr than myodystrophy. *J Neurosci Res* 2006;**84**:237–43.

102. Liu J, Ball SL, Yang Y, Mei P, Zhang L, Shi H, Kaminski HJ, Lemmon VP, Hu H. A genetic model for muscle-eye-brain disease in mice lacking protein O-mannose 1,2-N-acetylglucosaminyltransferase (POMGnT1). *Mech Dev* 2006;**123**:228–40.

103. Kurahashi H, Taniguchi M, Meno C, Taniguchi Y, Takeda S, Horie M, Otani H, Toda T. Basement membrane fragility underlies embryonic lethality in fukutin-null mice. *Neurobiol Dis* 2005;**19**:208–17.

104. Thornhill P, Bassett D, Lochmuller H, Bushby K, Straub V. Developmental defects in a zebrafish model for muscular dystrophies associated with the loss of fukutin-related protein (FKRP). *Brain* 2008;**131**:1551–61.

105. Martin-Blanco E, Garcia-Bellido A. Mutations in the rotated abdomen locus affect muscle development and reveal an intrinsic asymmetry in Drosophila. *Proc Natl Acad Sci USA* 1996;**93**:6048–52.

106. Haines N, Seabrooke S, Stewart BA. Dystroglycan and protein O-mannosyltransferases 1 and 2 are required to maintain integrity of Drosophila larval muscles. *Mol Biol Cell* 2007;**18**:4721–30.

107. Wairkar YP, Fradkin LG, Noordermeer JN, DiAntonio A. Synaptic defects in a Drosophila model of congenital muscular dystrophy. *J Neurosci* 2008;**28**:3781–9.

108. Ichimiya T, Manya H, Ohmae Y, Yoshida H, Takahashi K, Ueda R, Endo T, Nishihara S. The twisted abdomen phenotype of Drosophila POMT1 and POMT2 mutants coincides with their heterophilic protein O-mannosyltransferase activity. *J Biol Chem* 2004;**279**:42638–47.

109. Williamson RA, Henry MD, Daniels KJ, Hrstka RF, Lee JC, Sunada Y, Ibraghimov-Beskrovnaya O, Campbell KP. Dystroglycan is essential for early embryonic development: disruption of Reichert's membrane in Dag1-null mice. *Hum Mol Genet* 1997;**6**:831–41.

110. Brockington M, Muntoni F. The modulation of skeletal muscle glycosylation as a potential therapeutic intervention in muscular dystrophies. *Acta Myol* 2005;**24**:217–21.

111. Nguyen HH, Jayasinha V, Xia B, Hoyte K, Martin PT. Overexpression of the cytotoxic T cell GalNAc transferase in skeletal muscle inhibits muscular dystrophy in mdx mice. *Proc Natl Acad Sci USA* 2002;**99**:5616–21.

112. Xu R, Chandrasekharan K, Yoon JH, Camboni M, Martin PT. Overexpression of the cytotoxic T cell (CT) carbohydrate inhibits muscular dystrophy in the dyW mouse model of congenital muscular dystrophy 1A. *Am J Pathol* 2007;**171**:181–99.

113. Argov Z, Eisenberg I, Grabov-Nardini G, Sadeh M, Wirguin I, Soffer D, Mitrani-Rosenbaum S. Hereditary inclusion body myopathy: the Middle Eastern genetic cluster. *Neurology* 2003;**60**:1519–23.

114. Nishino I, Noguchi S, Murayama K, Driss A, Sugie K, Oya Y, Nagata T, Chida K, Takahashi T, Takusa Y, Ohi T, Nishimiya J, Sunohara N, Ciafaloni E, Kawai M, Aoki M, Nonaka I. Distal myopathy with rimmed vacuoles is allelic to hereditary inclusion body myopathy. *Neurology* 2002;**59**:1689–93.

115. Nonaka I, Noguchi S, Nishino I. Distal myopathy with rimmed vacuoles and hereditary inclusion body myopathy. *Curr Neurol Neurosci Rep* 2005;**5**:61–5.

116. Nishino I, Malicdan MC, Murayama K, Nonaka I, Hayashi YK, Noguchi S. Molecular pathomechanism of distal myopathy with rimmed vacuoles. *Acta Myol* 2005;**24**:80–3.

117. Eisenberg I, Avidan N, Potikha T, Hochner H, Chen M, Olender T, Barash M, Shemesh M, Sadeh M, Grabov-Nardini G, Shmilevich I, Friedmann A, Karpati G, Bradley WG, Baumbach L, Lancet D, Asher EB,

Beckmann JS, Argov Z, Mitrani-Rosenbaum S. The UDP-N-acetylglucosamine 2-epimerase/N-acetylmannosamine kinase gene is mutated in recessive hereditary inclusion body myopathy. *Nat Genet* 2001;**29**:83–7.

118. Hinderlich S, Salama I, Eisenberg I, Potikha T, Mantey LR, Yarema KJ, Horstkorte R, Argov Z, Sadeh M, Reutter W, Mitrani-Rosenbaum S. The homozygous M712T mutation of UDP-N-acetylglucosamine 2-epimerase/N-acetylmannosamine kinase results in reduced enzyme activities but not in altered overall cellular sialylation in hereditary inclusion body myopathy. *FEBS Lett* 2004;**566**:105–9.

119. Penner J, Mantey LR, Elgavish S, Ghaderi D, Cirak S, Berger M, Krause S, Lucka L, Voit T, Mitrani-Rosenbaum S, Hinderlich S. Influence of UDP-GlcNAc 2-epimerase/ManNAc kinase mutant proteins on hereditary inclusion body myopathy. *Biochemistry* 2006;**45**:2968–77.

120. Huizing M, Rakocevic G, Sparks SE, Mamali I, Shatunov A, Goldfarb L, Krasnewich D, Gahl WA, Dalakas MC. Hypoglycosylation of alpha-dystroglycan in patients with hereditary IBM due to GNE mutations. *Mol Genet Metab* 2004;**81**:196–202.

121. Salama I, Hinderlich S, Shlomai Z, Eisenberg I, Krause S, Yarema K, Argov Z, Lochmuller H, Reutter W, Dabby R, Sadeh M, Ben-Bassat H, Mitrani-Rosenbaum S. No overall hyposialylation in hereditary inclusion body myopathy myoblasts carrying the homozygous M712T GNE mutation. *Biochem Biophys Res Commun* 2005;**328**:221–6.

122. Ricci E, Broccolini A, Gidaro T, Morosetti R, Gliubizzi C, Frusciante R, Di Lella GM, Tonali PA, Mirabella M. NCAM is hyposialylated in hereditary inclusion body myopathy due to GNE mutations. *Neurology* 2006;**66**:755–8.

123. Galeano B, Klootwijk R, Manoli I, Sun M, Ciccone C, Darvish D, Starost MF, Zerfas PM, Hoffmann VJ, Hoogstraten-Miller S, Krasnewich DM, Gahl WA, Huizing M. Mutation in the key enzyme of sialic acid biosynthesis causes severe glomerular proteinuria and is rescued by N-acetylmannosamine. *J Clin Invest* 2007;**117**:1585–94.

124. Malicdan MC, Noguchi S, Nonaka I, Hayashi YK, Nishino I. A Gne knockout mouse expressing human GNE D176V mutation develops features similar to distal myopathy with rimmed vacuoles or hereditary inclusion body myopathy. *Hum Mol Genet* 2007;**16**:2669–82.

125. Malicdan MC, Noguchi S, Nishino I. Perspectives on distal myopathy with rimmed vacuoles or hereditary inclusion body myopathy: contributions from an animal model. *Lack of sialic acid, a central determinant in sugar chains, causes myopathy? Acta Myol* 2007;**26**:171–5.

126. Malicdan MC, Noguchi S, Nishino I. Recent advances in distal myopathy with rimmed vacuoles (DMRV) or hIBM: treatment perspectives. *Curr Opin Neurol* 2008;**21**:596–600.

127. Argov Z, Mitrani-Rosenbaum S. The hereditary inclusion body myopathy enigma and its future therapy. *Neurotherapeutics* 2008;**5**:633–7.

128. Leroy JG, Seppala R, Huizing M, Dacremont G, De Simpel H, Van Coster RN, Orvisky E, Krasnewich DM, Gahl WA. Dominant inheritance of sialuria, an inborn error of feedback inhibition. *Am J Hum Genet* 2001;**68**:1419–27.

129. Simpson MA, Cross H, Proukakis C, Priestman DA, Neville DC, Reinkensmeier G, Wang H, Wiznitzer M, Gurtz K, Verganelaki A, Pryde A, Patton MA, Dwek RA, Butters TD, Platt FM, Crosby AH. Infantile-onset symptomatic epilepsy syndrome caused by a homozygous loss-of-function mutation of GM3 synthase. *Nat Genet* 2004;**36**:1225–9.

130. Proia RL. Gangliosides help stabilize the brain. *Nat Genet* 2004;**36**:1147–8.

131. Bessler M, Hiken J. The pathophysiology of disease in patients with paroxysmal nocturnal hemoglobinuria. *Hematology Am Soc Hematol Educ Program* 2008:104–10.

132.  Szpurka H, Schade AE, Jankowska AM, Maciejewski JP. Altered lipid raft composition and defective cell death signal transduction in glycosylphosphatidylinositol anchor-deficient PIG-A mutant cells. *Br J Haematol* 2008;**142**(3):413–22.

133.  Savage WJ, Barber JP, Mukhina GL, Hu R, Chen G, Matsui W, Thoburn C, Hess AD, Cheng L, Jones RJ, Brodsky RA. Glycosylphosphatidylinositol-anchored protein deficiency confers resistance to apoptosis in PNH. *Exp Hematol* 2009;**37**:42–51.

134.  Maeda Y, Watanabe R, Harris CL, Hong Y, Ohishi K, Kinoshita K, Kinoshita T. PIG-M transfers the first mannose to glycosylphosphatidylinositol on the lumenal side of the ER. *Embo J* 2001;**20**:250–61.

135.  Almeida AM, Murakami Y, Layton DM, Hillmen P, Sellick GS, Maeda Y, Richards S, Patterson S, Kotsianidis I, Mollica L, Crawford DH, Baker A, Ferguson M, Roberts I, Houlston R, Kinoshita T, Karadimitris A. Hypomorphic promoter mutation in PIGM causes inherited glycosylphosphatidylinositol deficiency. *Nat Med* 2006;**12**:846–51.

136.  Almeida AM, Murakami Y, Baker A, Maeda Y, Roberts IA, Kinoshita T, Layton DM, Karadimitris A. Targeted therapy for inherited GPI deficiency. *N Engl J Med* 2007;**356**:1641–7.

137.  Topaz O, Shurman DL, Bergman R, Indelman M, Ratajczak P, Mizrachi M, Khamaysi Z, Behar D, Petronius D, Friedman V, Zelikovic I, Raimer S, Metzker A, Richard G, Sprecher E. Mutations in GALNT3, encoding a protein involved in O-linked glycosylation, cause familial tumoral calcinosis. *Nat Genet* 2004;**36**:579–81.

138.  Chefetz I, Sprecher E. Familial tumoral calcinosis and the role of O-glycosylation in the maintenance of phosphate homeostasis. *Biochim Biophys Acta* 2008. doi:10.1016/j.bbadis.2008.10.008.

139.  Strom TM, Juppner H. PHEX, FGF23, DMP1 and beyond. *Curr Opin Nephrol Hypertens* 2008;**17**:357–62.

140.  Ju T, Cummings RD. A unique molecular chaperone Cosmc required for activity of the mammalian core 1 beta 3-galactosyltransferase. *Proc Natl Acad Sci USA* 2002;**99**:16613–18.

141.  Berger EG. Tn-syndrome. *Biochim Biophys Acta* 1999;**1455**:255–68.

142.  Ju T, Cummings RD. Protein glycosylation: chaperone mutation in Tn syndrome. *Nature* 2005;**437**:1252.

143.  Buck KS, Smith AC, Molyneux K, El-Barbary H, Feehally J, Barratt J. B-cell O-galactosyltransferase activity, and expression of O-glycosylation genes in bone marrow in IgA nephropathy. *Kidney Int* 2008;**73**:1128–36.

144.  Ju T, Lanneau GS, Gautam T, Wang Y, Xia B, Stowell SR, Willard MT, Wang W, Xia JY, Zuna RE, Laszik Z, Benbrook DM, Hanigan MH, Cummings RD. Human tumor antigens Tn and sialyl Tn arise from mutations in Cosmc. *Cancer Res* 2008;**68**:1636–46.

145.  Hara T, Yamauchi M, Takahashi E, Hoshino M, Aoki K, Ayusawa D, Kawakita M. The UDP-galactose translocator gene is mapped to band Xp11.23-p11.22 containing the Wiskott-Aldrich syndrome locus. *Somat Cell Mol Genet* 1993;**19**:571–5.

146.  Gao XD, Tachikawa H, Sato T, Jigami Y, Dean N. Alg14 recruits Alg13 to the cytoplasmic face of the endoplasmic reticulum to form a novel bipartite UDP-N-acetylglucosamine transferase required for the second step of N-linked glycosylation. *J Biol Chem* 2005;**280**:36254–62.

147.  Shafi R, Iyer SP, Ellies LG, O'Donnell N, Marek KW, Chui D, Hart GW, Marth JD. The O-GlcNAc transferase gene resides on the X chromosome and is essential for embryonic stem cell viability and mouse ontogeny. *Proc Natl Acad Sci USA* 2000;**97**:5735–9.

148. Sparrow DB, Guillen-Navarro E, Fatkin D, Dunwoodie SL. Mutation of Hairy-and-Enhancer-of-Split-7 in humans causes spondylocostal dysostosis. *Hum Mol Genet* 2008;**17**:3761–6.

149. Sparrow DB, Chapman G, Wouters MA, Whittock NV, Ellard S, Fatkin D, Turnpenny PD, Kusumi K, Sillence D, Dunwoodie SL. Mutation of the LUNATIC FRINGE gene in humans causes spondylocostal dysostosis with a severe vertebral phenotype. *Am J Hum Genet* 2006;**78**:28–37.

150. Heinonen TY, Maki M. Peters'-plus syndrome is a congenital disorder of glycosylation caused by a defect in the beta1,3-glucosyltransferase that modifies thrombospondin type 1 repeats. *Ann Med* 2009;**41**:2–10.

151. Lesnik Oberstein SA, Kriek M, White SJ, Kalf ME, Szuhai K, den Dunnen JT, Breuning MH, Hennekam RC. Peters Plus syndrome is caused by mutations in B3GALTL, a putative glycosyltransferase. *Am J Hum Genet* 2006;**79**:562–6.

152. Sato T, Sato M, Kiyohara K, Sogabe M, Shikanai T, Kikuchi N, Togayachi A, Ishida H, Ito H, Kameyama A, Gotoh M, Narimatsu H. Molecular cloning and characterization of a novel human beta1,3-glucosyltransferase, which is localized at the endoplasmic reticulum and glucosylates O-linked fucosylglycan on thrombospondin type 1 repeat domain. *Glycobiology* 2006;**16**:1194–206.

153. Hess D, Keusch JJ, Oberstein SA, Hennekam RC, Hofsteenge J. Peters Plus syndrome is a new congenital disorder of glycosylation and involves defective Omicron-glycosylation of thrombospondin type 1 repeats. *J Biol Chem* 2008;**283**:7354–60.

154. Roseman S. Reflections on glycobiology. *J Biol Chem* 2001;**276**:41527–42.

155. Okajima T, Fukumoto S, Furukawa K, Urano T. Molecular basis for the progeroid variant of Ehlers-Danlos syndrome. Identification and characterization of two mutations in galactosyltransferase I gene. *J Biol Chem* 1999;**274**:28841–4.

156. Lindahl U. Further characterization of the heparin-protein linkage region. *Biochim Biophys Acta* 1966;**130**:368–82.

157. Faiyaz-Ul-Haque M, Zaidi SH, Al-Ali M, Al-Mureikhi MS, Kennedy S, Al-Thani G, Tsui LC, Teebi AS. A novel missense mutation in the galactosyltransferase-I (B4GALT7) gene in a family exhibiting facioskeletal anomalies and Ehlers-Danlos syndrome resembling the progeroid type. *Am J Med Genet A* 2004;**128A**:39–45.

158. Gotte M, Kresse H. Defective glycosaminoglycan substitution of decorin in a patient with progeroid syndrome is a direct consequence of two point mutations in the galactosyltransferase I (beta4GalT-7) gene. *Biochem Genet* 2005;**43**:65–77.

159. Quentin E, Gladen A, Roden L, Kresse H. A genetic defect in the biosynthesis of dermatan sulfate proteoglycan: galactosyltransferase I deficiency in fibroblasts from a patient with a progeroid syndrome. *Proc Natl Acad Sci USA* 1990;**87**:1342–6.

160. Dawson PA, Markovich D. Pathogenetics of the human SLC26 transporters. *Curr Med Chem* 2005;**12**:385–96.

161. Sugahara K, Schwartz NB. Defect in 3'-phosphoadenosine 5'-phosphosulfate formation in brachymorphic mice. *Proc Natl Acad Sci USA* 1979;**76**:6615–18.

162. ul Haque MF, King LM, Krakow D, Cantor RM, Rusiniak ME, Swank RT, Superti-Furga A, Haque S, Abbas H, Ahmad W, Ahmad M, Cohn DH. Mutations in orthologous genes in human spondyloepimetaphyseal dysplasia and the brachymorphic mouse. *Nat Genet* 1998;**20**:157–62.

163. Akama TO, Nishida K, Nakayama J, Watanabe H, Ozaki K, Nakamura T, Dota A, Kawasaki S, Inoue Y, Maeda N, Yamamoto S, Fujiwara T, Thonar EJ, Shimomura Y, Kinoshita S, Tanigami A,

Fukuda MN. Macular corneal dystrophy type I and type II are caused by distinct mutations in a new sulphotransferase gene. *Nat Genet* 2000;**26**:237–41.

164. Schmale 3rd GA, Conrad EU, Raskind WH. The natural history of hereditary multiple exostoses. *J Bone Joint Surg Am* 1994;**76**:986–92.

165. Zak BM, Crawford BE, Esko JD. Hereditary multiple exostoses and heparan sulfate polymerization. *Biochim Biophys Acta* 2002;**1573**:346–55.

166. Li H, Yamagata T, Mori M, Momoi MY. Association of autism in two patients with hereditary multiple exostoses caused by novel deletion mutations of EXT1. *J Hum Genet* 2002;**47**:262–5.

167. Esko JD, Selleck SB. Order out of chaos: assembly of ligand binding sites in heparan sulfate. *Annu Rev Biochem* 2002;**71**:435–71.

168. Bornemann DJ, Duncan JE, Staatz W, Selleck S, Warrior R. Abrogation of heparan sulfate synthesis in Drosophila disrupts the Wingless, Hedgehog and Decapentaplegic signaling pathways. *Development* 2004;**131**:1927–38.

169. Lin X, Wei G, Shi Z, Dryer L, Esko JD, Wells DE, Matzuk MM. Disruption of gastrulation and heparan sulfate biosynthesis in EXT1-deficient mice. *Dev Biol* 2000;**224**:299–311.

170. Stickens D, Zak BM, Rougier N, Esko JD, Werb Z. Mice deficient in Ext2 lack heparan sulfate and develop exostoses. *Development* 2005;**132**:5055–68.

171. Hennet T. From glycosylation disorders back to glycosylation: What have we learned? *Biochim Biophys Acta* 2008. doi:10.1016/j.bbadis.2008.10.006.

172. Freeze HH. Update and perspectives on congenital disorders of glycosylation. *Glycobiology* 2001;**11**:129R–143R.

173. Lehrman MA. Stimulation of N-linked glycosylation and lipid-linked oligosaccharide synthesis by stress responses in metazoan cells. *Crit Rev Biochem Mol Biol* 2006;**41**:51–75.

174. Panneerselvam K, Freeze HH. Mannose corrects altered N-glycosylation in carbohydrate-deficient glycoprotein syndrome fibroblasts. *J Clin Invest* 1996;**97**:1478–87.

175. Panneerselvam K, Etchison JR, Skovby F, Freeze HH. Abnormal metabolism of mannose in families with carbohydrate-deficient glycoprotein syndrome type 1. *Biochim Mol Med* 1997;**61**:161–7.

176. Korner C, Lehle L, von Figura K. Carbohydrate-deficient glycoprotein syndrome type 1: correction of the glycosylation defect by deprivation of glucose or supplementation of mannose. *Glycoconj J* 1998;**15**:499–505.

177. Fujita N, Tamura A, Higashidani A, Tonozuka T, Freeze HH, Nishikawa A. The relative contribution of mannose salvage pathways to glycosylation in PMI-deficient mouse embryonic fibroblast cells. *FEBS J* 2008;**275**:788–98.

178. Freeze H. Towards a therapy for phosphomannomutase 2 deficiency, the defect in CDG-Ia patients. *Biochim Biophys Acta* 2009. doi:10.1016/j.bbadis.2009.01.004.

179. Beck M. New therapeutic options for lysosomal storage disorders: enzyme replacement, small molecules and gene therapy. *Hum Genet* 2007;**121**:1–22.

180. Elson-Schwab L, Garner OB, Schuksz M, Crawford BE, Esko JD, Tor Y. Guanidinylated neomycin delivers large, bioactive cargo into cells through a heparan sulfate-dependent pathway. *J Biol Chem* 2007;**282**:13585–91.

181. Shang J, Gao N, Kaufman RJ, Ron D, Harding HP, Lehrman MA. Translation attenuation by PERK balances ER glycoprotein synthesis with lipid-linked oligosaccharide flux. *J Cell Biol* 2007;**176**:605–616.

182. Lehrman MA. Teaching dolichol-linked oligosaccharides more tricks with alternatives to metabolic radiolabeling. *Glycobiology* 2007;**17**:75R–85R.

183. Griggs RC, Batshaw M, Dunkle M, Gopal-Srivastava R, Kaye E, Krischer J, Nguyen T, Paulus K, Merkel PA. Clinical research for rare disease: opportunities, challenges, and solutions. *Mol Genet Metab* 2009;**96**:20–6.

184. Nicot AS, Toussaint A, Tosch V, Kretz C, Wallgren-Pettersson C, Iwarsson E, Kingston H, Garnier JM, Biancalana V, Oldfors A, Mandel JL, Laporte J. Mutations in amphiphysin 2 (BIN1) disrupt interaction with dynamin 2 and cause autosomal recessive centronuclear myopathy. *Nat Genet* 2007;**39**:1134–9.

185. Morrow EM, Yoo SY, Flavell SW, Kim TK, Lin Y, Hill RS, Mukaddes NM, Balkhy S, Gascon G, Hashmi A, Al-Saad S, Ware J, Joseph RM, Greenblatt R, Gleason D, Ertelt JA, Apse KA, Bodell A, Partlow JN, Barry B, Yao H, Markianos K, Ferland RJ, Greenberg ME, Walsh CA. Identifying autism loci and genes by tracing recent shared ancestry. *Science* 2008;**321**:218–23.

186. Matthijs G. Research network: EUROGLYCANET: a European network focused on congenital disorders of glycosylation. *Eur J Hum Genet* 2005;**13**:395–7.

187. Freeze HH. Congenital Disorders of Glycosylation: CDG-I, CDG-II, and beyond. *Curr Mol Med* 2007;**7**:389–96.

188. Blagoveshchenskaya A, Mayinger P. SAC1 lipid phosphatase and growth control of the secretory pathway. *Mol Biosyst* 2009;**5**:36–42.

189. Blagoveshchenskaya A, Cheong FY, Rohde HM, Glover G, Knodler A, Nicolson T, Boehmelt G, Mayinger P. Integration of Golgi trafficking and growth factor signaling by the lipid phosphatase SAC1. *J Cell Biol* 2008;**180**:803–12.

190. Faulhammer F, Konrad G, Brankatschk B, Tahirovic S, Knodler A, Mayinger P. Cell growth-dependent coordination of lipid signaling and glycosylation is mediated by interactions between Sac1p and Dpm1p. *J Cell Biol* 2005;**168**:185–91.

191. Morava E, Lefeber DJ, Urban Z, de Meirleir L, Meinecke P, Gillessen Kaesbach G, Sykut-Cegielska J, Adamowicz M, Salafsky I, Ranells J, Lemyre E, van Reeuwijk J, Brunner HG, Wevers RA. Defining the phenotype in an autosomal recessive cutis laxa syndrome with a combined congenital defect of glycosylation. *Eur J Hum Genet* 2008;**16**:28–35.

192. Lee SH, Takahashi M, Honke K, Miyoshi E, Osumi D, Sakiyama H, Ekuni A, Wang X, Inoue S, Gu J, Kadomatsu K, Taniguchi N. Loss of core fucosylation of low-density lipoprotein receptor-related protein-1 impairs its function, leading to the upregulation of serum levels of insulin-like growth factor-binding protein 3 in Fut8-/- mice. *J Biochem* 2006;**139**:391–8.

193. Ohtsubo K, Takamatsu S, Minowa MT, Yoshida A, Takeuchi M, Marth JD. Dietary and genetic control of glucose transporter 2 glycosylation promotes insulin secretion in suppressing diabetes. *Cell* 2005;**123**:1307–21.

# Index

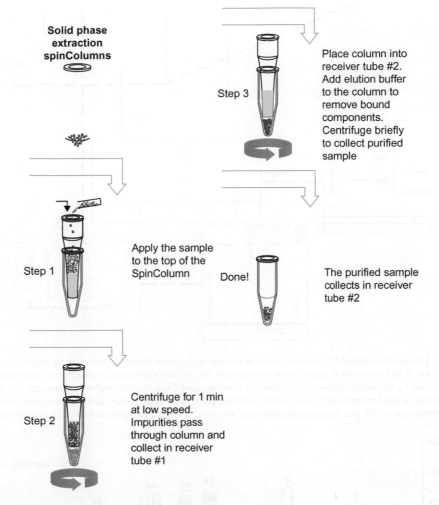

**Solid phase extraction spinColumns**

Step 1 — Apply the sample to the top of the SpinColumn

Step 2 — Centrifuge for 1 min at low speed. Impurities pass through column and collect in receiver tube #1

Step 3 — Place column into receiver tube #2. Add elution buffer to the column to remove bound components. Centrifuge briefly to collect purified sample

Done! — The purified sample collects in receiver tube #2

**PLATE 1** SpinColumn™ solid-phase extraction protocol for free glycans.

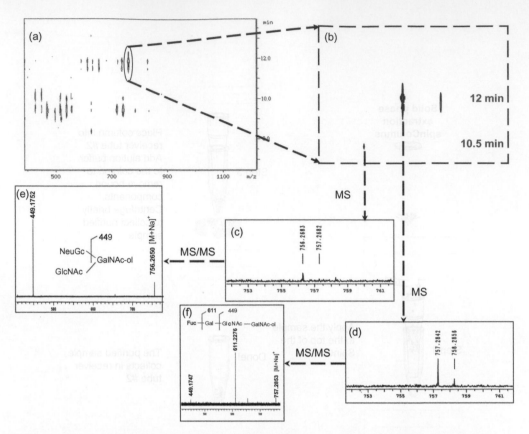

**PLATE 2** (a) Two-dimensional (2D) contour plot of a mixture of *O*-glycans cleaved from mucin; (b) zoomed 2D contour plot; (c and d) spectra of glycans with *m/z* values of 756 and 757 and their corresponding tandem mass spectra (e and f). An average resolution of ~30000 is demonstrated. Experimental conditions: cyano column, 26 cm; mobile phase, 2.4 mM ammonium formate buffer (240 mM, pH 3) and 0.2 mM sodium acetate in the mixture of acetonitrile/water (71:29, *v/v*); field strength, 600 V/cm; injection, 12 kV, 30 s. Reproduced from [58], with permission.

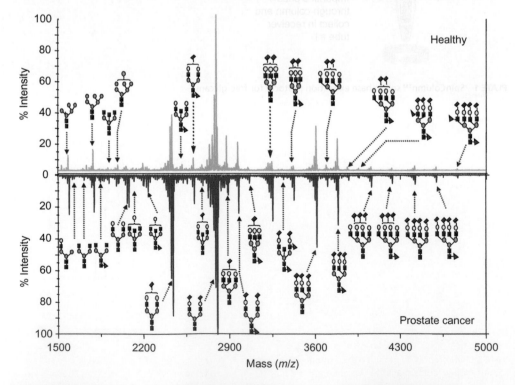

**PLATE 3** Matrix-assisted laser desorption/ionization (MALDI) mirror spectra of permethylated *N*-glycans derived from human blood serum of a healthy individual vs. a prostate cancer patient. Blue squares, *N*-acetylglucosamine; green circles, mannose; yellow circles, galactose; red triangles, fucose; pink diamonds, *N*-acetylneuraminic acid. Reproduced from [122], with permission.

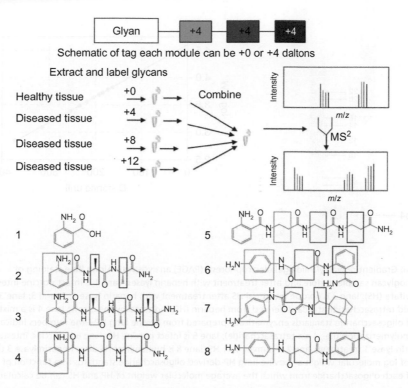

**PLATE 4** Representation of tetraplex tags and experimental flowchart and synthesized compounds indicating potential sites for isotopic modification. Reproduced from [130], with permission.

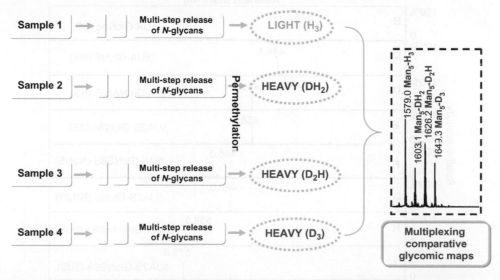

**PLATE 5** Representation of multiplexed comparative glycomic mapping through permethylation using stable isotopic methyl iodide reagents. This approach is an extension of C-GlycoMAP [131].

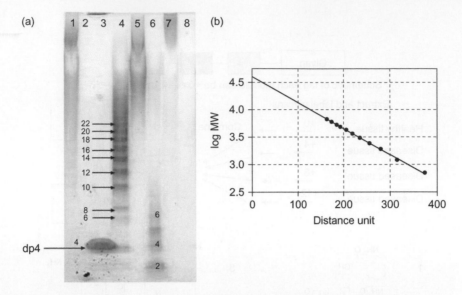

**PLATE 6** (a) Gradient polyacrylamide gel electrophoresis (PAGE) analysis with Alcian Blue staining of glycosaminoglycan samples before and after treatment with heparin lyases. Lane 1 is intact porcine intestinal heparan sulfate (HS); lane 2 is porcine intestinal HS after treatment with heparin lyase 1, 2, and 3; lane 3 is a hexasulfated tetrasaccharide standard derived from heparin (HP) indicated by the arrow; lane 4 is a mixture of HP-derived oligosaccharide standards enzymatically prepared from bovine lung HP—the numbers indicate their degree of polymerization (i.e., 4 is a tetrasaccharide); lane 5 is intact porcine HP; lane 6 is porcine intestinal HP after heparin lyase 1 treatment; lane 7 is human liver HS; lane 8 is human liver HS after heparin lyase 3 treatment. (b) A plot of log molecular weight of bovine lung HP-derived oligosaccharide standards as a function of migration distance of each oligosaccharide from which the average molecular weight of HP and HS can be calculated.

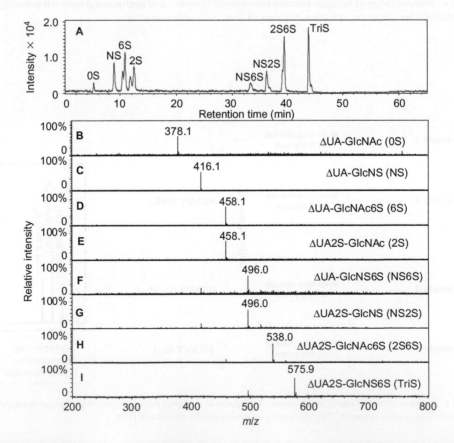

**PLATE 7** Heparin (HP)/heparan sulfate (HS) disaccharide analysis by liqiod chromatography/mass spectroscopy (LC-MS). (A) Extracted ion chromatography; (B–I) mass spectra of each disaccharide in negative mode.

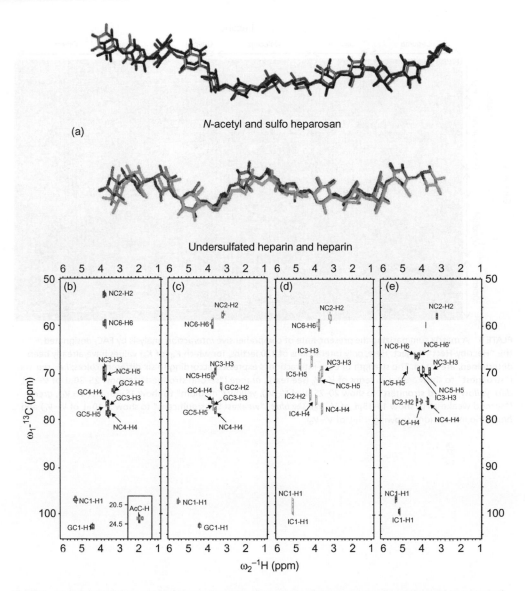

**PLATE 8** High-field NMR spectra and three-dimensional structures of chemo-enzymatic synthesized heparin (HP) and its precursors. (a) Three-dimensional structures of chemo-enzymatic synthesized HP and its precursors; (b–e) heteronuclear multiple quantum coherence of chemo-enzymatic synthesized HP and its precursors.

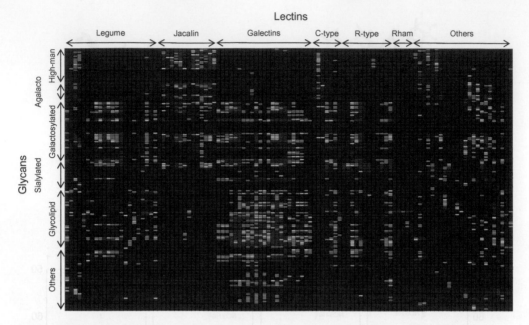

**PLATE 9** A matrix summarizing the present state of comprehensive *interaction* analysis by FAC, designated the "Hect-by-Hect" Project. Here, only interactions of 120 lectins, for which $K_d$ (or $K_a$) values have already been determined, are listed. The strength of each interaction is expressed according to six different colors; *i.e., red* ("strongest" to show >30 μL in terms of $V-V_0$ (see text)), *orange* ("second strongest" to show 25–30 μL of $V-V_0$), *dark yellow* ("third strongest" to show 20–25 μL of $V-V_0$), *yellow* ("medium" to show 15–20 μL of $V-V_0$), *green* ("second weakest" to show 10–15 μL of $V-V_0$), *sky blue* ("weakest but significant" to show 5–10 μL of $V-V_0$) and *blue* ("no interaction" showing <5 μL of $V-V_0$).

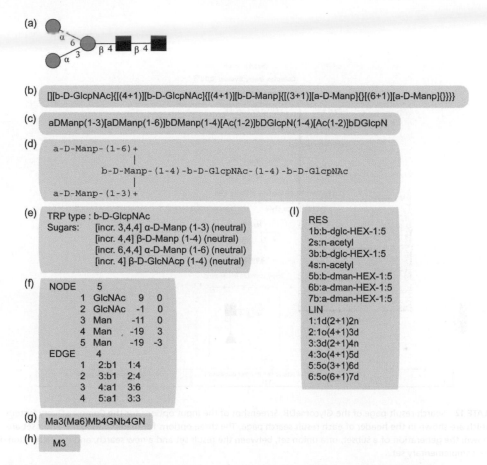

**(a)**

**(b)** [][b-D-GlcpNAc]{[(4+1)][b-D-GlcpNAc]{[(4+1)][b-D-Manp]{[(3+1)][a-D-Manp]{}{(6+1)][a-D-Manp]{}}}}}

**(c)** aDManp(1-3)[aDManp(1-6)]bDManp(1-4)[Ac(1-2)]bDGlcpN(1-4)[Ac(1-2)]bDGlcpN

**(d)**
```
a-D-Manp-(1-6)+
              |
              b-D-Manp-(1-4)-b-D-GlcpNAc-(1-4)-b-D-GlcpNAc
              |
a-D-Manp-(1-3)+
```

**(e)**
```
TRP type : b-D-GlcpNAc
Sugars:    [incr. 3,4,4] α-D-Manp (1-3) (neutral)
           [incr. 4,4] β-D-Manp (1-4) (neutral)
           [incr. 6,4,4] α-D-Manp (1-6) (neutral)
           [incr. 4] β-D-GlcNAcp (1-4) (neutral)
```

**(f)**
```
NODE    5
        1  GlcNAc   9    0
        2  GlcNAc  -1    0
        3  Man    -11    0
        4  Man    -19    3
        5  Man    -19   -3
EDGE    4
        1  2:b1   1:4
        2  3:b1   2:4
        3  4:a1   3:6
        4  5:a1   3:3
```

**(i)**
```
RES
1b:b-dglc-HEX-1:5
2s:n-acetyl
3b:b-dglc-HEX-1:5
4s:n-acetyl
5b:b-dman-HEX-1:5
6b:a-dman-HEX-1:5
7b:a-dman-HEX-1:5
LIN
1:1d(2+1)2n
2:1o(4+1)3d
3:3d(2+1)4n
4:3o(4+1)5d
5:5o(3+1)6d
6:5o(6+1)7d
```

**(g)** Ma3(Ma6)Mb4GNb4GN

**(h)** M3

**PLATE 10** Sequence formats used by different carbohydrate databases. Compilation of the different sequence formats used by the carbohydrate structure databases for the example of the *N*-glycan core. (a) Pictorial representation, as used by the CFG (http://www.functionalglycomics.org/static/consortium/Nomenclature.shtml), with reducing end at the right. (b) LINUCS encoding used in GLYCOSCIENCES.de. (c) BCSDB encoding. (d) ASCII 2D graph as employed in CCSD (CarbBank). (e) Notation used by GlycoBase of the Université des Sciences et Technologies de Lille. (f) KEGG Chemical Function format (KCF) used by KEGG. (g) Linear Code® used in the CFG database. (h) Oxford notation used in GlycoBase from NIBRT. (i) GlycoCT encoding used in EUROCarbDB and GlycomeDB.

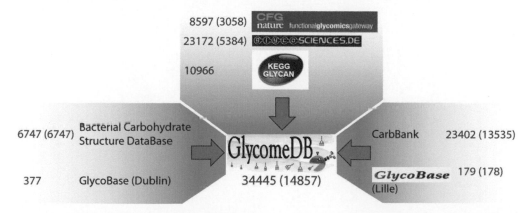

**PLATE 11** The GlycomeDB concept. GlycomeDB integrates and stores the structure and species information from seven open access carbohydrate structure databases. For GlycomeDB and for each database the number of unique sequences in the database specific format (as of November 2008) is given. The numbers in parentheses show the number of sequences with at least one species annotation in the database.

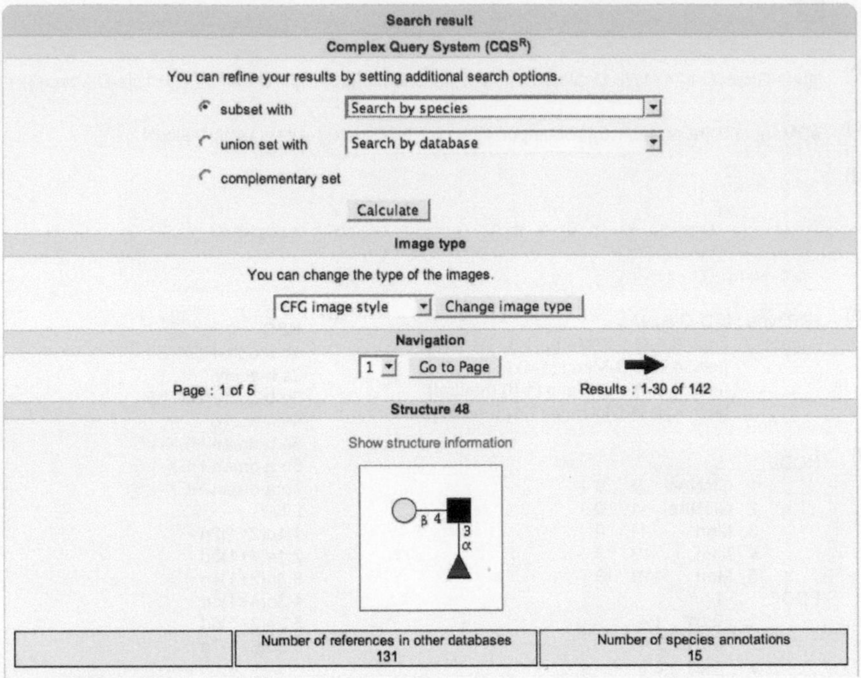

**PLATE 12** Search result page of the GlycomeDB. Screenshot of the input options for the Complex Query System which are shown in the header of each result search page. The three options for manipulating the result set are shown: the generation of a subset, or a union set, between the result set and a new search, and the calculation of the complementary set.

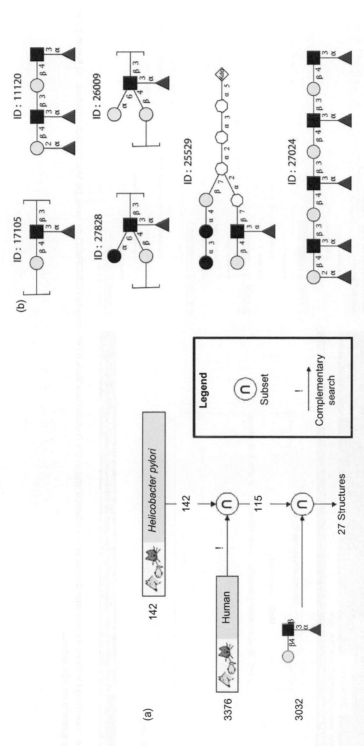

**PLATE 13** Example using the GlycomeDB Complex Query System (Case Study 1). Task: Retrieve glycan structures containing the Lewis x trisaccharide epitope that are found in *Helicobacter pylori* but not in *Homo sapiens*. (a) Workflow for the complex query in GlycomeDB: the results of the "species search" for *Helicobacter pylori* (142 structures) are refined using the Complex Query System by, first, a subset search for all non-human structures, and second, by a substructure search for the Lewis x epitope, resulting in 27 structures matching the search criteria. The numbers before each of the search symbols show the total numbers of structures in GlycomeDB matching the represented search criteria. The numbers in the middle represent the intermediate result after a subset operation. (b) Six of the 27 retrieved structures are shown with their GlycomeDB structure ID. Three of these are indicated as polysaccharides (enclosed with square brackets). In all structures the Lewis x trisaccharide can be seen.

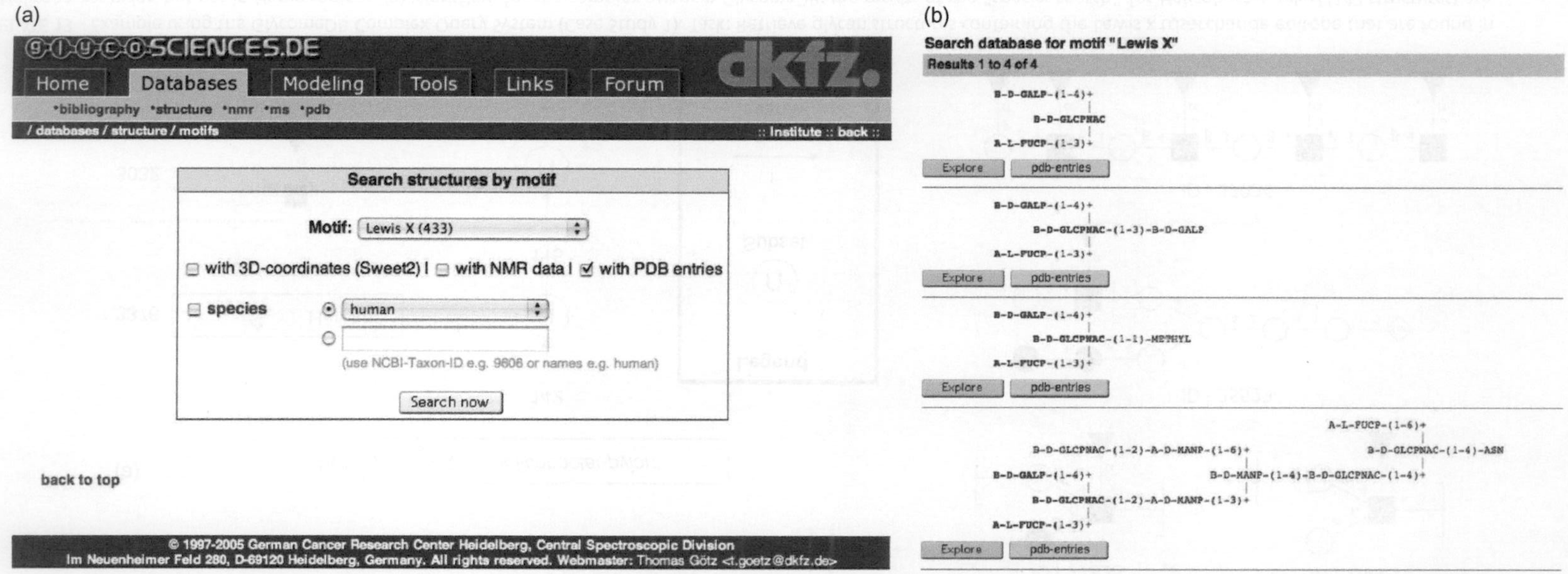

**PLATE 14** Searching for three-dimensional structures of carbohydrates in the Protein Data Bank (PDB) using GLYCOSCIENCES.de tools. (a) Motif search interface. (b) Results page offering access to the PDB entries.

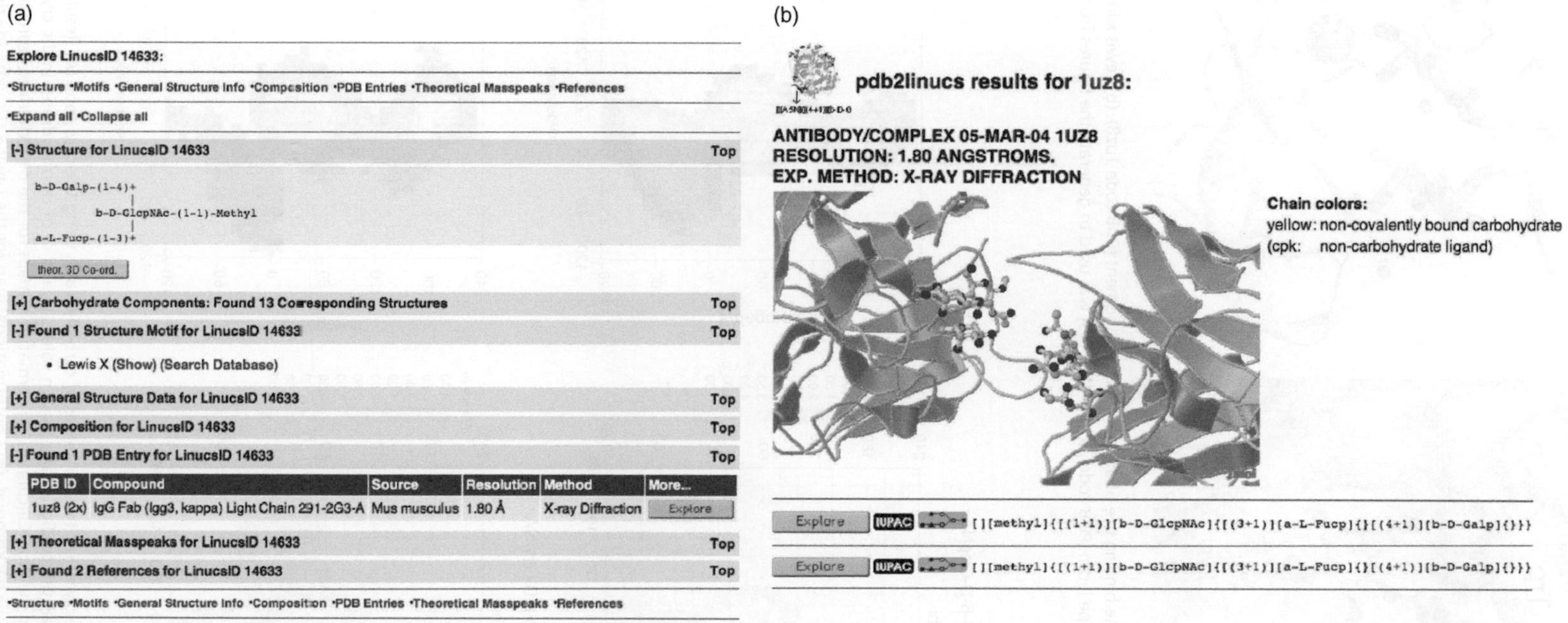

**(a)**

**Explore LinucsID 14633:**

•Structure •Motifs •General Structure Info •Composition •PDB Entries •Theoretical Masspeaks •References

•Expand all •Collapse all

[-] Structure for LinucsID 14633                                                    Top

b-D-Galp-(1-4)+
                    b-D-GlcpNAc-(1-1)-Methyl
a-L-Fucp-(1-3)+

    their 3D Co-ord.

[+] Carbohydrate Components: Found 13 Corresponding Structures        Top

[-] Found 1 Structure Motif for LinucsID 14633                       Top

• Lewis X (Show) (Search Database)

[+] General Structure Data for LinucsID 14633                        Top

[+] Composition for LinucsID 14633                                   Top

[-] Found 1 PDB Entry for LinucsID 14633                             Top

| PDB ID | Compound | Source | Resolution | Method | More... |
|--------|----------|--------|------------|--------|---------|
| 1uz8 (2x) | IgG Fab (Igg3, kappa) Light Chain 291-2G3-A | Mus musculus | 1.80 Å | X-ray Diffraction | Explore |

[+] Theoretical Masspeaks for LinucsID 14633                         Top

[+] Found 2 References for LinucsID 14633                            Top

•Structure •Motif •General Structure Info •Composition •PDB Entries •Theoretical Masspeaks •References

**(b)**

**pdb2linucs results for 1uz8:**

**ANTIBODY/COMPLEX 05-MAR-04 1UZ8**
**RESOLUTION: 1.80 ANGSTROMS.**
**EXP. METHOD: X-RAY DIFFRACTION**

Explore [IUPAC] [[methyl]{[(1-1)][b-D-GlcpNAc]{[(3-1)][a-L-Fucp]{[(4-1)][b-D-Galp]{}}}

Explore [IUPAC] [[methyl]{[(1-1)][b-D-GlcpNAc]{[(3-1)][a-L-Fucp]{[(4-1)][b-D-Galp]{}}}

**Chain colors:**
yellow: non-covalently bound carbohydrate
(cpk:   non-carbohydrate ligand)

PLATE 15  Results from a search for carbohydrate structures containing the Lewis x motif in the PDB. (a) Information page on the Lewis x trisaccharide structure at GLYCOSCIENCES. de. (b) pdb2linux display of the PDB entry 1uz8.

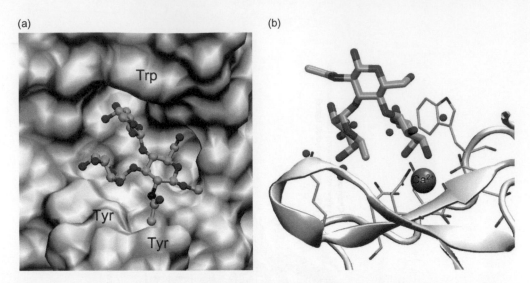

**PLATE 16** (a) Lewis x in the binding site of anti-Lewis x Fab fragment (PDB code 1uz8). (b) Lewis x in the binding site of scavenger receptor C-type lectin (PDB code 20×9). VMD [65] was used to generate the figures from the PDB files.

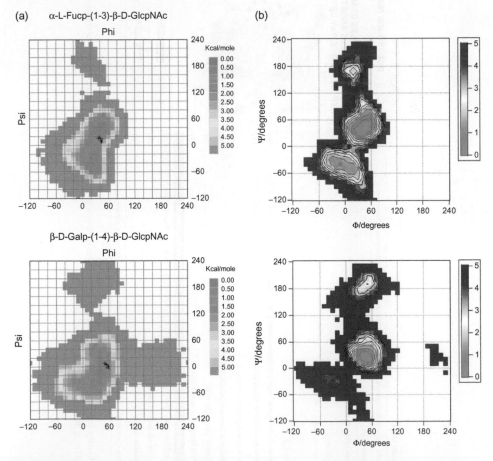

**PLATE 17** Analysis of the glycosidic torsion angles of carbohydrates using the *carp* tool. (A) Results of the *carp* analysis for PDB entry 1uz8. Glycosidic linkage torsion values ($\phi$ and $\psi$) of carbohydrates in the crystal structure are indicated by crosses in the plot. Calculated conformational maps for the disaccharides representing the linkages are displayed in the background. (b) Conformational map for the Lewis x trisaccharide derived from GlycoMapsDB.

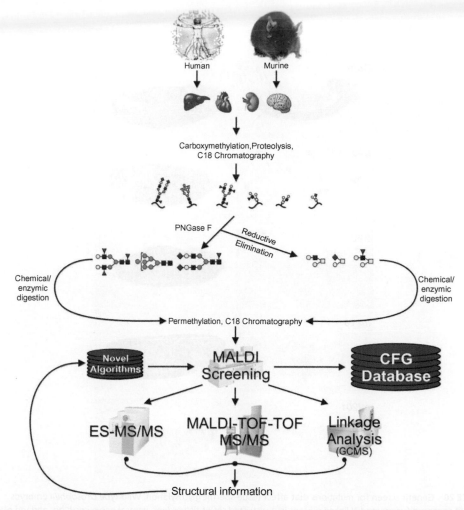

Human    Murine

Carboxymethylation,Proteolysis,
C18 Chromatography

PNGase F    Reductive
Elimination

Chemical/
enzymic
digestion

Chemical/
enzymic
digestion

Permethylation, C18 Chromatography

Novel
Algorithms

MALDI
Screening

CFG
Database

ES-MS/MS    MALDI-TOF-TOF
MS/MS

Linkage
Analysis
(GCMS)

Structural information

**PLATE 18** A simplified schematic representation of the glycomic strategy employed by Core-C for the Consortium for Functional Glycomics (CFG). Glycoproteins are extracted from human or murine tissues and purified before the release of the specific subpopulations of glycans. These glycan populations can then either be directly analyzed or subjected to chemical or enzymatic digestions in order to elucidate further structural information. The glycan pools are derivatized via permethylation in order to optimize the sensitivity of detection of the constituents and direct subsequent MS/MS fragmentations. A range of analytical techniques can be employed, but typically an initial screening by matrix-assisted laser desorption ionization (MALDI)-MS is performed as a first step. These results are used to select informed targets for further analysis by MS/MS techniques and supplementary experiments such as linkage analysis. Structural information gathered is then used to refine the original screening information. All data are deposited within the various databases curated by the CFG [15,16].

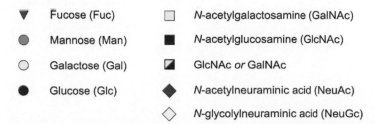

▼    Fucose (Fuc)              □    *N*-acetylgalactosamine (GalNAc)

●    Mannose (Man)             ■    *N*-acetylglucosamine (GlcNAc)

○    Galactose (Gal)           ◨    GlcNAc *or* GalNAc

●    Glucose (Glc)             ◆    *N*-acetylneuraminic acid (NeuAc)

                              ◇    *N*-glycolylneuraminic acid (NeuGc)

**PLATE 19** Symbol nomenclature as outlined by the Consortium for Functional Glycomics Nomenclature Committee (May 2004). Full documentation available from: http://glycomics.scripps.edu/CFGnomenclature.pdf.

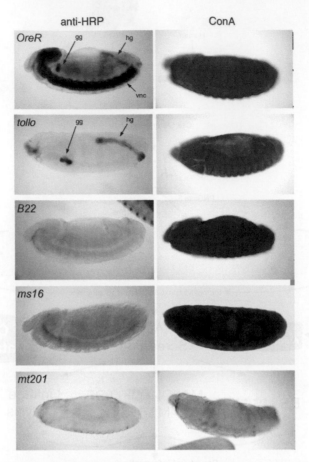

**PLATE 20** Genetic screen for mutations that affect tissue-specific glycosylation. Wild-type *Drosophila* embryos (*OreR*) express difucosylated *N*-linked glycans in a restricted set of tissues (*vnc*, ventral nerve cord; *gg*, garland gland; *hg*, hindgut). Examples of mutations that alter HRP-epitope expression without affecting the general glycosylation machinery have been previously identified (*tollo*) or generated by EMS-mutagenesis (*B22*, *ms16*). The integrity of the general glycosylation machinery in these mutants is revealed by staining with Concanavalin A, reporting the distribution of high- or pauci-mannose glycans. EMS-mutagenesis also produces mutations that affect both HRP-epitope expression and general glycosylation (*mt201*). The blue color in the *tollo* panel results from lacZ activity staining, which is carried on the mutant chromosome and used as a convenient marker of embryo genotype.

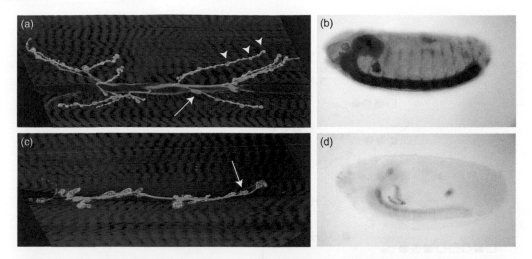

**PLATE 21** Decreased HRP-epitope expression is associated with altered neuromuscular junction morphology. (a,c) The neuromuscular junctions (green) at two muscles (red) in the body wall of a wild-type (a) or EMS-mutant (c) larva are shown. In the wild-type, small (arrowhead) and large (arrow) nerve terminal specializations (boutons) are detected. In the mutant, small boutons are missing and the overall complexity of junctional branching is reduced. (b,d) Embryos of the same genotypes as shown in (a) and (c) stained with anti-HRP antibody (b, wild-type; d, EMS-mutant).

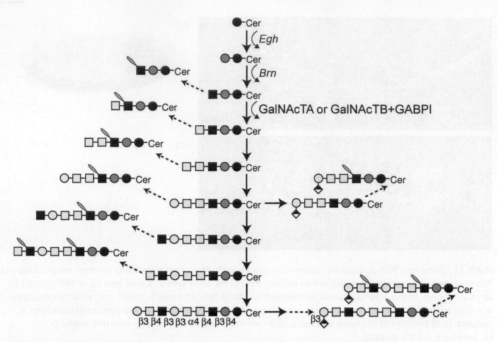

**PLATE 22** Glycosphingolipid glycan diversity of wild-type *Drosophila* embryos. The arthro-series glycosphingolipids are built by extension from a glucosylceramide core through the addition of Man by the Egghead mannosyltransferase (*Egh*), forming mactosylceramide (Manβ4Glcβ-ceramide). Brainiac (*Brn*), a GlcNAc transferase, forms the arthro-triaosylceramide, which is the first substrate for post-synthetic modification by phosphoethanolamine. Neutral glycans are frequently modified by the addition of phosphoethanolamine (orange ovals) on GlcNAc residues, forming zwitterionic structures. Acidic charge is contributed by GlcA linked to non-reducing terminal Gal residues on neutral or zwitterionic cores.

Printed and bound by CPI Group (UK) Ltd, Croydon, CR0 4YY

03/10/2024

01040312-0010